This is a book in the series

STUDIES IN PHYSICS AND CHEMISTRY

Consulting Editors: R. Stevenson and M. A. Whitehead

Additional volumes in preparation

STUDIES IN PHYSICS AND CHEMISTRY

Number 10

CRYSTAL STRUCTURES: A Working Approach

by

HELEN MEGAW

CRYSTAL STRUCTURES:
A Working Approach

HELEN D. MEGAW

**Lecturer in Physics,
University of Cambridge**

WITH DIAGRAMS EXECUTED BY J. BARBER

1973

W. B. SAUNDERS COMPANY, Philadelphia, London, Toronto

W. B. Saunders Company: West Washington Square
Philadelphia, PA   19105

12 Dyott Street
London, WC1A   1DB

833 Oxford Street
Toronto 18, Ontario

Crystal Structures: A Working Approach   ISBN 0-7216-6260-9

Print No.:   9   8   7   6   5   4   3   2   1

In recognition
of the work
of two outstanding crystallographers,
J. D. Bernal
and
W. H. Taylor

PREFACE

A knowledge of crystal structures—of the ways in which atoms are built together into regular three-dimensional patterns—underlies the whole modern development of solid-state science. This book is intended to provide a working foundation for anyone who wants to know about the architecture of crystals, not as a branch of formal theory, nor as an end in itself, but as a starting point for useful predictions about materials and for advances towards a deeper understanding of properties of matter.

This book is written for readers with no previous knowledge of crystallography but with a reasonable general background in the other physical sciences. It is meant to be useful to anyone, from first-year student to established research worker, who wants to equip himself with a knowledge of crystallography as a working tool. Each crystallographic concept is explained when it is first introduced. Formality and rigour are not sought for their own sake; on the other hand, the difficulties that often arise in practical applications are not dodged by limiting attention to the simplest examples—they are faced as they occur. The book is written on a do-it-yourself plan, with a step-by-step approach which shows how each principle formulated works out in practice.

This plan has decided the shape of the book. First comes a general introductory account of the concepts of crystal structure building, followed by a fairly thorough exposition of crystal symmetry and the use of symmetry notation. The development of the mathematical ideas is interrupted by a chapter in which concepts discussed up to that point are illustrated in a number of simple structures. Then come further chapters concerned with a selection of important and representative structures; a formal description of each is given, followed by discussion. Here the use of symmetry as a tool, which was developed in the first part of the book, is given practical application. The same approach is used in the final section, which deals with thermal effects—lattice vibrations, thermal expansion, and solid-state transitions not involving diffusion processes. The treatment is geometrical, not thermodynamic: emphasis is placed primarily on the similarities and differences between structures on either side of the transition, because without knowledge of these no theories of transition mechanisms can be very meaningful. The viewpoint is that of lattice dynamics rather than statistical mechanics, but formal theory is not used; the geometry of possible lattice modes rather than their frequencies is the subject of interest.

These final chapters approach more closely than the rest of the book to the front line of new advances of knowledge. Many of the reported facts and ideas have not yet been tested by time and by continued use, contrary to most of those in the earlier chapters. Examples are still multiplying, classifications changing, theories developing rapidly. What is already known, however, is enough to shape out lines of thought for the future, and that, rather than a definitive statement of the present position, is the aim of these chapters.

Both in developing new ideas and in making use of longer-established concepts, it is important for anyone thinking about crystal structures to be able to re-interpret them for his own purposes. Any structure can be looked at in a variety of ways. Sometimes ways that were originally helpful and that have therefore become traditional may not be the best for a particular later purpose. One aim of this book is to make possible more intelligent use of the original literature and of formal works of reference. Given the formal facts about a structure, its "crystal data" (symmetry, lattice parameters, and atomic position parameters), anyone with the proper skills can choose his own way of thinking about it, reconstruct the structure, explain it with more complete diagrams than have perhaps been published, calculate interatomic distances, look for specific features, and look for relationships to other structures that have hitherto gone unnoticed. The chapters in this book describing particular structures are intended to teach such skills, and the exercises at the ends of the chapters to give practice in using them. In particular, calculation of interatomic distances and bond lengths (to slide rule accuracy) is shown as an easy and important technique, once suitable diagrams have been constructed.

It is perhaps useful to add a few sentences about what the book does *not* attempt to do.

1. It is not concerned with the methods by which our knowledge of structures has been obtained, i.e., with methods of crystal structure analysis.

2. It is not concerned, in general, with physical or chemical properties, or how they follow from the particular atomic arrangements. Knowledge in this field is still very scanty (except for the very simplest structures). Only where the connection is very direct, as is true of ferroelectricity and of thermal expansion, has some discussion been included here.

3. It does not attempt to deal with imperfections in crystals, or with the ways in which crystallites are agglomerated together in the solid. Many macroscopic properties depend directly on these effects rather than on the crystal structure itself; but since their character in turn depends on the structure, a working knowledge of structures is essential in these fields too.

4. In describing structures and listing lattice parameters and position parameters, it does not aim at recording the most precise values available; rounded-off and approximate values are freely used, where they are sufficient for the immediate purpose. For more accurate values, the original literature or specialist compilations from it should be consulted. The same applies to measurements of other physical quantities such as thermal expansions and transition temperatures.

A classified bibliography is given at the end of the book, with a note of explanation about its contents. It is not intended to be comprehensive, but to serve as a general guide. Two books must, however, be mentioned here as well. Volume I of the *International Tables for X-ray Crystallography* (edited by N. F. M. Henry and K. Lonsdale) is essential to anyone working with crystal structures; it is the standard authority on geometrical symmetry, its nomenclature, and its application to periodic structures. R. W. G. Wyckoff's *Crystal Structures* (second edition) is the most valuable handbook for summarised descriptions of crystal structures (up to about 1960) and for references to the original literature concerning them. I have drawn heavily on both these books, to whose authors and compilers all crystallographers must be deeply indebted.

For the exercises at the ends of chapters, a formal set of answers is not supplied. In some, the answer is implied in the wording of the problem—a statement to be confirmed or verified. Others, on symmetry, can be checked by reference to the *International Tables*. Others again concern particular structures discussed in published papers; here the general correctness of answers can be judged by reference to papers listed in the bibliography. (The reader should bear in mind that the original authors may be using accurate values of parameters where the questions use rounded-off values.)

I hope that the present work may prove a useful tool for anyone interested in crystal structures, their interrelationships, and their bearing on solid-state behaviour.

HELEN D. MEGAW

CONTENTS

CHAPTER 15. PHASE TRANSITIONS   434

Chapter 1   INTRODUCTION

1.1   THE NATURE OF CRYSTAL STRUCTURES

Crystallography is the science concerned with the existence and character of extended arrays of atoms, the pattern formed by them, the forces involved in their stability, and the properties that follow in consequence of the spatial array of forces. As such, it is the fundamental science underlying our knowledge of solids.

Most solids are essentially crystalline, but the crystals composing them are not perfect. They have internal defects; they are fragmented and distorted by forces exerted on them by their neighbours; their limited size is itself a defect, because the outer layers necessarily differ from the interior. These effects have to be taken into account in considering the properties of real crystals. Nevertheless, it is right to begin with the study of perfect crystals, because only when the perfect structure is known can the nature of the defects be intelligently discussed. A perfect crystal is no more attainable than a perfect gas, but just as the concept of the perfect gas is the starting point for our understanding of actual gases, so the concept of a perfect crystal is for our understanding of actual solids.

A perfect crystal is characterised by its regular three-dimensional periodicity. Identical groups of atoms are repeated exactly at precisely equal intervals. This property of regularity is the antithesis of the property of randomness, which characterises a perfect gas. We may contrast the two in detail. A perfect gas has identical particles, infinitely small, with no forces between them, and no fixed positions in space, but with large and random velocities. A perfect crystal has identical groups of atoms, of finite size, held by mutual forces at fixed positions in space, at rest. The internal energy of the perfect gas is entirely kinetic, that of the perfect crystal is entirely potential. The perfect gas obeys exactly the statistical laws for a random assemblage of moving particles, and geometry plays no part; the perfect crystal obeys the geometric laws for a periodic arrangement of particles at rest, and randomness plays no part. Perfect gases are all alike

1

(except for the mass of their particles); perfect crystals can have an immense variety of periodic arrangements. It is important to realise from the outset that complexity of pattern is not a sign of imperfection. The complexity or distortion of the atomic group does not matter so long as it is exactly repeated with perfect periodicity.

The comparison can usefully be taken further. For actual gases, we often have to use a more realistic model in which the particles are allowed a finite small size and weak mutual attractions. In the same way, we often have to consider a crystal in which the atoms are not at rest but vibrating about mean positions. All actual crystals are imperfect in this way because of thermal movements (just as all actual gases are imperfect because of molecular size). Often, however, it is an entirely satisfactory approximation to ignore vibrations, treating the mean position of a vibrating atom as the centre of the atom at rest. We shall use this approximation through most of the book, leaving aside the question of thermal vibrations, their causes and consequences, to be considered in the two final chapters.

1.2 THE GEOMETRY OF CRYSTALS

Crystals are distinguished by their geometric regularity. Two factors contribute to this: first, the existence of a periodic repetition of the same atom or group of atoms in the same orientation, operating three-dimensionally over indefinitely long distances; and second, the existence of symmetry relations between atomic positions within the group or in repetitions of it in different directions. All crystals must possess periodic repetition—it is what makes them crystals; most crystals have additional symmetry. In a broad sense, the periodicity itself is a symmetry operation; but for the moment we want to consider it separately from the other kinds of symmetry.

The short account of symmetry in the next few paragraphs is intended for readers to whom the ideas are new. It is an informal one meant to serve as an introduction to the fuller treatment in subsequent chapters.

Symmetry is of course not confined to crystals, and it is helpful to think of examples from daily life before attempting formal definition.

The most easily recognised symmetry element, as well as the commonest, is the *mirror plane;* it is what we usually think of when we call an everyday object "symmetrical." For an object possessing a mirror plane of symmetry, one half is the mirror image of the other: for example, an ordinary chair, or (to a good approximation) our own bodies. An ordinary rectangular or oval table has two vertical mirror planes intersecting at right angles. Objects with three mirror planes at right angles can easily be constructed, but they are less common in everyday use, because most useful objects have features such as legs to stand on, or lids or doors to open, which make them one-sided in at least one dimension in space.

A less obvious but very important kind of symmetry element is the *rotation axis*. If an object having an *n*-fold rotation axis is turned about its axis, it presents an identical appearance to the viewer *n* times in a complete rotation of 360°. An obvious illustration is the rotating display stand sometimes used for selling paperback books or postcards. A square table has a fourfold rotation axis (in addition to its two mirror planes); a three-legged table is generally designed to have a threefold rotation axis. Many flowers possess fivefold rotation axes; a gentian is a particularly beautiful example, because of the absence of the mirror planes which, in flowers like a buttercup or a wild rose, are combined with the fivefold axis.

Another important element, the *centre of symmetry*, can be illustrated as follows. If one puts one's two hands together, palm to palm and thumb to thumb, in the ordinary way, the combination has a mirror plane of symmetry. Now let them rotate, keeping palm to palm, till the right thumb lies alongside the left little finger, and vice versa. The two hands, in their new position, are related by a centre of symmetry.

This same example illustrates the idea of *inversion axes*. We start with a left hand only. Suppose it is acted on by a centre of symmetry to become a right hand, in the thumb-to-little-finger configuration, and immediately rotated by a twofold axis perpendicular to the palm, so that it ends up in the thumb-to-thumb configuration. The combined operation of a centre plus a twofold axis thus has the same result as a mirror plane, and is equivalent to it. If we think of it in this way, however (as it is sometimes useful to do), we call it a twofold *inversion axis*.

Operations of these kinds—mirror planes, rotation axes, inversion axes, centres of symmetry—can be combined in various ways, as we have seen in some of our examples. The combinations are called *groups of point symmetry*, or more simply *point groups*.

The point groups that can occur in crystals are discussed in Chapter 6. They are restricted in number because not all symmetry operations which are possible in everyday objects can occur in crystals. A flower can have a fivefold rotation axis; a crystal cannot. The only axes possible for a crystal have $n = 2, 3, 4$, or 6 (unless, for formal convenience, we include the case of $n = 1$—see §6.4). The reason for this limitation is the periodic character of crystals; whereas a periodic repetition of fourfold or sixfold axes, all identical in their relations to each other, is a possibility, the same is not true for fivefold axes. (A proof of this is given in §7.2.) Point groups do exist containing rotation axes with $n = 5, 7, 8$, or more, but such groups cannot describe the overall symmetry of any crystal. This does not mean, of course, that we can never find parts of a crystal—parts cut out in imagination from the whole—which possess one of these "forbidden" symmetries; but such local symmetries disappear when we scan over more extended regions, and do not affect our argument.

Periodic repetition is itself a symmetry operation, but of a different kind; it cannot be demonstrated in finite objects, since, to be perfect, it must go on and on and never come to a boundary. Hence, in one sense, no crystal is ever perfect; but in practice most crystals are so large compared with the length of their repeat unit that the effects of the boundaries are negligible. It is therefore useful to begin by considering infinitely extended crystals, and legitimate to apply the results of our treatment to the majority of actual crystals, keeping in mind the possibility that modifications may be necessary when the crystals dealt with are exceptionally small. We shall not have to consider any such special cases in this book.

The directions in space along which periodic repetition occurs provide us with natural axes of reference for a geometrical description of the crystal—a necessary starting point for further study. This aspect is considered in Chapter 3. Once we have set up such rational axes, we can define directions and positions in the crystal, construct a nomenclature, and deduce geometrical laws, consequent on the periodicity, which govern angular relationships between the faces and between the edges of actual crystals. This is the subject matter of Chapter 5. The observed rational relationship between directions could by itself have provided an adequate basis for choosing sensible axes of reference, without an *a priori* knowledge of the periodic structure, but the argument is less direct, and harder to set out, than when we can assume such knowledge.

It is however important to remember that, historically, the regularities of external form of crystals were being observed and studied before there was any possibility of knowledge of the internal structure, and for long remained our only evidence of the existence of an underlying periodicity. The *law of constancy of angles* had its earliest experimental formulation in the work of Nicolaus Steno (Niels Stensen) published in 1669. The *law of rational indices* (cf. §5.3) was discovered a hundred years later by R. J. Haüy, who was led to look for it by his hypothesis that a crystal such as calcite could be built up by the periodic repetition of identical very small blocks. (Robert Hooke, in 1665, had suggested that crystal shapes could be constructed by a packing of "bullets or globules," but had not carried the idea further.)

The difficulty in any attempt to advance from Haüy's theory was that of knowing what, physically speaking, were the unit building blocks. Haüy's little rhombohedra might be a reasonable postulate for calcite, but were not so obvious for other crystals, nor could they easily be reconciled with the ideas of atomic and molecular chemistry that began to develop in the nineteenth century. Leaving this problem aside, two other lines of advance were open. On the one hand, the *law of rational indices* provided a firm foundation for the collection and correlation of experimental facts about crystal geometry; research of this kind was actively pursued. On the other hand, the idea of a three-dimensional periodicity took shape as an interesting postulate, capable of mathematical development. The existence of fourteen essentially different kinds of three-dimensional periodicity—now known as the fourteen *Bravais lattices*—was established by A. Bravais in 1848. Ways of combining these with other symmetry operations were investigated over the next fifty years by L. Sohncke, E. S. Federov, A. M. Schoenflies, and W. Barlow, as a result of whose work the existence of 230 independent combinations or *space groups* was proved.

The two lines of advance—observation-based morphological crystallography on the one hand, abstract mathematical logic on the other—could not help each other much, though naturally there were attempts to correlate them. Evidence from chemistry and physics was indirect and hard to evaluate. Complete and elegant though the mathematical treatments were, their links with experiment were weak: they remained essentially an abstract system, whose relevance to actual crystals was arguable. This situation lasted until the discovery of X-ray diffraction in 1912, which we shall consider further in §1.4.

Nowadays the mathematical treatments are fully accepted, and it is recognised that all possible crystal symmetries can be logically derived from a few simple postulates. The symmetry of crystals can be, and sometimes is, treated as a part of Group Theory. This does not mean, however, that all *crystal structures* are predictable by Group Theory—far from it. Group Theory concentrates attention on the operations of symmetry without considering the character and placing of the atoms on which they operate, i.e., the nature of the *asymmetric unit*. Our concern in this book is not with symmetry for its own sake but with how the kinds of structure actually found are conditioned by symmetry; we need to know how to *use* the laws of symmetry as a tool. For this reason, it is better not to follow the approach of formal Group Theory, but to develop its concepts in more usable form. Hence, in the chapters on symmetry which follow, dealing separately with lattices (Chapter 6), point groups (Chapter 7), and space groups (Chapter 8), knowledge is built up gradually from examples relevant to the situation in actual crystals, with the complexity of the examples increasing step by step. When, with experience, we have acquired a good working knowledge of what to expect in any given circumstances—in fact, a "feel" for the subject—we are equipped not only to work out for ourselves the implications of all the simpler, and commoner, symmetries, but also to make intelligent use of the standard reference books for details of the more complicated cases.

Here we may ask: why should we *want* to be able to use the laws of symmetry? Such skills may be necessary for those who are engaged in crystal structure analysis; of what importance are they once the structures are known?

There are several answers. First, we may consider a technical one. We need to be able to give concise but exact descriptions of crystal structures, and to understand such descriptions when given by other people. It is of course impossible to record the position of every atom in the crystal; even for the position of every atom in the translation-repeat unit it would often be tedious (though it is done in Chapter 4 for simple structures, since the chapters dealing with symmetry do not come till later). If we need only state the positions of all atoms in the asymmetric unit, and the symmetry operations required to derive all others from them, the description is shortened.

A second and more important use is that knowledge of the laws of symmetry allows us to recognise equalities and make comparisons. Interatomic distances related by symmetry are necessarily equal; so are interatomic forces. If one distance or force is changed—for example, by thermal expansion, or by the substitution of a chemically different atom—symmetry requires that the others must be changed in a corresponding way. Otherwise symmetry is lost, the structure as a whole changes, and a transition takes place. (Such transitions will be discussed in Chapter 15.) Again, symmetry-equivalent atoms must make equal contributions to the energy of the structure. In many ways, in the descriptions of structures in Chapters 11 to 15, we shall use arguments based on symmetry for understanding not only the geometry of the structures but also their physical and chemical characters on the atomic scale.

Thirdly, we must remember that the symmetry of a crystal does not only affect the positional arrangement of its atoms, and its external shape; it also governs the distribution of interatomic forces, and hence affects all physical and chemical properties. The macroscopically observable properties necessarily result from some sort of averaging process over all the structural units in the crystal, and must have a symmetry closely related to the structural symmetry. With such properties, however, we shall not be generally concerned in this book.

When discussing the applications of symmetry to actual examples, it is important to remember that different axes of reference may legitimately be chosen for the same crystal on different occasions, and we need to be able to transform readily from one set to another. This will be dealt with in Chapter 10.

Chapter 9 is a short digression, in which the ideas and methods of Chapters 7 and 8 are applied to the more complicated problem where the atoms operated on by symmetry have not only position but also some directional property (such as magnetic spin) which may be represented by a + or − sign. The chapter is a brief introduction to the subject of black-and-white symmetry, for which there are increasing applications in crystal physics, notably in connection with magnetic structures.

One fundamental question sometimes suggests itself in any discussion of crystal symmetry: Why should a structure show any particular symmetry—or, indeed, why should it show regularity of repetition? For very simple structures we may feel that the answer is intuitively obvious, but for more complex ones that is not so. To say that perfect repetition is often thermodynamically the most stable arrangement is merely to restate the fact, not to explain it. The question has been too little considered in the past, and it cannot be answered except in vague terms. We accept it at present as a fact of observation. Perhaps the sort of experimental work that is now going on, to which Chapters 14 and 15 serve as an introduction, may soon lead us to a better understanding of this important topic.

1.3 CRYSTAL CHEMISTRY

It was pointed out in §1.2 that Group Theory has no direct concern with the nature of the asymmetric unit; it imposes no requirements, and sets no limit to its size or complexity. We could invent as many structures as we pleased by devising different asymmetric units, unhindered by any geometrical restrictions. For real structures, however, there are restrictions of a different kind: the atoms of the unit must be so placed, relative to each other and to the symmetry elements which repeat them, that *chemically* satisfactory environments are provided for each, i.e., that for each atom the distances and directions of its neighbours are compatible with what is required by the interatomic forces or bonds. This is a very important consideration—so important that it is unsatisfactory to go far with discussion of other aspects of crystal symmetry until we have seen what near-neighbour geometrical relations are imposed by the nature of the interatomic forces.

This point of view is therefore taken up in Chapter 2. We do not need to go deeply into chemical theory for the purpose. Our concern is to know what sorts of groups are stable and of common occurrence, and how they are linked to other such groups. For this, empirical evidence from crystal structure analysis is of primary importance. On the whole, stable groups tend to have rather simple symmetries, irrespective of (though sometimes influenced by) the symmetry of the crystal as a whole. This can be explained in a general way in terms of basic laws of mechanics and electrostatics applied to a quantum-theory picture of the atom. Such concepts are extremely helpful in predicting atomic arrangements. They represent a good first approximation to the actual situation. At a more sophisticated level, however, chemical theory is much less useful. It tends to concentrate attention on the limited classes of examples with which it can at present deal, and to turn it away from those which require a new approach. For many of the kinds of structures to be dealt with in this book, observation is still ahead of theory—as it has been so often in the past. If, with the help of the simpler and more basic concepts, we can sort out the observed facts and present them in a coherent and self-consistent picture, the ground will be prepared for more radical advances of rigorous theory. This is the aim of the second half of the book.

The question at the root of all crystal chemistry is: Why do the observed structures exist, rather than others we might have thought of with the same chemical compositions? Rarely, if at all, can this be answered quantitatively, but qualitatively we can often give very good reasons. The discussion can be in terms either of *energies* or of *forces*—just as it could be for any macroscopic structure, such as a hinged framework or a girder bridge.

It is often easier to work with energies, because they are scalars and can be added directly without considering spatial directions. Moreover, in chemical reactions, changes of energy can be measured macroscopically as heat, and linked fairly easily with thermodynamic theory; hence, on the whole, emphasis in chemistry has generally been on energies. Macroscopic measurements, however, always represent a *summation* of atomic-scale contributions from the different structure-building units, and do not necessarily throw light on structural details. For insight into these structural details, it is generally most helpful to consider forces rather than energies. Forces are vectors, and have direction; the angles at which they act are important physical features. The geometrical arrangement of forces is intimately linked with the geometrical arrangement of interatomic vectors, and the two need to be considered together in any discussion of actual structures.

1.4 METHODS OF STRUCTURE ANALYSIS

Though we are not directly concerned, in this book, with the methods by which our knowledge of crystal structures is derived, it is desirable here to give an outline of them, because the picture we can make of a crystal—the things we know about it, and the things we have to put aside as unknown—depend very much on what our experimental techniques are capable of discovering.

The point was made in §1.2 that, for long after the concept of periodicity in crystals had been put forward, and during the decades when its mathematical consequences were being developed, there was no means of finding out experimentally the nature of the structural building units. That they were molecules might have been a reasonable hypothesis; but the concept of a molecule was itself still an abstraction, useful for describing reaction processes quantitatively, but giving no assurance that it corresponded closely to any actual entity in the solid.

Even this hypothesis would, as we now know, have been misleading in dealing with structures such as the alkali halides, which (like most of the structures to be described in this book) have *no* chemical molecules in the old sense. Though ideas of close packing of atoms had been used by Barlow, between 1880 and 1900, to predict correctly some of the simpler structures we shall describe in Chapter 4, to most of his fellow chemists the idea of the abolition of molecules as discrete units was a great stumbling block—and continued to be so even after the facts had been demonstrated by structure analysis. In any case, this kind of prediction could only be made for very simple structures. Over the field of solid-state structure as a whole, there was no generally accepted theory, and no way by which speculations could be put to the test.

Modern crystallography therefore dates from 1912, when Max von Laue's discovery of X-ray diffraction by crystals opened up a new chapter.

The experiment, devised by Laue and carried out by W. Friedrich and P. Knipping, was in fact undertaken to find out more about the nature of X-rays, and its success meant new opportunities for advance in that direction. But it also provided a most powerful tool for the study of crystals, and it was quickly recognised as such and developed with very great success by W. H. Bragg (Sir William Bragg) and his son, W. L. Bragg (Sir Lawrence Bragg.)

Essentially, periodicity in a crystal means that it can serve as a three-dimensional diffraction grating for radiation of suitable wavelength. If we know the wavelength, the directions of the diffracted beams indicate the lengths of the periodic repeats, and their intensities indicate the distribution of the diffracting matter associated with a repeat unit. For X-rays, the scattering underlying the diffraction process is done by electrons; hence, a study of X-ray diffraction teaches us about the distribution of electrons. However, the electrons are not to be thought of as individual points, but as spread-out clouds of *electron density* surrounding the nucleus of each atom, densest near the nucleus, thinner towards the periphery. We nearly always assume them to be spherical about the nucleus. Modern work can sometimes detect deviations from this, where the variation of density with distance from the nucleus depends on direction. Such effects are of importance in providing evidence for theories of interatomic bonding, or suggesting new developments of theory. They are relevant to the discussion of interatomic distances in §2.4. Any more detailed theoretical treatment is, however, outside the scope of this book; for most of the time, it is legitimate to think of the atoms as spheres.

A new experimental method came into use about 1950 when neutron beams of suitable wavelength and sufficient intensity began to be available. Neutrons are diffracted by atomic nuclei, which are effectively points, not spread out like the electron clouds. (Neutrons are also diffracted by electrons in orbitals with unpaired spins, but structure investigations using this property have been much less common.) Now while the fraction of incident X-rays scattered by an atom depends on the number of its electrons, i.e., its atomic number, the fraction of incident neutrons scattered by a nucleus has no such simple dependence. Hence while with X-rays it is hard to locate light atoms accurately in the presence of heavy ones—hydrogen being particularly elusive—with neutrons the light atoms may show up just as well. Conversely, of course, some atoms are harder to locate accurately by neutrons. Moreover, the techniques of neutron diffraction often involve more elaborate apparatus than do those of X-ray diffraction, and are less flexible for adapting to different problems; and of course, if we are interested in electron distribution, they are no help to us. Nowadays, it is generally recognised that the two methods of investigation must be regarded as complementary.

Electron diffraction is less generally useful, though it has made valuable contributions. Electrons are diffracted by the periodic distribution of electric potential; but the fraction of the incident beam diffracted by an atom is so large—much larger than its value for X-rays or neutrons—that only thin layers of a crystal can be studied. On the whole, electron diffraction has been more used to investigate imperfections than perfect structures.

Other techniques have their part to play—for example, spectroscopic methods, nuclear magnetic resonance, nuclear quadrupole resonance—but none of them are so generally applicable. They are of most use when a good deal is known about the structure by diffraction methods but some ambiguities are left. If too little is known to start with, the danger of drawing wrong conclusions through oversimplification is a very real one; it has not always been avoided in the past.

1.5 SIZE, HOMOGENEITY, AND PERFECTION

Crystal structures are defined on the assumption that the pattern of atoms repeats perfectly to infinity. How does this match up to reality?

Size is generally not a matter for concern. A crystal even a few microns in diameter generally contains a very large number of repeat units—so large a number that the atoms in the surface layers, where the pattern inevitably becomes modified, represent a tiny fraction of the whole and can be ignored. Very small crystals, where size is important, will not be considered in this book.

Crystals grown without any special precautions, or found naturally, are commonly matted together in a variety of orientations, sometimes interspersed with crystals of different composition, and subject to all sorts of mechanical stresses, which may distort them considerably. For structure determinations, however, one must find or grow good crystals, homogeneous and free from mechanical distortion. (Fortunately, they do not need to be large—0.1 mm or less is quite large enough for X-ray study.) It is such crystals whose structures will be described and discussed in this book. The knowledge gained from them is a necessary background if one has to deal with less perfect materials later on.

When we refer to material as homogeneous, we mean that it does not vary systematically over the region we are investigating. Crystal grains that are obviously inhomogeneous can generally be avoided. But what if inhomogeneities are on the atomic scale? The best we can do, if we cannot avoid such imperfections, is to average them out, and assume that all repeat units are equally affected by them. This is done automatically in interpreting diffraction results.

Suppose for example that, out of a symmetry-related set of atoms, one of every p atoms is missing, selected at random. Then we may usefully treat the structure as if it were a perfect one with every site of the set occupied by a new kind of atom whose scattering power is $(p - 1)/p$ times that of the original. This implies that the chance of finding an imperfection in any given repeat unit is the same, and is independent of the position of the unit.

The randomness of distribution of the imperfections is the condition that allows us to average in this way.

Suppose it were not so. Suppose that the imperfections occurred at regular intervals—for example, that after every $p - 1$ perfect repeats the pth unit had one particular atom missing, or had it replaced by a chemically different atom. The new periodicity, p times as large as the original, would now form the element of the diffraction grating, and the vacancies or substituted atoms would no longer constitute imperfections; they simply give rise to a new form of order.

In practice, however, it is sometimes hard to detect these longer periodicities, and if we failed to do so we should probably deduce a random distribution of the vacancies or the foreign atoms. Though there is a distinction in principle between disorder and a complex order, we cannot always expect a clear answer in practice.

The same kind of argument holds good when the imperfections are not vacancies or foreign atoms but slight displacements of atoms of like species from their average position. To take a simple example, if atoms which are otherwise symmetry-equivalent may be displaced by $\pm\delta$ along a particular line, and the choice of $+$ or $-$ sign is random, diffraction methods show us half-strength atoms at each of the two possible sites, provided δ is not too small. (If it is too small, the two half-atom peaks are not resolved, but merge into one elongated atom.) On the other hand, if $+$ and $-$ signs alternate regularly, the new structure is perfect, but has a periodicity twice that of the original. Only if we overlook the new periodicity shall we think that the atoms are disordered.

The fact that atoms in a crystal are not at rest but undergoing thermal motion means that at any instant they have a great variety of small displacements in different directions. The electron density deduced from diffraction evidence shows a distribution slightly spread out round the mean position. The spreading can be ignored when we only want to know mean positions, and we shall do so until we inquire into thermal effects in Chapter 14.

We do not normally consider structures imperfect because their atoms are undergoing thermal vibrations—we treat them as perfect structures of rather spread-out atoms. Similarly, it is helpful to describe structures as *quasiperfect* if they have other small imperfections distributed completely at random, so far as can be told from the experimental evidence. A quasiperfect structure may be defined as one which has perfect sets of sites, some of which are occupied at random by two or more different kinds of atoms (counting a vacancy as a kind of atom—a *zero atom*). We refer to the average content of any site of a symmetry-equivalent set as an *average atom*. The quasiperfect structure is thus like a perfect structure except that, in one or more of its sets of symmetry-equivalent sites, ordinary chemical atoms are replaced by average atoms. This is a useful concept when it comes to considering solid solutions.

Single crystal domains are regions within which the crystal has a perfect or quasiperfect structure. It is such structures with which we shall be concerned. Effects which do not depend on the structure but on the *texture*—the way in which small domains are sometimes fitted together into a large mosaic—are outside the scope of the book.

1.6 TWINNING

Another kind of imperfection of frequent occurrence is *twinning*. A twin is formed by two pieces of perfect structure which have grown together in a rational orientation (i.e., an orientation determined by the geometry of the lattice). The perfect regions are known as *twin components*. Usually the two pieces meet in a well-defined plane, the *composition plane*, which is usually also a rational plane of the structure. Sometimes twinning is repeated, so that the so-called crystal is really a mosaic of small pieces of perfect structures; in such a case the perfect pieces or components are often called *twin domains* and the process is called *polysynthetic twinning*.

As usual, our main attention in this book will be given to the perfect structure within a single component or domain. There are, however, two general points about twinning which are relevant for us.

(a) If the arrangement of atoms in a crystal very nearly satisfies some symmetry operation, such as a mirror plane, then the twinning operation is often such that the two components are related by that symmetry operation. This is sometimes known as *pseudosymmetric twinning*.

(b) When pseudosymmetric twinning occurs repeatedly, forming very small domains—as it often does—it is not always easy to work with a single domain. The average structure includes both orientations, and thus possesses the symmetry element that the individual components lacked. Under such circumstances it is often difficult to tell whether the structure deduced from direct observation is a true structure or an average structure over two kinds of twin component; if it is the latter, the true structure has lower symmetry than that directly observed.

1.7 POLYMORPHISM AND PHASE TRANSITIONS

A compound of given chemical composition may possess different perfect structures under different conditions. With change of temperature, for example, it may change from one structure to another. Since, in solids, the structure characterises the chemical phase, changes of structure represent solid-state phase transitions. Examples will be dealt with in Chapter 15. It is not only change of temperature, of course, which brings about such transitions; change of pressure, or occasionally of electric field, may also be effective.

More unexpected, perhaps, is the fact that sometimes different structures for the same compound can exist stably at the same temperature. This phenomenon was recognised from crystal morphology—from inspection and measurement of the external forms of crystals—long before structure determination was possible; it was called *polymorphism*. The early examples, of course, were limited to materials which were readily available in well-developed crystals, and therefore mainly to minerals; nowadays we know that it is much more widespread.

Polymorphism was once regarded as something of a difficulty for formal thermodynamics: if one structure was more stable than the other, how could they continue to coexist indefinitely at room temperature? The answer, in general terms, may be put as follows. Growth processes and transition processes in solids do not necessarily have the quality of randomness that is presupposed by conventional thermodynamics. The choice between two structures may be determined by some factor which affects their relative energies at the growing stage; once formed, the chosen structure is separated from the other by an energy barrier which may be too high to overcome at room temperature.

Examples of polymorphic forms of the same material can be found in §4.10–11, §11.7–8, §11.11–14, and in later chapters. Transitions between polymorphic forms are discussed in Chapter 15.

1.8 ISOMORPHISM

The term *isomorphous* was originally applied to crystals which had identical symmetry and very nearly the same interfacial angles between their macroscopically developed faces. It was realized that the very similar external geometry must imply very similar atomic arrangements and hence very similar chemical roles for the constituent atoms.

Not all crystalline materials, of course, will necessarily show good external faces. A definition of isomorphism in terms of structures is therefore capable of wider application, and at the same time it can be made more exact: *Two substances are said to be strictly isomorphous if they have identical symmetry and if their atomic positions can be fully described with the same set of arbitrary parameters—the actual numerical values of the parameters differing only slightly for the two substances.* This is in some ways stricter than the older definition, and certainly stricter than some later modified usages. Another word sometimes preferred for the effect is "isostructural," but it is not free from the danger of vagueness. It seems better to keep the strict definition of isomorphism, and refer to pairs of structures which do not satisfy it as "related structures" rather than "isomorphous structures."

Sometimes chemical similarity is specified as a necessary condition of isomorphism. With the preceding geometrical definition this need not be stated separately, because close numerical similarity between a whole set of arbitrary parameters does not occur unless the atoms involved are playing chemically similar roles.

We are left, of course, with the difficulty encountered with the original definition: how closely alike must the numerical values be, to satisfy us? It is not easy to give a general answer and to formulate rigid rules, but with actual examples there is usually no problem.

Some examples will be found in Chapter 4, and many more in Chapters 11, 12, and 13.

1.9 SOLID SOLUTIONS

The idea of solid solution is based on macroscopic observation. If two compounds (for example, potassium chloride and potassium bromide) can enter in arbitrary proportions into the formation of a single crystal with a single set of physical properties intermediate between those of the two pure constituents, analogy with liquid solutions suggests that this is a solid solution. The effect most commonly occurs for end members which are isomorphous.

This suggests that we might be able to give a more clear-cut definition in structural terms: *A solid solution is a material in which sites of a symmetry-equivalent set are occupied at random by atoms of two or more chemical species, in continuously variable proportions.* This means that a solid solution can be treated as a quasiperfect structure, built from average atoms, in which the two or more chemical species occur in proportions specified by the composition. Sometimes, as in the case of KCl and KBr, a solid solution can exist over the complete range of composition between simple end members; sometimes, as in the case of NaCl and KCl, it is limited to the immediate neighbourhood of the two end members. *

It is clear from our definition that solid solutions of a particular series must be strictly isomorphous with one another over the whole range of composition possible to them. If, with change of composition, the symmetry changes, the solid solution must be considered to belong to a new series.

Solid solutions can occur even when the two pure end members have different structures. This, which has sometimes been found puzzling in the past, need not be a difficulty if we realise that what is involved is partial replacement of atoms in a hypothetical host structure by a different species of atom; whether atoms of either species could by themselves build up a stable structure of the same kind is irrelevant.

---

* A given pair of end members may show different equilibrium effects at different temperatures—complete randomness at high temperatures and segregation into two phases, which may be isomorphous, at low temperatures. The definition of solid solution does not necessarily imply an equilibrium state; at room temperature, it is often true that a material of arbitrary intermediate composition between two end members is a homogeneous solid solution or a two-phase mixture according to the rate at which it has been cooled from a high-temperature equilibrium state.

Any two species of atom capable of replacing one another at random in symmetry-equivalent sites must of course resemble each other fairly closely in their chemical nature, but they need not have the same valency. The balance of positive and negative valencies must obviously be maintained, but this can be done by a double set of substitutions on different sites. For example, Al^{3+} can replace Si^{4+} in one set to the extent that Ca^{2+} replaces Na^+ in another. (Here again we note that if a vacancy can occur in a sometimes-occupied set of sites, it counts for our present purpose as a zero atom.) Very complicated solid solutions can easily be formed, if suitable ingredients are present, and this commonly happens in the growth of natural minerals.

Our definition of a solid solution requires that the different kinds of atoms which can replace one another should be distributed *at random* in the sites of the symmetry-equivalent set. This is an important condition. In practice, the distribution is often not random but ordered. The result is, strictly speaking, not a solid solution but a new compound with its own structure—a new single phase. The new structure may, however, in its turn serve as the host structure for a solid solution. Suppose, for example, that the original structures are AX and BX, and in the new ordered structure A and B alternate, giving the composition ABX_2. If the same structure can exist with composition $A_{1+n}B_{1-n}X_2$, in which the alternate sites are still recognisable as A-rich and B-rich, we must regard the material as a solid solution of AX and ABX_2 rather than AX and BX.

In practice, there may be intermediate states between complete randomness and complete order of occupancy, and these may depend on the temperature at which the material was formed and any subsequent annealing. It is thus not always easy to know whether a single-phase material of intermediate composition is a solid solution between its end members (random disorder) or an intermediate compound (ordered). It is generally legitimate in such cases to assume disorder as a working hypothesis unless and until clear-cut evidence for ordering is available.

More important than valency in deciding whether two atomic species can replace one another at random is the question of atomic (or ionic) size (dealt with in Chapter 2). In general, the smaller the size misfit, the greater the tendency to randomness, and the wider the range of true solid solution; the greater the size misfit, the greater the tendency either to ordering or to separate growth of two phases of different composition. This is because each guest atom in a site of the symmetry-equivalent set S exerts stresses on its neighbours, which in consequence tend to change their position slightly, and since each change in turn affects the next neighbour, it will ultimately affect (to a greater or lesser extent) the environment surrounding the next site of set S, even if this is some distance from the site with the guest atom. According to the particular structure concerned, the modification of the environment may make it less favourable or more favourable for another foreign atom. If *less* favourable, the available foreign atoms will be spaced out as far from one another as possible; assuming there is a high proportion of them, a new structure with a longer periodic repeat is formed. If *more* favourable, the sites occupied by the foreign atoms tend to clump together, giving (in the extreme case) little crystallites of the pure "guest" compound.

Solid solutions are of great practical importance, because they provide fairly predictable means of modifying physical properties for desired purposes. On the other hand, their structures are generally (though not invariably) the same as those of known stoichiometric compounds, and there is no need to single them out for separate treatment in this book. It must be remembered, however, that materials which, when first worked with, are classified as solid solutions may later turn out to represent new structures with interesting properties of their own.

1.10 USE OF APPROXIMATIONS

It is clear from earlier sections—and will be amply illustrated in later chapters—that we sometimes have to be content with incomplete knowledge of a structure. In so far as what we know gives us an approximation to the true structure it is of value. Suppose that at an early stage of investigation we think a structure is cubic, and later it turns out not to be exactly so, or suppose we think it has random disorder of substitution and later it turns out to be ordered with a more complex ordering pattern, the early picture needs only slight modification to make it fit the new evidence. For many purposes, the extra precision may not be needed. Even if we know a structure departs slightly from true cubic symmetry, for example, it may be good enough for our particular interest of the moment to treat it as cubic.

This sounds paradoxical when we are dealing with symmetry. Symmetry is an exactly defined concept; mathematically speaking, a symmetry element is either present or absent, and there is no halfway house. But in everyday life we know that there can be approximations to symmetry. Our own bodies are the most obvious examples—it is useful to think of the right and left sides as mirror images, even though the relationship is not exact and the differences may be very important for some purposes. In the same way, in crystals, atoms may sometimes be very close to symmetry-related positions, even though their actual positions are not symmetry-related. True, the laws of symmetry are mathematically rigorous, and will be treated as such in their development in Chapters 3 and 5 to 10. But it is only in so far as they express actual relationships between the positions of atoms and interatomic links that they influence observable properties. If by introducing (in imagination, and in our diagrams) very slight changes in atomic positions we can introduce a new symmetry element, this idealised structure of higher symmetry is obviously a sensible one to consider as a first approximation. In such a case the actual structure is said to be *pseudosymmetric*. The idealised structure may be sufficiently precise for our purpose; if it is not, the departures involved in changing to the actual structure can be introduced as a second approximation. We must be careful, of course, even when working with the idealised structure, to begin by recording the true structure if we know it—or at least to note that the true symmetry is less than in our working assumption.

In the same way, it is a completely permissible first step to treat an average structure as if it were a perfect or quasiperfect structure. The process of sorting out the complexities, if any, over which an average has been taken can come later. We must remember, of course, that there may be important local differences in atomic environment depending on whether an observed structure is perfect or averaged over small distinguishable domains, and take care in any theories we may construct while the matter is unresolved.

The principle of introducing simplifying assumptions at an early stage and saving complexities to be dealt with at a second approximation is an important one in all aspects of crystallographic study—as, indeed, in other branches of science. We shall see further examples in Chapter 2, when considering atomic sizes. It does not mean, of course, that the first approximation will always be adequate: the differences to be considered at the second approximation may contain the vital information for some purposes. It does, however, give us an adaptable way of approaching the subject which is extremely powerful intellectually in helping us to see what is going on. A sensible first approximation does not have to be discarded as wrong when new details become known; it is merely noted as less precise, and therefore not to be used when precision is needed, though remaining adequate for most of its original purposes.

1.11 CRYSTAL STRUCTURE DIAGRAMS AND MODELS

A word may be said here about the conventions of crystal structure diagrams. Such diagrams are really maps, and like all maps they can be drawn in different ways to emphasise different aspects of the real thing. In Chapter 2, where simple configurations of atoms are described, the illustrations are drawn in four ways in Figure 2-2 to familiarise the reader with some of the alternative conventions in common use. They are as follows:

(i) Ball-and-spoke diagram, in perspective. Circles mark the centres of atoms, and joining lines represent bonds.

(ii) Coordination polyhedra, in the same perspective. The lines represent anion-anion edges; anions are taken to be at the corners of the polyhedra, and the central cations are omitted.

(iii) Packing diagram, in the same perspective. The ions are shown as spheres, with radii to scale (the cation being sometimes hidden by its surrounding anions.)

(iv) Ball-and-spoke diagram, projected on the horizontal plane of the earlier figures. Solid lines represent bonds, and broken lines anion-anion edges. Atoms or bonds coinciding in projection are shown with double lines, and otherwise heavy lines indicate the upper part of the group, light lines the lower. Heights of atoms above or below the central plane are marked (in arbitrary units.)

Later in the book, we shall generally make little use of perspective diagrams, which are much less helpful for serious study than are projections and sections. The latter can of course be used to show packing spheres, or outlines of coordination polyhedra, or atomic centres and interatomic bonds, just as required. The reason for preferring projections and sections is exactly the same as that which makes the architect and engineer rely on scale plans and elevations rather than perspective views. A projection can be drawn to scale and annotated with full information about heights, so that the reader with a little practice can reconstruct and visualise the complete structure from any viewpoint, and work on it for himself. A perspective diagram is more subjective, because lengths and angles cannot easily be found from it, and the reader has to accept the viewpoint provided, instead of being given the means to inspect the structure from any viewpoint he wishes. This difficulty also applies to stereoscopic views. Perspective views and stereoscopic views are helpful for beginners and can be useful to display particular features of structures at any stage, but they cannot replace projected diagrams as a working tool.

The use of models is of course helpful, especially the models one makes for oneself. As with two-dimensional maps, one can choose whether to use large spheres to show packing, or smaller balls and spokes to show bonds, or prefabricated coordination polyhedra to show how they join up. To see a two-dimensional diagram one has drawn translated into the corresponding three-dimensional model can be illuminating. But a model cannot be altered as easily as a drawing if it becomes desirable to modify it or make comparisons or call attention to new aspects; and even the best-made model is generally inadequate for quantitative work, for which one has to turn back to two-dimensional diagrams. A complete description of a structure ought therefore to include at least one projection, drawn to scale and carefully annotated, as a working diagram for the reader. Stereoscopic views and photographs of models ought to take second place as optional extras.

Chapter 2 INTERATOMIC FORCES AND STRUCTURE BUILDING

2.1 THE NATURE OF INTERATOMIC FORCES

A stable structure is one which has the lowest free energy. We have seen in §1.2 that it is generally a good approximation to consider a structure with all atoms at rest, and we shall do so throughout this chapter. Then the free energy consists entirely of potential energy.

The potential energy of a crystal is the sum of the potential energies of every pair of atoms in it. To each such energy there corresponds an interatomic force. As we noted in §1.3, it is often more helpful to discuss problems in terms of forces rather than energies.

We begin by considering a single pair of atoms in equilibrium—a diatomic molecule. Obviously, there must be both attractive and repulsive forces between them—attractive to keep them together, repulsive to prevent them from coalescing. At the equilibrium separation, the attractive and repulsive forces must be equal in magnitude; at greater distances, the attractive forces must predominate, at smaller distances, the repulsive. We may expect both kinds of force to decrease with increasing distance, but the repulsive forces must decrease more rapidly. This is illustrated in Figure 2-1.

The same argument applies to pairs of nearest-neighbour atoms in a crystal if we can assume that all such pairs are identical and all more distant pairs have negligible interactions. This is true only in special cases, but we shall see how it can be modified to apply to others. The nearest-neighbour attractions are the *interatomic bonds*.

It is customary and useful to classify attractive forces into four main types:

    (i) metallic,
    (ii) ionic (electrostatic, or Coulomb),
    (iii) homopolar (covalent),
    (iv) van der Waals.

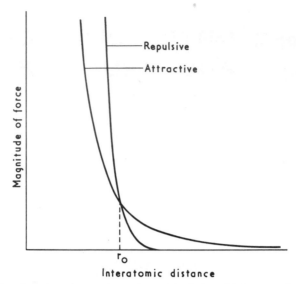

Figure 2-1 Variation of attractive and repulsive forces with interatomic distance in an isodesmic structure. The forces are equal in magnitude and opposite in sign at r_0, the equilibrium distance.

We need not concern ourselves greatly with the theory of their origin, but it is important to emphasise the empirical differences between them.

Metallic forces can be pictured as due to the "electron gas" of conduction electrons shared between all the atoms. The forces are fairly weak, and are essentially *undirected* or *central*. An example of a structure with only metallic bonds, all alike, is copper.

Ionic forces are electrostatic forces between atoms which have gained or lost electrons. The ions are treated as spherical distributions of charge, which are electrostatically equivalent to point charges at the centres. The forces are therefore *central*. They are directly proportional to the charge, and hence to the chemical valency; they are inversely proportional to the square of the interatomic distance. The slow decrease with distance means that electrostatic attractions or repulsions between more widely separated pairs of ions are not negligible, and must be taken into account in the energy summations; this, however, can be done in a routine way for simple structures. An example of a structure with only ionic bonds, all alike, is sodium chloride, NaCl.

Covalent forces are due to specific interactions between the wave functions of the atoms involved. They decrease more rapidly with distance than do ionic forces; hence it is rarely necessary to consider anything but nearest neighbours. The forces are essentially *directed*; in other words, the bonds from an atom to its various neighbours make particular angles with one another which are determined by the wave functions of the atoms involved. A given species of atom can make more than one geometrically-distinct set of bonds, depending on the electron orbitals called into play: for example, carbon may have four bonds directed to the corners of a regular tetrahedron (as in diamond, or in methane, CH_4), or three bonds directed to the corners of an equilateral triangle (as in benzene, or in a carbonate ion). In a crystal, the interbond angles may be slightly distorted from their ideal values, because of interaction with other structural forces. Nevertheless, the directed character of covalent bonds is their most vital property for the crystallographer. An example of a structure with only covalent bonds, all alike, is diamond.

Van der Waals forces (sometimes called *dispersive forces*) are present between any pair of atoms, and give weak attractions inversely proportional to the seventh power of the distance. They are undirected. In discussions of structure building, van der Waals forces may be neglected when stronger forces are present between the atoms being considered; but if there are no other direct or indirect links, they provide the essential cohesive forces. An example of a solid containing only van der Waals bonds is argon; the low melting point is of course directly related to the low forces of cohesion.

This classification of bond types, like most classifications, is over-simplified. It distinguishes theoretically-important concepts; but few examples are as clear-cut as those we have mentioned. We must consider some of the actual complications.

Firstly, many crystalline solids contain more than one type of bond. It often happens that covalent forces bind a finite group of atoms into a *molecule*, and the molecules are held together by nothing stronger than van der Waals forces—as for example in solid benzene. Again, a group linked by covalent forces may constitute a *molecule-ion*, connected to other ions by electrostatic forces, as for example the cholesteryl group in cholesteryl iodide or the carbonate group in calcium carbonate. The distribution of bonds of different strength may be very strongly anisotropic, as in cadmium iodide, where ionic or partly ionic forces build up a layer stretching to infinity in two dimensions, while successive layers are held together by van der Waals forces only. Structures such as these, with bonds of wholly different types, are called *heterodesmic*. The very simple structures considered in the previous paragraphs, in which all bonds are alike, are *isodesmic*. Structures in which all bonds are generally similar in kind, though not identical, are called *homodesmic*. This kind of classification, put forward by J. D. Bernal in the early 1930's, underlies all our thinking about the relations between crystal structure and properties.

Secondly, it is now known that many bonds do not fit simply into one of the four categories, but are in some respects intermediate. One important example is the *hydrogen bond*, whose weakest link (which determines its effective strength) may be thought of as intermediate between electrostatic and van der Waals; we shall consider it later in Chapter 13. Again, many bonds are intermediate between pure ionic and pure covalent; they are *semipolar*. They have the directed character of covalent bonds, but their angles are much more easily distorted from the ideal by the influence of neighbouring atoms. Semipolar bonds are very common and very important in the structures to be dealt with in this book, and some further consideration is needed at this point.

Theoretically, in a pure ionic bond the cation has transferred all its valence electrons to the surrounding ions, thereby completing their rare-gas shell. Suppose this happens, and suppose that in the resulting compound the equilibrium interionic distance is such that the shells (whose radii can be calculated from the wave functions of the atoms) do not overlap; then the ionic character is unambiguous. On the other hand, suppose that, starting the same way, we find that the shells of cations and anions overlap; then one cannot really say to which atom the overlapping fraction of the electron cloud belongs, nor how far it is to be considered as shared. This situation is most likely to occur when the "cation" has a high valency. An alternative description is then possible in which only a proportion of the valence electrons is transferred to the "anion," the remainder taking part in a covalent bond, which may even have double-bond character. Modern chemical theories all recognise the importance of intermediate states between the purely ionic and the purely covalent (though their approaches and terminologies may differ). It is thus unrealistic (however convenient at times) to think in terms of fully-charged cations of high valency, such as Si^{4+}.

2.2 INTERATOMIC DISTANCES: PREDICTION OR MEASUREMENT?

Our reason for trying to classify and describe interatomic forces was to enable us to predict the kinds of structures they could produce. In particular, we need to think about interatomic distances or *bond lengths* between atoms linked by attractive forces. It is easiest to begin by considering compounds which are wholly or mainly ionic.

We must choose between two approaches. On the one hand, we may use what we know of the structure of the atom, or its wave function, to predict interatomic distances; we shall have to make some postulate about its state of ionisation, and we may have to restrict ourselves to the low-valency ions which are most easily calculable. On the other hand, we may use the interatomic distances observed in a small number of representative structures, in conjunction with simple empirical rules, to predict not only interatomic distances in other compounds but the actual structure types to be expected in them. With this latter approach, we make deductions from *variations* in bond distances about variations in bond character, and no assumption is needed initially, nor has any been made, as to the degree of ionic character in the bond.

Historically, the latter approach was initiated by W. L. Bragg in 1920, and developed in the following decade, notably by Bragg and his coworkers at Manchester, and by V. M. Goldschmidt and his coworkers at Oslo. It has proved immensely helpful to our understanding of solids, and a stimulus to new work.

Towards the end of the same decade, L. Pauling put forward a more theoretical treatment, based on the assumption that the atoms are ionised and carry a charge represented by their chemical valency. Though his treatment included other arbitrary assumptions, which could only be justified for particularly simple structures, it too proved very successful and stimulating.

The success of Pauling's approach has, however, sometimes misled later writers into thinking that all structures of the kind studied by him and by Goldschmidt are fully ionic; or, alternatively, that Goldschmidt's approach is in disagreement with the evidence that has since accumulated for their partly covalent character. In fact, "cations" and "anions" in the empirical or Goldschmidt approach are simply concise expressions for "the more electropositive atoms" and "the more electronegative atoms."

We shall begin by considering the Goldschmidt approach, and later return to Pauling's.

2.3 IONIC RADII: THE EMPIRICAL APPROACH

One vitally important consequence of the work of Bragg and of Goldschmidt was the clear meaning it gave to the concept of ionic (or atomic) radius: *a radius which belongs to a particular species of atom, which is unchanged from one compound to another, and which can be used, in conjunction with corresponding values for other atoms, to predict interatomic distances approximately.* (Atoms in different states of chemical valency are considered as different species.) This idea is so easy to picture and to use that nowadays one often tends to forget that it needed to be justified empirically in the first instance—and indeed was seriously attacked when it was first put forward. We must look at the evidence on which it is based.

Before doing so, we pause for a moment to notice that *metallic radii*, the radii of atoms in metals and alloys, may differ from the ionic or atomic radii with which we are concerned, and which apply in compounds where the atoms differ enough in electronegativity to be distinguished as cations and anions. We shall not have many examples of metallic structures in this book, and therefore will not discuss metallic radii further, except to note that they are generally found by taking half the closest distance of approach between like atoms in metallic structures.

To illustrate the character of ionic radii, consider the compounds listed in Table 2-1, which (except for those with asterisks) are isomorphous (isostructural) and which are also isodesmic. Beginning with NaCl, we arbitrarily assign a radius of 0.98 Å to Na^+, and hence by subtraction obtain 1.83 Å for Cl^-; from KCl, we obtain 1.31 Å for K^+; from KBr, 1.98 Å for

Table 2-1 Interatomic Distances in Alkali Halides (in Å)

	Li	Na	K	Rb	Cs
F	2.01	2.31	2.66	2.82	3.00
Cl	2.57	2.81	3.14	3.27	3.56*
Br	2.75	2.98	3.29	3.43	3.71*
I	3.00	3.23	3.53	3.66	3.95*

* These compounds have the CsCl structure; all others have the NaCl structure.

Br^-; and from NaBr, 1.00 Å for Na^+—a value very close to our original assumption. Alternatively, we could have used the distances in NaCl, NaBr, and KCl, to predict the distance in KBr, correct to within 0.02 Å. If we extend the scheme to include all the alkali halides, the greatest disagreement is only 0.06 Å. Hence, each kind of atom has a nearly constant radius throughout the whole series of its compounds.

The scheme can be widened still further by considering other structures, not necessarily isomorphous with these, until a characteristic radius is allotted to every species of ion. Small discrepancies are to be expected between values for the same ion deduced by different chains of reasoning, not so much because of experimental error as because of real differences in nearest-neighbour interatomic distances due to the effects of other fairly-near-neighbour atoms. These will be discussed later; one can correct for some of them, but even so, differences as great as about 0.05 Å may remain. Setting up a self-consistent set of empirical radii must therefore involve taking some more or less arbitrary average between the different values, within these fairly narrow limits.

The consistency of the results (within these limits) is proof of the validity of the idea, but it does not necessarily confirm our chosen value for the Na^+ radius; any other choice over quite a wide range would have done equally well. We have to find some criterion to fix one radius, and then the rest will follow.

The earliest estimate of this kind (made by A. Landé in 1920) was based on the assumption that in LiI the iodine atoms are in contact, since I^- must be very much larger than Li^+, having 54 electrons in its rare-gas core as compared with 2 for Li^+. More generally, it could be argued that, if ions remain spherical and of constant radius, we are likely, in studying a large number of compounds of a particular large ion, to find some in which they touch each other; hence the smallest observed distance between like ions will be a measure of their diameter. From Bragg's work on oxides and silicates, he thus deduced, in 1927, a value of 1.35 Å for the radius of O^{2-} The same value was obtained by an independent method using the measured interatomic distances in MgSe, MnSe, CaSe, and CaO; the equality of the values for MgSe and MnSe was taken to mean that the Se^{2-} atoms were in contact, and not pushed apart by the smaller Mg^{2+} and Mn^{2+} atoms; and from the Se^{2-} radius thus determined, values for Ca^{2+} and O^{2-} were found from CaSe and CaO respectively, by subtraction.

Another quite different method of finding the value of an ionic radius depends on the fact that the ionic volume is proportional to the polarisability,* which determines the molar refractivity, and this can be measured. Measurements of molar refractivity by J. A. Wasastjerna in 1932 led to radii of 1.32 Å for O^{2-} and 1.33 Å for F^-. Later work confirmed the value for F^-, but suggested that the value for O^{2-} should be higher. The general agreement between radii found by this method and those found by the independent method outlined in the previous paragraph provides satisfactory confirmation of the validity of the concept.

The first fairly complete set of radii for common ions was compiled by Goldschmidt in 1926, using Wasastjerna's values for O^{2-} and F^- in conjunction with interatomic distances measured in a comprehensive series of structure determinations undertaken for the purpose in his own laboratory. These are the *Goldschmidt radii*. As a result of later work, additions and minor corrections to the set have been made and are incorporated in the list in Table 2-2, but no complete reassessment of the evidence as a whole was made until fairly recently. We shall consider some of the new work later. None of it represents a sufficient advance, either in theoretical understanding or in empirical removal of discrepancies, to justify a radical revision of the set of accepted radii.

In treating ionic radii as strictly additive, we imply that the contact distance between the two ions is independent of any other forces exerted on them by other parts of the structure; it means in effect that the ions are considered incompressible, i.e., that the repulsive force is zero outside a particular value of the radius, and infinite for smaller values. Our picture is of a perfectly rigid sphere. It is obvious that this cannot be strictly true; we must allow our spheres some elasticity. We may, however, rightly expect the repulsive force to increase extremely rapidly as we decrease the interatomic distance below its usual equilibrium value. The rigid-atom model is thus a sensible approximation, which we can use with confidence provided we do not demand too much precision.

There is, however, one correction which needs to be mentioned at this stage, though discussion will be left until later. The cation-anion distance

---

* Polarisability is a word with confusingly different meanings. It refers to the deformation of the electron cloud of an atom or ion in an electric field. If the field changes each atom into a dipole, the polarisability is defined as the ratio of the dipole moment per unit volume (*or* per atom) to the field acting on it, which is deduced from the uniform applied field. This is the sense in which it is used at this point in the text. Qualitatively, however, in the discussion of structures, it is used in relation to the deformation of the atom as a whole in the field of one or more neighbours; the field is generally far from uniform, and the deformation is into something more complicated than a single dipole. See §2.9 for further discussion.

Table 2-2 Ionic Radii in Å*

+

	G	P	A	SP/6	SP/4	SP/12
Li^+	0.78	0.60	0.68	0.74	0.59	—
Na^+	0.98	0.95	0.97	1.02	0.99	—
K^+	1.33	1.33	1.33	1.38	—	1.60
Rb^+	1.49	1.48	1.47	1.49	—	1.73
Cs^+	1.65	1.69	1.67	1.70	—	1.88
Cu^+	—	0.96	0.96	—	—	—
Ag^+	1.13	1.26	1.26	1.15	—	—
Au^+	—	1.37	1.37	—	—	—
Tl^+	1.49	1.44	1.37	1.50	—	1.76

2+

	G	P	A	SP/6	SP/4	SP/12
Be^{2+}	0.34	0.31	0.35	—	0.27	—
Mg^{2+}	0.78	0.65	0.66	0.72	0.58	—
Ca^{2+}	1.06	0.99	0.99	1.00	—	1.35
Sr^{2+}	1.27	1.13	1.12	1.13	—	1.40
Ba^{2+}	1.43	1.35	1.34	1.36	1.42	1.60
Ra^{2+}	—	—	1.43	—	—	—
Zn^{2+}	0.83	0.74	0.74	0.75	0.60	—
Cd^{2+}	1.03	0.97	0.97	0.95	0.80	1.07
Hg^{2+}	1.12	1.10	1.10	1.02	0.96	1.14
Mn^{2+}	0.91	0.80	0.80	{0.67L / 0.83H}	—	—
Fe^{2+}	0.83	0.75	0.75	{0.61L / 0.78H}	0.63	—
Co^{2+}	0.82	0.72	0.72	{0.65L / 0.74H}	—	—
Ni^{2+}	0.78	0.70	0.69	0.69	—	—
Cu^{2+}	—	—	0.72	0.73	—	—
Ge^{2+}	—	—	0.73	—	—	—
Sn^{2+}	—	—	0.93	—	—	—
Pb^{2+}	1.32	—	1.20	1.18	—	1.49

3+

	G	P	A	SP/6	SP/4	SP/12
B^{3+}	—	—	0.23	—	0.12	—
Al^{3+}	0.57	0.50	0.51	0.53	0.39	—
Sc^{3+}	0.83	0.81	0.81	0.74	—	—
Y^{3+}	1.06	0.93	0.92	0.90	—	—
La^{3+}	1.22	1.15	1.14	1.06	—	1.32
Ga^{3+}	0.62	0.62	0.62	0.62	0.47	—
In^{3+}	0.92	0.81	0.81	0.79	—	—
Tl^{3+}	1.05	0.95	0.95	0.88	—	—
V^{3+}	0.65	0.66	0.74	0.64	—	—
Cr^{3+}	0.64	0.64	0.63	0.61	—	—
Mn^{3+}	0.70	0.62	0.66	{0.58L / 0.65H}	—	—
Fe^{3+}	0.67	0.60	0.64	{0.55L / 0.64H}	0.49	—
Co^{3+}	—	—	0.63	{0.52L / 0.61H}	—	—
As^{3+}	—	—	0.58	—	—	—
Sb^{3+}	—	—	0.76	—	—	—
Bi^{3+}	—	—	0.96	1.02	—	—
Ce^{3+}	1.18	—	1.07	1.01	—	—
Gd^{3+}	1.11	—	0.97	0.94	—	—
Pu^{3+}	—	—	1.08	1.00	—	—

Sets of radii are given in columns under the initials of their authors, as follows: G—Goldschmidt, summarised in *Internationale Tabellen zur Bestimmung von Kristallstrukturen*, 1935 (with a later revised value for Zr^{4+}); P—Pauling, mainly from *The Nature of the Chemical Bond*, 1939; A—Ahrens, *Geochim. Cosmochim. Acta*, **2**: 155, 1952; SP—Shannon and Prewitt, *Acta Cryst.*, **B25**: 925, 1969, with minor modifications *Acta Cryst.*, **B26**: 1046, 1970. Their values have been rounded off to 0.01 Å.

All radii except the SP values are for coordination number 6. For SP values the coordination number is noted at the top of the column; e.g., SP/4 refers to 4-coordination.

The letters L and H following some of the radii refer to the low-spin and high-spin state of the atoms concerned.

4+

	G	P	A	SP/6	SP/4
Si⁴⁺	0.39	0.41	0.42	0.40	0.26
Ti⁴⁺	0.64	0.68	0.68	0.60	—
Zr⁴⁺	0.77	—	0.79	0.72	—
Hf⁴⁺	—	—	0.78	0.71	—
Ge⁴⁺	0.44	0.53	0.53	0.54	0.40
Sn⁴⁺	0.74	0.71	0.71	0.69	—
Pb⁴⁺	0.84	0.84	0.84	0.77	—
V⁴⁺	0.61	—	0.63	0.59	—
Mn⁴⁺	0.52	—	0.60	0.54	—
Nb⁴⁺	0.69	—	0.74	0.69	—
Mo⁴⁺	0.68	—	0.70	0.65	—
Ce⁴⁺	1.02	1.01	0.94	0.80	—
Th⁴⁺	1.10	—	1.03	1.00	—
U⁴⁺	1.05	—	0.97	—	0.73
Pu⁴⁺	—	—	0.93	—	—

5+

	G	P	A	SP/6	SP/4
P⁵⁺	—	0.34	0.35	—	0.17
V⁵⁺	0.65	0.59	0.59	0.54	0.35
Nb⁵⁺	—	0.70	0.69	0.64	0.32
Ta⁵⁺	—	—	0.68	0.64	—
As⁵⁺	—	0.47	0.46	0.50	0.33
Sb⁵⁺	—	0.62	0.62	0.61	—
Bi⁵⁺	—	0.74	0.74	—	—

— and 2 —

	G	P	A	SP/2	SP/3	SP/4	SP/6	SP/8
F⁻	1.33	1.36	1.33	1.28	1.29	1.31	1·33	—
Cl⁻	1.81	1.81	1.81	—	—	—	—	—
Br⁻	1.96	1.95	1.96	—	—	—	—	—
I⁻	2.20	2.16	2.20	—	—	—	—	—
O²⁻	1.32	1.40	1.40	1.35	1.36	1.38	1.40	1.42
S²⁻	1.74	1.85	1.96	—	—	—	—	—

6+

	G	P	A	SP/6	SP/4
S⁶⁺	—	0.29	0.30	—	0.12
Cr⁶⁺	—	0.52	0.52	—	0.30
Mo⁶⁺	—	0.62	0.62	0.60	0.42
W⁶⁺	—	—	0.62	0.60	0.42
Re⁶⁺	—	—	—	0.52	—
U⁶⁺	—	—	0.80	0.75	0.48

7+

	G	P	A	SP/6	SP/4
Mn⁷⁺	—	0.46	0.46	—	0.26
Re⁷⁺	—	—	0.56	0.57	0.40
I⁷⁺	—	0.50	0.50	—	—

This supplementary table additionally gives the variation with coordination number of anions, according to SP.

For the Ahrens radius of O²⁻, a contraction of 0.05 to 0.10 Å is to be applied when the cation is highly polarising, i.e., small and of high valency.

* Radii for less common cations and coordination numbers in addition to those quoted here may be found in the original papers by Ahrens and by Shannon and Prewitt referred to above.

does depend to some extent on the number of anion neighbours possessed by the cation, known as its *coordination number* (see §2.5). Two courses are open to us. *Either* we may treat chemically similar cations with different co-ordination numbers as different kinds of cations for the purpose of finding empirical radii, *or* we may turn to theory to find a correction factor giving the change of radius with coordination number. The latter was done by Pauling for use with his semi-empirical 'crystal radii,' discussed in §2.4. Correction factors can be used equally well with the Goldschmidt radii, and with these available the Goldschmidt standard set has been drawn up to give the interatomic distance directly when the coordination number is 6. Approximate values for important coordination numbers are given in Table 2-3; others can be interpolated if necessary.*

Table 2-3 Correction Factors for Different Coordination Numbers

COORDINATION NUMBER	RATIO OF INTERATOMIC DISTANCE TO SUM OF STANDARD RADII
4	0.94
6	1.00
8	1.04
12	1.10

More recently, however, the separate empirical evaluation according to coordination number has been carried out by R. D. Shannon and C. T. Prewitt. Some of their radii (discussed further in §2.4) are given in Table 2-2. They introduced two additional modifications. They assumed that the anion could change its radius with a change in coordination number, just as the cation does, though to a lesser extent; and they also assumed that for paramagnetic ions the radius depended on the spin state, and therefore gave different values for high spin and low spin. On the whole, their radii give very good predictions of interatomic distances, but when discrepancies occur they are not always easy to explain. Some examples are given in Tables 2-4 and 2-5.

In accepting different radii for different coordination numbers, we have already departed from the rigid-ion model. Further consideration of the limitations of this model will be left till later in the chapter.

---

* The use of these correction factors is not invariably helpful. For example, Pauling points out that for alkaline-earth fluorides with cation coordination number 8 a better prediction can be made by using the radius for coordination number 6 (cf. Table 2-4).

2.4 IONIC RADII: THEORETICAL TREATMENTS

All theoretical treatments yet available use some empirical evidence as a starting point. The best-known treatment is that of Pauling. It begins with the postulate that the effective radius is the radius of the outermost shell of electrons, which is inversely proportional to the positive charge attracting these electrons to the centre. The effective charge is less than the charge on the nucleus because the inner electrons shield the outer ones from it. We may therefore write

$$\text{Radius} = \frac{C_n}{Z - S} \tag{2.1}$$

where Z is the nuclear charge, S a screening constant, and both C_n and S have constant values for all ions with the same rare-gas shell. The quantity S was evaluated by Pauling partly by calculation, partly from experimental results of work on molecular refractivity and X-ray spectroscopy.

Though the values of the C_n's are unknown, the ratio of the radii of ions with the same rare-gas shell can be found, since they have the same C_n. For example,

$$\frac{r_K}{r_{Cl}} = \frac{Z_{Cl} - S}{Z_K - S} \tag{2.2}$$

Using the measured interatomic distance in KCl, the individual radii of K^+ and Cl^- can then be found. A similar procedure can be used for the pairs Na^+F^-, Rb^+Br^-, and Cs^+I^-.

Since the constants C_n can now be deduced, "univalent radii" can now be calculated for all other atoms with complete rare-gas shells, i.e., radii which would predict the interatomic distances in compounds such as MgO (isomorphous with NaCl) if the electrostatic attractive force between Mg and O were e^2/r^2 as in NaCl. It is then assumed that the ratio of the "crystal radii" to the "univalent radii" is the ratio of the theoretical equilibrium distances in two similar structures differing only in their electrostatic forces. The Born theory (see §2.7) allows this ratio to be calculated. If the numerical value of the ionic charge in the second structure is z, the ratio of the electrostatic energies is z^2; then, using equation 2.13, the ratio of the r_0's is $z^{-2/(n-1)}$, or about 0.85 for $z = 2$, $n = 10.0$ (n is here the exponent in the expression for the Born repulsion, and not, as in the previous paragraph, the serial number of the rare-gas shell.) Interatomic distances calculated from the "crystal radii" agree satisfactorily with experimental values.

This treatment deals only with ions with complete rare-gas shells; other radii have to be found empirically with these as starting point.

An important extension was made later by L. H. Ahrens. After reviewing the Pauling radii and making minor adjustments, he showed that theoretically the ionisation potential of an atom (a measurable quantity) should be proportional to its radius, and that experimentally this was confirmed by the smooth variation of the Pauling radii with ionisation potentials. Hence the radii of other ions, not included in the original set, could be predicted from their ionisation potentials. In this way, selecting best weighted averages, a complete table was drawn up (cf. Table 2-2).

The same kind of interpolation was used by Shannon and Prewitt for ions where direct evidence from structures of pure materials was not available. Values obtained in this way are included in Table 2-2.

Table 2-4 Comparison of Interatomic Distances and Radius Sums for Some Representative Simple Structures

	OBSERVED DISTANCE IN Å	PREDICTED DISTANCE IN Å		
		PAULING	GOLDSCHMIDT	SHANNON AND PREWITT
Mg—O {in MgO (6, 6)	2.10	} 2.05	2.10	2.12
{in Mg(OH)$_2$ (6, 3)	2.10			2.08
Ca—O {in CaO (6, 6)	2.40	} 2.39	2.38	2.40
{in Ca(OH)$_2$ (6, 3)	2.36			2.36
Sr—O in SrO (6, 6)	2.57	2.53	2.59	2.56
Ba—O {in BaO (6, 6)	2.76	2.75	2.75	2.76
{in BaSnO$_3$ (12, 6)	2.91	3.02	3.02	3.00*
Zn—O {in ZnO (4, 4)	1.98	2.01	2.02	1.98
{in Zn(OH)$_2$ (6, 3)	2.19	2.14	2.15	2.10
Al—O {in Al$_2$O$_3$ (6, 3)	1.91	1.90	1.89	1.89
{in CaAl$_2$Si$_2$O$_8$ (4, 3)	1.75	1.79	1.78	1.77*
Si—O in SiO$_2$ (4, 2)	1.60	1.70	1.61	1.61
Mg—F in MgF$_2$ (6, 3)	2.04	2.01	2.11	2.01
Ca—F in CaF$_2$ (8, 4)	2.35	2.45†	2.49†	2.43
Ba—F in BaF$_2$ (8, 4)	2.68	2.86†	2.86†	2.73

The cation coordination number, followed by the anion coordination number, is shown in round brackets.

In cases where cation-oxygen distances are not equal by symmetry, their mean value is given.

For the Pauling and Goldschmidt radii, radius sums have been multiplied by the appropriate correction factor of Table 2-3 to give the predicted distance.

* It would be preferable to modify Shannon and Prewitt's method and, for the anion coordination number, count only strongly-bound cation neighbours. This would reduce the anion coordination number to 2, and the predicted distances to 2.95 and 1.76 Å respectively.

† Correction for coordination number has added 0.10 and 0.11 Å to the Ca—F and Ba—F distances respectively.

Table 2-4 shows the sort of agreement that can be obtained between observed and predicted interatomic distances in representative structures with fairly simple atomic arrangements. It is reasonably satisfactory but far from perfect. Many more examples will be found in Chapter 4, and in Chapters 11 to 15, where structures are described in detail.

In working with any of these tables, we must remember that they are valid only to the approximation to which it is legitimate to treat the atoms as nearly rigid spheres—completely rigid except for the empirical correction for coordination number. It is dangerous to assume that meaningful calculations based on them can be done with very high precision.

In the remainder of this section we consider some of the difficulties involved in the use of any set of radii. (Readers to whom the ideas are new should skip this section at a first reading, and come back to it later.)

Pauling suggested that at least part of the discrepancies between observed and predicted distances is due to the *double repulsion effect*. This will be discussed more fully in §2.8; here we need only note that when the anions round a cation approach each other closely, the cation-anion distance will be greater than in structures where there is no anion-anion contact. The effect can be quite large—as much as 10 percent of the cation-anion distance in extreme cases. Pauling discussed it quantitatively for the alkali halides, and obtained a new set of radii from which, not by simple addition but by means of a rather complicated function based on the potential energies of attractive and repulsive forces (cf. §2.7), much more precise predictions of interatomic distances could be obtained. This, however, is not only too difficult a process for general everyday use, but it is not even applicable with safety to structures other than the alkali halides. Qualitatively, however, the effect must be taken into account. We must bear in mind that when radii have been derived, theoretically or empirically, from structures where there is no anion-anion contact, they will predict too small a cation-anion distance for other structures where such contact occurs; conversely, if they are derived empirically from structures with such contacts (which are very common where there are small high-valent cations) they will predict too great a cation-anion distance for structures lacking contacts. This effect may explain many apparent discrepancies.

A second difficulty, noted by Ahrens, is that there are 'shrinkage effects' due to polarisation, notably for distances involving O^{2-}. This could perhaps be thought of as a different way of looking at the problem of double repulsion rather than as a completely new effect. No theoretical rules were suggested for calculating the magnitude of the shrinkage or correcting for it; one merely chooses a value of the O^{2-} radius deduced for a similar structure from the observed cation-oxygen distance and a known cation radius, and continues to use that value for similar compounds whose cation radius is to be determined. Shannon and Prewitt's way of meeting the same difficulty was to give a different radius to each kind of O^{2-} according to its coordination number. The effects are illustrated in Table 2-5 for structures

Table 2-5 Comparison of Interatomic Distances and Radius Sums for Some Oxides of High-Valent Cations, Showing "Shrinkage"

	OBSERVED DISTANCE IN Å	RADIUS SUM IN Å		
		AHRENS	GOLDSCHMIDT	SHANNON AND PREWITT
Perovskite Family				
Ti—O { in CaTiO$_3$	1.92			
in SrTiO$_3$	1.95	2.08	1.96	2.00 (1.92)
in BaTiO$_3$	2.00			
Nb—O { in NaNbO$_3$	1.99	2.10	—	2.04 (1.96)
in KNbO$_3$	2.01			
Sn—O in BaSnO$_3$	2.06	2.11	2.06	2.09 (2.01)
Rutile Family				
Ti—O in TiO$_2$	1.94	2.08	*	1.96
Sn—O in SnO$_2$	2.06	2.11	*	2.06
V—O in VO$_2$	1.93	2.03	*	1.95
Nb—O in NbO$_2$	2.01	2.14	*	2.06

All cations considered have coordination number 6.

In cases where cation-oxygen distances are not equal by symmetry, a mean has been taken.

Anion coordination numbers are as follows: perovskites, Shannon and Prewitt's count, 6; modified method, 2 (see footnote to Table 2-4, with resulting radius sum in round brackets); rutiles, 3.

* The interatomic distance is used directly with the standard O^{2-} radius of this set to evaluate the cation radius; hence it is not an independent check.

of the perovskite family and the rutile family, in both of which the coordination number of the cation concerned is 6. With the Pauling-Ahrens radii, the apparent shrinkage is in the range 0.05 to 0.15 Å. The Shannon and Prewitt predictions are better than those of Ahrens for the rutile structures, but they are still not very good for the perovskite structures. This method of evaluating the coordination number of the oxygen ion by counting all its cation neighbours is physically much less realistic than counting only the strongly bound ones, in cases when the bonds are very unequal. Adopting the latter method, we find improved agreement. Shannon and Prewitt's allowance for the coordination of the anion is an empirical recognition of the fact that, if the bond is not purely ionic but partly covalent, its length is affected by other bonds to the same cation. We shall come back to this, from a slightly different point of view, in §2.10.

A third point to bear in mind is that, when the oxygen neighbours of a cation are not all equidistant from it, we may be wrong in assuming that the average cation-oxygen distance is always the same as when the individual distances are equal. The discussion in §2.8 suggests that the average distance can sometimes be increased by making the individual distances unequal. This effect is illustrated in Table 2-5 by the differences in Ti—O distance between $CaTiO_3$ and $SrTiO_3$ on the one hand and $BaTiO_3$ on the other. (The structures will be discussed in §11.2, §12.6, and §12.10.) None of the existing theories take account of this point.

Other recent work has been more radical in its approach. It assumes that the ionic radius is the radius of the sphere at whose surface the electron density decreases to zero (or to an arbitrarily small value). With this definition, both theory and experiment suggest a value of about 1.25 Å for the O^{2-} radius.

It is, however, not obvious that we should revise the whole set of ionic radii to fit in with this result. We have no assurance that individual radii based on electron cloud sizes are constant from one structure to another—indeed, when we remember the varying amounts of covalent character possible, it seems most unlikely that they *should* be constant—nor can we assume that they will be as valid for anion-anion contacts as for cation-anion contacts. Again, when polarisation occurs and the electron cloud becomes nonspherical (cf. §2.9), the effect on its "radius" is neither easily pictured nor easily calculated. Discrepancies between observed and predicted interatomic distances as a whole, such as those in Tables 2-4 and 2-5, and, in particular, differences between the apparent radius of the same cation in different structures, are no more easily explained with radii based on electron cloud size than with the older, accepted, sets.

The property that makes the Goldschmidt and Pauling radii, and others closely derived from them, so useful in crystal structure work is that the radius assigned to an anion for its anion-cation contacts is, to a good approximation, the same as that for its normal anion-anion contacts. This was the property implicitly assumed by Bragg in his first approach to the subject; it is the property that allows geometrical discussion of structures in terms of the close-packing of spheres, occurring locally or throughout space—a topic we shall consider further in §2.5, 4.5, and 4.18. From one point of view, it must be regarded as a lucky accident that systems based (like Goldschmidt's and Pauling's) on such an indirectly related effect as molar refractivity have this property. On the other hand, it is just *because* they have this property that they are so intuitively acceptable and so useful, and have won such wide recognition.

The fact that, with these radii, anion-anion distances are sometimes less than twice the anion radius is no difficulty, because this effect only occurs when the anions are pulled together indirectly through their strong bonds to a common cation. Knowing that the rigid-sphere model is only a first approximation, we are easily able to move from it to a second approximation in which the ions are allowed to be compressible, and their radii are therefore dependent (though only slightly) on structural forces. We shall see in §2.9 that polarisation and electron cloud distortion are associated with strong interatomic forces; by treating them as equivalent to an elastic distortion of the atom, we are often able to discuss structural details in terms of simple statics.

The advantage in using empirical rather than theoretical radii for the prediction of interatomic distances is that they have already taken into account the average effect both of local stresses and of polarisation (cf. §2.8 to §2.10). They are designed to give us a correct picture of the usual packing distances of the atoms, without any need to go into details about electron distribution or the degree of covalency in the bonds. Deviations from their predictions indicate something unusual about the structure, and can draw attention to important new effects. If we tried to work in this way with theoretically derived radii, we should first have to be sure that we had identified, and made corrections for, all the usual stresses and polarisations—a rather difficult task.

The advantages of theoretical derivations of radii are twofold: they give us confidence, because of their general agreement with the empirical treatments, that both are on the right lines; and they suggest ways of tackling discrepancies.

2.5 COORDINATION POLYHEDRA

We now consider what groupings of atoms we may expect from the rigid-sphere model. Cations tend to be surrounded by anions, and anions by cations. It is most convenient to think of the group of anions round a cation, even though some or all of the anions may be shared by more than one cation. To begin with, we shall consider only cases where the packing is fairly regular, and all the cation-anion distances are nearly equal; this will need reconsideration later. The group of anions is the *coordination polyhedron*, and the number of anions in it is the *coordination number*.

The coordination numbers for different cation-anion combinations may be predicted with the help of Goldschmidt's packing principle, which can be expressed as follows: *The number of anions surrounding a cation tends to be as large as possible, subject to the condition that all the anions touch the cation.* This principle, based on experimental evidence, may also be shown to be compatible with what we should expect from energy considerations (see §2.8). The requirement that the anions must touch the cation is sometimes called the "no-rattling rule." It implies that the overall architecture of the structure is determined by the $A - X$ bonds rather than the $A - A$ or $X - X$ interactions. The principle is only a guide and not an inviolable rule, but it enables us to predict and classify a great variety of structures which can be thought of as at least partly ionic, and draws attention to cases which appear

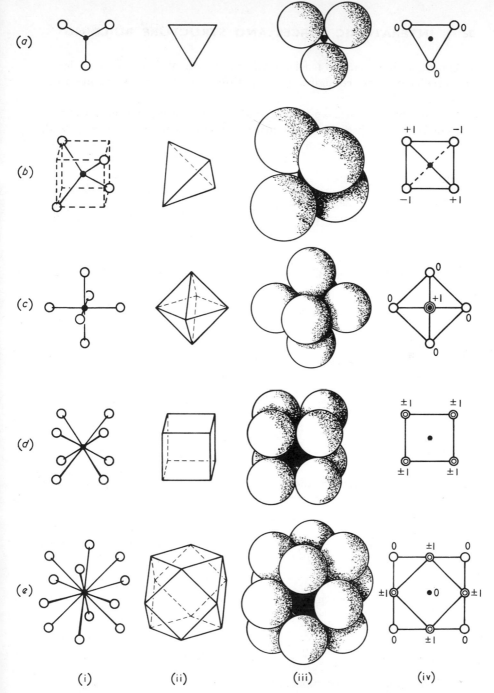

Figure 2-2 Coordination polyhedra. For each coordination number, the figure is illustrated in perspective: (i) as a ball-and-spoke arrangement, to show interatomic bonds; (ii) as an outline of the polyhedron, showing anion-anion edges; (iii) as a packing diagram, showing the relative sizes of the atoms; also (iv) in plan, projected on the horizontal plane of the other diagrams (except for (a), where the whole figure is in a plane). Each row represents a different coordination number: (a) 3-coordination (triangle); (b) 4-coordination (tetrahedron); (c) 6-coordination (octahedron); (d) 8-coordination (cube); (e) 12-coordination (cuboctahedron).

to contradict it as being in need of further thought. We shall use it to derive the predicted stable configurations for different values of the radius ratio $r_A : r_X$.

In the first instance we consider only those coordination numbers that are compatible with highly symmetrical configurations, namely, 3, 4, 6, 8, and 12, because the polyhedra they form are not only the simplest to describe but also the most important physically. They are illustrated in Figure 2-2.

(a) 3-coordination: the triangle.

The anion-anion distance is the edge of an equilateral triangle. Hence

$$\frac{r_X}{r_A + r_X} = \cos 30° = \frac{\sqrt{3}}{2}, \qquad \therefore \ \frac{r_A}{r_X} = \frac{2}{\sqrt{3}} - 1 = 0.155 \quad (2.3)$$

(b) 4-coordination: the tetrahedron.

Consider the anions as occupying four alternate corners of a cube of side a, with the cation at its centre. The anion-anion distance is the face diagonal of the cube. Hence

$$r_X = \frac{\sqrt{2}}{2} \cdot a, \qquad \text{and} \qquad r_A + r_X = \frac{1}{2}\sqrt{3} \cdot a$$

$$\therefore \ \frac{r_A}{r_X} = \sqrt{\frac{3}{2}} - 1 = 0.221 \tag{2.4}$$

(c) 6-coordination: the octahedron.

The anion-anion distance is the octahedron edge.

$$r_A + r_X = \frac{1}{\sqrt{2}} \cdot 2r_X; \qquad \therefore \ \frac{r_A}{r_X} = \sqrt{2} - 1 = 0.414 \quad (2.5)$$

(d) 8-coordination: the cube.

The anion-anion distance is the cube edge.

$$r_A + r_X = \frac{\sqrt{3}}{2} \cdot 2r_X; \qquad \therefore \ \frac{r_A}{r_X} = \sqrt{3} - 1 = 0.732 \quad (2.6)$$

(e) 12-coordination: the cuboctahedron.

The anion-anion distance is an edge of the small square in the projection; the cation-anion distance is the semidiagonal of the large square. Hence

$$r_A + r_X = 2r_X; \qquad \therefore \ \frac{r_A}{r_X} = 1 \tag{2.7}$$

There are certain comments to be made on these results.

(i) The calculated ratios $r_A:r_X$ are in each case the *lowest* values satisfying the Goldschmidt principle for the particular coordination. Groups become progressively less stable as the radius ratio increases above this value.

(ii) There is generally good agreement between the predicted and observed coordination numbers. Some cation-anion combinations, however, occur in different coordinations in different structures—or even in different parts of the same structure—especially those for which $r_A:r_X$ is near the limiting value for the larger number. (See Table 2-6 for examples.)

(iii) Other polyhedra with the same coordination numbers may occur. The most important are probably the trigonal prism (instead of the octahedron) for 6-coordination and the square antiprism (instead of the cube) for 8-coordination. (See Figure 2-3.)

(iv) Unless symmetry operators (Chapter 6) are present, the coordination polyhedra need not be perfectly regular; the $A—X$ bond lengths may differ slightly from each other, and so may interbond angles. Generally, if the ratio $r_A:r_X$ is close to the lower limit for one of the configurations in Figure 2-2, irregularities are small, but if it is much above the limit the irregularities may be quite large—sometimes so large that it is hard to determine the coordination unambiguously. In such cases, the number of really close neighbours is often much lower than would be predicted from the radius ratio; but if the investigator is more tolerant and decides to count all atoms within a distance only slightly larger than normal as neighbours, the number is greater than predicted. In such cases, attempts to be dogmatic and define a true coordination number are physically meaningless.*

(v) Radius ratios appropriate to coordination numbers 5, 7, 9, 10, and 11 can be estimated roughly by interpolation between those for 3, 4, 6, 8, and 12. For example, we might expect 5-coordination for a ratio of 0.3, and 7-coordination for one of 0.6. However, none of the former set of numbers can give regular polyhedra, nor are they associated with point symmetry operators demanding equality of bond lengths. Hence irregularities are to be expected, prediction is uncertain, and the ambiguities mentioned in the preceding paragraph are likely to occur.

(vi) A coordination number of 12 is allowed only if the cation is at least as large as the anion, which (as seen from Table 2-2) is unusual. This kind of coordination is difficult to fit into an ionic structure, unless there is a second kind of cation with a smaller coordination number.

The general conclusion to be drawn is that, while clear-cut meaning can be attached to coordination polyhedra in some cases, there are others

---

* In the literature, difficulties have sometimes been due to the assumption that the term coordination number has a unique meaning in such circumstances.

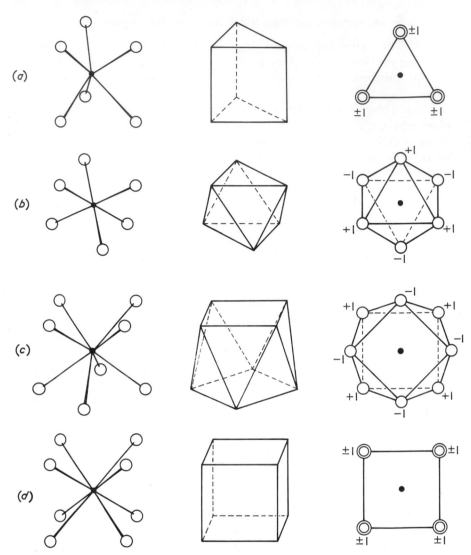

Figure 2-3 Other forms of 6-coordination and 8-coordination: (a) 6-coordination, trigonal prism; (b) 6-coordination, octahedron, drawn with triad axis vertical for comparison with (a); (c) 8-coordination, square antiprism; (d) 8-coordination, cube as in Figure 2-2(d), for comparison with (c). Drawings are constructed as in Figure 2-2, except that the packing diagrams are omitted.

in which definitions are arbitrary or impossible. Polyhedra that are well-defined geometrically are generally also stable, and are important as structure-building units; those that are ill-defined geometrically play a much less predictable role.

Table 2-6 shows the radius ratios and coordination numbers of certain useful cation-anion combinations.

Table 2-6 Common Coordination Polyhedra and Their Radius Ratios (using Goldschmidt Radii)

TETRAHEDRA 4	OCTAHEDRA 6	CUBES 8	CUBOCTAHEDRA 12
Ideal 0.22	Ideal 0.41	Ideal 0.74	Ideal 1.00
BeF_4 0.26	AlO_6 0.43	CaF_8 0.80	KO_{12} 1.01
BeO_4 0.26	TiO_6 0.50	CaO_8 0.81	BaO_{12} 1.08
SiO_4 0.30	MgO_6 0.59	KO_8 1.01	
AlO_4 0.43	MgF_6 0.59		
	NaO_6 0.74		
	CaO_6 0.81		

2.6 ELECTROSTATIC VALENCE AND THE LINKAGE OF POLYHEDRA

Polyhedra may be linked together by sharing anions in common; two or more polyhedra may share a corner, or two polyhedra may share an edge, or a face; and each may be linked in the same or a different fashion to other neighbours. Plenty of examples will be found in later chapters.

A very useful survey of the possibilities of linkage can be made, starting from the simple assumption that all the atoms are fully ionised. The success of this treatment, formulated by Pauling, must not be taken to mean that the starting assumption is correct; it is merely a good enough approximation for our purpose.

Pauling put forward an idea for the estimation of bond strength. He defined *electrostatic valence* (or *electrostatic bond strength*) as the ratio of the valency of the cation to its coordination number. Examples are given in Table 2-7. The concept is best thought of in terms of Faraday's lines of

Table 2-7 Electrostatic Valences of Representative Cation-Anion Configurations

CATION	Na^+	K^+	Ca^{2+}	Mg^{2+}	Al^{3+}	Al^{3+}	Ti^{4+}	Si^{4+}	Nb^{5+}	P^{3+}	P^{5+}	S^{6+}
Number and species of anion neighbours	$6Cl^-$	$8O^{2-}$	$8F^-$	$6O^{2-}$	$6O^{2-}$	$4O^{2-}$	$6O^{2-}$	$4O^{2-}$	$6O^{2-}$	$4O^{2-}$	$4O^{2-}$	$4O^{2-}$
Electrostatic valence	$\frac{1}{6}$	$\frac{1}{8}$	$\frac{2}{8}=\frac{1}{4}$	$\frac{2}{6}=\frac{1}{3}$	$\frac{3}{6}=\frac{1}{2}$	$\frac{3}{4}$	$\frac{4}{6}=\frac{2}{3}$	$\frac{4}{4}=1$	$\frac{5}{6}$	$\frac{3}{4}$	$\frac{5}{4}$	$\frac{6}{4}=\frac{3}{2}$

force. If the charge on the cation is ze (where z is the valency and e the electronic charge), there are $4\pi ze$ lines of force leaving it. If it is surrounded and completely enclosed by n equivalent anions, each of these receives

$4\pi ze/n$ lines of force, and hence an induced charge of $-ze/n$. The electrostatic valence is thus the fraction of the cation charge involved in the formation of one cation-anion bond.

If the cation environment is less ideal, the concept loses its exactness. For example, if there are large gaps between the anions, so that part of the lines of force pass through, to end on more distant anions, the Coulomb attraction is slightly less than would be predicted from the electrostatic valence. When polyhedra are irregular, we must count all anions of the enclosing shell as neighbours, but recognise that different A—X bond lengths are associated with different bond strengths—the shorter the stronger— though there is no way of calculating their relative values. However, when the irregularity is not too great, the simple definition of electrostatic valence can be extremely useful as a starting point, though it must not be pressed too far in detailed arguments.

Just as one cation-anion bond uses only part of the charge on the cation, so it often uses only part of the charge on the anion. Then the anion can form bonds with other cations, thus providing linkage between polyhedra. The possibilities can be derived from Pauling's *electrostatic valence rule*, which states: *The sum of the electrostatic valences of bonds to any given anion is numerically equal to the negative valency of the anion.* According to this, for example, since each of the four O^{2-} anions round Si^{4+} receives a Si—O bond of electrostatic valence 1, it may simultaneously be linked to a second similar Si^{4+}, *or* to two octahedrally coordinated Al^{3+}, *or* to three octahedrally coordinated Mg^{2+}. The bond summations in these three cases are $1 + 1$, $1 + 2 \times 1/2$, and $1 + 3 \times 1/3$ respectively, totalling 2 in each case. Other examples are given in later chapters.

Of course, the rule is not always obeyed exactly. We cannot expect it to be any more reliable (except perhaps by an averaging effect) than are our estimates of individual bond strengths, which are defined from the coordination number, and therefore are inaccurate when the polyhedra are irregular. Nevertheless, it provides a valuable guide to the overall linkages of many structures.

Even when the electrostatic valence is greater than unity, as for the tetrahedral groups $P^{5+}O_4^{2-}$ and $S^{6+}O_4^{2-}$, the rule gives reasonable results. Since the bond strengths here are 5/4 and 6/4 respectively, the tetrahedra cannot share corners with others of the same kind. On the other hand, $P^{3+}O_4^{2-}$ tetrahedra can and do share corners, forming chains of overall composition $(PO_3)_n^{3-}$. It is of course unrealistic to suppose that the central atoms of these tetrahedra are fully ionised cations, but it is interesting to see that a treatment devised for purely ionic bonds can be extrapolated so successfully to cases where the bonds are much more nearly covalent.

It is, however, among these compounds of high electrostatic valence that the prediction of interatomic distances from the sum of ionic radii becomes least accurate, and the meaning to be attached to numerical values of ionic radii least certain (cf. §2.4).

2.7 THE BORN THEORY

In complete contrast to the Goldschmidt treatment is that put forward at about the same time by Max Born. Whereas Goldschmidt's emphasis was on ionic radii derived as directly as possible from measured distances, Born began by setting up theoretical expressions for the potential energy, from which radii could ultimately be deduced. Whereas Goldschmidt's approach was concerned with three-dimensional geometry and packing, Born's largely ignored geometry. Whereas Goldschmidt's treatment could be applied to structures of many independent parameters, Born's applied to structures defined by a single parameter, the interatomic distance. While the Goldschmidt treatment has been successful in predicting an immense variety of complex structures, the Born treatment has allowed detailed quantitative study of a few simple, but very important, types.* The two treatments are thus complementary. The Born theory, now to be described, provides a background of ideas which may be useful even when its conditions are not strictly fulfilled.

Born used the postulates put forward by G. Mie in 1903, that the potential energy of any pair of atoms is the sum of parts associated with the different kinds of forces between them; that the forces, and therefore the energies, can be expressed as simple inverse powers of the interatomic distance; and that the energy of the crystal is the sum of the energies of all interacting pairs of atoms. Writing, for any pair of atoms,

$$V_{\text{pair}} = -\frac{a}{r^m} + \frac{b}{r^n} \qquad (2.8)$$

where the first term is associated with the attractive force and the second with the repulsive force, we see that, since the repulsive force falls off more rapidly, $n > m$; and if the attractive force is electrostatic, $m = 1$. For summing over all pairs of atoms, we must either assume that all the constants a, b, m, n are the same for each pair, or, if $m = 1$, we may take b and n as the same for each pair and a as proportional to the known product of the two ionic charges concerned. We further assume that the separation of any pair is directly proportional to the separation of the nearest-neighbour pair, the constants of proportionality being fixed by the particular structure type. The energy can then be written in the form

$$V = -\frac{A}{r^m} + \frac{B}{r^n} \qquad (2.9)$$

---

* Born proceeded to develop his approach for moving atoms in a way which has been most fruitful for studies of lattice dynamics (mentioned qualitatively in Chapter 14) but *not* for the prediction of crystal structures. We are concerned here only with the introductory theory dealing with atoms at rest.

where V represents the energy per nearest-neighbour pair (i.e., the total energy divided by the number of such pairs), and r is now the nearest-neighbour distance. In calculating A and B from a and b, we note that, because n is large, only nearest-neighbour interactions contribute to B. If m is fairly large, the same may be true for A; but if the attractive force is electrostatic, and $m = 1$, summation will certainly have to go out to much more distant neighbours, and will include electrostatic repulsions between like ions as well as attractions between unlike ions. Geometrically, the calculations for $m = 1$ are tedious, but can be done once and for all for any structure type fulfilling the condition that all interatomic distances depend only on the single variable r, a condition which requires that the structure must be cubic. The results are usually expressed by giving the *Madelung constant* for the structure type, defined as the ratio of the net electrostatic force per formula unit* to the electrostatic attractive force between a pair of nearest neighbours.

Similar calculations can be made for simple noncubic structures which have only one parameter besides r, namely, the axial ratio; the Madelung constant is then a function of the axial ratio.

Even for isodesmic structures, the energy equation 2.9 is only an approximation to the truth. Two modifications have been introduced on theoretical grounds. Quantum-theory treatment suggests that the repulsive potential should be an exponential function of the form

$$V_{\mathrm{rep}} = B' \exp \left[-r/\rho \right] \qquad (2.10)$$

where B' and ρ are new constants. This expression is of course very different from B/r^n for small values of r, but such values are physically unattainable. The two expressions agree in predicting a rapid decrease of energy to a negligible value outside a narrow "contact range" of distances, and immediately within that range they can be made to match closely by an appropriate choice of constants. The difference between the two in numerical work is very small. The other modification theoretically required is the addition of a further term on the right-hand side, inversely proportional to r^6, to allow for the van der Waals interaction. Since neither modification makes sufficient difference to be worth the trouble involved for most purposes, we shall continue to use the simpler expression, equation 2.9.

We can re-express V, the net cohesive energy per bond, in a way which will allow us to calculate the force constant of the bond. We use the fact that, at equilibrium, V has its minimum value V_0 (which is negative in sign), and

---

* The term formula unit is a general one, which can be used in connection with structures where there are no discrete molecules; it specifies the set of atoms named in the formula, regardless of their positions in space or bonding relations. The *mole* then comprises N formula units, where N is Avogadro's number. For sodium chloride, the formula unit is obviously NaCl. In other cases, in which there are alternative ways of writing the formula, for example, ABX_n or $A_2B_2X_{2n}$, either choice is permissible for the formula unit, provided the mole is chosen to correspond.

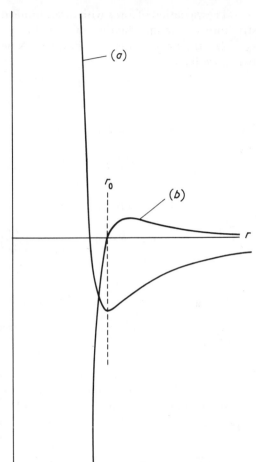

Figure 2-4 Variation with interatomic distance of (a) the energy, (b) the force acting on an atom, for an isolated pair of atoms. The force is the gradient of the energy. The equilibrium interatomic distance is given by the minimum of the energy curve and the zero of the force curve. (These curves have been drawn assuming the attractive and repulsive energies to be A/r^2 and B/r^9 respectively; any other reasonable assumptions would give very similar diagrams.) Notice how the magnitude of the force increases much more rapidly with distance for compressions (here shown as negative forces) than for extensions (positive forces).

therefore $\partial V/\partial r$, the interatomic force, is zero. This is shown graphically in Figure 2-4, where the curves have been drawn for $m = 2$, $n = 9$, and $B/A = 1000$. It gives us values for the equilibrium distance r_0 and energy V_0. Analytically we have

$$\frac{mA}{r_0^{m+1}} - \frac{nB}{r_0^{n+1}} = 0 \qquad (2.11)$$

Hence

$$\frac{mA}{nB} = r_0^{m-n} \qquad (2.12)$$

and

$$V_0 = -\frac{A}{r_0^m}\left(\frac{n-m}{n}\right) \qquad (2.13)$$

Comparison of equations 2.13 and 2.9 shows that, at equilibrium, the cohesive energy per bond V_0 is a fraction $(n-m)/n$ of the net electrostatic energy per bond.

The variation of force with interatomic distance is found by allowing the structure to expand slightly, so that, for a stretching force δF, r_0 becomes $r_0 + \delta r$; the force constant is $(\partial F/\partial r)_0$. Now by differentiating equation 2.9 twice, we have

$$\frac{\partial^2 V}{\partial r^2} = -\frac{m(m+1)}{r^{m+2}} A + \frac{n(n+1)}{r^{n+2}} B \qquad (2.14)$$

Hence, using equation 2.13,

$$\left(\frac{\partial^2 V}{\partial r^2}\right)_0 = \frac{mA}{r_0^{m+2}} (n-m) = -\frac{mnV_0}{r_0^2} \qquad (2.15)$$

Since, by Taylor's theorem,

$$\delta F = F - F_0 = \frac{\partial V}{\partial r} - \left(\frac{\partial V}{\partial r}\right)_0 = \left(\frac{\partial^2 V}{\partial r^2}\right)_0 \delta r \qquad (2.16)$$

it follows that

$$\left(\frac{\partial F}{\partial r}\right)_0 = \left(\frac{\partial^2 V}{\partial r^2}\right)_0 = -\frac{mnV_0}{r_0^2} \qquad (2.17)$$

If we know m and n, this relates the force constant to the bond energy and the bond length.

We now want to relate equation 2.9 to the macroscopic quantities lattice energy U (cohesive energy per mole) and compressibility κ. We need to know the volume per formula unit in terms of nearest-neighbour distance r, and the number of bonds per formula unit. Let them be αr^3 and β respectively. Then

$$U = N\beta V \qquad (2.18)$$

where N is Avogadro's number, and the molar volume v is therefore $N\alpha r_0^3$. The compressibility is defined as

$$\frac{1}{\kappa} = v_0\left(\frac{\partial^2 U}{\partial v^2}\right)_0 \qquad (2.19)$$

But

$$\frac{\partial U}{\partial v} - N\beta\left[\frac{\partial V/\partial r}{\partial v/\partial r}\right] - \frac{N\beta}{3N\alpha}\left[\frac{\partial V/\partial r}{r^2}\right] \qquad (2.20)$$

and

$$\left(\frac{\partial^2 U}{\partial v^2}\right)_0 = \frac{N\beta}{3N\alpha}\left[\frac{1}{3N\alpha r^2}\left(\frac{1}{r^2}\frac{\partial^2 V}{\partial r^2} - \frac{2}{r^3}\frac{\partial V}{\partial r}\right)\right]_0 = \frac{\beta}{9N\alpha^2 r_0^4}\left(\frac{\partial^2 V}{\partial r^2}\right)_0 \qquad (2.21)$$

since $(\partial V/\partial r)_0 = 0$. Hence, substituting in equation 2.19 for v_0 and $(\partial^2 U/\partial v^2)_0$,

$$\frac{1}{\kappa} = \frac{\beta}{9\alpha r_0}\left(\frac{\partial^2 V}{\partial r^2}\right)_0 \qquad (2.22)$$

or, using equation 2.17,

$$\frac{1}{\kappa} = \frac{\beta}{9\alpha r_0}\left(\frac{\partial F}{\partial r}\right)_0 \tag{2.23}$$

Thus,

$$\frac{1}{\kappa} = -\frac{\beta\, mn V_0}{9\alpha r_0^3} \tag{2.24}$$

The quantities α and β are constants for the structure type, easily calculated from its geometry. Hence if we know the compressibility we can deduce the force constant in the bond from equation 2.23. Moreover, if we know the lattice energy, r_0, and m, we can predict n from equation 2.13 (since A is related to the Madelung constant A' by a factor involving only universal constants and the cation valency), and hence we can find the compressibility from equation 2.24. In general, for ionic crystals, if we know the structure type, r_0, and any one of the quantities n, U_0, or κ, we can from these equations predict the other two. Strictly speaking, the compressibilities used should be those at absolute zero. Since, however, applications of the theory generally involve simplifying assumptions about the structures (sometimes explicit and sometimes implicit) which may introduce quite large errors, it is often more practical to use it only in an order-of-magnitude way, for which room-temperature values are good enough. It can then be quite helpful in cor-relating physical characteristics with structures over a wider range of materials.

Much detailed work has been done on the alkali halides, and for them satisfactory agreement has been found between predicted and experimental values of the lattice energy. The exponent n has values between 5 and 12, depending on the number of core electrons; from equation 2.13, it can be seen that the lattice energy is not sensitive to uncertainties in n. Some simple compounds in which other evidence suggests that the bonds have considerable homopolar character give good agreement when their lattice energies are calculated as if they were ionic; it is not permissible to reverse the argument and take this as evidence for their ionic character.

2.8 STRESSED AND UNSTRESSED BONDS

As stated in the last section, it is only to very simple structures that the Born theory as outlined there applies. In actual structures we cannot assume that no other forces act on a cation-anion pair except their mutual attraction and repulsion and (if the attraction is electrostatic) electrostatic interactions with more distant neighbours.

Consider, for example, a cation A at the centre of a square of anions X (Figure 2-5). The equilibrium of the group as a whole depends on the

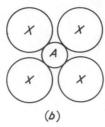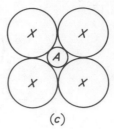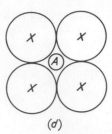

(a)　　　　　　(b)　　　　　　(c)　　　　　　(d)

Figure 2-5 Hypothetical example of the same coordination polyhedron (a square) with the same radius sum but different radius ratios. Circles A are the cations; circles X, the anions.

repulsive stresses in X—X as well as the algebraic sum of the Born attraction and repulsion in each A—X bond; obviously, the net force in A—X is *not* zero, but must be a tension, holding the whole together against the X—X repulsions. These X—X repulsions constitute the double repulsion effect mentioned in §2.4, and they must be allowed for in deducing interatomic distances from the Pauling radii. They are made up of two parts: an electrostatic part proportional to $4e^2/(r_{X-X})^2$ and a covalent part varying over a very short distance near $r = r_X$, where r_X is the effective radius of X. The electrostatic part is small, but varies slowly with distance, and can be allowed for in calculations; the covalent part can be negligible or very large according to the relative sizes of the ions. The examples in Figure 2-5 are all drawn for the same value of length A—X, but with decreasing values of r_A/r_X. In diagrams (a) and (b), covalent repulsions are small; (c) is the limiting case at which they suddenly become very large. In (d) the X—X repulsions control the size and shape of the group; the abnormally long A—X bonds are a clear indication of tensile strain, and hence of tensile stress. The Goldschmidt packing rule (§2.5) is a statement of the observed fact that situations like (d) rarely occur. It might therefore be re-expressed as follows: *Coordination polyhedra in which the A—X bonds are in too great tension tend to be unstable.* We must include the qualification "too great" because a structure in which moderate stresses of tension and compression are balanced against one another may well be more stable (in the sense that it needs greater energy to disrupt it) than one in which all stresses are zero. Since small strains in bonds can correspond to great stresses, and ionic radii are not known very exactly, we cannot know with certainty just what stress can be tolerated. The rule (like so many others) is a general guide rather than an absolute criterion. It is all the more difficult to test because there are many cases where, without independent means of finding the cation radius, we cannot tell whether the situation corresponds to that in Figure 2-5(c), as we are apt to assume, or to Figure 2-5(d), as it may well do.

It is however worth considering theoretically what should happen if the cation is slightly too small. Assuming that the same group of anions is retained, we find that at first (as expected) the cation remains central, but with further decrease in size it moves off-centre. The Born-theory treatment of this is as follows:

Suppose the unstressed A—X length is r_0, and the two atoms X_1 and X_2 are held fixed at a separation $2r_0 + s$ (Figure 2-6). Then if A is at a distance $r_0 + x$ from X_1, it is at a distance $r_0 + s - x$ from X_2. The oppositely directed forces F_1 and F_2 on A due to X_2 and X_1 are shown in Figure 2-7, for two values of s. They must of course be equal, for equilibrium; hence x must correspond to an intersection of the two curves. But will equilibrium be stable? The condition for stable equilibrium is that for any change δx the net force is in the direction opposite to δx, i.e., that dF/dx is negative, where F is the resultant of the two separate forces. Now because the force curve F_1 was derived from an inverse power law with high n, its slope gradually decreases with increasing x, and F_2 is its mirror image. Hence, for fairly small values of s, as in Figure 2-7(a), the symmetrical position c is stable; but for greater separations, as in Figure 2-7(b), the symmetrical position is unstable, and the atom A has the choice of two off-centre positions, B and D, equivalent to each other.

In a sense, the "no-rattling rule" is still satisfied. The cation does not move freely over the larger space available to it inside its polyhedron of X's, but is located at a particular site other than the geometrical centre.

Empirically, this off-centring is observed in many structures containing small high-valent cations in octahedral coordination, e.g., in those with TiO_6 and NbO_6 groups. It is not easy to predict the relative magnitudes of the off-centre displacements from the tabulated "ionic" radii, because the latter are not known accurately enough for the purpose; empirically, the displacement increases in the order Ta^{5+}, Ti^{4+}, Nb^{5+}, W^{6+}, V^{5+}, though the ionic radius of Ti^{4+} appears as less than that of Nb^{5+} in all the sets in Table 2-2,* and the differences of radii between other members of the series are neither large nor consistent. For very large displacements, such as those of V^{5+}, some of the O's of the original octahedron are so far from the cation that they may not be counted as neighbours, and the fact that the groups can be interpreted as extreme members of the series may be overlooked.

It would be satisfactory if we could calculate the degree of misfit—the separation s in Figure 2-7—at which off-centring sets in. There are, however, two reasons why it is difficult or impossible to do so. First, as already mentioned, we have no good independently known values of cation radii from bonds which are not stressed, and therefore we cannot measure the

---

* A possible explanation is based on the effect mentioned in §2.4. Empirical radii are all evaluated from mean interatomic distances within a tetrahedron or octahedron, whether the individual distances are equal or unequal. But, for the same cation, the mean is greater if they are unequal, because of the nonlinearity of the forces. A cation like Nb^{5+}, which is almost always off-centre, thus appears empirically to be larger than one like Ti^{4+}, which is often central. If cation radii in such cases were deduced from the shortest cation-oxygen distances, a different picture would appear. Compare, for example, the shortest distances in $BaTiO_3$ and $KNbO_3$, which are 1.90 and 1.86 Å respectively, with the octahedral mean distances given in Table 2-5.

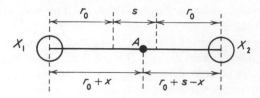

Figure 2-6 Separation of anions X_1 and X_2 at more than twice the normal cation-anion distance r_0.

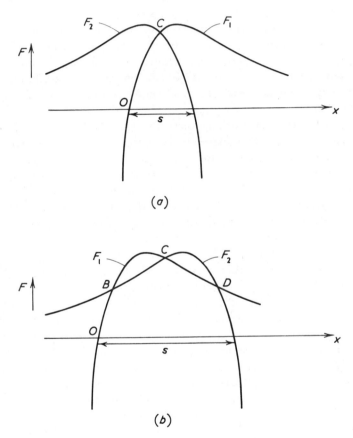

Figure 2-7 Forces F_1 and F_2 on cation A due to anions X_2 and X_1. Force F_1 is considered positive if it acts from left to right, force F_2 if it acts from right to left. At the origin O the cation is at its normal distance r_0 from X_1, but at a distance $r_0 + s$ from X_2. (a) Small separation s; equilibrium position of cation is at C. (b) Large separation s; equilibrium positions of cation are at B and D.

misfit. Secondly, the argument has been carried out in terms of point atoms. In fact, the atoms are deformable clouds of electrons (as we shall consider more generally in the next section, §2.9). When A approaches X_1, for example, it distorts the electron cloud of X_1, and its own electron cloud is distorted by X_1; negative charges tend to concentrate on the side of X_1 near A, and positive charges on the opposite side; the net electrostatic attraction between A and X_1 increases much more than if they were point atoms. This enhances the stability of the off-centre position. At the same time, it alters the form of the attraction and repulsion potentials in an unpredictable way. Thus, though our argument in terms of point atoms and central forces remains qualitatively correct, and provides a useful guide to what is happening, it cannot be used for actual calculations.

The situation in actual crystals may be even more complicated than in our hypothetical example. Thus, the off-centring of a cation in a square of anions need not be along a diagonal, as we assumed for simplicity; it might be parallel to an edge, or lie in a more general direction. Moreover, the anion polyhedron may be subject to forces from other atoms in the crystal; we shall have many examples later where the most strongly bonded polyhedra form a continuous network with other cations filling the interstices between them and capable of exerting forces on them. In fact, we must revise our earlier statement that cation-anion bonds are responsible for the architecture of crystals; indirectly this is still true, but in many cases where anions are in contact in polyhedron edges, it is the rigid polyhedron which is the structural building block, with the cation merely serving to hold the polyhedron together.

All this complication does not mean that we cannot hope to understand energy relations in actual structures. It *does* mean that we must be careful to concentrate first on the most important features, the most strongly bonded units, and then allow for the effect of weaker forces as correction terms. The real contribution of the Born theory to our understanding of crystals remains qualitative rather than quantitative.

2.9 POLARISATION

The discussion in the last paragraph drew attention to the deformation of the electron cloud of an ion by the electrostatic field of a neighbouring ion. This effect is known as *polarisation*. It is easiest to picture for an isolated pair of atoms. Each deforms the other, but the anion cloud is generally larger and the outer shell less tightly bound to its core than is the cation cloud, and so the anion is more easily deformed. We are therefore usually concerned with the *polarising power* of the cation, the *polarisability* of the anion. In general, small highly-charged cations have high polarising power; indeed, for atoms with completed rare-gas shells, the electrostatic valence is a measure of the polarising power. Atoms with unfilled d-shells have greater polarising power than would be expected from their valence; this point will be considered later.

There is danger here of confusion with the terminology of electrostatic field theory. *Polarisation* in electrostatics usually refers to the separation of + and − charges in a homogeneous medium; it may be defined as dipole moment per unit volume. Assuming that the dipoles are created by an electric field, the ratio of polarisation to field is the polarisability. Polarisation and field are both vectors; if the medium is anisotropic and the polarisability depends on direction, it will therefore be a tensor. The difficulty in translating this macroscopic theory into atomic terms is that there is nothing obvious corresponding to a dipole. Moreover, we are not here concerned with any externally-applied fields which can be varied at will to find the resulting polarisation and hence the polarisability. The field acting on one ion is that due to all neighbouring ions. For example, in sodium chloride, the field at any point within the Cl^- ion is due (in addition to the contribution of the Cl^- ion itself) to the six neighbouring Na^+ ions, as well as more distant neighbours; it is not uniform everywhere within Cl^-, but is symmetrical about the centre, though not radially symmetrical, and varies rapidly with distance from the centre. It is possible, if one wishes, to think of each elementary unit of the electron cloud as drawn out into a dipole: one might perhaps think of the whole as a set of six dipoles, each lying within the Cl^- ion and directed along one Na^+–Cl^- bond.

In a symmetrical case like NaCl, the effect may be calculable, but when the atomic environments are less symmetrical it is much more difficult, especially if the atoms cannot be considered fully ionic. The character and extent of the polarisation in any one bond depends not only on the atoms directly linked by the bond but also on the polarising power of other neighbours, and on the directions of their closest approach. In fact, not only the polarisation but also the polarisability of an atom depends on direction. In this connection, interesting experimental evidence about the electronic polarisability of O^{2-} comes from measurements of the refractive index of oxides. (Electronic polarisability refers to the electron cloud deformation when the atomic nucleus remains at rest). The polarisability varies from one compound to another, and is approximately proportional to the volume available to the oxygen atom; i.e., it is greater when the oxygen atom has more room for lateral distortion. This seems to imply that the electron cloud can be distorted more easily in directions in which it is not restricted by the close approach to other atoms. If this nonspherical distortion is true for the electric charge within the atom, it may also be true for the function expressing the repulsive force. We must therefore be prepared for the possibility that a highly polarised atom may behave, for packing purposes, as if it were no longer a sphere.

There is another way of considering polarisation which is helpful in understanding its role in crystal structures. Deformation of the electron cloud of one atom by the field of its oppositely charged neighbour always increases their mutual attraction and decreases the distance between them. If they are already fairly close together, this decrease of distance causes or increases overlap between their wave functions, and hence increases the covalent interaction. Without going into details, we may always think of increasing polarisation, in a series of compounds, as equivalent to increasing covalent character in their interatomic bonds. Conversely, even when we know that interatomic bonds are appreciably covalent, we may include them in the general account of structures derived in terms of simple electrostatic forces if we add the proviso that the atoms concerned are highly polarising or polarisable. For example, when atoms have unfilled *d*-shells, our earlier statement that they are highly polarising is merely a different way of saying that their tendency to form covalent bonds is important. This way of using the concept of "polarisation" as equivalent qualitatively to "degree of covalent character" allows the extension of what is sometimes thought of as the ionic theory of structures very successfully to structures which are certainly far from ionic. To understand the implications, however, we must begin at the other end and consider what we should expect of structures involving mainly covalent bonds.

2.10 COVALENT BONDS

The theory of covalent bonds is dealt with thoroughly in most books of structural chemistry, and will not be repeated here. The feature with which we are chiefly concerned is their *direction*. If a number of atoms X are grouped regularly round an atom A, it may be possible for the wave functions of all the atoms to recombine into a new arrangement, giving a set of A—X bonds. The stability of any bond system depends partly on the number and nature of the orbitals provided by A and X, partly on the electrons available to fill them, and partly on the orientations of the A—X directions and the lower limit set to the A—X distance by the effective "radii." Since the most stable bond systems, in general, are the symmetrical ones—in particular, the tetrahedral and octahedral systems—the results will in many cases not be very different from those predicted by simple packing considerations. Even the interatomic distances will generally not be grossly different for ionic and covalent bonds, since the cores of the atoms, inside the electrons used for covalent bonding, are very similar to ions with completed rare-gas shells. It is therefore often possible to explain an observed atomic group equally well by either theory, as far as its geometry is concerned. In such cases, we conclude that both ionic and covalent forces contribute to its stability, but we cannot tell which kind predominates. If we need to know more about the character of the force, we must either think carefully about small differences in interatomic distance or else turn to investigation of physical or chemical properties which depend on the difference.

There is, however, one way in which the difference between ionic and covalent forces shows up structurally. It depends on their directedness. Ionic forces are central forces, and if cation A is enclosed in a tetrahedron of four anions X, it does not matter (as far as the AX forces are concerned) that the tetrahedron should be regular; on the other hand, a tetrahedral covalent bond system has its AX bonds at the exact tetrahedral angle. Deformation of the regular tetrahedron by other structural forces would be easy for ionic AX bonds, but very difficult for covalent AX bonds. Deformation does occur, of course; in organic chemistry, which deals mainly with covalent bonds, the other forces are often collectively referred to as "steric effects". For partly covalent bond systems, we may expect bond angles approximating to the ideal covalent, but with much greater departures from it than would be likely for pure covalent bonds.

The difficulty in applying this criterion to many important structures arises when the ionic packing is too good. If cation A is exactly the right size to fit in a tetrahedron of anions X which all touch each other, any possibility of decreasing any of the angles $X—A—X$ is ruled out because of the $X—X$ repulsions. Figure 2-5 illustrates this for a hypothetical example: in (a) or (b) distortion to something other than a square would be easy, if the $A—X$ bonds were ionic, but in (c) it would be prevented. Of course, the $X—X$ bonds are not absolutely incompressible, but the forces needed are of the same order of magnitude as those involved in distortion of covalent bond angles. Thus a close-packed structure such as (c) might be either ionic or covalent, but a very open structure which was nevertheless symmetrical, such as (a), would probably be covalent.

An example is provided by ZnS. From Table 2-2, if we take the ionic radii for Zn^{2+} and S^{2-}, the radius ratio is 0.47, well exceeding the limit for 6-coordination (§2.5), whereas in the actual structure (§4.14) the coordination is tetrahedral; moreover, the S's are not close-packed, but tetrahedrally coordinated by Zn. On the other hand, all tetrahedra are regular. The openness of the structure therefore suggests covalent rather than ionic bonding, and this is confirmed by the physical properties of the material. It is important, however, to remember that we are not choosing between two wholly different kinds of bonding, but estimating the character of the bond as lying somewhere between the two extremes, nearer to one than the other.

There are other examples in which covalent bond character can be recognised by the geometry of an atom's environment. The Zn^{2+} atom is tetrahedrally coordinated in ZnO, though the radius ratio is even larger than for ZnS. The Cu^{2+} atom often has a square coordination of four near neighbours, with two others at slightly greater distances perpendicular to the plane of the square. The Pb^{4+} atom has four bonds in a one-sided arrangement; the anions form the square base of a flat pyramid, with Pb^{4+} at the apex. Perhaps the most important example of all is the oxygen atom, which must be considered in more detail.

Many oxides are based on a close-packing of oxygen atoms and the environments of the O atoms are well-balanced electrostatically. For oxides of small high-valent cations, however, this is not always true. In silicates, for example, if SiO_4 tetrahedra share corners, an O atom common to both can have no other cation neighbours. Electrostatically, therefore, one would expect the two Si—O bonds to be collinear. But in fact straight-line Si—O—Si configurations are extremely rare in silicates; the angles found range from roughly 100° to 160°. Quartz (§11.14) is a clear-cut example. There are no obvious packing considerations to explain the distortions from 180°. On the other hand, chemical theory shows that covalently-bonded oxygen tends to have two bonds at an angle of about 110°. The observed angles, and their consistency from one silicate to another, are thus most naturally explained as evidence for partly covalent bonding. In structures where the electrostatic bond strength is less than 1, we expect the covalent character in general to be less, but not to disappear. This strong but varying tendency to covalent bonding in oxides is the cause of the high electronic polarisability of oxygen, and its variability from one compound to another.

Modern studies of bonds associated with high-valent "cations," starting from the assumption of covalent character and using chemical bond theory, show that their lengths are significantly dependent on the other bonds to the "anion." An example in which the effect is well-marked occurs in silicates, where Si—O bonds have been found to differ by 0.08 Å according to whether the oxygen is shared between the two Si atoms ("bridging") or has only Na atoms as its other neighbours ("nonbridging"), the bonds being shorter in the latter case. Most differences are less than this, and they are often smoothed out, at least partly, when the interatomic distances are averaged over the tetrahedron. It thus remains true that the sum of the ionic radii provides a first approximation to the interatomic distance, but that more detailed chemical theory may help towards an improved second approximation; and its general approach can, in any case, indicate qualitatively the sort of corrections to be expected.

2.11 PACKING STRUCTURES AND LINKAGE STRUCTURES

It is useful for many purposes to distinguish between *packing structures* and *linkage structures*. Packing structures are typically ionic, with a specific volume not much greater than that required for the anions alone, with bonds of not very unequal strengths linking each anion to a number of cations fairly uniformly distributed in space. Linkage structures are typically semipolar, and very open, with high specific volume; the strongly bonded polyhedra may form cages or enclose cavities capable of holding other atoms or ions, but the bond strengths of the cavity cations are much less than those of the framework cations. The polyhedra of packing structures tend to share many edges or faces; those of linkage structures generally share corners only.

Packing structures tend to have high symmetry; linkage structures often have lower overall symmetry, but their polyhedra remain very regular. Very often, of course, a given structure may be discussed equally well in terms of its packing or its linkage, but the extreme cases can be so different that it is well to remember that both exist. Typical examples are sodium chloride (§4.7) for a packing structure, quartz (§11.14) for a linkage structure.

Linkage is so important for silicates that it has been used as a basis for classification. The classification scheme, developed by W. L. Bragg, who first analysed many of the structures on which it is based, is as follows (with names for the classes as suggested by H. Strunz).

(a) Structures with separate SiO_4 tetrahedra (nesosilicates)
(b) Structures with two tetrahedra sharing one corner (sorosilicates), or several tetrahedra sharing two corners joining up in a ring (cyclosilicates)
(c) Structures with all tetrahedra sharing two corners to form infinite chains, or with a small proportion sharing a third corner, giving double chains (inosilicates)
(d) Structures with all tetrahedra sharing three corners, to form infinite sheets (phyllosilicates)
(e) Structures with all tetrahedra sharing all corners, to form three-dimensional frameworks (tectosilicates)

This provides a valuable starting point for consideration of such materials. It is, however, oversimplified in two ways, which can be misleading if it is used unthinkingly. First, one must count AlO_4 tetrahedra (and perhaps tetrahedra containing other high-valent cations) as if they were SiO_4 tetrahedra when they play the same role in the structure, and the question of when to do so is sometimes ambiguous. Second, in emphasising the SiO_4 tetrahedra we must not forget the polyhedra of other cations forming links between the units built from SiO_4 tetrahedra. Indeed, sometimes (especially with isolated SiO_4 tetrahedra) the framework is really determined by the octahedra round the other cations, and it is better to think of these in the first instance, with SiO_4 tetrahedra merely providing links. One example is the structure of Mg_2SiO_4 discussed in §11.9; many others occur among silicates investigated in the 1950's by N. V. Belov and his coworkers. Looked at in this way, the packing features of such structures become more important, and the linkage features less so.

2.12 SUMMARY

The concepts which are important in discussing structures are as follows:

(a) Ionic (or atomic) radii predict interatomic contact distances.
(b) Packing of ions as rigid spheres determines the coordination number (the number of anions surrounding a cation).

(c) Coordination polyhedra tend to be regular; the tetrahedron and the octahedron are the commonest and most important.

(d) The character of the interatomic forces can generally not be deduced from the geometry of the polyhedra. Semipolar bonds intermediate between ionic and covalent are common.

(e) Structures with covalent and partly covalent bonds forming continuously linked networks can conveniently be considered as ionic structures with varying degrees of polarisation of their ions.

(f) Rigidity of bond angles is characteristic of covalent bonds but not of ionic bonds.

(g) Many structures can easily be described in terms of packing of rigid spheres, but in others close-packing only operates to form polyhedra round high-valent cations, which are linked by their corners to give a more open framework.

PROBLEMS AND EXERCISES

1. Considering the alkali halides (except the Cs compounds), compare the observed interatomic distances (Table 2-1) with those predicted by the different sets of atomic radii (Table 2-2). Which set of radii gives best agreement?

2. If ionic radii were perfectly additive, the sums of interatomic distances in pairs of compounds $A_1X_1 + A_2X_2$ and $A_1X_2 + A_2X_1$ would be equal. Test this relation for the alkali halides (except the Cs compounds), using Table 2-1. Show that the discrepancy is greatest when X_1 is F. Can you detect any other regularities in the discrepancies? (It is convenient to define A_1 as having a lower atomic number than A_2, and X_1 lower than X_2.)

3. Look up the lattice parameters of MgO, MnO, CaO, and the corresponding sulphides and selenides, all of which have the rock salt structure (cf. §4.7). Assuming that in some of them the cations may be too small to touch all the anions of their polyhedron (in spite of §2.5), estimate, without using any other information, the radii of Se, S, O, Ca, Mn, and Mg. Compare the values with those listed in Table 2-2.

4. The ammonium ion, NH_4^+, often behaves like a spherical ion, giving structures isomorphous with those of corresponding compounds of alkali metals. NH_4I has the rock salt structure (cf. §4.7), with the coordination number 6. The cube edge a is 7.244 Å. Using Table 2-2, deduce the effective radius of NH_4^+, and predict its coordination number in the other alkali halides. Are these predictions fulfilled?

5. Using Table 2-2, predict the interatomic distances for the following pairs of atoms, with the cation coordination numbers mentioned. For co-ordination numbers other than 6, compare results using (a) Goldschmidt radii, with the correcting factors of Table 2-3, (b) Shannon and Prewitt radii for the specified coordination. Al—O, 6, 4; Mg—O, 6, 4; Ca—F, 6, 8; Ba—O, 6, 12; Na—O, 6, 8; K—O, 6, 8, 12; Ti^{4+}—O, 6, 4; V^{5+}—O, 6, 4; Zn—O, 6, 4; Zr—O, 6, 4, 8.

6. In each of the examples in Problem 5, calculate the anion–anion edge length of the polyhedron, assuming it to be regular. Which of the polyhedra do you consider likely to be stable arrangements?

7. With oxygen as anion, what coordination numbers would you predict for the following cations: Zn^{2+}, Fe^{2+}, Ni^{2+}, Cu^{2+}, Y^{3+}, La^{3+}, Ga^{3+}, Ge^{4+}, Zr^{4+}, Sn^{4+}, Mn^{4+}, V^{5+}, Nb^{5+}, Ta^{5+}, Cr^{6+}, Mo^{6+}, W^{6+}, U^{6+}?

8. Two different expressions may be used for the repulsive force: that in equation 2.10, and the second term of equation 2.9. Show that these can be made very nearly equal when r is close to r_0, and find the connections between B' and ρ, B and n, to make this true.

9. (a) Find the connection between the quantity A in the text and the Madelung constant A' for an ionic structure, using S.I. units.

   (b) Write down expressions for (i) the lattice energy, (ii) the compressibility, in terms of n, r_0, the cation valency, constants for the structure type, and universal constants.

10. For the following, use values of r_0 from Table 2-1.
    (a) RbCl has the rock-salt structure (§4.7), with $A' = 1.75$, 6 bonds per formula unit, and volume $2r_0^3$ per formula unit; $m = 1$ and $n = 9.5$. Find its lattice energy, its compressibility, and the force constant in the RbCl bond.

    (b) The following crystals all have the rock salt structure. Use their compressibilities to estimate n and predict the lattice energies; compare the latter with the experimental values given.

	COMPRESSIBILITY	LATTICE ENERGY
NaCl	42.6×10^{-12} $m^2 N^{-1}$	0.765×10^6 J mole^{-1}
KCl	56.2×10^{-12} $m^2 N^{-1}$	0.685×10^6 J mole^{-1}
RbBr	79.3×10^{-12} $m^2 N^{-1}$	0.640×10^6 J mole^{-1}

    (c) Assuming the crystals in the following table to be ionic, predict for each the lattice energy, compressibility, and force constant in the bond. Compare your predictions for lattice energy and compressibility with their observed values, and comment on them in relation to corresponding quantities in (b).

Table 2-8

REFERENCE TO STRUCTURE TYPE		A'	α^*	β	r_0	n	U_{obs}	κ_{obs}
					(Å)		$(10^6 \text{ J mole}^{-1})$	$(10^{-12} \text{ } m^2 N^{-1})$
CsCl	§4.8	1.76	1.54	8	3.56	10.5	0.65	59.4
MgO	§4.7	1.75	2.00	6	2.10	7.0	3.8	7.2
BeO†	§4.10	1.64	3.08	4	1.65	6.0	4.4	—

* The derivation of α can be carried out as an exercise after reading Chapter 4.
† BeO is not cubic, but its bonds are very nearly equal.

11. A regular octahedron of oxygen atoms has edge length l. The central cation B is displaced towards one corner by a small fraction x of the centre-to-corner distance. Calculate the six B—O distances.

    Repeat for the case in which the displacement is a small fraction of the centre-to-mid-edge length, and again when it is a small fraction of the centre-to-mid-face length.

12. The structure of scheelite, $CaWO_4$, consists of isolated WO_4 tetrahedra held together by Ca; all Ca atoms have the same set of neighbours, and so have all the O atoms. Pauling's valence rule is exactly satisfied. What can you conclude about the coordination of Ca and of O?

13. The mineral thaumasite contains isolated $Si(OH)_6$ groups, and the OH groups can be assumed to be effectively spherical, with the same radius as O^{2-}. The three OH groups forming one face of the octahedron have Si—OH bond lengths of 1.73 Å, the three forming the opposite face 1.83 Å. Comment on this arrangement.

14. The change of spin state in Fe^{2+} results in a change of radius and hence allows rearrangement of the position of neighbouring atoms. Calculate the displacement relative to the cation of an oxygen atom Q, not a nearest neighbour, in the following hypothetical model.

    In the low-spin state, the six nearest neighbours, $ABCDEF$, form a regular octahedron, with Q touching the three atoms ABC. In the high-spin state, the three atoms DEF touch each other and the cation; so do DEA, EFB, FDC; and Q continues to touch ABC. The atoms may all be assumed spherical, with a constant radius of 1.40 Å for O.

    (This mechanism of changing spin state is responsible for the uptake and release of oxygen by haemoglobin in the blood, associated with breathing—though the neighbours of the cation are not isolated O ions, and the geometry of the arrangement is less simple than in our hypothetical example.)

Chapter 3   LATTICES AND LATTICE COMPLEXES

3.1   PERFECT PERIODIC STRUCTURES

A crystal is essentially a three-dimensional periodic structure, i.e., a structure in which the unit of pattern repeats with perfect regularity. In order to study the geometrical aspects, we assume that all atoms are at rest, and that the crystal is infinite in extent. In fact, all atoms are undergoing thermal motion, but we may allow for this by taking a time average over all the positions occupied by an atom during its thermal vibration, and using this as if it were an actual atom at rest. Again, no crystal is infinite, but most crystals examined are very large compared with the size of their unit of pattern; our discussions will apply to regions far enough inside the crystal to be unaffected by what happens on the surface.

An infinite periodic structure can be described completely and rigorously in terms of a *lattice* and a *lattice complex*. The lattice is made up of a set of translations, and describes how the repetition occurs; the lattice complex is the object or group of objects on which the repetition acts. It is easiest to discuss them separately.

3.2   LATTICES, TRANSLATION VECTORS, AND AXES OF REFERENCE

A *lattice* is a set of points which have identical surroundings.* It may be thought of as a scaffolding (of points, not lines) to which the actual components of the structure may be attached; or alternatively as a transparent

---

* Unfortunately, a different meaning of the word "lattice" has become common in solid-state literature; there it is often used to mean "structural array of atoms," or simply "structure," and confusion has often resulted. There is no simple remedy for the confusion, except to be aware of the danger, and to avoid using the word in this second sense in any context where it could be misunderstood. In this book, the word will only be used in the first or strictly geometrical sense defined in the text.

Figure 3-1 Two-dimensionally periodic pattern. (Reproduced, with permission, from the work of M. C. Escher in *Symmetry Aspects of M. C. Escher's Periodic Drawings*, by C. H. MacGillavry. A. Oosthoek, Utrecht, The Netherlands, 1965.)

net or graph (of points, not lines) which may be laid down on the structure to indicate the repeats. This can be illustrated in two dimensions; it is very easy, after one has grasped the idea in two dimensions, to extend it mentally to three.

Figure 3-1 shows a periodic pattern, from the work of the Dutch artist M. C. Escher. To construct its lattice, lay a piece of tracing paper on it, and put a dot to mark a chosen point of the pattern—for example, the eye of a white horse—every time it recurs. Then, if you move the paper parallel to itself, so that one dot comes above a different point—for example, a white wing tip—every other dot is also above a white wing tip. The dots are the lattice points. Wherever the paper is laid (provided its orientation is not changed), the surroundings of each lattice point are seen to be identical. (We must except, of course, the bits cut off at the edges—a truly periodic pattern would not be cut off but continue indefinitely.)

It is convenient to describe a lattice with the help of *translation vectors*.

A translation vector is any vector joining lattice points (for example, in Figure 3-1 the eyes of two white horses). It can be thought of in two ways. In an existing structure, it represents a displacement which leaves the environment completely unchanged; for building up a structure, it is an operator generating new lattice points, always at the same distance and in the same direction from preexisting ones.

Any three nonparallel vectors can generate a lattice. The two-dimensional lattice in Figure 3-2 can be generated by a_1 and b_1. On the other hand, the choice of translation vectors to generate a given lattice is not unique; the pair a_2 and b_2 would have been just as effective (though less convenient to draw). Not all pairs are, however, equally satisfactory; for example, by using a_1 and b_2 we can generate only half the points of the original lattice.

Having made a particular choice of three translation vectors (or in our example, two) we complete the parallelepiped (or, two-dimensionally, the parallelogram) with these vectors as sides. This is known as the *unit cell* of the lattice.

The directions of the translation vectors give us our *axes of reference* for the description of the lattice and the structure. It is necessary to choose senses for them, and this is conventionally done so that the axes are right-handed, and preferably so that all three interaxial angles are equal to or greater than 90°. (With some choices of unit cell, where all three angles differ from 90°, the latter condition cannot be adhered to.) The axes x, y, and z are associated with cell edges a, b, and c respectively; indeed, sometimes they are referred to as the a, b, and c axes.

Since the choice of translation vectors is not unique, the choice of unit cell is not unique, nor is the choice of axes of reference. *The lattice is the same, regardless of how we choose our axes, but our descriptions of it (and sometimes our name and symbol for it) differ according to our choice of axes.* This is an important point to remember. It becomes especially relevant when we consider later (Chapter 7) the effect of symmetry on the lattice, which sometimes makes one particular choice of axes so much more convenient than any other that we forget a choice exists, and sometimes leaves us hesitating between two choices, each with its own advantages for different purposes. (The conventional rules for choosing are given in the Appendix to Chapter 10.)

Let us consider the unit cells given by our three pairs of translation vectors (Figure 3-3).

The cells given by a_1, b_1 and a_2, b_2 are of equal area; the former would be the better choice, because of its shorter edges and lesser obliquity, unless special features in the lattice complex favoured the latter. Each of them is associated with one lattice point. (The easiest way to check this is to imagine the unit cell as a transparent box and to displace it slightly with respect to the original grid, as shown by dotted lines in the bottom left-hand corner of the figure; obviously it encloses only one of the grid points.) A unit cell associated with only one lattice point is called *primitive*.

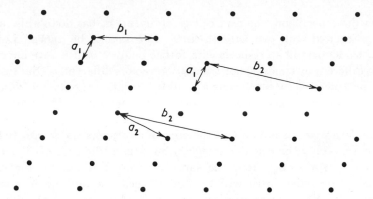

Figure 3-2 Two-dimensional lattice, showing four different translation vectors, $\mathbf{a}_1$, $\mathbf{b}_1$, $\mathbf{a}_2$, $\mathbf{b}_2$.

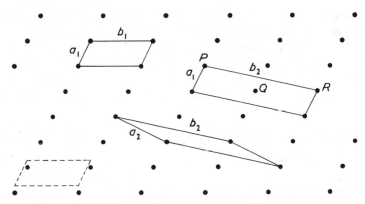

Figure 3-3 Two-dimensional lattice, showing several different choices of unit cell, using the translation vectors of Figure 3-2.

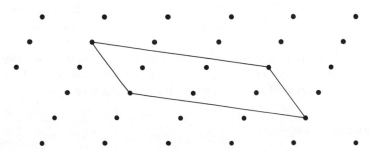

Figure 3-4 The same two-dimensional lattice, showing a possible but very unusual (and probably very inconvenient) choice of unit cell.

On the other hand, the unit cell given by $\mathbf{a}_1$, $\mathbf{b}_2$ has double the area, and is associated with two lattice points. It is said to be *centred*. Lattice point Q is identical in all respects with lattice points P and R—we have not made it different by choosing our translation vectors differently—but since it cannot be generated by the vectors $\mathbf{a}_1$ and $\mathbf{b}_2$ operating repeatedly from our origin P, we have to give other instructions for finding it. By saying that the lattice with translation vectors $\mathbf{a}_1$ and $\mathbf{b}_2$ is centred, we mean that there are lattice points not only at the ends of all vectors $m\mathbf{a}_1 + n\mathbf{b}_2$ (where m and n are positive or negative integers) but also at the ends of all vectors $m\mathbf{a}_1 + n\mathbf{b}_2 + \frac{1}{2}(\mathbf{a}_1 + \mathbf{b}_2)$ from the same origin. It is exactly the same lattice as the primitive lattice with translation vectors $\mathbf{a}_1$ and $\mathbf{b}_1$. A lattice is thus not in itself either primitive or centred; to describe it as primitive or centred is only meaningful *after* we have chosen axes of reference. All lattices can, if one so wishes, be described using primitive unit cells. For some lattices, as we shall see later in Chapter 7, nonprimitive unit cells are a much more obvious and convenient choice, and the lattices are named accordingly, the name *primitive lattice* (and the corresponding lattice-symbol p in two dimensions or P in three dimensions) being kept for cases where the primitive unit cell is usually the most convenient.

More complicated kinds of centring are possible than those shown in Figure 3-3; for example, a unit cell might be chosen as in Figure 3-4. There are, however, relatively few occasions when descriptions of this kind are likely to be useful, and no special nomenclature has been developed for them.

3.3 LATTICE CENTRING IN THREE DIMENSIONS

In three dimensions there is a greater variety of important kinds of lattice centring than the one kind we considered in detail in two dimensions.

In illustrating these (Figures 3-5 to 3-8) it is generally necessary—as in much crystal structure work—to include more than one unit cell, so as to show the surroundings of the lattice points under discussion. In the unit cell in the illustration, all the angles are right angles, but that is merely for convenience of drawing; the arguments and conclusions are exactly the same if all the angles are oblique. Lattice points which coincide with unit cell corners are shown as solid circles, and additional lattice points due to the centring operation as open circles, but this is merely to emphasise the different methods of deriving them—in fact, if they *are* lattice points, the open circles must be identical in all respects with the solid ones.

(i) One-Face Centring

In this type of centring (Figure 3-5) (sometimes called side centring or base centring) one of the three independent faces of the parallelepiped has a lattice point at its centre (and so of course has the opposite face, and all others

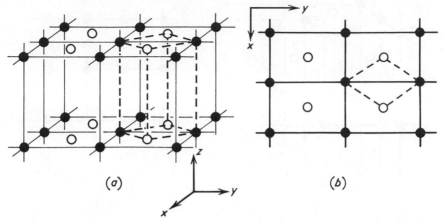

Figure 3-5 One-face-centred lattice. (a) Perspective diagram: solid lines outline unit cells of centred lattice; broken lines outline primitive unit cell of same lattice; black circles and open circles mark identical lattice points. (b) Same lattice in projection down z axis; all points are at height zero.

related to the first by translation vectors). In the diagram, the ab face is centred, which means that the centring point is at a vector distance $\frac{1}{2}\mathbf{a} + \frac{1}{2}\mathbf{b}$ from the corner. The lattice is called C-face-centred, and given the symbol C. Lattices centred on the bc or ca faces are A-face-centred or B-face-centred, with the symbol A or B respectively.

(ii) All-Face-Centring

In this type of centring (Figure 3-6) (sometimes simply called face centring, when there is no risk of confusion with one-face centring) all three independent faces of the parallelepiped are centred: i.e., there are lattice points at the ends of vectors $\frac{1}{2}\mathbf{a} + \frac{1}{2}\mathbf{b}$, $\frac{1}{2}\mathbf{b} + \frac{1}{2}\mathbf{c}$, $\frac{1}{2}\mathbf{c} + \frac{1}{2}\mathbf{a}$ from the corners. There are thus four lattice points per unit cell. To obtain a primitive unit cell, Figure 3-6(d), a completely different set of axes of reference is needed. An all-face-centred lattice has the symbol F.

One might at first sight expect that we could have a two-face-centred lattice. An attempt to construct this by centring the ab and ca faces is shown in Figure 3-7. Inspection of the array of points so obtained shows that it is not a lattice, because the environments of all points are no longer alike. A point such as P has four near neighbours in the vertical plane, Q has four near neighbours in the horizontal plane, while R has sets in both vertical and horizontal planes.

It is useful to look back at Figure 3-6(a) and check that each point, whether it is a solid or an open circle, has the same set of neighbours—twelve in each case, four in each of the three principal planes. The same sort of check should be applied to the lattices of Figures 3-5(a) and 3-8(a).

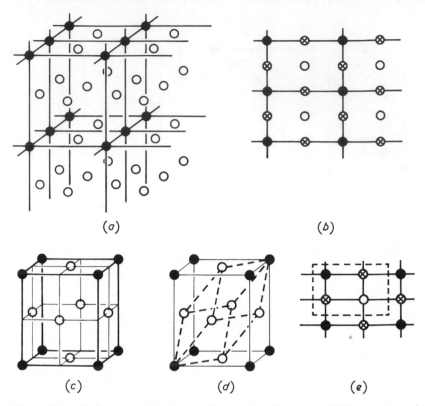

(a)

(b)

(c)

(d)

(e)

Figure 3-6 All-face-centred lattice. (a) Perspective diagram: solid lines outline unit cells of the centred lattice; black circles and open circles mark identical lattice points (the open circles lying at face centres). (b) Same lattice in projection down the z axis: points marked with crosses are at height 1/2, the others are at height zero. (c) Perspective diagram of one-face-centred unit cell, outlined by heavy lines, with construction for the insertion of face-centring points shown by weaker lines. (d) Perspective diagram of one unit cell of the face-centred lattice, outlined by solid lines, showing also a primitive unit cell, outlined by heavy broken lines. (e) Projection of one unit cell, showing displacement of origin (as indicated by broken lines) to demonstrate that there are four lattice points per cell.

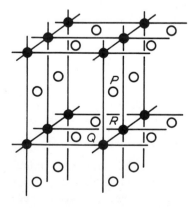

Figure 3-7 Perspective diagram of a periodic structure with the same unit cell as in Figure 3-6(a), of which two faces are centred. Note that black circles and open circles are now not identical; i.e., the structure is not a lattice.

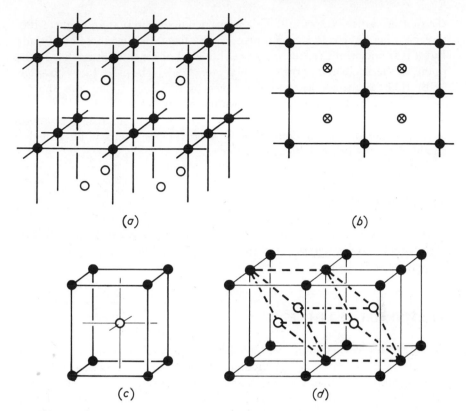

Figure 3-8 Body-centred lattice. (a) Perspective diagram: solid lines outline unit cells of centred lattice; black and open circles mark identical lattice points. (b) Same lattice in projection down the z axis; points marked with crosses are at height 1/2, the others are at height zero. (c) Perspective diagram of one body-centred unit cell, outlined by heavy lines, with construction for the insertion of body-centring point shown by weaker lines. (d) Perspective diagram of four unit cells of the body-centred lattice, outlined by solid lines, showing also a primitive unit cell, outlined by heavy broken lines.

(iii) Body Centring

In this type of centring there is a lattice point at the centre of each unit cell, i.e., at the end of a vector $\frac{1}{2}\mathbf{a} + \frac{1}{2}\mathbf{b} + \frac{1}{2}\mathbf{c}$ from the corner (Figure 3-8). The symbol for a body-centred lattice is I (from the German, *Innenzentrierung*).

(iv) Centring of Trigonal Lattices

As in the two-dimensional illustration of Figure 3-4, in three dimensions, many other kinds of centring are possible if the axes of reference are chosen in unusual ways. None of them are useful enough to have been given special notation, except for the one example now to be considered.

Suppose that the lattice possesses a threefold symmetry axis (see §6.3) and that we wish to use this as our z-axis of reference, with x and y axes along

the shortest lattice vectors in the plane perpendicular to it at an angle of 120°
to each other. We could have a primitive lattice satisfying these conditions.
But if it is not primitive, the only possible kind of centred lattice has lattice
points at the origin and at the ends of vectors* $\frac{2}{3}\mathbf{a} + \frac{1}{3}\mathbf{b} + \frac{1}{3}\mathbf{c}, \frac{1}{3}\mathbf{a} + \frac{2}{3}\mathbf{b} + \frac{2}{3}\mathbf{c}$.
With such a lattice, by choosing axes of reference along the shortest
vectors we should obtain a primitive unit cell, which is a rhombohedron.
For this reason it is generally called a rhombohedral lattice, but with
our first-chosen axes of reference it might more usefully be called a
rhombohedrally-centred trigonal lattice.

3.4 LATTICE PARAMETERS

When our axes of reference have been chosen, the unit cell is described by
specifying its edge lengths a, b, c, and its interaxial angles, α, β, γ. These are
known collectively as the *lattice parameters*.

3.5 SYMMETRY OF LATTICES

So far (with one exception) we have been concerned with lattice properties
that are independent of any particular symmetry. Lattices can, as we know,
possess symmetry; then certain choices of axes of reference may become so
advantageous that they are universally adopted, and unit cells are defined
accordingly. Two examples, which are particularly important, will be
mentioned here.

(i) A cubic lattice has a unit cell with the symmetry of a cube, which
means that its three edges are equal and its three angles are 90°. Cubic
lattices may be primitive, all-face-centred, or body-centred, but not one-face-
centred (because symmetry requires the three faces ab, bc, and ca to be
equivalent).

(ii) A hexagonal lattice has a unit cell which is a right prism with a 60°
rhombus as base;† this means that edges a and b are equal, and enclose an
angle of 120°, while c (of arbitrary length) is at right angles to both a and b.

It is important to remember that the shape of the unit cell does *not*
determine the symmetry of the lattice; instead, the symmetry imposes
restrictions on the possible shapes, which we shall consider more fully later
(Chapter 7). All that we need notice at this stage is that certain important
and commonly occurring symmetries give rise to unit cells of the shapes

---

* Lattice points at the origin and the ends of vectors $\frac{1}{3}\mathbf{a} + \frac{2}{3}\mathbf{b} + \frac{1}{3}\mathbf{c}, \frac{2}{3}\mathbf{a} + \frac{1}{3}\mathbf{b} + \frac{2}{3}\mathbf{c}$,
would give the same kind of lattice rotated by 180° about z; the orientation given in the text
is the conventional choice (see §7.6).

† Some books show the unit cell of a hexagonal lattice as a hexagonal prism. This is
wrong; the unit cell of a hexagonal lattice, like that of any other lattice, is a parallelepiped,
and a hexagonal prism is made up of three unit cells.

described; granted this, we can use them in Chapter 4 when we describe particular structures.

Enumeration of the possible types of lattice, and their conventional unit cells, will be left till after we have considered symmetry in Chapter 6.

3.6 THE LATTICE COMPLEX

The lattice is not itself the structure; it is only a set of reference points describing the repetition properties of the structure. Associated with each lattice point is the *lattice complex*, which may be anything from a single atom to a group of large molecules or a part of a continuously bonded network. It may be defined as a piece of structure which is just sufficient, when repeated by the operation of the lattice, to build up the complete, infinite structure.

As with translation vectors, so with the lattice complex, the choice to be made for describing any given structure is not unique. In Figure 3-9, several choices are shown, each of which when repeated by the lattice vectors shown at the bottom right will build up the whole pattern. When a structure has isolated motifs, or clearly defined molecules, it is natural to choose the lattice complex in a way which keeps them as entities (as, for example, in Figure 3-9(a)); but it would be permissible (if there were any advantage in it) to

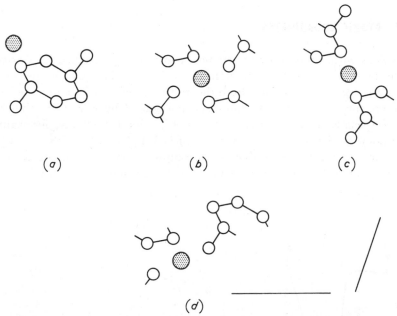

(a)   (b)   (c)

(d)

Figure 3-9 Different choices of lattice complex for the same hypothetical structure, with lattice vectors as shown by the straight lines at bottom right. In (a) and (b) the lattice complex fits within the unit cell; in (c) and (d) the translation-repeat unit is a different shape.

take half of one motif and the complementary half of another, separated by the lattice translation vector—like the picture in a poorly adjusted television screen in which the bottom half appears above the top half. Some variants of this kind are shown in Figure 3.9(b), (c), and (d). There is no geometrical requirement that we should draw our boundaries for the lattice complex through lines of physical weakness, or empty spaces of the structure, though it is often easier to visualise the structure if we do. Another convenient choice is to take the contents of the primitive unit cell with the origin at the centre of the six-membered ring, as in Figure 3.9(b); this makes it easy to check that we have neither left anything out nor included anything twice, but involves the risk that we may sometimes overlook important interatomic bonds which cut across the cell boundaries.

It is convenient to introduce the term *translation-repeat unit* for the volume (or in two dimensions the area) containing the lattice complex. For example, if we choose the lattice complex of Figure 3.9(c), a corresponding translation-repeat unit is given by Figure 3.10, the unit cell being the parallelogram obtained by joining the heavy dots. Translation-repeat units for any structure are the smallest volumes (or areas), of identical shape and contents, which, put together without change of orientation, fill all space. The contents of a translation-repeat unit constitute the lattice complex. As we have seen, the translation-repeat unit has the volume of the primitive unit cell, but it may have any shape—it does not need to be a parallelepiped.

3.7 ATOMIC COORDINATES

Positions within the unit cell are most conveniently specified by Cartesian coordinates which are fractions of the unit cell edge. Thus a point specified by x, y, z is at a vector distance from the origin (a corner of the unit cell) with components xa, yb, zc. The numbers represented by x, y, z are the *atomic position parameters*. They may, but need not, be simple fractions; for example, the point at the body centre of any cell is $1/2, 1/2, 1/2$. Values of x, y, z greater than unity are possible, but are unnecessary in specifying the structure, because the existence of the lattice tells us that for every point x, y, z

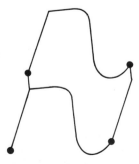

Figure 3-10   Translation-repeat unit corresponding to Figure 3-9(c); black dots mark corners of the unit cell.

there are others at $1 + x, y, z$; $2 + x, y, z$; and so forth. Negative values of x, y, z are commonly used; this is equivalent to choosing as lattice complex a set of atoms grouped around the origin rather than a set confined within the unit cell.

In some early work on crystal structures, notably many of Bragg's classical papers, it was customary to list atomic parameters as θ_1, θ_2, θ_3, corresponding to our $2\pi x$, $2\pi y$, $2\pi z$. Values reported as $180°$, $0°$, $90°$, for example, would now be written $1/2$, 0, $1/4$.

A centred lattice imposes requirements for relations between atomic co-ordinates. Thus, if a lattice is body-centred, it implies that for every atom (or electron or molecule) at x, y, z there is another identical one at $1/2 + x$, $1/2 + y$, $1/2 + z$. Provided we have stated that the lattice is body-centred, it is therefore unnecessary to write down the coordinates of both atoms; either set is implied by the other. We may write the body-centring operation as $(0, 0, 0; 1/2, 1/2, 1/2) +$, to show how the complete set of lattice-related atoms is derived from any one of them. Corresponding relations for the other lattice centrings are given in Table 3-1.

Table 3-1 Lattice-Centring Operators

KIND OF OPERATION	SYMBOL	OPERATORS	NUMBER OF LATTICE POINTS PER UNIT CELL
One-face centring	A	$(0, 0, 0; 0, \frac{1}{2}, \frac{1}{2}) +$	2
	B	$(0, 0, 0; \frac{1}{2}, 0, \frac{1}{2}) +$	2
	C	$(0, 0, 0; \frac{1}{2}, \frac{1}{2}, 0) +$	2
All-face centring	F	$(0, 0, 0; 0, \frac{1}{2}, \frac{1}{2}; \frac{1}{2}, 0, \frac{1}{2}; \frac{1}{2}, \frac{1}{2}, 0) +$	4
Body centring	I	$(0, 0, 0; \frac{1}{2}, \frac{1}{2}, \frac{1}{2}) +$	2
Rhombohedral centring* (with trigonal symmetry only)	R	$(0, 0, 0; \frac{2}{3}, \frac{1}{3}, \frac{1}{3}; \frac{1}{3}, \frac{2}{3}, \frac{2}{3}) +$	3

* In conventional or *obverse* setting, see §7.6.

3.8 COMPLETE DESCRIPTION OF STRUCTURE

For a complete description of a structure made up of atoms of known species, we need to be able to specify the lattice parameters and the atomic co-ordinates of all atoms within the unit cell. We then have a complete and objective geometrical description of the structure, and can record it for other people to study without tying them to our particular way of looking at it. In doing so we have, of course, chosen an origin and particular axes of reference, but if the reader prefers a different choice he has all the information needed to allow him to make the change. (Ways of handling changes of axes of reference are discussed in Chapter 10.)

The existence of lattice centring, as we have seen, allows us to abbreviate our statement of the atomic coordinates. Further abbreviation is often possible when we know and record the symmetry; this is embodied in the space-group symbol, to be discussed later (Chapter 8). However, it is not necessary to know anything about the symmetry before using the description, provided all the atomic coordinates as well as the lattice parameters are recorded. We can therefore postpone formal consideration of symmetry till after we have described a number of structures which provide examples of its effects.

PROBLEMS AND EXERCISES

1. (a) Construct the two-dimensional lattice of Figure 3-1, assuming that colour does not matter, i.e., that only the outlines are part of the structure. (b) Choose a unit cell in three different ways, each containing only one lattice point per cell. Find the lattice parameters in each case. (c) Can you find a rectangular unit cell? How many lattice points does it contain? (d) How would the lattice type be named, conventionally?

2. Construct the periodic structure corresponding to the lattice vectors and lattice complex of Figure 3-9. (Use tracing paper; make a grid corresponding to the lattice vectors; lay it in the correct orientation with a lattice point superposed on any chosen point in the lattice complex; move the next lattice point to coincide with the same point of the lattice complex; repeat.) Extend the pattern far enough to show how each of the diagrams in Figure 3-9(a) to (d) can be picked out from it.

3. Construct a structure which would be formed by operating on the lattice complex of Figure 3-9(a) with a rectangular lattice.

4. With the help of suitable tracings, show that Figure 3-10 is a translation-repeat unit with the same lattice as Figure 3-9, and that it can be so placed in the periodic structure as to enclose the lattice complex of Figure 3-9(c).

5. Construct a translation-repeat unit, with the same lattice, containing the lattice complex of Figure 3-9(d).

6. In Figure 3-1, if we consider the black and white horses as different, the most simply chosen lattice complex consists of one white horse plus one black horse. Find a translation-repeat unit which does not break up the white horse into separate parts, but, subject to this, has as simple an outline as possible. Show that the separate parts of the black horse within this translation-repeat unit can be pieced together to give the complete horse.

7. By measurement of Figure 3-9(a), find the atomic coordinates of all atoms, assuming that the origin is taken at the large shaded atom. Repeat for Figures 3-9(b), (c), and (d), with the same choice of origin, and show how the coordinates are related to those from (a). Alternatively, derive the coordinates for (b), (c), and (d), by use of the lattice principle, from those obtained from (a).

Chapter 4 SOME SIMPLE STRUCTURES

4.1 INTRODUCTION

In this chapter, we select a number of simple structures, important not only for their common occurrence and for the wide use in physics and technology of the materials possessing them, but also because they are simple illustrations of many important structure-building principles that apply to more complicated structures to be considered later.

We shall need to consider each structure in at least two ways: first, an objective geometrical description,* of the kind explained in Chapter 3, and second, a description in terms of atomic groupings and interatomic bonds, of the kind discussed in Chapter 2. Sometimes it is helpful to have more than one description under either heading to emphasise different features of the crystal architecture. For any useful understanding of solids, it is necessary to be able to change from one description to another, according to one's immediate purpose, and to grasp the relationship between them.

It is possible to read these descriptions without knowing more about symmetry than that it is the cause of the particular shapes of the unit cells recorded. For convenience of later reference, however, the space group of each structure is noted.

4.2 COPPER

Space group: $Fm3m$ (No. 225)
Lattice: Cubic, all-face-centred, with side of cube $a = 3.60$ Å
Lattice complex: Cu atom at $(0, 0, 0)$

---

* One common mistake is to think that specification of the lattice describes the structure adequately. It is worse than useless—it is positively confusing—to specify the lattice unless one also specifies the lattice complex either explicitly or implicitly.

The face-centring operation gives four atoms per unit cell. This can most easily be seen in Figure 4-1(a) by imagining the outline of the cell shifted so that its corner lies just outside the atom M, originally at $(0, 0, 0)$; the cell then includes M, N, I, F, and excludes all the other atoms shown.*

Contact distances between nearest neighbours are shown in Figure 4.1(b). They lie along cube face diagonals, and their length is $\sqrt{2}a/2$, or 2.54 Å. Each atom has twelve equidistant neighbours, some of them outside the volume shown in Figure 4-1(a) to (g).

An alternative way of looking at this structure is extremely useful. Figure 4-1(f) shows that there are planes of atoms such as $DBEHKG$ perpendicular to the body diagonal MC; the next plane (not outlined) contains $AFJNLI$; and there must be corresponding planes through M and C (containing atoms in adjacent unit cells) since atoms M and C are identical in all respects with the others. A projection of these planes down the MC direction is shown in Figure 4-1(h). Successive planes are separated by a distance $(1/3)MC = \sqrt{3}a/3$. Each atom is in contact with six others in the same plane—the atoms in the same plane, in fact, constitute a *close-packed layer*—Figure 4-1(i). The sixfold symmetry of the layer is obvious; we shall come back to this later.

Though the layers are all identical, successive layers are staggered. The layer through B has three-sided hollows between the atoms, which may be divided into two sets with projections at $CPQ \ldots$ and $AFIJNL \ldots$ in Figure 4-1(i). Atoms of the next layer fit into one of these sets, $CPQ \ldots$, and make in their turn two new sets of hollows, with projections at $AFIJNL \ldots$ and $DBEGHK \ldots$, of which the former is chosen by the third layer. Continuing to build upwards, the fourth layer repeats the first, with atoms projecting at $DBEGHK \ldots$, and every third layer gives an exact repeat in the vertical direction, the vertical separation between atoms being the lattice translation vector MC. The layers fit together as closely as they can: the structure is *close-packed*.

Though we chose MC as our direction of projection, similar results could have been obtained using JE, or AK, or DL. This means that if we constructed a cube containing a large number of unit cells and bevelled off the corners, we should find on each corner a close-packed plane of atoms like that shown for a single unit cell in Figure 4-1(g). By contrast, sections parallel to the cube faces show a rather open square packing, as in Figure 4-1(c) and (e).

Cubic close-packed structures are extremely important in themselves, and also as a starting point for the understanding of many other structures. It is essential to be able to think of them as viewed either along a cube edge, as in

---

* An alternative method of counting, commonly used, is as follows: There are eight atoms such as A, each shared by eight cubes, total 1; and six atoms such as B, each shared by two cubes, total 3; grand total 4. This is not incorrect, but is clumsy, and is misleading in its suggestion that atoms such as A and B are intrinsically different; it is not recommended.

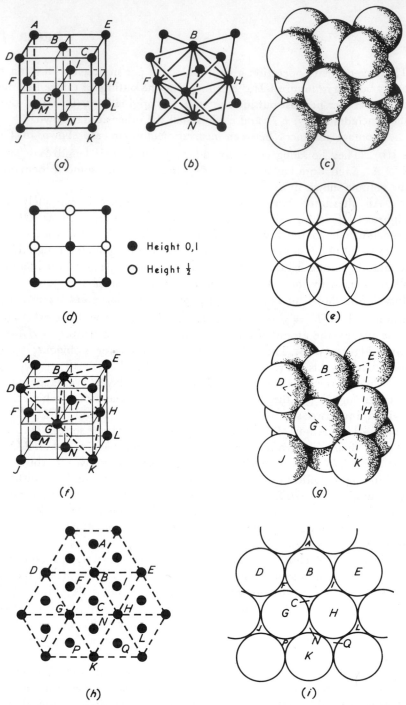

Figure 4-1 Structure of copper. (a) Perspective view of unit cell, showing positions of atomic centres. (b) Same view, showing nearest-neighbour contacts. (c) Same view, showing packing. (d) Projection of atomic centres on *xy* plane. (e) Projection on same plane, showing packing: heavy circles are atoms at heights 0 and 1, light circles at height 1/2. (f) Same perspective view as (a), showing plane of atoms perpendicular to body diagonal. (g) Same as (c), except for removal of corner atom C, to show packing in plane *DBEHKG*. (h) Projection of (f) down cube body diagonal *CM*. (i) Same as (h), showing packing of one layer of atoms *DBEHKG*, and projection of centres of others.

Figure 4-1(d) or (e), or along a cube body diagonal as in (h) or (i), and to move from one viewpoint to another as required for any particular purpose.

Other examples of the cubic close-packed structure* include many metals, notably aluminium, nickel, platinum, lead, silver, and gold, and some solid phases of the rare gases, neon, argon, krypton, and xenon.

Clearly the cubic close-packed structure can only occur in elements (that is, if we exclude disordered solids with quasiperfect structures, discussed in §1.5), and only in those elements in which the forces are central rather than directed (since directed-bond systems involving twelve neighbours are unlikely). Metallic bonding, as in copper, and van der Waals bonding, as in argon, satisfy this requirement.

4.3 IRON

> Space group: $Im3m$ (No. 229)
> Lattice: Cubic body-centred, with side of cube $a = 2.86$ Å
> Lattice complex: Fe atom at $(0, 0, 0)$

There are two atoms per unit cell, as shown in Figure 4-2(a). Contact distances between nearest neighbours lie along cube body diagonals as shown in Figure 4-2(b), and their length is $\sqrt{3}a/2$ or 2.50 Å. Each atom has eight neighbours. The packing in the plane of the cube face is in square formation, but the atoms are not in contact—see Figure 4-2(e).

It is, of course, possible to project this structure down the cube body diagonal, but it does not lead to particularly interesting results.

The structure exemplified by iron is known as the *monatomic body-centred cubic structure*. (Often the adjective monatomic is omitted, but strictly it should not be, because there are many other structures with cubic body-centred lattices.)

Other examples of this structure include the metals molybdenum, tungsten, and sodium. Iron is capable of existing in a different structure at high temperatures.

4.4 MAGNESIUM

> Space group: $P6_3/mmc$ (No. 194)
> Lattice: Primitive hexagonal; unit cell with sides $a = 3.20$ Å, $c = 5.20$ Å
> Lattice complex: (i) Mg atoms at $(0, 0, 0)$ and $(1/3, 2/3, 1/2)$, as shown in Figure 4-3(c),
> (ii) With alternative choice of origin, Mg atoms at $\pm(2/3, 1/3, 1/4)$, as shown in Figure 4-3(d)

---

* It is sometimes called "the face-centred cubic structure." This is an unfortunate name, overlooking the fact that there are many other structures with face-centred cubic lattices. The particular feature of this one is that its lattice complex consists of a single atom. A correct but clumsy name would be "the face-centred cubic structure with a monatomic lattice complex."

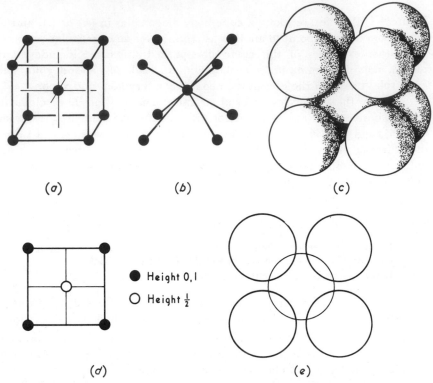

(a) (b) (c)

● Height 0,1

○ Height $\frac{1}{2}$

(d) (e)

Figure 4-2 Structure of iron. (a) Perspective view of unit cell, showing positions of atomic centres. (b) Same view, showing nearest-neighbour contacts. (c) Same view, showing packing. (d) Projection of atomic centres on xy plane. (e) Projection on same plane, showing packing; heavy circles are atoms at heights 0 and 1, light circle at height 1/2.

Again there are two atoms per unit cell—see Figure 4-3(a)—but in contrast to the two previous structures, the points occupied by atoms *do not constitute a lattice*. Within the ab plane, the atoms form a close-packed layer *exactly* like that of copper, and two successive layers are superimposed—Figure 4-3(e)—like the A and B layers of Figure 4-1(h); there the resemblance ends. The third layer lies vertically above the first, and there is no necessary geometrical relation (as there was in the cubic structure) between the separation of layers and the interatomic distance within a layer. The separation of layers is $\frac{1}{2}c$. Each atom has twelve neighbours, as shown in Figure 4-3(b)—six in its own plane, three above, and three below. The shortest distance between atoms in successive layers is

$$\sqrt{\left(\frac{2}{3}\cdot\frac{\sqrt{3}}{2}\,a\right)^2 + \left(\frac{1}{2}\,c\right)^2} \qquad (4.1)$$

If this were equal to a, the shortest distance within the layer, the packing would be as good as in the cubic arrangement. Evaluating the expression, we find that c/a should be $\sqrt{(8/3)} = 1.633$.

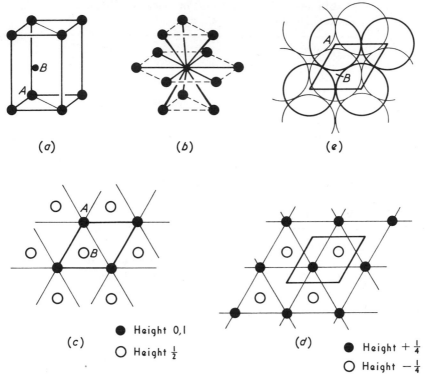

(a)   (b)   (e)

(c)   (d)

● Height 0,1
○ Height ½

● Height +¼
○ Height −¼

Figure 4-3 Structure of magnesium. (a) Perspective view of unit cell, showing positions of atomic centres. (b) Similar view, including additional atoms, showing nearest-neighbour contacts. (Broken lines join neighbours in same plane.) (c) Projection on *xy* plane, showing atomic centres: unit cell (heavy outline) chosen with origin at *A*. (d) Same projection: unit cell (heavy outline) chosen with different origin. (e) Projection on same plane, showing packing; heavy circles are at height 0 and 1, light circles at height 1/2.

For magnesium, it can be seen from Table 4-1 that the requirement is well met: magnesium is in fact a good example of *hexagonal close-packing*. The same is true for most other metals with structures of the same geometrical description. However, it is possible while keeping this description to have structures with different *c/a* ratios, which are not close-packed. Examples are zinc and cadmium, whose parameters are also given in Table 4-1. The individual layers in these metals are close-packed, but the layers are more widely separated than would be expected from packing under the influence of central forces. Some of the rare-gas solids are isomorphous with Mg.

4.5 COMPARISON OF CUBIC AND HEXAGONAL CLOSE-PACKING

The possibilities of close-packing may be restated as follows. If the sites occupied by atoms in a single close-packed layer are denoted by *A* (Figure 4-4), the two close-packed ways of superimposing a second identical layer

Table 4-1 Axial Ratios in Hexagonal-Close-Packed Metals

ELEMENT	a (Å)	c (Å)	c/a
Be	2.28	3.58	1.57
Mg	3.20	5.20	1.62
Ti	2.92	4.67	1.60
Zr	3.23	5.14	1.59
Os	2.73	4.31	1.58
Zn	2.66	4.94	1.86
Cd	2.97	5.61	1.89

place its atoms directly above either B or C, and either way gives the same results (except for a rotation of the structure by 60°). Suppose sites B are used. A third identical layer may then be placed with its atoms over either A or C. If A is chosen, the unit cell has been completed with two layers, and the sequence can continue $ABABABAB$ . . .; this is the hexagonal close-packed sequence. Alternatively, if C is chosen, the next repetition of the same operation brings the atoms back to A; the unit cell has been completed with three layers, and the sequence is $ABCABCABC$ . . .—the cubic close-packed sequence.

We can see that the cubic-close-packed array of sites constitutes a lattice by noting the translation vectors needed to generate one layer from the next. Let the two (equal) shortest vectors at 120° to each other *in* the layer be called **a** and **b**, and the vector perpendicular to the layer (the cube body diagonal of Figure 4-1) be called **c**. (In future we shall refer to these as defining the *hexagonal axes of reference of the face-centred cubic lattice*.) Then the $ABCABC$ . . . sequence is generated by repeated operation of the vector $(2/3)\mathbf{a} + (1/3)\mathbf{b} + (1/3)\mathbf{c}$. By contrast, the $ABAB$ . . . sequence needs alternate operation of the two different vectors $(2/3)\mathbf{a} + (1/3)\mathbf{b} + (1/2)\mathbf{c}$

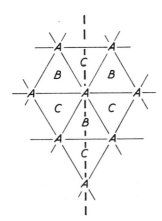

Figure 4-4 Sites of cubic close-packing, projected on *xy* plane. The broken line is the trace of the plane used as plane of projection in Figure 4-5.

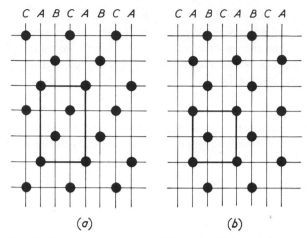

Figure 4-5 Close-packed arrays, projected down the short a vector on the plane normal to it and to the xy plane. The lines marked A, B, C are projections of those that projected into the correspondingly marked points in Figure 4-4. Projection of the unit cell is shown by heavy outline. (a) Cubic close-packing; (b) hexagonal close-packing.

and $(1/3)\mathbf{a} + (2/3)\mathbf{b} + (1/2)\mathbf{c}$. For lack of a better name, we may call the latter set of vectors the hexagonal-close-packing operator, and apply it, in later examples, to lattice complexes consisting of more than a single atom. For this operator there is no necessary geometrical relation between the lengths of a and c once the physical requirement of equal interatomic distances ceases to hold. For the cubic face-centred lattice, on the other hand, c/a must remain equal to $\sqrt{6}$; otherwise the structure ceases to be cubic.

Figure 4-5 shows the projection of the cubic and hexagonal close-packed structures on the plane normal to one of the shortest intralayer vectors, whose trace is shown by the broken line in Figure 4-4. The unit cells are outlined; the different values of c, corresponding to three layers in (a) and two layers in (b), should be noted.

More complicated sequences of layers than these two are possible, for example, $ABCBABCBA \ldots$ or $ABCBCABCBCA \ldots$. They are of much less general importance, and can be discussed individually as need arises.

It is useful to consider the character of the largest empty spaces in a close-packed structure. They are located halfway between layers, and their number is equal to the number of atoms in the layer. Consider the projection on the plane of the layer (Figure 4-4). Interstices between the two adjacent layers whose atoms project at B and C appear on the projection at points A. They are equidistant from six atoms—three atoms B below and three atoms C above—and the six form a regular octahedron. Such *octahedral interstices* could contain spheres of radius $0.20a$ without disturbing the close-packing.

Smaller interstices are found at the centres of tetrahedra formed by three atoms of layer B and one of layer C, or three of layer C and one of layer B. The former set, projecting at C, are at a height of one-quarter the interlayer distance above the plane through atoms B; the latter, projecting at B, are at three-quarters the interlayer distance above the same plane. These *tetrahedral interstices* could accommodate spheres of radius $0.11a$ without disturbing the close-packing.

Since the nature of the interstices is completely determined by a pair of close-packed layers, it is the same for cubic-close-packed and hexagonally-close-packed structures; only in the relative positions of interstices between successive pairs of atomic layers do the two types of packing differ.

4.6 DIAMOND

Space group: $Fd3m$ (No. 227)
Lattice:* Cubic face-centred; unit cell is a cube, with side $a = 3.56$ Å
Lattice complex: (i) 2 C atoms at $(0, 0, 0)$ and $(1/4, 1/4, 1/4)$
(ii) With different choice of origin, 2 C atoms at $\pm(1/8, 1/8, 1/8)$

There are eight atoms in the unit cell, each of the two in the lattice complex giving rise to four by the lattice centring operation.

The first choice of origin has been used in Figure 4-6, though for many other purposes the second choice is more convenient. Figure 4-6(c) could be converted to illustrate the second choice by moving the square outline of the unit cell so that its corner is midway between two atoms, i.e., at height $1/8$ midway between the origin and point P; in (d) and (e) the diagrams would be unchanged, but $3/24$ would have to be subtracted from each marked height.

The contact distance is along the cube diagonal, and is equal to $\sqrt{3}a/4$, or 1.54 Å. Each C atom has four neighbours, the bonds being directed towards the corners of a regular tetrahedron, as shown in Figure 4-6(b). The structure is a very open one: only four of the octants of the unit cell have atoms at their centres $(1/4, 1/4, 1/4; 3/4, 3/4, 1/4; 3/4, 1/4, 3/4; 1/4, 3/4, 3/4)$, and there would have been plenty of room for atoms in the other four. In fact, the structure is controlled by the need for tetrahedral bonds at every carbon atom, which is inconsistent with closer packing.

It is interesting to project this structure down the cube body diagonal, as in Figure 4-6(d). We get an arrangement exactly like that of Figure 4-1(h), except that there is an additional atom vertically above every atom of that figure, separated from it in height by $1/4$ of the repeat distance ($3/4$ of the

---

* It is incorrect to use the expression "diamond lattice." Writers who do so generally mean the diamond *structure;* the *lattice* of diamond is the same as that of Cu or NaCl. To say that "there are two interpretating face-centred lattices" is again a misuse of the word lattice; a lattice is a set of points, not of atoms, and the lattice of a structure is a property of the structure, not a piece of it.

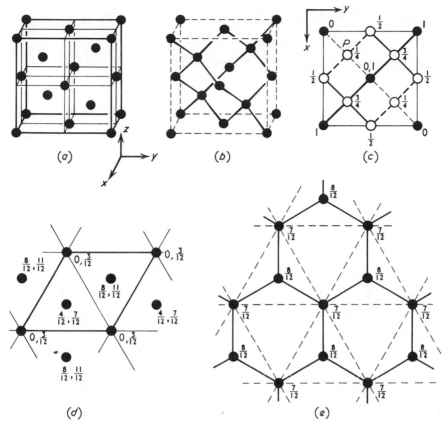

Figure 4-6 Structure of diamond. (a) Perspective view of the unit cell, with positions of atomic centres. (b) Same view, showing interatomic bonds. (c) Projection of the unit cell on the *xy* plane, showing centres of atoms and interatomic bonds; bonds in the upper half of the unit cell are shown by solid lines, in the lower half by broken lines. (d) Projection down a cube body diagonal; heights of atoms are marked. (e) Same projection as (d), with one puckered layer of atoms picked out to show hexagonal rings. The broken lines outlining triangles round points marked 8/12 are the bases of tetrahedra pointing upwards, with vertices at 11/12. Similar tetrahedra could have been shown round points marked 7/12, pointing downwards with vertices at 4/12.

interlayer distance). The heights of all atoms are marked in Figure 4-6(d). From this we can see that there are puckered layers, consisting of hexagonal rings of carbon atoms. One such is shown in Figure 4-6(e). The vertices of tetrahedra centred at height 8/12 point upwards, and those centred at 7/12 point downwards, connecting this puckered layer to others above and below.

The diamond structure is also possessed by silicon, with $a = 5.42$ Å, and germanium, with $a = 5.62$ Å. The structure demands strong tetrahedral bonds, irrespective of atomic size; hence it is only formed by elements with a tendency to form tetrahedral covalent bonds.

4.7 SODIUM CHLORIDE (ROCK SALT)

Space group: *Fm3m* (No. 225)
Lattice: Face-centred cubic; unit cell a cube with side $a = 5.63$ Å
Lattice complex: (i) Na^+ ion at $(0, 0, 0)$, Cl^- ion at $(1/2, 1/2, 1/2)$,
(ii) With alternative choice of origin,
Cl^- at $(0, 0, 0)$, Na^+ at $(1/2, 1/2, 1/2)$

Each unit cell contains four ions of each kind. (To check, imagine the out-line of the unit cell in Figure 4-7(a) displaced slightly with respect to the atoms, as before—cf. §4.2.) In other words, there are four *formula units* of NaCl per unit cell.

The contact distances are $r_{Na} + r_{Cl} = a/2 = 2.81$ Å. Each Na^+ ion is surrounded by six Cl^- ions at the corners of an octahedron; each Cl^- ion is

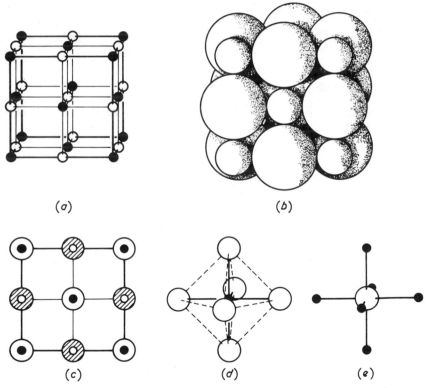

(a)   (b)

(c)   (d)   (e)

Figure 4-7 Structure of NaCl. (a) Perspective view of the unit cell, showing positions of atomic centres; black circles are Na, open circles Cl. (b) Same view, showing packing—radii drawn to scale. (c) Projection on plane of cube face. Small circles are Na, larger circles Cl; black and shaded circles are at heights 0 and 1, open circles at height 1/2. (d) Perspective view of the environment of the Na atom. Solid lines show Na—Cl bonds, broken lines the outline of the coordination polyhedron. (e) Perspective view of the environment of the Cl atom. Solid lines show Na—Cl bonds.

similarly surrounded by six Na^+ ions, as shown in Figure 4-7(d) and (e). There is no question of one Na^+ belonging to one particular Cl^- to make a molecule; instead, we might picture the structure as having each one-sixth of Na^+ linked to the neighbouring one-sixth of Cl^-.

In terms of coordination polyhedra, we see that all edges of each $NaCl_6$ octahedron are in common with a neighbouring octahedron; all the octahedra are, of course, exactly alike. In terms of packing, we see that the array of Cl^- atoms is slightly expanded from a close-packed formation, the nearest Cl–Cl distance being $\sqrt{2}a/2$, or 3.98 Å, instead of the 3.62 Å expected for close-packing; this expansion is needed to make the largest interstices (bounded by octahedral groups of Cl^-) large enough to accommodate Na^+. The radius ratio r_{Na}/r_{Cl} is 0.54 as compared with the lower limit of 0.41 for 6-coordination.

Clearly Pauling's electrostatic valence rule is exactly obeyed, because each of the six bonds between Na^+ and its neighbours has strength 1/6, and six such bonds reach each Cl^-.

It is interesting to project this structure down the cube body diagonal. We then see that the close-packed planes of Cl^-, corresponding exactly to the planes of Cu atoms in Figure 4-1(i), are interleaved by planes containing only Na^+. Using the notation of §4.5, the sequence of layers is as follows:

$$Cl^- \text{ in } A, \quad Na^+ \text{ in } C, \quad Cl^- \text{ in } B, \quad Na^+ \text{ in } A, \quad Cl^- \text{ in } C, \quad Na^+ \text{ in } B$$

The sodium chloride structure (sometimes called the $B1$ structure from the original classification in the Strukturbericht) is an extremely common one. Compounds possessing it include most of the alkali halides (except CsCl, CsBr, and CsI); oxides, sulphides, selenides and tellurides of Mg, Ca, Sr, and Ba (except MgTe); oxides of transition metals, such as FeO (commonly nonstoichiometric); many intermetallic compounds; and LiH. In some of these the 6-coordination would be predicted from the radius ratio, as explained in §2.5; in others, even among the nearly ionic compounds, it would not. The common tendency for covalent bonds to be octahedral (cf. §2.10) may help to explain some of the anomalies. It is interesting to notice that in CsF and RbF the cation is larger than the anion.

4.8 CAESIUM CHLORIDE

Space group: $Pm3m$ (No. 221)
Lattice:* Primitive cubic; unit cell with side $a = 4.11$ Å

---

* It is a common mistake to refer to the lattice of this structure, or to the structure itself, as body-centred. It would only be body-centred *if* Cs^+ and Cl^- were identical, which of course is not true. It can, however, be treated *as if* it were body-centred for any purpose for which the distinction between the two kinds of atoms either cannot be made in practice or is unimportant.

Lattice complex: (i) Cs^+ at $(0, 0, 0)$,

                     Cl^- at $(1/2, 1/2, 1/2)$

             (ii) With alternative choice of origin,

                     Cl^- at $(0, 0, 0)$, Cs^+ at $(1/2, 1/2, 1/2)$

Each unit cell contains one atom of each kind, i.e., one formula unit. The contact distances, shown in Figure 4-8(d) and (e), lie along the body diagonal of the cube, giving an effective radius sum $r_{Cs} + r_{Cl} = \sqrt{3}a/2 = 3.56$ Å.

Each Cs^+ ion is surrounded by eight Cl^- ions at the corners of a cube. This can be seen directly from Figure 4-8(d), where the choice of origin has been made as in description (ii); or, less directly, by drawing more of the structure than is shown in Figure 4-8(a), where the choice of origin has been made as in description (i). Similarly, each Cl^- ion is surrounded by eight Cs^+ ions at the corners of a cube; this can be seen directly from Figure 4-8(a) or 4-8(e), or less directly by extending Figure 4-8(d). The coordination polyhedra, which are cubes, share square faces with each other.

We might expect the effective radii to be rather greater than for the same ions in 6-coordinated environments, and in fact we find their sum, 3.56 Å, to be greater than that predicted by the uncorrected Goldschmidt values 1.70 and 1.80 Å. The radius ratio is 0.92, as compared with the predicted lower limit of 0.74 for 8-coordination. As in NaCl, the electrostatic valence rule is exactly obeyed, though here each bond has strength 1/8.

The structure is sometimes called Type $B2$, from the early classification in the Strukturbericht.

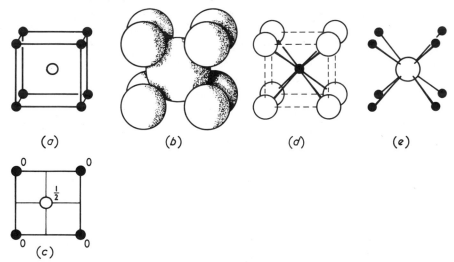

(a) (b) (d) (e)

(c)

Figure 4-8 Structure of CsCl. (a) Perspective view of the unit cell, showing the position of atomic centres; small black circles are Cs, large open circle Cl. (b) Same view, showing packing—radii are drawn to scale. (c) Projection on the plane of a cube face; atoms are marked as in (a). (d) Perspective view of the environment of the Cs atom. Solid lines show Cs–Cl bonds, broken lines the outline of the coordination polyhedron. (e) Perspective view of the environment of the Cl atom. Solid lines show Cs–Cl bonds.

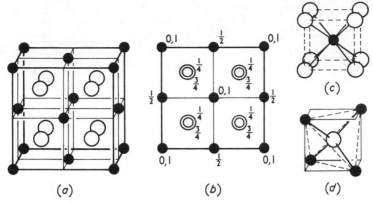

Figure 4-9 Structure of CaF_2. (a) Perspective view of the unit cell, showing the position of atomic centres; small black circles are Ca, large open circles F. (b) Projection on the plane of a cube face; atoms are marked as in (a). (c) Perspective view of the environment of the Ca atom. Solid lines show Ca—F bonds, broken lines the outline of the coordination polyhedron. (d) Perspective view of the environment of the F atom. Heavy solid lines show Ca—F bonds, broken lines join the tetrahedron of Ca atoms surrounding F atoms, and weak solid lines outline the octant of the unit cell.

Other examples of this structure type are TlCl, the bromides and iodides of Cs and Tl, and many intermetallic compounds. The compounds NH_4Cl and NH_4Br are closely related, but must be discussed separately because of problems presented by the position of the hydrogen atoms (cf. §15.24).

4.9 CALCIUM FLUORIDE

> Space group: $Fm3m$ (No. 225)
> Lattice: Cubic face-centred; unit cell with side $a = 5.45$ Å
> Lattice complex: Ca^{2+} at $(0, 0, 0)$;
> $\qquad$ 2 F^- at $\pm(1/4, 1/4, 1/4)$

The unit cell contains four Ca^{2+} and eight F^- atoms, i.e., four formula units of CaF_2.. Each Ca atom is surrounded by eight equidistant F atoms, arranged at the corners of a cube as shown in Figure 4-9(c); the environment is exactly like that of Cs in CsCl. Each F atom, however, is only 4-coordinated, its Ca neighbours being at the corners of a regular tetrahedron, as indicated in Figure 4-9(d). The contact distance is $\sqrt{3}a/4$, giving 2.36 Å, as compared with the Goldschmidt radius sum of 2.39 Å. If this had been corrected for 8-coordination, according to the general rule of Table 2-3, the predicted distance would have been much too great. The radius ratio r_{Ca}/r_F is 0.80, as compared with the lower limit of 0.74 for 8-coordination.

The application of Pauling's rule is a little less trivial than in the last two examples. Each bond from Ca has strength 2/8, and each F receives four such bonds, adding up to 1, which is its negative charge.

Isomorphous materials include SrF_2, BaF_2, ThO_2, and UO_2.

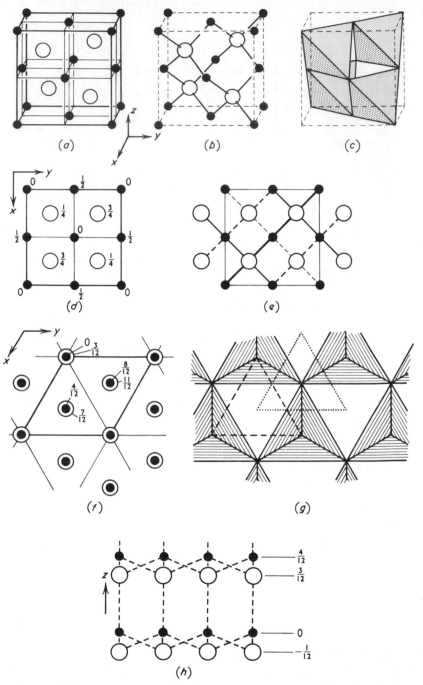

Figure 4-10 Structure of zinc blende. (a) Perspective view of the unit cell, showing the centres of atoms; small black circles are Zn, larger open circles S. (b) Same view, showing interatomic bonds. (c) Same view, showing the outlines of tetrahedra of Zn surrounding S. (d) Projection on the xy plane. Atoms are distinguished as in (a); the heights marked are

4.10 ZINC BLENDE (OR SPHALERITE)—THE FIRST STRUCTURE OF ZnS

Space group: $F\bar{4}3m$ (No. 216)
Lattice: Cubic face-centred; unit cell side $a = 5.41$ Å
Lattice complex: Zn at $(0, 0, 0)$; S at $(1/4, 1/4, 1/4)$

There are four Zn and four S atoms in the unit cell, i.e., four formula units (Figure 4-10). Each Zn atom is surrounded by four S atoms at the corners of a regular tetrahedron, and each S atom is surrounded by four Zn atoms. The close relation to the diamond structure (§4.6) is obvious; if we built a model of this structure and then made all the Zn and S atoms in it identical, it would be a model of diamond. The contact distance is $\sqrt{3}a/4$ or 2.34 Å, as compared with the Goldschmidt radius sum, 2.57 Å. Correction for 4-coordination, according to Table 2-3, reduces the predicted distance to 2.42 Å. The radius ratio is 0.48, as compared with the lower limit of 0.41 for 6-coordination and 0.22 for 4-coordination.

Though it is very artificial to discuss this structure in terms of fully ionised atoms, it is worth noting that Pauling's rule is exactly obeyed. Each bond from Zn has strength 2/4, and four such bonds reach each S, adding up to 2.

The structure of zinc blende is extremely important because it is possessed by many materials with valuable semiconducting properties. It is therefore desirable to consider it in detail from several aspects.

Like diamond, it is a rather open structure, though less so because of the disparity in size between the two kinds of atom. This suggests that its stability is partly due to a tendency to form tetrahedrally directed bonds.

It is interesting to project this structure, as we did that of diamond, down the cube body diagonal (Figure 4-10(f)). Again we get puckered rings, but now Zn and S alternate. Three S atoms at height $-1/12$, shown as open circles marked 11/12 in Figure 4-10(f), form an equilateral triangle, with Zn, at height zero, slightly above its centre. The next puckered ring brings S at height 3/12 immediately above this Zn, completing the tetrahedron around it. There is thus a layer of tetrahedra linked by their corners, all parallel and all pointing upwards, as shown in Figure 4-10(g). The top corner is also a

fractions of cube edge. (e) Same projection, showing interatomic bonds. Bonds in the upper half of the cell are shown by heavy solid lines, in the lower half by broken lines. The diagram has been extended to include enough atoms outside the unit cell to display the environment of Zn atoms. (f) Projection down the cube body diagonal. Atoms are distinguished as in (a); the heights marked are fractions of the cube body diagonal, with origin the same as in (a). Hexagonal axes of reference are used in this and the two following diagrams, (g) and (h). (g) Same projection as (f), showing the tetrahedra of S surrounding Zn. The first layer of tetrahedra (surrounding Zn at height zero) is shown in full; the triangular bases of one tetrahedron of the second layer and one of the third are shown by broken and dotted lines respectively. (h) Projection on the yz plane. Atoms are distinguished as in (a); distances of layers from the plane $z = 0$ are given as fractions of c. (The full height of the unit cell is not shown.) Broken lines are projections of interatomic bonds.

corner of the triangular bases of the next layer, one of which is shown by broken lines. This layer does not project on top of the first; its base triangles, and its Zn atoms, are related to those of the first as the *B* positions of cubic close-packing are to the *A*. A third layer of tetrahedra is above the second, in the *C* position; the base of one triangle of this layer is shown by dotted lines. The fourth layer repeats the first.

What we have been doing here is to apply the cubic close-packing operator—the cubic face-centred lattice—to the lattice complex, which is the pair of atoms Zn and S separated by a distance 2.34 Å along the (hexagonal) *c* direction.

All the tetrahedra point upwards. The axis down which they were projected is thus *uniterminal* (Figure 4-10[h]). Projection down any one of the other three body diagonals of the cube would have given the same result, provided that the positive direction was taken as that making an angle of about 110° with the positive direction of the first.

The tetrahedra are linked into a continuous three-dimensional framework by sharing each corner between two tetrahedra.

We could equally well have described the structure in terms of tetrahedra of Zn atoms surrounding central S atoms, as seen in Figure 4-10(b) and (c). Geometrically the only difference in the description would be that, for the same choice of axial directions, all tetrahedra would point downwards. Physically, it is easier to think in terms of large S anions grouped round the smaller Zn cations.

Here there is a distinction from the diamond structure. In diamond the two sets of tetrahedra contain the same atoms, and since one points up and the other down, both senses of each axis are equivalent—the axes are not uniterminal.

A wide variety of materials possess the zinc blende structure. Some representative examples are BeS, CuF, CdS (first form), GaAs, InSb.

4.11 WURTZITE—THE SECOND STRUCTURE OF ZnS

Space group: $P6_3mc$ (No. 186)
Lattice: Hexagonal primitive; unit cell with sides $a = 3.81$ Å, $c = 6.23$ Å
Lattice complex: Zn atoms at $(2/3, 1/3, 0)$, $(1/3, 2/3, 1/2)$;
    S atoms at $(2/3, 1/3, u)$, $(1/3, 2/3, 1/2 + u)$, with $u \simeq 3/8$.
(Any other choice of origin which adds the same amount to all z coordinates would be equally permissible.)

There are two formula units of ZnS in the unit cell (Figure 4-11).

The easiest way of understanding this structure is to realise that it is related to the zinc blende structure as hexagonal close-packing is to cubic close-packing. If we take the lattice complex of zinc blende—a Zn atom and

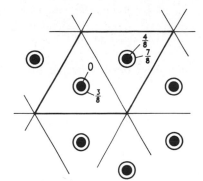

Figure 4-11 Structure of wurtzite. Projection on the *xy* plane; small black circles are Zn, larger open circles S. The unit cell is shown with a heavy outline. Heights of atoms are marked as fractions of *c*. Compare with Figure 4-10(f).

an S atom separated in this case by 2.33 Å—and use the hexagonal close-packing operator defined in §4.5, we get the wurtzite structure.*

The Zn–S distance along the *c* axis is not necessarily exactly equal to the Zn–S distances inclined to the axis, but in fact there is little difference between them, and the tetrahedra are therefore very nearly regular.

There are only two layers of tetrahedra in the *c*-repeat distance, instead of three as in zinc blende, but all tetrahedra point in the same direction, as they do in zinc blende, and the axis is therefore uniterminal. The projection of zinc blende in Figure 4-10(h) would do equally well for wurtzite if the separations of the layers were read as multiples of 1/8 instead of multiples of 1/12, since the differences between the two structures concern the heights of atoms above the plane of projection, which are not marked (but which can be seen by comparing Figures 4-11 and 4-10(f)).

Like hexagonal close-packed structures but unlike diamond, wurtzite has a unique *c* axis; there are no other equivalent axes in the structure. Other materials possessing this structure are ZnO, BeO, CdS (second form), MgTe, and NH_4F (considering the NH_4 group as a spherical atom—a point discussed in more detail in §15.24).

Recent accurate work has shown that in ZnO at room temperature the axial ratio is 1.6021 instead of the ideal 1.6330, the parameter *u* is 0.3825 instead of 3/8, and the Zn–O distances parallel to *c* are 0.019 Å longer than the others—a difference which is just significant compared with experimental error.

4.12 COPPER–GOLD ALLOY, Cu₃Au †

Space group: *Pm3m* (No. 221)
Lattice: Primitive cubic; unit cell with side $a = 3.74$ Å

---

* There is a high-pressure form of carbon, the rare mineral lonsdaleite, found in meteorites, whose structure is related to that of wurtzite as diamond is to zinc blende, i.e., it can be derived from the wurtzite structure by replacing both Zn and S by C.

† This structure is of particular interest in connection with the possibilities of disorder in alloys. Here we are concerned only with the perfect structure.

Lattice complex: Au at $(0, 0, 0)$;

3 Cu at $(0, 1/2, 1/2)$, $(1/2, 0, 1/2)$, $(1/2, 1/2, 0)$

1 formula unit of Cu_3Au per unit cell.

Each Au atom has twelve Cu neighbours, as shown in Figure 4-12(b) and (c). Each Cu has four Au neighbours (at the corners of a square) and eight other Cu neighbours. The interatomic distance is $\sqrt{2}a/2 = 2.65$ Å.

The geometrical interest of this structure lies in its relation to the structure of copper. The pattern of sites occupied by atoms is exactly the same in both; they differ in the distinction between the occupying atoms. (In just the same way, diamond and zinc blende have the same pattern of sites, but differ in the species of atoms occupying them.) The size of the unit cell is the same for Cu and Cu_3Au, but the lattices are different; the number of lattice points per cell is 1 for Cu_3Au, as compared with 4 for Cu.

The structure of Cu_3Au illustrates the distinction between a lattice and an array of atomic sites: its lattice is primitive, but its array of sites is face-centred cubic. This structure has sometimes been described as a 'superlattice' or 'superstructure', but geometrically there is no reason for making a category to distinguish it from the other structures considered in this chapter.

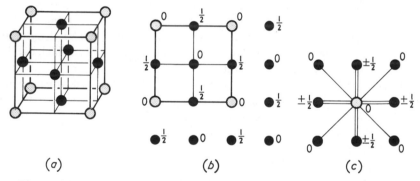

(a) (b) (c)

Figure 4-12 Structure of Cu_3Au. (a) Perspective view of the unit cell (heavy outline), showing the centres of atoms; black circles are Cu, hatched circles Au. (b) Projection on a cube face. The unit cell is shown with a heavy outline. Atoms are distinguished as in (a); heights above the cube face are marked. (c) Part of the projection in (b), showing the environment of the Au atom; the double lines are two bonds that coincide in projection.

4.13 IRON–ALUMINIUM ALLOY, Fe_3Al*

Space group: $Fm3m$ (No. 225)

Lattice: Cubic face-centred; unit cell with side $a = 5.79$ Å (2×2.90 Å)

Lattice complex: 3 Fe at $(0, 0, 0)$, $(1/2, 0, 0)$, $(1/4, 1/4, 3/4)$;

                 Al at $(1/4, 1/4, 1/4)$

4 formula units of Fe_3Al per unit cell

* The description applies to the perfect structure—see footnote to Cu_3Au.

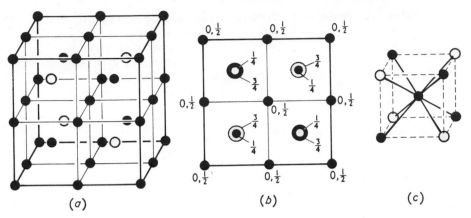

Figure 4-13 Structure of Fe₃Al. (a) Perspective view of the unit cell (heavy outline), showing the centres of atoms; black circles are Fe, hatched circles Al. (b) Projection of the unit cell on a cube face. Atoms are distinguished as in (a), except that atoms at height 3/4 are shown larger than the others. Heights above the cube face are marked. (c) Perspective view of the environment of the Fe atom at (1/2, 1/2, 1/2). Atoms are distinguished as in (a). The piece of structure shown is the central part of that in (a); solid lines are interatomic contacts, and broken lines the outline of the polyhedron of neighbours of the central Fe.

In this structure (Figure 4-13), each Al atom has eight Fe neighbours at the corners of a cube, and the same is true for two-thirds of the Fe atoms. Each of the other Fe atoms has four Al and four Fe neighbours; taken alone, the set of Al neighbours or Fe neighbours forms a tetrahedron, but together they form a cube, as shown in Figure 4-13(c). The interatomic distance is $\sqrt{3}a/4 = 2.50$ Å.

The geometrical interest of this structure is in its relation to cubic body-centred on the one hand, and to CsCl on the other. Applying the lattice operator to the Fe atom at (1/2, 0, 0), we find other Fe atoms at (0, 1/2, 0) and (0, 0, 1/2), which, with that at (0, 0, 0) and its lattice-related points, occupy all the corners of the cubes constituting octants of the unit cell. At the centres of four of these octants are Al, and at the centres of the other four, Fe. The structure could thus be derived from the cubic body-centred structure of Fe by selecting every alternate unit cell systematically and replacing its body-centring atom by Al.

Like Cu₃Au, this structure illustrates the distinction between a lattice and an array of atomic sites. The lattice here is cubic face-centred, but the array of sites is cubic body-centred. This structure, too, has been called a superlattice or superstructure, but again there is no need to make a special category for it.

4.14 ARRAY OF OCCUPIED SITES

It is useful to consider some of the other simple structures in terms of their overall array of occupied sites (Table 4-2). The number of formula

Table 4-2 Patterns of Occupied Sites in Some Simple Structures

ARRAY OF SITES	STRUCTURE	NUMBER OF FORMULA UNITS PER UNIT CELL OF ARRAY
Cubic face-centred	Cu	4
	Cu$_3$Au	1
Cubic body-centred	Fe	2
	CsCl	1
	Fe$_3$Al	$\frac{1}{2}$
Cubic primitive	NaCl	$\frac{1}{2}$
Diamond	Diamond, C	8
	Zinc blende, ZnS	4
Hexagonal close-packed	Mg	2
ABAC packing sequence	NiAs	2

units per unit cell of the array of sites (i.e., of what would be the structure if all atoms were identical) is also given. If it is a fraction, the unit cell of the actual structure must be a multiple of that of its array of sites.

It is worth emphasising, because confusion sometimes results from not doing so, that just as the CsCl structure has a body-centred array of sites, so the NaCl structure has a *primitive* array of sites.

4.15 NICKEL ARSENIDE, NiAs*

Space group: $P6_3/mmc$ (No. 194)
Lattice: Primitive hexagonal; unit cell sides $a = 3.60$ Å, $c = 5.01$ Å, with $c/a = 1.39$
Lattice complex: 2 Ni at $(0, 0, 0)$, $(0, 0, u)$
2 As at $(2/3, 1/3, u)$, $(1/3, 2/3, 1/2 + u)$, with $u = 1/4$

The structure is as illustrated in Figure 4-14(a) and (b). Each Ni atom is surrounded by six As atoms in an octahedral arrangement, as shown in Figure 4-14(d). The octahedron is not regular but flattened: the height is only 1.2 times the projection of the NiAs distance, instead of 1.4 times as in a regular octahedron. This allows the approach of Ni atoms above and below, and so we must count them as neighbours of the central Ni, which has thus a more complicated environment than is found in any structure considered up to this point. The Ni–As distance is $\sqrt{(a^2/3 + c^2/16)} = 2.43$ Å, the Ni—Ni distance $c/2 = 2.50$ Å. Each As atom is surrounded by six Ni atoms at the corners of a trigonal prism.

---

* Many disordered variants of this structure are found in actual materials, even in nickel arsenide itself at nonstoichiometric compositions. The perfect or idealised structure is described here.

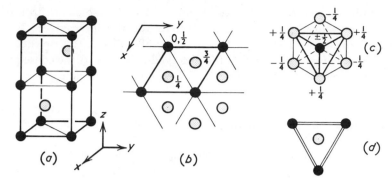

Figure 4-14 Structure of NiAs. (a) Perspective view of the unit cell (heavy outline), showing the centres of atoms; black circles are Ni, hatched circles As. (b) Projection on the xy plane; the unit cell is shown with a heavy outline. Atoms are distinguished as in (a); heights are marked. (c) Projection on the xy plane of the environment of the Ni atom, with heights marked; atoms are distinguished as in (a). The As atoms at $\pm 1/4$ form a flattened octahedron; addition of Ni atoms at $\pm 1/2$ makes it into a rhombohedron. (d) Projection on the xy plane of the environment of the As atom. Atoms are distinguished as in (a); heights are marked. Double lines indicate the edges of the trigonal prism which coincide in projection.

If we consider the centres of the As atoms alone, they form an array like hexagonal close-packing slightly compressed in the direction perpendicular to the layers. (Compare the c/a ratio of 1.39 with its ideal value of 1.63 for close-packing.) Into the octahedral interstices described in §4.5 we can insert Ni atoms; there are as many such interstices as atoms, and so by filling all of them we achieve the composition NiAs. This is a somewhat artificial way of deriving the structure, because the Ni atoms are in fact about as large as the As atoms, if, as seems reasonable, we use metallic rather than ionic radii; even if we use ionic radii, the Ni atoms are not so much smaller than the As atoms that they can fit into the interstices without pushing the As atoms further apart. The array of As atoms is very far from close-packed, as can be seen by comparing the a value of 3.6 Å with the As (metallic) atomic diameter of 2.5 Å; we might better refer to it as an *expanded hexagonal packing*. The usefulness of this way of thinking in terms of interstices in an expanded packing becomes apparent when we need to compare the NiAs structure with other structures, including imperfect structures not considered here. For example, if only half the octahedral interstices were occupied by metal atoms—those at height zero, but not at height 1/2—we should have the CdI_2 structure (§4.16).

An alternative way of considering it can also be helpful. If we take into account both Ni and As in considering the packing, the sequence may be written, in the notation of §4.5, as $(A)B(A)C(A)B(A)C\ldots$, the round brackets marking the planes occupied by Ni.

A property of this structure which is not possessed by cubic structures, but was previously noted for the structure of Mg (§4.4), is the arbitrariness of its c/a ratio. An increase in c/a would not only make the octahedron of As atoms surrounding Ni more regular, but would also increase the Ni—Ni length relatively to Ni—As. To put it thus is of course to reverse cause and effect; we may really deduce that the observed axial ratio is the consequence of the relative magnitude of the Ni—Ni equilibrium distance, and hence throws light on the character of the Ni—Ni interaction. In this chapter, however, we are concerned with the consequences and not the causes, but we may note that the Ni—Ni distance is that characteristic of metallic packing of Ni.

Other materials possessing the NiAs structure have different c/a ratios, ranging from VS ($c/a = 1.73$) through FeS ($c/a = 1.68$) and NiS ($c/a = 1.55$) to FeSb ($c/a = 1.25$); correspondingly, the ratio of the metal-metal distance to that between unlike atoms ranges from about 1.17 for VS to 0.96 for FeSb.

4.16 CADMIUM IODIDE, CdI_2

Space group: $P\bar{3}m1$ (No. 164)
Lattice: Primitive hexagonal; unit cell sides $a = 4.24$ Å, $c = 6.84$ Å, with $c/a = 1.61$
Lattice complex: Cd at $(0, 0, 0)$;
2 I at $\pm(2/3, 1/3, u)$, with $u \simeq 1/4$
1 formula unit of CdI_2 per cell

This is a layer structure (Figure 4-15) consisting of planes of Cd atoms (parallel to x and y) sandwiched between planes of I atoms; the sandwich is the repeat unit in the z direction.

Each Cd atom is surrounded by six I atoms forming a slightly flattened octahedron. Each I atom also has six neighbours arranged in an irregular octahedron, three Cd atoms on one side of it (above or below) and three I atoms on the other, rather farther away. The Cd—I distance is 2.99 Å; the I—I distances are 4.24 Å in the xy plane, 4.21 Å for other edges of the octahedron round Cd and for contacts with the next CdI_2 layer. (If u were rather less than 1/4, the next-layer contact distances would be greater than 4.21 Å, and the other edges less.)

If we consider I atoms alone, they form a hexagonally packed array, as can be seen from Figure 4-15(a); strictly, it is an expanded array, but it is not greatly enlarged from a close-packed one. The octahedral interstices between pairs of I planes are occupied by Cd atoms as illustrated in Figure 4-15(b). Figure 4-15(c) shows a projection in the yz plane. The lines marked $PRTV$ represent planes of I atoms. (Individual atomic positions within the planes are not marked.) All the interstices in the intervening planes Q and U are occupied by Cd atoms, whereas all those in S are empty. The repeat distance along z is PT.

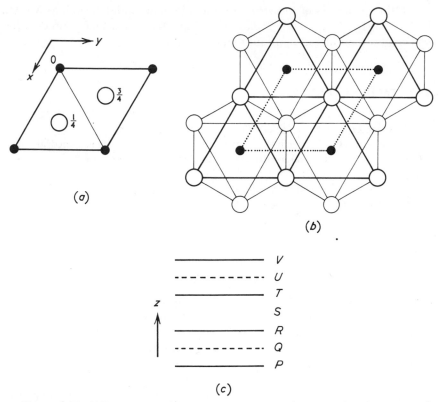

Figure 4-15 CdI$_2$ structure. (a) Projection on the xy plane, showing the centres of atoms; small black circles are Cd, large open circles I. Heavy lines outline the unit cell. Heights of atoms are marked as fractions of c. (b) Same projection, showing the outline of octahedra round Cd. Atoms are distinguished by small black circles and large open circles as in (a), but heavy open circles and lines now show atoms and interatomic distances in the plane at $z = 3/4$ and lighter open circles and solid lines show parts of the structure below this; the dotted line is the outline of the unit cell. (c) Projection on a plane through the z axis, showing planes occupied by I (solid lines) and Cd (broken lines).

If, instead of thinking of Cd atoms as filling interstices, we treat them as equivalent to I atoms and consider the array of occupied sites as a whole, we can represent it as a close-packing with every fourth plane of atoms missing. If such empty planes are indicated by O, the sequence is $(A)BOC(A)BOC\ldots$, the round brackets (parentheses) indicating the planes occupied by Cd. This is illustrated in the projection on the yz plane in Figure 4-16(e).

CdI$_2$ is an example of a *layer structure*, because the Cd—I bonds that link the atoms into a two-dimensionally infinite layer are much stronger than the I—I bonds, which are the only ones holding layers together in the z direction.

For the CdI$_2$ structure as for the NiAs structure, variations in the c/a ratio can occur in different materials, but in practice the deviations from close-packing are much less for the CdI$_2$ structure.

Other examples of this structure include many halides, such as $MgBr_2$, $FeBr_2$, MgI_2, CaI_2, and PbI_2; also $Mg(OH)_2$, $Ca(OH)_2$, and other hydroxides (considered more fully in §13.9).

4.17   CADMIUM CHLORIDE, $CdCl_2$

> Space group: $R\bar{3}m$ (No. 166)
> Lattice: Rhombohedrally-centred hexagonal (i.e., rhombohedral using hexagonal axes of reference), with lattice operator
> $(0, 0, 0; \ 2/3, 1/3, 1/3; \ 1/3, 2/3, 2/3)+$
> Unit-cell sides $a = 3.85$ Å, $c = 17.46$ Å. $c/a = 4.54$ ($= 3 \times 1.51$)
> Lattice complex: Cd at $(0, 0, 0)$;
> $\qquad$ 2Cl at $\pm(2/3, 1/3, u)$, with $u \simeq 1/12$

This, like CdI_2, is a layer structure (Figure 4-16); the layers are the same in both structures, but they pack differently.

The environment of Cd is the same as previously; each Cd has six Cl neighbours arranged around it in an octahedron as shown in Figure 4-16(b). On the other hand, though Cl still has three Cd neighbours on one side and three Cl on the other, the six taken together constitute a trigonal prism, as shown in Figure 4-16(c), and not an octahedron. The Cd—Cl distance is 2.66 Å; Cl—Cl distances are 3.85 Å in the xy plane, 3.63 Å in the other octahedron edges, and 3.38 Å between layers.

We may usefully derive the structure in two ways. In the first way, taking a $CdCl_2$ layer like the CdI_2 layer, we add a second similar layer so that its Cd atoms lie vertically above the upper Cl atoms of the first layer, and a third so that its Cd atoms lie vertically above the upper Cl atoms of the second. The fourth layer is vertically above the first. This process resembles the operation of cubic close-packing, described in §4.5. Unlike the case of zinc blende (§4.10) where we used the same operation, it does not result in a cubic structure, because the lattice complex of $CdCl_2$ does not have the requisite symmetry. The process is nevertheless a lattice operation, summarised by the lattice vectors noted at the beginning of this section. In consequence, the $CdCl_2$ structure is related to the CdI_2 structure as a rhombohedrally centred hexagonal lattice is to a primitive hexagonal lattice.

The second way of deriving the structure is to construct the close-packing sequence, as we did for CdI_2, marking the layer of unoccupied interstices between Cl atoms as O and the layer occupied by Cd atoms with round brackets. This gives $(A)BOA(B)COB(C)AOC(A) \dots$ Its projection on the yz plane is shown in Figure 4-16(d), with the corresponding projection of CdI_2 in Figure 4-16(e) for comparison. The packing of the layers in both structures can be seen, and also the 3:1 ratio of the heights of the unit cells.

Other examples of this structure are $MgCl_2$, $FeCl_2$, ZnI_2, and a second form of PbI_2. It is less common than the CdI_2 structure.

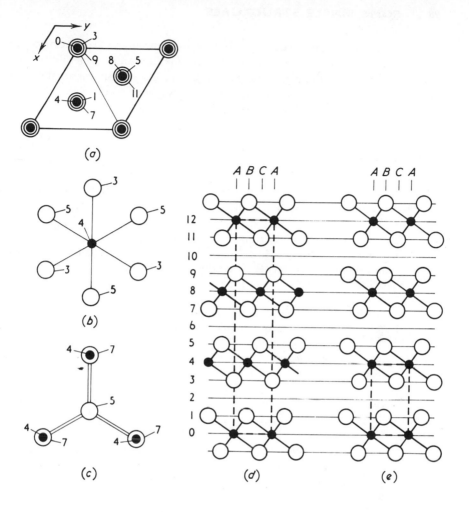

Figure 4-16 $CdCl_2$ structure. (a) Projection on the xy plane, showing the centres of atoms; small black circles are Cd, large open circles Cl. One Cd atom and two Cl atoms always coincide in projection; their heights are marked in multiples of $\frac{1}{12}c$. (b) Part of the same projection, showing the environment of the Cd atom. Heights are marked in multiples of $\frac{1}{12}c$. The lines are CdCl bonds. (c) Part of the same projection, showing the environment of the Cl atom; double lines represent Cd—Cl and Cl—Cl bonds, which coincide in projection. (d) Projection on the yz plane. Atoms are distinguished as in (a). Heavy lines are Cd—Cl bonds; fine lines are traces of xy planes at heights $n/12$, numbered to match (a), (b), and (c). The vertical lines marked A, B, C correspond to the nomenclature in Figures 4-4 and 4-5. The unit cell is outlined with a heavy broken line. (e) Diagram corresponding to (d) drawn for the CdI_2 structure.

As so often happens when structures based on cubic close-packing and hexagonal close-packing can exist separately for the same or closely related compounds, intermediate disordered forms can occur. In these, the AX_2 layers are essentially the same, but the packing sequence is irregular (cf. §4.5). We shall not consider them further.

4.18 SUMMARY

The structures described in this chapter have provided simple illustrations of very important principles. They have introduced us to:

the idea of *close-packed arrays*, cubic or hexagonal (§4.2 to 4.5);

that of *expanded packings*, which have the same pattern, though the distances are too great for contact between the atoms (§4.15);

the filling in of interstices of such arrays by atoms of a different kind (§4.7 and 4.15 to 4.17);

the sharing of edges of coordination polyhedra (§4.7);

the distinction between an array of occupiable sites and the distribution of the occupying atoms (§4.12 to 4.14);

the existence of *linkage structures* dependent on bond directions rather than packing (§4.6 and 4.10 to 4.11).

In addition, they have provided examples of some of the concepts discussed in Chapters 1 and 2:

the dependence of coordination number (and hence of structure type) on radius ratio (§4.7 to 4.9);

Pauling's electrostatic valence rule (§4.7 to 4.10);

isomorphism (almost every section);

polymorphism (§4.10 to 4.11).

Geometrically, we have seen the usefulness of projections of cubic structures down their triad axes, and their description by means of hexagonal axes of reference (§4.2, 4.5, 4.6, 4.10); and the extension of the idea of the cubic and hexagonal close-packing operators to structures which are not close-packed (§4.10, 4.11, 4.16, 4.17).

On the other hand, we have had examples (§4.4, 4.15) where structures that are geometrically isomorphous may in fact differ considerably in their axial ratios and interatomic distances, and hence in their physical character.

Simple structures of metals and a few intermetallic compounds have been included (§4.2 to 4.4, 4.12 to 4.13, 4.15), and in one case (§4.15) we have touched briefly on the need for a set of radii other than the ionic radii of Table 2-2 for interpreting them.

Most of the structures described in this chapter are so symmetrical that in stating the unit-cell size we have also specified the interatomic distances;

the few examples (§4.11 and 4.15 to 4.17) where a single arbitrary parameter is needed to define the lattice complex still represent very special cases. This simplicity is not representative of crystal structures, and if we limit ourselves to such types we shall not obtain a sufficiently general picture of the interplay of interatomic forces. Other examples, however, must wait until we have developed the ideas of crystal symmetry. We shall consider them in Chapter 11 and later chapters.

In other respects, most of the general principles of this chapter will be illustrated again and again in structures to be described later. We shall not, however, include any more complicated examples of intermetallic structures, though modern work has shown the existence of an immense variety of these. Our chief concern will be with "ionic" and semipolar compounds where there are big differences of electronegativity that allow us to think in terms of "cation" and "anion," along the lines outlined in Chapter 2.

PROBLEMS AND EXERCISES

1. Obtain a convenient set of equal spheres (marbles, round beads, silver cake decorations, or tapioca) and glue them together in close-packed layers. Stack them together to form a cubic close-packed structure. Show that, by suitably breaking off corners, similar close-packed layers inclined to the horizontal can be found. Find also the cube faces.

   Change the stacking of the layers to form a hexagonal close-packed structure. Can you find any nonhorizontal close-packed layers?

2. Illustrate cubic and hexagonal close-packing by using sheets of paper or metal with circular holes cut in a hexagonal array. (Such sheets can sometimes be bought as metal or decorative paper; if not, they can be made by using a paper punch to punch holes at corners of a hexagonal net just large enough to prevent the paper from tearing between the holes.)

3. If atoms are assumed to be equal spheres, filled with electrons out to the packing radius, what fraction of the unit cell is empty space (a) in cubic close-packing, (b) in hexagonal close-packing, (c) in body-centred cubic packing?

4. For the metals in Table 4-1, calculate the interatomic distances between neighbouring atoms (a) in the same layer, (b) in adjacent layers.

5. Prove that, for a cubic close-packed array of atoms with cube edge a, there are two kinds of interstices, capable respectively of holding atoms of radius about $0.14a$ in octahedral coordination and atoms of radius

about $0.08a$ in tetrahedral coordination, without disturbing the original array. Find the ratios of these radii to the radius of the close-packed atoms.

6. Find (a) the radius of an atom which could hypothetically be fitted into the open spaces in diamond, assuming all atoms to be spheres, (b) the fraction of empty space in diamond, with the assumptions of Question 3.

7. Assuming all electron density to be contained within spheres defined by the packing radii, how does the empty space in crystals with the rock salt structure depend on the radius ratio?

8. For which of the structures listed in §4.7 as isomorphous with rock salt would this structure have been predicted by radius-ratio considerations? Comment on the others, noting the predicted structure, the magnitude of the discrepancy, and any special factors which may favour the observed structure.

9. Project the unit cell of zinc blende on a plane perpendicular to the body diagonal of the cube, and mark the heights of all atoms above the lowest corner of the cube, expressing them as fractions of this body diagonal. Add atoms from adjacent unit cells of the periodic structure which project within the same boundaries and mark their heights. Outline the tetrahedra with S as corners, and discuss their linkage, both in the horizontal plane and in other planes inclined to the horizontal.

   How many ZnS_4 tetrahedra are there in the translation-repeat unit with the same volume as the cube?

10. Show that if, in the wurtzite structure, all tetrahedra are regular, with the cations at their centres, the axial ratio is 1.633 and the u parameter 3/8.

    If c/a differs from the ideal by a factor $1 + x$, obtain expressions in terms of x for (a) the ratio of the sloping and horizontal anion-anion edges of the tetrahedra, (b) the value of u that makes all cation-anion distances equal.

11. If the volume per formula unit is expressed as αr_0^3, where r_0 is the nearest-neighbour distance, evaluate α for the rock salt, caesium chloride, and wurtzite structures (assuming the wurtzite structure to have ideal parameters).

12. Obtain expressions for interatomic distances in the nickel arsenide structure (a) between like atoms, Ni—Ni, (b) between unlike atoms, Ni—As, in terms of the lattice parameters a and c/a.

    Assuming that Ni—Ni contacts give the Ni diameter, find the As radius.

13. Apply the method of Question 12 to the compounds in the following table, which are all isomorphous with NiAs, to find radii for all the atoms. Average over atoms of the same kind to obtain a set of radii appropriate for use with this sort of compound; estimate limits of error.

Hence predict the lattice parameters of FeSe, CoS, and NiTe, all with the NiAs structure (cf. R. W. G. Wyckoff: *Crystal Structures*).

	a (Å)	c (Å)		a (Å)	c (Å)
FeS	3.45	5.67	NiS	3.42	5.30
FeTe	3.80	5.65	NiSe	3.66	5.33
FeSb	4.06	5.13	NiAs	3.60	5.01
CoSe	3.61	5.28	NiSb	3.94	5.14
CoTe	3.89	5.36	MnTe	4.12	6.70
CoSb	3.87	5.19	MnAs	3.72	5.70
			MnSb	4.12	5.78

Chapter 5 DIRECTIONS OF PLANES AND LINES

5.1 INTRODUCTION

This chapter summarises important geometrical relations and explains the notation by which information about the directions of planes and lines is concisely expressed.

For a periodic structure, we shall be concerned with *rational directions* and *rational planes*, i.e., directions and planes which are parallel to translation vectors.

5.2 THE EXTERNAL FORM OF CRYSTALS

Though solids are usually crystalline, this does not mean that they always grow as recognisable crystals. The texture of the solid depends not only on the crystal structure but on the conditions of growth, which are rarely uniform enough to allow the development of large pieces of perfect structure. For modern crystallography, this is not too serious; very small single crystals, either grown as such or cut from larger polycrystalline samples, are adequate for structure determination. Historically, however, crystallography began with the study of external forms, and the results of such study laid the foundations from which structural work could develop.

If a crystal grows in uniform conditions, it may be expected to develop faces which are parallel to rational lattice planes. Preference for one plane rather than another depends on the nature of the lattice complex—in particular, on the inclination of interatomic bonds to the plane in question—as well as on the rate at which atoms or groups of atoms can be brought up from the surrounding liquid. Indeed, one can sometimes make a good guess at the structure from a knowledge of the external faces preferred.

Greater uniformity of conditions will be needed to ensure that all preferred faces grow at the same rate rather than merely that each remains a plane face. We may therefore expect to find regularities in the directions of faces rather than in their positions or sizes, and this in fact happens.

Historically, the observation of regularities came before knowledge of the structure, and provided the starting point from which theories of the structure could be derived. The *law of constancy of angles* is the first fundamental law of crystallography; it was put forward in 1669 by Nicholaus Steno (Niels Stensen). It may be stated as follows: *The interfacial angles of a crystal are characteristic of the kind of crystal, however its external shape may vary in other respects.*

5.3 THE LAW OF RATIONAL INDICES: MILLER INDICES

The second fundamental law of crystallography, also based on observation, is the *law of rational indices.* It may be stated as follows: *If directions parallel to three edges of a well-developed crystal are chosen as axes of reference, and if the intercepts of any face on the axes are in the ratio* a : b : c, *then the ratio of the intercepts of any other face can be written as* a/h : b/k : c/l, *where* h, k, l *are fairly small integers (positive or negative) or zero.* This formulation, and the use of the indices h, k, l, was developed and extensively applied by W. H. Miller, Professor of Crystallography at Cambridge, in the second quarter of the nineteenth century. It is possible to choose h, k, l so that they have no common factor, and this is always done when we are concerned with the crystal as a finite object. The integers thus defined are the *Miller indices** of the face, or of the plane parallel to it; the symbol is written (hkl) in round brackets (parentheses). The plane used to define the *axial ratios* a:b:c, the *parametral plane*, is (111).

It is important to notice that neither the law nor the definition of Miller indices makes any assumption about the internal structure of the crystal, though the law can be so readily explained from the idea of a lattice that it led to a belief in the existence of lattices long before there were any experimental means of investigating them. Assuming a lattice, the law merely states that crystal faces are parallel to rational lattice planes. The definition of indices is rewritten from this standpoint: *A plane parallel to a rational plane of the lattice defined by vectors* **a**, **b**, **c** *has Miller indices* (hkl) *when its intercepts on the axes of reference are in the ratio* a/h : b/k : c/l *where* h, k, l *are integers (or zero) with no common factor.*

The index zero implies that the intercept is infinite, i.e., that the plane is parallel to the axis in question.

Figure 5-1 indicates the use of Miller indices. Let OA, OB, OC be equal (or proportional) to a, b, c, respectively, and let $OB' = OB/2$, $OB'' = OB/3$, $OC' = OC/2$, $OC'' = OC/3$, $OA'' = OA/3$. Then the plane ABC is (111), ABC' is (112), ABC'' is (113), $A''B'C''$ is (323); but $A''B''C''$, which is parallel to

---

* *Miller-Bravais indices* are 4-digit indices using a modified form of the same principle, applicable in certain special circumstances. (See Appendix 1 to this chapter.)

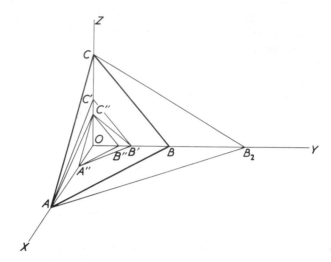

Figure 5-1   Intercepts of planes on axes of reference.

ABC, is not (333) but (111), since *Miller indices are concerned only with direction, not with position*. By the same argument, AB_2C is (212).

The law of rational indices, and the definition of indices, *holds good for any choice of rational directions as axes of reference*. The axes need not be orthogonal, and they need not be specially related to any symmetry directions of the crystal. The numerical values of the indices of course depend on the particular choice of axes of reference; we shall see later (Chapter 10) how to convert from one set to another.

There is one important difference between the use of external edges and of lattice vectors to define the axes of reference. Suppose we choose the same directions in each case. If our definition is in terms of lattice vectors, the lengths *a*, *b*, *c*, are thereby fixed as being the shortest vectors in these directions, and therefore the parametral plane (111) is fixed. If our definition is in terms of external edges, we have a free choice of parametral plane, and may happen to choose one that gives a different axial ratio—for example, AB_2C in Figure 5-1. For description and discussion of the structure such a choice is unsatisfactory, because if *OA* and *OC* are primitive vectors **a** and **c**, OB_2 is twice the primitive vector **b**, and this would be confusing.

Very many crystals, especially crystals of minerals, were described from their external form before X-ray diffraction became available. It is not surprising that the choice of parametral plane was sometimes unfortunate, and had to be revised when the unit-cell measurements were made. This, of course, means that all the faces had to be re-indexed to correspond. On reading old descriptions, one must therefore check whether the axial ratios are the same as in a modern description, and if not one must convert the indices before using them. What *is* rather unexpected—but satisfying—is the high proportion of crystals for which the morphological choice of parametral plane has proved to be the correct choice from the lattice standpoint.

When symmetry is present, faces occur in sets such that each member of the set is derived from any other by a symmetry operation. Such a set is called a *form*, e.g., the six faces of a cube, the four faces of a regular tetrahedron, or the three faces of an open-ended triangular prism. If the axes of reference are chosen suitably with respect to symmetry directions, the faces of a form have indices which are permutations of positive or negative values of the same three integers, according to rules easily derived from the symmetry (cf. Chapter 6). It is then often convenient to write the form as {*hkl*}, using curly brackets (braces), to mean all the planes which can be derived by the symmetry of the crystal from the plane named: e.g., {100} may be used for the six faces of a cube, (100), ($\bar{1}$00), (010), (0$\bar{1}$0), (001), (00$\bar{1}$).

5.4 DIRECTIONS OF LINES: ZONES AND ZONE INDICES

A *zone* is a set of planes all parallel to one direction, the *zone axis*. The zone axis is not located anywhere in particular; it is simply a direction. Any rational direction in the crystal, being the intersection of two rational planes, can be considered as a zone axis, and for this reason the indices specifying a direction are commonly called *zone indices*. Their definition is as follows: if a line through the origin passes also through the point with Cartesian coordinates ua, vb, wc, then its direction has the zone indices [*uvw*]. Note the square brackets used with this definition. A form of symmetry-related directions is denoted by $\langle uvw \rangle$.

Several useful relations about zones may be noted.

(i) Planes are in the same zone if their normals are in the same plane.

(ii) A plane (*hkl*) is in the zone [*uvw*] if

$$hu + kv + lw = 0 \qquad (5.1)$$

(iii) If planes $(h_1k_1l_1)$ and $(h_2k_2l_2)$ lie in the zone [*uvw*], then

$$u = k_1l_2 - k_2l_1, \quad v = l_1h_2 - l_2h_1, \quad w = h_1k_2 - h_2k_1 \qquad (5.2)$$

(iv) The zone axis [*uvw*] generally does not coincide with the normal to plane (*uvw*), but does so in crystals with cubic symmetry and in other crystals if it is a symmetry direction.

5.5 FAMILIES OF PLANES AND INTERPLANAR SPACING: BRAGG INDICES

For a lattice, we are concerned not only with the directions of planes but with the interplanar spacings—the distances between parallel planes through successive lattice points.

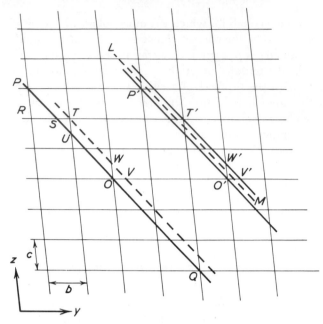

Figure 5-2 Indexing of families of planes (see text).

Consider the example in Figure 5-2. This is a projection (not necessarily orthogonal) down the x axis on to the yz plane, and PQ is a plane parallel to the x axis. Plane PQ goes through some lattice points, and its other intersections with unit cell edges always divide them in simple ratios; for example, $RS = 2RT/3$. The lattice point T which approaches closest to PQ without actually lying in it is separated from it by $TS = (1/3)b$ in the y direction and $TU = (1/2)c$ in the z direction. Hence if the origin is taken at a lattice point O in PQ, the next plane parallel to PQ passes through T and makes intercepts $OV = (1/3)b$ and $OW = (1/2)c$. Both PQ and TV are then parallel to a plane with indices 032.

But we know more than that about these planes. We know that their separation d is the perpendicular distance from T to PQ, and that all successive planes in this direction through lattice points have the same separation d. Planes PQ and TV are in fact members of a *family of planes*, characterised by their interplanar spacing as well as their direction.

More generally, the planes whose traces are PQ and TV are inclined to the x axis as well as to the y and z axes. The length of the intercept on OX is most easily seen by looking at one of the other principal projections, on the xy or xz plane. Like the intercepts on OY and OZ, it must be a submultiple of the cell edge, giving an integral index h.

What happens if we try to construct families of planes in which h, k, l have a common factor? Obviously the direction will not be affected, because it depends only on the *ratio* of the intercepts. The distance from the origin to the nearest plane of the family will however be $1/n$th of that for which the indices have been divided by the common factor n. This is illustrated in the right-hand part of Figure 5-2. Taking the origin at O', the plane LM, the nearest to O' of the family 064, makes axial intercepts half those made by the plane $W'V'$, and therefore lies at half the distance from the origin. But this means that it cannot go through any lattice points, because we had already drawn $T'W'V'$ through the points nearest to $O'P'$. This represents a general result: if we require all the planes of a family to pass through lattice points, we can only use families whose indices hkl have no common factor. This restriction rarely applies, however; much more often, we need to be able to take into consideration families whose spacings are submultiples of those passing through lattice points. Then if a family with indices hkl has spacing d, a family parallel to it with spacing d/n is described by indices nh, nk, nl.

Indices used to specify the interplanar spacing of a family of planes as well as their direction are known as *Bragg indices* (or *diffraction indices*,) because of their importance in treatments of three-dimensional diffraction). A family of parallel equidistant planes has Bragg indices hkl if, when the origin is taken on one of them, the next makes axial intercepts $a/h, b/k, c/l$. Conventionally, Bragg indices are written without brackets of any kind and the following points are to be noted about them. Like Miller indices, they presuppose a choice of axes of reference and axial ratios; unlike Miller indices, they also presuppose a choice of absolute values for a, b, c. They may or may not have a common factor; for example, families with Bragg indices 021, 042, 063, all have the same direction as the plane (021). They are not needed for describing crystals, but they are extremely valuable in dealing with diffraction by crystals.

Geometrical relationships between families of planes are much more easily handled if each family can be represented by a point. This is done with the help of the reciprocal lattice, described in §5.7.

When the axial lengths and Bragg indices are known, calculation of d_{hkl} is purely geometrical. Formulae are given in Appendix 3.

5.6 THE SPHERICAL PROJECTION AND THE STEREOGRAPHIC PROJECTION

In order to show relationships between the directions of different planes, it is much more convenient to be able to represent each plane as a point. This is done in two steps, first representing the plane by its normal, and then representing this normal as a point.

The *spherical projection* is constructed as follows. Take any point inside the crystal as origin, and from it draw normals to all the faces. Draw a sphere round the origin of arbitrary radius. The point where each normal cuts the sphere is the spherical projection of the face. This is illustrated in Figure 5-3(a) and (b) for a rectangular block ABC with a face q bevelling one edge and a face p bevelling one corner.

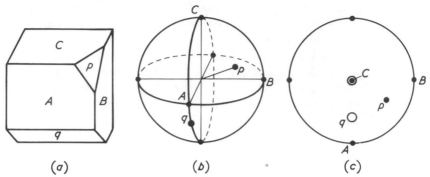

Figure 5-3 (a) Perspective view of hypothetical crystal; faces *A, B, C* are at right angles to each other. (b) Spherical projection of (a)—perspective view. (c) View of (b) from above; points in lower hemisphere are shown by circles.

The next step is to obtain a two-dimensional representation of the sphere, and this, of course, can be done in various ways. An orthogonal projection on a diametral plane is an obvious method, but in practice not a useful one for quantitative work. For discussion and illustration, a sketch of the sphere, viewed from above, is often all that is needed, without specifying the method of projection; this is shown in Figure 5-3(c). For quantitative work, the commonest and most useful method is the *stereographic projection*.

The construction of the stereographic projection, shown in Figure 5-4, is as follows. Let the diameter *NS* be drawn perpendicular to the diametral plane *OAB*, which is chosen as the plane of projection—Figure 5-4(a). The point *p*, whose projection is required, is joined to *S*. The point *P*, where *pS* cuts the equatorial plane *OAB*, is the stereographic projection of *p* —see Figure 5-4(c) and (b).

For a point such as *q* in the southern hemisphere, the projection of *Q* constructed as described lies outside the equatorial circle, as shown in Figure 5-4(d). It is however more usual to join points in the southern hemisphere to the north pole, and distinguish them on the projection with an open circle, labelled *Q'* in Figure 5-4(b) and (d).

The stereogram is shown in Figure 5-4(b). The lines *OA, OP* are projections of arcs of the great circles *NAS, NpS* in (a), and the angle *ω* between them is equal to the angle between the planes *NAS, NpS*. The length *OP* is $r \tan \varphi/2$ where $NOp = \varphi$—see Figure 5-4(c). *P* can thus be plotted graphically from a knowledge of *ω* and *φ*, the angular coordinates in space of the line *Op*.

A stereographic projection has the following useful properties.

(i) A great circle projects as an arc of a circle cutting the equatorial circle at opposite ends of a diameter: for example, in Figure 5-4(e) the great circle *AmA'* becomes the arc *AMA'*. If the plane of the great circle includes *NS*, its projection is a diameter of the circle, for example *NA q S* in Figure 5-4(a) and *AA'* in Figure 5-4(b).

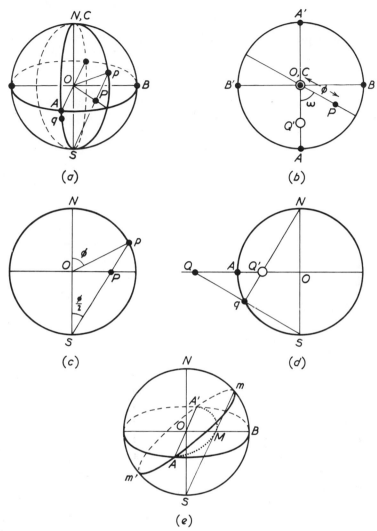

Figure 5-4 Construction of stereographic projection. (a) Perspective view of spherical projection and stereographic projection on plane AOB; projection of p is at P. (b) Stereographic projection. (c) Vertical section of (a) in plane containing p, showing relation of OP to the angular coordinate φ. (d) Vertical section of (a) in plane containing A and q, showing alternative methods of projection of q from pole N or S. (e) Perspective view of spherical projection of a zone $mAm'A'$ and stereographic projection $A'MA$ of the part $A'mA$. The construction on the stereogram is made by finding the projection of m at M, and drawing an arc of a circle through A, A', and M.

(ii) A small circle on the sphere projects as a circle, but the projection of its centre is not the centre of the projected circle. (This fact will not be used in this book, and is not illustrated.)

(iii) Arcs representing great circles in projection intersect at the same angles as do the planes of the great circles in space; the projection is thus angle-true.

(iv) Great circles represent zones. Hence if two faces $(h_1k_1l_1)$, $(h_2k_2l_2)$ lie on the same great circle, any other face on the same great circle has indices which are linear combinations of these, $(nh_1 + mh_2, nk_1 + mk_2, nl_1 + ml_2)$. Notice in Figure 5-3(a) the zone $CA\,q$ with parallel edges, and in Figure 5-4 (a) and (b) the great circle $NA\,q\,S$ and the diameter $CQ'A$.

Stereographic projections are very useful when angles between important directions in the crystal have to be calculated; graphical methods can be used, or, alternatively, guidance as to the most efficient tactics for trigonometrical calculation can be obtained by inspection. It must be remembered, however, that stereographic projections deal *only* with directions, and include no information about interplanar spacings. Miller indices and not Bragg indices are used.

5.7 THE RECIPROCAL LATTICE

We may record facts about interplanar spacings as well as directions of families of planes in the following way. Consider families 100, 010, 001 parallel to the unit-cell faces. From a fixed origin draw lines perpendicular to these planes, of lengths inversely proportional to their spacings. The vectors so obtained are $\mathbf{a}^*$, $\mathbf{b}^*$, $\mathbf{c}^*$ with magnitudes

$$a^* = \frac{\lambda}{d_{100}}, \qquad b^* = \frac{\lambda}{d_{010}}, \qquad c^* = \frac{\lambda}{d_{001}} \tag{5.3}$$

where λ, the constant of proportionality, may be given any convenient value. It is commonly taken either as unity, when the starred quantities have dimensions $(\text{length})^{-1}$; or (when diffraction effects are to be examined) as the wavelength of the radiation used, and then the starred quantities are dimensionless.

The vectors $\mathbf{a}^*$, $\mathbf{b}^*$, $\mathbf{c}^*$ serve as the translation vectors of a new lattice, the *reciprocal lattice*.

It follows from our definition that a point on the x^* axis, at a distance ha^* from the origin, corresponds to a family of planes parallel to 100 and with spacing d_{100}/h, i.e., to the family with Bragg indices $h00$. More generally, writing hkl for the point at $h\mathbf{a}^* + k\mathbf{b}^* + l\mathbf{c}^*$, its distance s_{hkl} from the origin can be shown to equal λ/d_{hkl} (see Appendix 2 to this chapter); hence any point hkl represents a family of planes with hkl for its Bragg indices.

Figures 5-5(a)(i) and (b)(i) represent sections of direct lattices whose third axis is perpendicular to the paper. Figures 5-5(a)(ii) and (b)(ii)

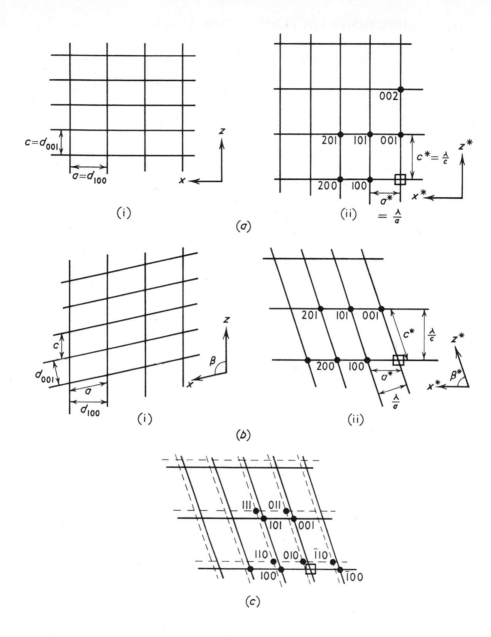

Figure 5-5 Sections of direct and reciprocal lattices. (a) Axes orthogonal: (i), direct lattice; (ii), reciprocal lattice. (b) Axes monoclinic, $\alpha = \gamma = 90°$: (i), direct lattice; (ii), reciprocal lattice. (c) Reciprocal lattice with all angles oblique: section at height of first layer (broken lines) projected orthogonally on to plane of zero layer (solid lines).

represent the corresponding reciprocal lattices. The example in Figures 5-5(b)(i) and (b)(ii) calls attention to the important fact that, for non-orthogonal axes, d_{100} and d_{001} are not identical with **a** and **c** respectively, in either magnitude or direction.

If the third axis in direct space is not perpendicular to the plane of the paper, d_{100} and d_{001} do not lie in the same plane as **a** and **c**. The plane x^*z^* in reciprocal space is thus not the same as the plane xz in direct space. The direction normal to x^*z^* is y; the direction normal to xz is y^*. Figure 5-5(c) shows the orthogonal projection on x^*z^* of two successive layers $h0l$, $h1l$, in such a case. Each layer is displaced relative to the one below it by a constant vector depending on the interaxial angles α and γ.

It is important to notice the reciprocal character of the two lattices. If we had originally only been given the reciprocal lattice vectors, we could have used them to construct the direct lattice. Interplanar spacings in direct space correspond to vectors from the origin to points in reciprocal space, while interplanar spacings in reciprocal space, such as λ/a in Figure 5-5(b)(ii), correspond to vectors from the origin to lattice points in direct space.

The great importance of the reciprocal lattice lies in the fact that reciprocal space is identical with diffraction space, and each reciprocal-lattice point can be associated with the direction of a diffraction maximum of a perfect crystal. We shall not be concerned with that aspect in this book. It is also useful for displaying the relationships between lattice parameters and interplanar spacings, and either allowing graphical methods of calculation or providing sketch diagrams from which the required geometrical formulae can be readily derived.

5.8 SUMMARY

Miller indices (hkl) define the directions of planes (without locating their positions); Bragg indices hkl define the directions and interplanar spacing of families of parallel equidistant planes and name the points of the reciprocal lattice. Zone indices $[uvw]$ define directions of lines. Forms—symmetry-related sets of planes or lines—are written as $\{hkl\}$ and $\langle uvw \rangle$ respectively.

Stereographic projections provide convenient ways of displaying, and measuring graphically, the relations between directions of planes and lines. The reciprocal lattice serves the same purpose for the directions and periodic separations of families of planes and lines. In both, the point representing the plane represents more directly the direction of its normal.

APPENDIX 1 MILLER–BRAVAIS INDICES

Miller-Bravais indices should almost always be used with hexagonal axes of reference in preference to Miller indices. In principle they are like Miller

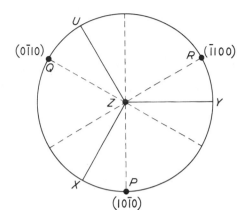

Figure 5-6 Stereogram illustrating Miller-Bravais indices. Hexagonal axes of reference are in directions X, Y, Z, and the fourth (redundant) axis in direction U.

indices, but include a fourth index i referring to a redundant axis in the xy plane. The advantage of this is as follows. If there is a trigonal or hexagonal axis Z (Figure 5-6) acting through an origin O (at the centre of the sphere of projection), it operates on the line OX to produce identical lines OY and OU. By choosing OX and OY as axes of reference and ignoring OU, we are obscuring the symmetry relations. Using Miller indices, planes P and Q would be (100) and (0$\bar{1}$0) respectively, while the symmetry-related plane R would be ($\bar{1}$10). If we let the index i refer to the axis OU, then the plane ($hkil$) makes axial intercepts in the ratio $a/h : a/k : a/i : c/l$. The three planes P, Q, and R become (10$\bar{1}$0), (0$\bar{1}$10), ($\bar{1}$100) respectively. The planes whose traces are shown by the broken lines are (11$\bar{2}$0), (1$\bar{2}$10), and ($\bar{2}$110); these are the planes normal to the axes of reference. The rules are that, for any plane ($hkil$), (i) $h + k + i = 0$, (ii) operation of the triad axis corresponds to permutation of the first three indices, giving ($hkil$), ($kihl$), and ($ihkl$).

Because, for crystals of this kind, rhombohedral axes of reference are often chosen, for which ordinary Miller indices are most convenient, confusion could result if 3-figure indices were used without distinction when hexagonal axes are chosen. In fact, when writers say that they are using Miller indices, they generally imply that they have chosen rhombohedral axes. Hence it is desirable, when using hexagonal axes, always to indicate the existence of the third equivalent direction in the horizontal plane; if insertion of the value of i would distract the reader's attention, its place can be indicated by a dot.

APPENDIX 2 PROPERTIES OF THE RECIPROCAL LATTICE

In vector notation, the reciprocal lattice is defined by the relations

$$\mathbf{a} \cdot \mathbf{a}^* = \mathbf{b} \cdot \mathbf{b}^* = \mathbf{c} \cdot \mathbf{c}^* = \lambda,$$

$$\mathbf{a} \cdot \mathbf{b}^* = \mathbf{b} \cdot \mathbf{a}^* = \mathbf{a} \cdot \mathbf{c}^* = \mathbf{c} \cdot \mathbf{a}^* = \mathbf{b} \cdot \mathbf{c}^* = \mathbf{c} \cdot \mathbf{b}^* = 0 \quad (5.4)$$

Let **s** be the vector from the origin to the point hkl; then

$$\mathbf{s} = h\mathbf{a}^* + k\mathbf{b}^* + l\mathbf{c}^* \tag{5.5}$$

We require to find the relation between the reciprocal-space vector **s** and the interplanar spacing d_{hkl}.

Let **u** be unit vector in the direction of **s**; then

$$\mathbf{u} = \frac{1}{s}(h\mathbf{a}^* + k\mathbf{b}^* + l\mathbf{c}^*) \tag{5.6}$$

Now consider a line in direct space with the same direction as **u**. The first plane of the family hkl makes axial intercepts $\mathbf{a}/h$, $\mathbf{b}/k$, $\mathbf{c}/l$. The projection of $\mathbf{a}/h$ on the direction of **u** is $(1/h)\mathbf{a} \cdot \mathbf{u}$, which is $(1/h)(1/s)(h\mathbf{a} \cdot \mathbf{a}^*)$ or λ/s, from the above definition. The projections of $\mathbf{b}/k$ and $\mathbf{c}/l$ on **u** have the same value. Hence the direction of **u** must be the normal to the plane hkl, and λ/s must be the distance of the plane from the origin, i.e.,

$$s_{hkl} = \lambda/d_{hkl} \tag{5.7}$$

APPENDIX 3 OTHER GEOMETRICAL RELATIONS

(i) Volume of Unit Cell

In the most general case,

$$V = abc[\sin^2 \alpha + \sin^2 \beta + \sin^2 \gamma - 2(1 - \cos \alpha \cos \beta \cos \gamma)]^{\frac{1}{2}} \tag{5.8}$$

When $\alpha = \gamma = 90°$

$$V/abc = \sin \beta \tag{5.9}$$

When $\alpha = \beta = \gamma$

$$V/abc = (2 \cos^3 \alpha - 3 \cos^2 \alpha + 1)^{\frac{1}{2}} \tag{5.10}$$

(ii) Interplanar Spacing

In the most general case,

$$\frac{1}{d^2} = \left(\frac{abc}{V}\right)^2 \left[\sum \frac{h^2}{a^2} \sin^2 \alpha + 2 \sum \frac{hk}{ab} (\cos \alpha \cos \beta - \cos \gamma)\right] \tag{5.11}$$

(where there are three terms in each summation, obtained by permutation from the first).

When $\alpha = \gamma = 90°$, this reduces to

$$\frac{1}{d^2} = \frac{1}{\sin^2 \beta} \left[\frac{h^2}{a^2} + \frac{k^2 \sin^2 \beta}{b^2} + \frac{l^2}{c^2} + 2\frac{hl}{ac}(1 - \cos \beta) \right] \qquad (5.12)$$

When all the axes are orthogonal, it becomes

$$\frac{1}{d^2} = \frac{h^2}{a^2} + \frac{k^2}{b^2} + \frac{l^2}{c^2} \qquad (5.13)$$

(a convenient form to remember)

(iii) Reciprocal Lattice Vectors

In the most general case,

$$a^* = \frac{abc}{V}\frac{\lambda}{a} \sin \alpha \qquad (5.14)$$

$$\cos \alpha^* = \frac{\cos \beta \cos \gamma - \cos \alpha}{\sin \beta \sin \gamma} \qquad (5.15)$$

When $\alpha = \gamma = 90°$, we have

$$a^* = \frac{\lambda}{a \sin \beta}, \qquad b^* = \frac{\lambda}{b}, \qquad c^* = \frac{\lambda}{c \sin \beta}, \qquad \beta^* = \pi - \beta \qquad (5.16)$$

(iv) Formulae for Spherical Triangles

For a general spherical triangle ABC with sides a, b, c opposite to angles A, B, C,

$$\cos c = \cos a \cos b + \sin a \sin b \cos C \qquad (5.17)$$

$$\frac{\sin a}{\sin A} = \frac{\sin b}{\sin B} = \frac{\sin c}{\sin C} \qquad (5.18)$$

These simplify if any of the sides or angles is $90°$.

PROBLEMS AND EXERCISES

1. Figure 2-2(c) gives a perspective drawing of an octahedron. If the axes are taken with OZ vertical on the page, OY pointing to the right, and OX coming out from the page pointing slightly to the left, what are the Miller indices of all eight faces?

   With the same set of axes, what are the Miller indices of the faces of the tetrahedron in Figure 2-2(b)?

2. A crystal of quartz has the shape of a hexagonal prism capped with a hexagonal pyramid, the edges between prism A and pyramid faces B lying in a plane perpendicular to the axis of the prism, which is taken as the z axis of reference. The x and y axes of reference are chosen so that they bisect the angles between the normals to prism faces.
   Make a sketch of the crystal, marking the axes of reference. Show that the set of prism faces constitutes a zone, and that there are other zones each including one prism face, its adjacent pyramid face, and the hypothetical face which would be perpendicular to the prism axis.

   What are the indices of the prism faces (a) using Miller indices appropriate to the above axes of reference, (b) using Miller-Bravais indices? Mark them on your sketch.

3. A crystal of brookite (one form of TiO_2) has three faces A, B, C, mutually at right angles, chosen as (100), (010), (001); two faces, l, m, are in the [100] zone, two faces, x, y, are in the [010] zone, and a face t is in the [001] zone. The angles between the face normals are as follows: Al 22.6°, Am 40.1°, Ax 60.7°, Ay 74.3°, Bt 27.9°. Choosing either l or m as (110) and either x or y as (101), determine the axial ratios $a:b:c$ and the indices of the other faces. Repeat for the other three possibilities.

   The conventional choice of mineralogists takes m as (110) and x as (102). Repeat the work for this choice.

4. Draw a sketch stereogram of the quartz crystal of Question 2. Take one pyramid face as $(10\bar{1}1)$ and assign indices to all the rest.

   The angle between the normal to the pyramid face and the z axis is 51.6°. What is the axial ratio c/a? [Note: find first the ratio of the pyramid plane's intercept on the z axis to that on the normal to the prism face, and then notice the ratio of the latter to that on the x axis].

5. A prism with its axis horizontal has faces whose normals are at 0°, 10°, 20° $\cdots$ 90° to the vertical. Construct its stereogram. (A tracing of this can be used as a chart for measuring angles from the vertical on other stereograms constructed with the same radius.)

6. Show that, for a cubic crystal, the angle (001) $\wedge$ (111) is $\cos^{-1}$ (1/3). Using this result, construct a stereogram of the cuboctahedron of Figure 2-2(e), indexing all the faces.

7. Construct an accurate stereogram of the quartz crystal of Question 4.

8. Show that any face in the same zone as faces (hkl) and (pqr) must have Miller indices given by

$$(mh + np)/f, \qquad (mk + nq)/f, \qquad (ml + nr)/f$$

where m, n are integers and f is the highest common factor of the three numerators.

9. Using the result of Question 8, insert the following faces on the cubic stereogram of Question 6: (110), $(1\bar{1}0)$, (101), (011), (112), (211), (210). (Use relation (i) of §5.6, for constructing zones on the stereogram.)

10. A quartz crystal is like that described in Question 2, except for an additional small face s at the corner between two prism faces A and two pyramid faces B. If the faces are lettered so that A_1A_2, B_1B_2 are adjacent, with the edges A_1B_1 and A_2B_2 horizontal, then s lies in the zones A_1B_2 and A_2B_1. Insert it on the stereogram of Question 7. What are its indices?

    Calculate the angle between the normal to s and the z axis, using the known value of the axial ratio c/a.

    Check this result by using a tracing of the standard stereogram of Question 5 to measure the angle (approximately).

11. For the crystal of Question 10, calculate the angle A_1s, i.e., the angle between the normals to these two faces. (Find a spherical triangle on the stereogram, preferably a right-angled one with A_1s as one side and its other sides already known, and use the formulae of Appendix 3(iv).)

12. On the stereogram of the cubic crystal of Question 9, show (by using spherical triangles) that $(101) \wedge (011)$ is $60°$.

    Insert (201), and find the value of $(201) \wedge (211)$.

13. (a) Construct several unit cells of a two-dimensional lattice with $a = 2$ cm, $b = 4$ cm, interaxial angle $110°$. Find by measurement the interplanar spacings of families of planes 10, 0, 1.

    (b) Construct the corresponding reciprocal lattice, using a scaling constant $k = 10$.

    (c) On different parts of your direct lattice, insert families of planes 12, 2, $\bar{2}$. Find by measurement the direction and magnitude of the interplanar spacings. Check the magnitude by calculation.

    (d) Construct the reciprocal points corresponding to 12, 2, $\bar{2}$, and show that they coincide with points of the lattice constructed in (b).

Chapter 6 SYMMETRY AND ITS APPLICATION TO FINITE OBJECTS

6.1 THE SYMMETRY OF CRYSTALS

Our ideas of symmetry start from the appreciation of geometrical shapes. Mathematically, they can be systematised as a branch of group theory. Since, however, we want to be able to apply the concepts to actual crystal structures, we shall begin with simple ideas and illustrate them from cases to be met in practice, and shall only quote the results of mathematical theory when we want to generalise from our examples.

An object possesses certain symmetry if, after the application of a particular operation, it looks exactly as it did before, and continues to do so however often the operation is repeated. For example, a hexagonal prism looks exactly the same after rotation through 60°. The operation must be repeated six times to bring back the original position—we have to suppose this position can be recognised, though in fact any kind of marking would destroy the exact symmetry—and it is therefore called 6-fold rotation symmetry.

In this chapter we consider the various kinds of symmetry operators which can occur in solid objects. Some of them are excluded when the object is a periodic structure with a lattice (*or* a crystal obeying the law of rational indices, which we saw depends on its possession of a lattice); we shall not consider them in detail.

Symmetry operations may occur in combination. A set of symmetry operations which, however often they are used and in whatever order, lead to a finite number of settings of the object, is known as a *symmetry group*. If we demand that one point remain unmoved throughout all the operations, we have a point symmetry group. For the symmetry operators possible in a crystal, there are only 32 possible combinations of this kind; i.e., there are 32 *point groups*. These are discussed in §6.6 to 6.8 and 6.10.

Since a crystal is characterised by the directions of its face normals, all of which can be drawn through one point, its symmetry must belong to one of the 32 point groups. Crystals are divided into 32 *crystal classes*, each class corresponding to a particular point group.

It is convenient and usual to group the 32 classes into six (or sometimes seven) *crystal systems*. This is explained in §6.6, 6.8, and 6.10 and is summarised in §6.12.

Next it is necessary for us to examine what symmetry may be combined with lattices. This is done in Chapter 7. We find that there are 14 possible distinct combinations, the 14 *Bravais lattices*.

Three-dimensional structures consist, as we have seen, of a lattice operating on a lattice complex. New symmetry operations can occur in such structures. Symmetry groups of this kind are called *space groups*, and there are 230 of them. They are considered in Chapter 8.

6.2 NOTATION

The crystal classes have been given descriptive names, but these have not proved generally helpful and we shall not trouble with them. On the other hand, a concise symbolic notation defining the ingredients of the point group is a necessary tool.

The earliest notation which is still current is that of A. M. Schoenflies, put forward about the end of the nineteenth century. Devised originally for point groups, it could only be extended to space groups by adding an arbitrary serial number, which did not give any information about the translation symmetry. In the 1920's, crystallographers were looking for something more helpful, and finally in 1930 an international committee recommended a system based on the work of C. Hermann and C. Mauguin. This, the *Hermann-Mauguin notation*, is now widely used by most crystallographers. We shall adopt it in this book, explaining it step by step as we go along. It is fully recorded and used in the *International Tables for X-ray Crystallography*, the authoritative reference book which embodies internationally accepted conventions.

The Hermann-Mauguin notation is specially designed for space groups, and very simple rules allow the point group to be derived from it. One fact needs to be noted from the outset, that *the same point group (or space group) may have different symbols according to the choice of axes of reference*, and this is sometimes an obstacle to those unfamiliar with the principles of the notation. We shall deal later (Chapter 10) with the rules for transformation from one form to another. In cases where there could be confusion, it is sometimes helpful to add either the Schoenflies symbol or the serial number of the space group as listed in the *International Tables*—but as an addition to, and not a replacement of, the Hermann-Mauguin symbol.

6.3 SYMMETRY OPERATIONS OF THE POINT GROUP

Symmetry operators may be considered under four headings: the centre, the mirror plane, rotation axes, inversion axes. (Later, we shall see that *formally* we need only consider the two last-named.) Instead of starting with a symmetrical object and examining its symmetry, it is easiest to start with an asymmetric object—for example, a comma, with its tail sticking up from the paper or down into it—and see what each symmetry operator generates from it. The diagrams have to be two-dimensional, but in space the tail of the comma may point up or down from the paper and this is indicated by the letters *u* or *d*.

Centre of Symmetry (Figure 6-1)

If there is a centre of symmetry at $(0, 0, 0)$, it operates on any point (x, y, z) to give an identical point at $(-x, -y, -z)$—commonly written in crystallographic work as $(\bar{x}, \bar{y}, \bar{z})$. The comma-shaped object is turned into another, equidistant from the centre, but inverted; note that the tail of one sticks up, that of the other down.

Mirror Plane (Figure 6-2)

If there is a mirror plane at $z = 0$, it operates on any point (x, y, z) to give an identical point at $(x, y, \bar{z})$. Both commas are equidistant from the plane; now both tails stick up.

It might be difficult to explain a mirror plane of symmetry if we were not so familiar with it in everyday life. Familiarity is not only due to our experience with mirrors, but also arises because our bodies possess an approximate plane of symmetry. The mirror symmetry operation is in fact that of relating a right hand to a left hand.

This calls attention to the property of *inversion* possessed by mirror symmetry. An asymmetric object and its mirror image cannot be made to coincide, however they are displaced. Each is *inverted* with respect to the other.

The operation of a centre of symmetry also produces inversion; i.e., it turns a right hand into a left hand. A pair of hands, placed palm to palm with the fingers of each on the wrist of the other, are related by a centre of symmetry between the palms.

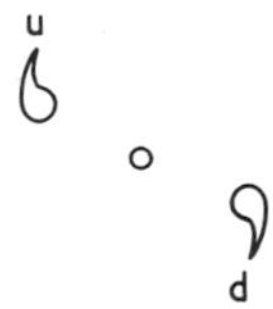

u

Figure 6-1 Operation of centre of symmetry (marked by small open circle). Letters *u* and *d* indicate whether, in three dimensions, the tail of the comma is up or down.

Figure 6-2 Operation of mirror plane of symmetry (trace of plane marked by heavy line).

Rotation Axes

A rotation axis of n-fold symmetry operates on any point to produce another like it equidistant from the axis, with its perpendicular to the axis in the same plane as that of the original point, at an angle of $2\pi/n$ to it.

In general, n may have any integral value except 1. For example, a 5-fold rotation axis is possessed by many common flowers, such as the wild rose and the buttercup. If n is infinite, the symmetry is cylindrical. We shall only be concerned, however, with n-values of 2, 3, 4, or 6, because they are the only values compatible with the existence of a lattice (cf. §7.2).

These rotation axes are illustrated in Figure 6-3(a) to (d), with the axis perpendicular to the plane of the paper. Names and symbols are as follows:

2-fold axis—diad   ●
3-fold axis—triad   ▲
4-fold axis—tetrad   ■
6-fold axis—hexad   ⬢

It is often necessary to show a diad axis *in* the plane of the paper; this is illustrated in Figure 6-3(e). Note the difference between this and Figure 6-2;

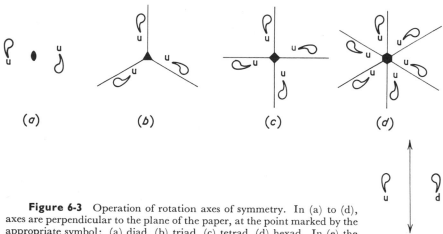

(a)   (b)   (c)   (d)

(e)

Figure 6-3 Operation of rotation axes of symmetry. In (a) to (d), axes are perpendicular to the plane of the paper, at the point marked by the appropriate symbol: (a) diad, (b) triad, (c) tetrad, (d) hexad. In (e) the diad axis is in the plane of the paper, marked by the line with arrows. Note the up or down direction of the tails of the comma.

they would be the same in projection, but three-dimensionally one tail sticks upward, the other downward. The same kind of comparison can be made between Figure 6-3(a) and Figure 6-1.

None of the commas produced by rotation are inverted; each can be made identical with the rest by giving it the correct orientation and position.

It is sometimes convenient to use the idea of a 1-fold symmetry axis. The definition then requires that an object should be in an identical state after a rotation of 360° But this is always true; hence *any* direction is a direction of 1-fold symmetry. However, in constructing a notation for the concise description of symmetry, it may be helpful to be able to state that a particular direction is *not* a symmetry direction; this can be done by referring to it as a 1-fold axis. It is rather like using zero as a digit in our ordinary arithmetical notation.

Inversion Axes

An inversion axis is a little more difficult to envisage than a rotation axis. Two steps are involved in each operation. First there is a rotation, but the object is not left in that state—it is immediately inverted through a centre.

Figure 6-4(a) illustrates the operation of an inversion tetrad. It rotates the object at A through 90° to B and inverts it through the centre of inversion (which acts like a centre of symmetry) to C. Repetition of the same operation, as shown in Figure 6-4(b), generates in turn D, E, and a repetition of A. If A and D are above the centre of inversion, C and E are an equal distance below it.

To each kind of rotation axis there corresponds an inversion axis. However, if using any of the other crystallographically possible inversion axes we carry out the same kind of procedure as we did for the 4-fold axis, we find that the set of objects generated (Figure 6-5) could equally well have been

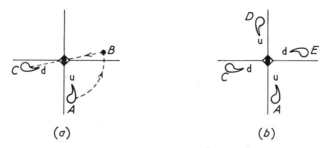

(a) (b)

Figure 6-4 Operation of inversion tetrad axis, (perpendicular to the paper, at the point marked by the symbol). (a) Single operation, repeating the original object A at C. Arrows show how the operation may be visualised. (b) Result of the completed group of operations: object A is repeated at C, D, and E.

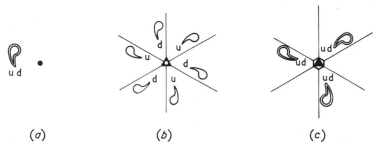

$$(a) \qquad\qquad (b) \qquad\qquad (c)$$

Figure 6-5 Operation of other inversion axes (perpendicular to the paper, at the points marked). (a) Inversion diad. The double comma represents the projection of two objects related by reflection in the plane of the paper. (b) Inversion triad. (c) Inversion hexad. Double commas as in (a).

derived from a combination of some of the other (simpler) symmetry elements as follows:

(a) 2-fold inversion axis: equivalent to a mirror plane, whose normal coincides with the axis;

(b) 3-fold inversion axis: equivalent to a 3-fold rotation axis plus a centre of symmetry;

(c) 6-fold inversion axis: equivalent to a 3-fold rotation axis plus a mirror plane whose normal coincides with the axis.

It is only the inversion tetrad which cannot be matched by some other combination. Inspection of Figure 6-4(b) shows that it *included* a rotation diad (relating A to D, C to E) but also something more (the relation of A to C or to E) which cannot be matched by anything else. The simplest example is a regular tetrahedron, which possesses an inversion tetrad along any line joining midpoints of opposite edges.

By analogy, a 1-fold inversion axis is simply equivalent to a centre of symmetry, and it is sometimes convenient to use this description. Like the 1-fold rotation axis, it has *no* direction associated with it.

For interest, it may be noted that there are also inversion axes of 5, 7, 8, 9 . . .-fold symmetry. When n is odd, such an axis always gives rise to $2n$ equivalent points, which could alternatively be produced by combining an n-fold rotation axis with a centre. When n is an odd multiple of 2, it gives n points, alternatively produced by combining an $n/2$-fold rotation axis with a mirror plane perpendicular to it. When n is a multiple of 4, there is no other combination of symmetry elements which can replace the inversion axis.

The advantage of using inversion axes, even when they can be matched by combinations which are easier to envisage, is that they make possible much simpler classification and description of symmetry groups. Since a mirror plane is equivalent to a 2-fold inversion axis, and a centre of symmetry to a 1-fold inversion axis, *all* symmetry elements can be classified as either rotation axes or inversion axes; this allows a simpler formal scheme. In discussing individual cases, however, it is generally simpler to work (when possible) with rotation axes, mirror planes, and centres of symmetry.*

6.4 NOTATION FOR SYMMETRY ELEMENTS

Rotation axes are symbolised by their value of n, i.e., by the figure 2, 3, 4, or 6. The corresponding inversion axes are symbolised by $\bar{2}$, $\bar{3}$, $\bar{4}$, or $\bar{6}$. The *normal* to a mirror plane is symbolised by m; when this has the same direction as an n-fold rotation axis, the combination is written n/m. Though $\bar{2}$ means the same as m, the latter symbol is conventionally preferred, unless there are special reasons. The symbol 1 is used, when needed, for a direction of no symmetry, and $\bar{1}$ when *only* a centre of symmetry is present. From the discussion in §6.3, we also note that although $\bar{3}$ cannot conveniently be written in any other way, $\bar{6}$ may be (and often is) written as $3/m$.

6.5 COMBINATION OF SYMMETRY OPERATORS

It is easy to derive some of the simpler point groups, which will illustrate the principles for the rest.

(i) Combination of a Rotation Diad Axis and an Inversion Diad Axis in the Same Direction

This is shown in Figure 6-6, with PQ as the symmetry direction. The inversion diad along PQ is equivalent to a mirror plane RS perpendicular to PQ

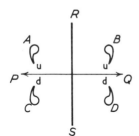

Figure 6-6 Combination of rotation diad and mirror normal lying in the same direction PQ.

---

* Early workers used rotation-reflection axes in preference to inversion axes. The operation of these consisted of rotation followed by reflection in the plane perpendicular to the axis. The 4-fold rotation-reflection axis is the same as the 4-fold inversion axis, but the 3-fold rotation-reflection axis is the same as the 6-fold inversion axis, and the 6-fold rotation-reflection axis is the same as the 3-fold inversion axis. Rotation-reflection axes are perhaps easier to think about individually than are inversion axes, but they are more confusing in their application to symmetry groups. Unfortunately, they are embodied in the Schoenflies notation for the point groups; and though their use has been abandoned by crystallographers since the adoption of the Hermann-Mauguin notation, they are sometimes still used in group-theory treatments and by theoretical chemists. This can be confusing if the reader is not forewarned.

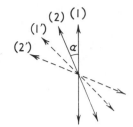

Figure 6-7 Combination of two rotation diads (1) and (2) lying in the plane of the paper at an angle α to each other.

through the centre of inversion. Its operation on object A gives B; then operation of the rotation diad on A and B gives C and D respectively. Since D is also the result of the operation of the mirror plane on C, further use of the same symmetry elements leads to nothing new. Inspection shows, however, that A and D are related by a centre of symmetry, as are B and C. We draw the general conclusion that the combination of a rotation diad and an inversion diad in the same direction implies additionally the existence of a centre of symmetry at the centre of inversion. Together, these symmetry elements form a point group.

(ii) Combination of Two Rotation Diads at an Angle to One Another

We can show that only certain angles are allowed. In Figure 6-7, let the solid lines be diad axes, at an angle α to each other. Operation of (2) on (1) gives $(1')$; operation of this on (2) gives $(2')$, and so forth. The angle between two symmetry-related axes is 2α. Only if 2α is an exact submultiple of 2π will a closed group be formed; otherwise, repetition goes on and on till the plane of the paper is a continuum of diad axes. But if diads (1) and (2) are repeated to give n of each kind, an n-fold symmetry axis has been produced perpendicular to the paper. Since, for crystallographic point groups, $n = 2, 3, 4$, or 6, it follows that $\alpha = \pi/n = 90°, 60°, 45°$, or $30°$.

(iii) Combination of Two Rotation Diads When No Higher Symmetry Is Present

It follows from the argument in (ii) that the diads must here intersect at 90°. This is shown in Figure 6-8. Operation with (1) on object A gives B;

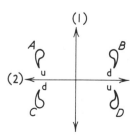

Figure 6-8 Combination of two rotation diads at right angles.

operation with (2) on A gives C. Operation with (2) on B gives D, *and so does operation with* (1) *on* C. Repeated operation with either axis gives nothing new.

However, inspection shows that A and D are related by a diad axis perpendicular to the paper, as are B and C. This third axis at right angles to the other two is a necessary consequence of the existence of the first two. In other words, *if there are two rotation diads at right angles to each other, there will be a third at right angles to both.* Together they constitute a point group.

(iv) Combination of Two Mirror Planes (i.e., of Two Inversion Diads) When No Higher Symmetry Is Present

An argument like that in (ii) shows that the plane normals must intersect at 90°. A construction like that in (iii) shows that *when two mirror plane normals, or inversion diads, are at right angles, there is a rotation diad at right angles to both.*

This result and that in (iii) can be stated together as follows: *When there are two directions of diad symmetry at right angles, there is a third direction of at least diad symmetry at right angles to both.*

(v) Combination of Rotation Diads and Mirror Planes

Suppose we add a mirror plane to the point group discussed in (iii): its normal must coincide with one of the three rotation diads. Then, allowing this to operate on the group of objects in Figure 6-8, we find that there are mirror plane normals along the other two rotation diads, and also a centre of symmetry.

Whatever combination of rotation and inversion diads we start with, we can never get any other point groups than those described.

6.6 POINT GROUPS OF THE TRICLINIC, MONOCLINIC, AND ORTHORHOMBIC SYSTEMS

We can now list all possible point groups which can be formed using symmetry elements for which n is not greater than 2. This is done in Table 6-1. They are classified into *systems* according to the number of their symmetry directions—*triclinic* if there is no symmetry direction, *monoclinic* if there is 1, *orthorhombic* if there are 3 (which must, as we have seen, be at right angles). Their stereograms are shown in Figure 6-9.

6.7 POINT-GROUP NOTATION

The point-group symbol names the symmetry elements necessary to construct the group. Several features should be noticed.

Table 6-1   Point Groups with 1-Fold or 2-Fold Operators Only

SYSTEM	NUMBER OF SYM- METRY DIRECTIONS	CHARACTER OF SYMMETRY OPERATORS	POINT GROUP
Triclinic	None	No symmetry	1
		Centre of symmetry	$\overline{1}$
Monoclinic	One	Rotation diad axis	2
		Mirror plane normal	m
		Rotation diad axis *and* mirror plane normal coinciding in direction; also a centre of symmetry	$2/m$
Orthorhombic	Three, at right angles	Three rotation diads	222
		Two mirror plane normals, and one rotation diad at right angles to them	$\{mm2\}$
		A rotation diad and a mirror plane normal coinciding in each of the three directions; also a centre of symmetry	mmm

There is no need to record the presence of a centre of symmetry, except when it is the only symmetry element (i.e., in point group $\overline{1}$). In all other point groups, its presence can be deduced from the other symmetry elements.

When there is more than one symmetry direction (i.e., for all point groups except in the triclinic and monoclinic systems) the order in which the

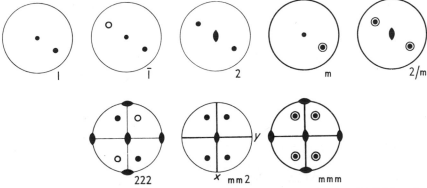

Figure 6-9   Stereograms of all point groups possessing no symmetry axis of higher than 2-fold symmetry. Point-group symbols are given. Mirror planes with normals in the plane of projection are shown by heavy lines along diameters; a mirror plane coinciding with the plane of projection is shown by a heavy line along the circumference. Projections of directions of diad axes are marked with the symbol for a diad perpendicular to the paper.

symbols are written carries information. In the orthorhombic system it is presupposed that the axes of reference coincide with the symmetry directions; then the three symbols refer to the directions of the x, y, and z axes respectively. The point group consisting of two mirror planes and a diad axis is written $2mm$, $m2m$, or $mm2$, according to whether we choose to call the rotation diad OX, OY, or OZ. If we do not wish to decide on the axes of reference, we may put the symbol (in any of its three forms) in curly brackets (braces), e.g., $\{mm2\}$; this means "the point group which, by a suitable choice of axes of reference, could have the symbol $2mm$." (Notation for point groups with axes of 3, 4, and 6-fold symmetry will be discussed in §6.8 and 6.10.)

For all point groups there is a *standard form* under which they are listed in the *International Tables for X-ray Crystallography*. This, however, is only for convenience of reference and comparison. Generally the choice of axes of reference for a crystal is dictated by reasons other than the character of the symmetry elements, and one has to put up with the consequence that its point-group symbol is not in the standard form. For example, because the standard form of one point group is $mm2$, it would be wrong either to insist that for every crystal in that class the z axis must be chosen along the rotation diad, or to write its point group as $mm2$ when the rotation diad is *not* the z axis.

Point-group symbols must name all the symmetry elements necessary to construct the group, but they may name more than are necessary. For example, it follows from §6.5 that 22 would be the same point group as 222, and $m2$ or $2m$ or mm would be the same as $\{mm2\}$. The two-symbol abbreviated form might be used for any of these. On the other hand, mmm is already an abbreviated form, and cannot be further abbreviated. Since each of the symmetry directions is not only a mirror plane normal but also a rotation diad, the point group could be written in full as $2/m\ 2/m\ 2/m$, and this could be abbreviated to $2/m\ 2/m$ or even to $2/mm$, because the operators named are enough to generate the rest of the group. However, for this group it is generally easiest and most elegant to name the three mirror planes and let them generate the rest.

The usual names of point groups are thus often compromises which tell us more than the minimum symmetry needed to generate the group, but less than the full symmetry present. To write the full symmetry would often be unnecessarily tedious and clumsy; but to write the minimum symmetry cannot be done in a unique way, and very heavily abbreviated forms would be much harder to remember and use. In the *International Tables for X-ray Crystallography* a compromise based on working experience has been adopted, and the forms given there are now accepted as standard. For the orthorhombic system, the three-figure symbol is accepted.

The only safe way of understanding symbols written in non-standard form, without memorising elaborate rules, is to be able from the symmetry operators named to draw a set of equivalent points, like those in Figures 6-1 to 6-5, and thence deduce any other symmetry present.

6.8 CUBIC POINT GROUPS

We have now considered all point groups whose symmetry directions are not more than 2-fold. Let us next consider the question: can we have more than one triad axis?

It is possible to prove, either by group theory or by spherical trigonometry, that if there is more than one triad there must be four triads, making angles of 109°28′ with each other, i.e., directed from the centre towards the four corners of a regular tetrahedron. Inspection shows that this formation has additional symmetry, namely, three diad axes bisecting the angles between each pair of triads (Figure 6-10). In the tetrahedron, these diads are the lines joining midpoints of opposite edges, and it is easy to show that they are at right angles to one another. They therefore make convenient axes of reference. Because they are equally inclined to a triad, unit lengths along them are equal. The unit cell of such a crystal must be a cube.

All crystals possessing four triad axes are classified as belonging to the *cubic system*. The point group we have just described has the least symmetry that it *can* have while remaining cubic. It does not even have the full symmetry of the regular tetrahedron, which has inverse tetrads where this lowest point group has rotation diads. This means that we should be able to find ways of bevelling off the corners of a regular tetrahedron which maintain the triad axes but destroy the inverse tetrads. Other symmetry elements can be added, giving higher point groups.

There are a total of five cubic point groups. The possible symmetry directions are the cube edge, the cube body diagonal, and the cube face diagonal. The symbol for the point group names the kind of symmetry operator present in each of these directions, in the order specified. The groups are as listed in Table 6-2. As with the orthorhombic symbols,

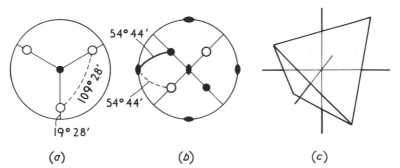

(a)   *(b)*   *(c)*

Figure 6-10 Minimum symmetry of the cubic system. (a) Stereographic projection down one triad axis, showing three other triads (open circles). (b) Stereographic projection down one diad axis, showing two other diads (lens-shaped symbols) and four triads (black circles and open circles). (c) Perspective view of regular tetrahedron, showing three diad axes at right angles as bisectors of pairs of opposite edges.

Table 6-2 Cubic Point Groups

CONVENTIONAL SYMBOL	ALTERNATIVE FORM OF SYMBOL	REMARKS
$m3m$	$4/m\ 3\ 2/m$	Maximum possible symmetry. Rotation tetrads and mirror plane normals along cube edges; rotation diads and mirror plane normals along cube face diagonals; centre of symmetry
$m3$	$2/m\ 3\ 1$	Symmetry planes with normals along cube edges, but no tetrads, and nothing along face diagonals; centre of symmetry,
432	43	Axes only; no mirror planes; no centre.
$\bar{4}3m$		The symmetry of the regular tetrahedron. No centre
23	231	Minimum possible symmetry. No centre.

abbreviated forms are permissible, and are sometimes the conventional ones. Note the presence of tetrad axes in $m3m$ and their absence in $m3$. Since cubic crystals are characterised by their triad axes, all the point groups must have 3 in the second place.

6.9 OTHER POINT GROUPS WITH AXES OF n-FOLD SYMMETRY ($n > 2$)

It can be shown that there are no other point groups with more than one n-fold axis when n is 3, 4, or 6. If n is 5, however, there are other possibilities. A point group can be constructed with 20 triad axes and 10 pentad axes. This cannot be applied to any crystal as a whole, because 5-fold axes cannot generate a lattice; but there is nothing to prevent its application to a finite object or configuration of atoms. Some very large molecules and some configurations of atoms in metals do have this kind of symmetry. It is no more difficult for a lattice of any given symmetry to operate on an object of this kind than it is to operate on a lattice complex with no symmetry at all.

6.10 POINT GROUPS OF THE TETRAGONAL AND HEXAGONAL (AND TRIGONAL) SYSTEMS

The remaining point groups all have one and only one axis of more than 2-fold symmetry. Those with 4 or $\bar{4}$ are classed in the *tetragonal system*. Those with 3, $\bar{3}$, 6, or $\bar{6}$ offer considerably more difficulty, and different classifications have been used in the past. It is now generally thought best to treat them all as belonging to the *hexagonal system* (since they can all be

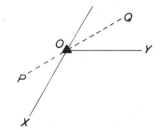

Figure 6-11 Directions to which point-group symbols refer in the hexagonal (and trigonal) system. OX and OY are axes of reference, and PQ the external bisector of the angle XOY.

conveniently described with hexagonal axes of reference) but to assign those with 3 or $\bar{3}$ to the *trigonal division* of the system. (See footnote dealing with rotation-reflection axes, p. 124 which are a cause of confusion here.)

In the point-group symbol, the first figure refers to the unique axis (always chosen as the z axis). The second figure refers to the direction perpendicular to it that we choose for the x axis. Since the y axis is always produced by the operation of the z-axis symmetry on the x axis, it is identical with x, and the third symbol can be used for a different direction in the xy plane, namely, *the external bisector of the angle XOY*. In the tetragonal system, it is the direction at 45° to OX; in the hexagonal system, it is at 30° (or 90°) to OX (Figure 6-11). The symbol implies a choice of axes of reference such that, if there are any symmetry directions in the XY plane, they are either along OX, or along the external bisector of XOY, or along both. If OX is not a symmetry direction, we use the figure 1 to indicate it (see §6.3).

We can now derive all the point groups of these systems.

Point groups 4, $\bar{4}$, 3, $\bar{3}$, 6, $\bar{6}$ need no comment, except to note that $\bar{6}$, though it can also be written $3/m$, should be assigned to the hexagonal division of the hexagonal system. Adding a mirror-plane normal in the same direction as the unique axis, we get new groups $4/m$ and $6/m$. (Of the others, $\bar{4}/m$ and $\bar{3}/m$ are identical with these, while $3/m$ and $\bar{6}/m$ are the same as $\bar{6}$.) Adding a diad in the x direction gives us groups which can be written 42, $\bar{4}2$, 32, $\bar{3}2$, 62, $\bar{6}2$, but these abbreviated forms do not show up some important features. In groups with even rotation axes, new diads are generated in the third direction (the external bisector of XOY), so we can usefully write them as 422, 622, while in 32 nothing new is generated, and to mark the difference the symbol should preferably be written 321. In groups with inversion axes, mirror-plane normals are generated; with even axes, the mirror-plane normals lie in the third direction and the symbols are $\bar{4}2m$, $\bar{6}2m$, while for $\bar{3}2$ the mirror-plane normal coincides with the diad, and the symbol is $\bar{3}\,2/m\,1$.

Next, consider adding mirror-plane normals in the XOY plane (instead of diad axes). For the inversion axes we find that the diad axes are generated; the groups are $\bar{4}m2$, $\bar{6}m2$, and $\bar{3}\,2/m\,1$, which repeat those found previously (though with a rotation of axes of reference relative to the symmetry elements

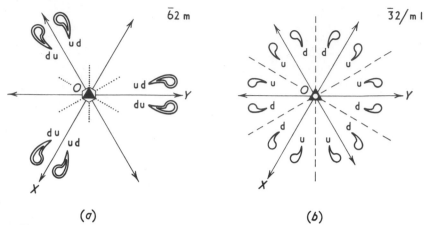

$$\bar{6}2m \qquad \bar{3}\,2/m\,1$$

(a) **(b)**

Figure 6-12 (a) Point group $\bar{6}2m$. Axes of reference are OX and OY. Rotation diads are in the directions of the solid lines; mirror plane normals are in the directions of the dotted lines. (b) Point group $\bar{3}\,2/m\,1$. Axes of reference are OX and OY. Rotation diads *and* mirror plane normals are in the directions of the solid lines; broken lines are *not* symmetry directions. (They represent traces of mirror planes, not mirror plane normals.)

for $\bar{4}m2$ and $\bar{6}m2$). For the rotation axes we find new groups $4mm$, $6mm$, and $3m1$.

Finally, we add a mirror-plane normal along the unique axis as well as at right angles to it, and obtain two new groups $4/m\,2/m\,2/m$ and $6/m\,2/m\,2/m$, commonly shortened to $4/mmm$ and $6/mmm$.

This completes the list of point groups: seven in the tetragonal system, seven in the hexagonal division, and five in the trigonal division of the hexagonal system.

The two groups $\bar{6}2m$ and $\bar{3}\,2/m\,1$ are a cause of much trouble to the inexperienced. They are shown in Figure 6-12. The key to working with them is to remember that m in the symbol defines the direction of the mirror-plane normal and *not* the trace of the mirror plane.

In $\bar{6}2m$, the inversion axis $\bar{6}$ gives the three double commas (cf. Figure 6-5(c)). The diad along OX, acting on these, has generated a mirror plane whose *trace* lies along OX and therefore whose *normal* lies at 90° (with another at 30°) to OX.

By contrast, in $\bar{3}2$ the trace of the mirror plane is at 30° (and 90°) to OX, and hence the plane normal coincides with OX. The point group may be written as $\bar{3}2$, $\bar{3}m$, $\bar{3}\,2/m$, or $\bar{3}\,2/m\,1$; the last-named is to be preferred. If we had chosen our axes at right angles to those marked OX and OY, the point group would have to be written $\bar{3}\,1\,2/m$.

6.11 APPLICATIONS OF POINT-GROUP SYMMETRY TO COORDINATION POLYHEDRA

The commonest coordination polyhedra are regular solids with high point-group symmetry.

The Tetrahedron

The regular tetrahedron (point group $\bar{4}3m$) has the following elements (Figure 6-13):

> Four triad axes, each joining a corner to the midpoint of an opposite face
> Three inversion tetrads, each joining midpoints of two opposite edges
> Six mirror planes, each going through one edge and the midpoint of the opposite edge

Viewed in projection down its tetrad axis, the regular tetrahedron appears as a square, and opposite edges (top and bottom) lying parallel to the plane of projection can be seen at right angles (Figure 6–13[d]). In projection down its triad axis, it appears as an equilateral triangle (Figure 6–13[e]). Geometrical relations can most easily be seen by drawing the surrounding cube, as in Figure 6-14(a). We then have:

> Cube edge $= a$
>
> Tetrahedron edge $= l = \sqrt{2}a$
>
> Centre-to-corner distance $= r = \sqrt{3}a/2$
>
> Height of tetrahedron (corner to midpoint of opposite face) $= 2a/\sqrt{3}$
>
> Distance between midpoints of opposite edges $= a$
>
> Angle between two centre-to-corner bonds $= 109°28'$
>
> Angle between cube edge and centre-to-corner bond $= 54°44'$

Distortion of the tetrahedron by pulling it out along one tetrad axis destroys all the triads and reduces the other tetrads to diads, but retains the mirror

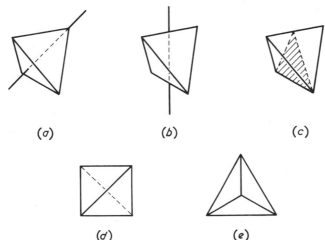

(a)   (b)   (c)

(d)   (e)

Figure 6-13 Symmetry of the regular tetrahedron. (a) Perspective view, showing direction of a triad axis. (b) Same, showing direction of inversion tetrad (which includes a diad). (c) Same, showing mirror plane (shaded). (d) Projection down inversion tetrad. (e) Projection down triad.

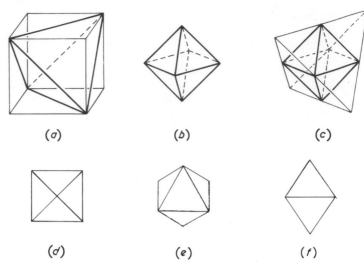

(a)   *(b)*   *(c)*

(d)   *(e)*   *(f)*

Figure 6-14 Relation of cube, regular octahedron, and regular tetrahedron. (a) Perspective view of tetrahedron inscribed in cube. (b) Perspective view of octahedron. (c) Perspective view of octahedron shown in (b) inscribed in tetrahedron shown in (a). (d) Projection of octahedron down tetrad axis. (e) Projection of octahedron down triad axis. (f) Projection of octahedron down diad axis.

planes; it becomes $\bar{4}m2$ or $\bar{4}2m$ according to one's choice of axes. Distortion by pulling it out along one triad destroys all other axes but leaves the planes; it becomes $3m1$ or $31m$.

The Octahedron

The regular octahedron (point group $4/m\ 3\ 2/m$ or $m3m$) has the full symmetry of the cube. It is shown in Figure 6-14(b). The triad axes join midpoints of opposite faces, tetrad axes join opposite corners, diad axes join midpoints of opposite edges. Mirror planes run through opposite edges, and also through the medians of opposite triangular faces; there is a centre of symmetry.

The relation to the tetrahedron can be seen by joining the midpoints of tetrahedron edges as in Figure 6-14(c). Four of the octahedron faces, chosen so that they meet each other only at corners and not at edges, coincide with faces of the tetrahedron; the other four would form a similar tetrahedron, inverted through a centre from the first.

Projected down a tetrad axis, the octahedron gives a square (Figure 6-14[d]); projected down a triad axis, its outline is a hexagon (Figure 6-14[e]).

Geometrical relations are as follows:

Height, corner to corner $= a$

Edge length $= l = a/\sqrt{2}$

Centre-to-corner distance $= r = a/2$

Distance between midpoints of opposite edges $= a/\sqrt{2}$

Distance between midpoints of opposite faces $= a/\sqrt{3}$

Angle between centre-to-corner bonds $= 90°$

Angle between triad axis and centre-to-corner bond $= 54°44'$

The Cube

The cube (point group $4/m\ 3\ 2/m$ or $m3m$) needs little more comment. Geometrical relations are as follows:

Edge $= a$

Centre-to-corner distance $= r = \sqrt{3}a/2$

Angle between centre-to-corner bonds $= 70°32'$

Angle between cube edge and centre-to-corner bond $= 54°44'$

Distortion of a cube along a tetrad axis gives a square prism, with point group $4/mmm$. Distortion along a triad axis (a body diagonal) gives a rhombohedron with point group $\bar{3}\ 2/m\ 1$; all three edges remain equal, but the angles, though equal, are no longer 90°.

Distortion of a cube along a face diagonal turns the face concerned into a rhombus, (Figure 6-15, face $ABCD$), while the two nonparallel faces, $ADHE$ and $DCGH$, remain squares; or if dimensional changes perpendicular to AC are allowed, they become rectangles. The new figure has lost all its triad and tetrad axes and the planes of symmetry parallel to the original cube faces, but it has retained the diagonal planes $ACGE$ and $BDHF$. In consequence, it is orthorhombic, but its symmetry directions do not coincide with the cube edges, our original axes of reference. Its point group, with the new axes, is mmm; if we prefer to go on using axes parallel to what were the cube edges, we have to write the point group as $\{mmm\}$ (cf. §6.7).

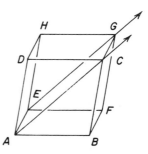

Figure 6-15 Distortion of a cube by elongation along a face diagonal.

6.12 SUMMARY

(i) Crystals of the cubic system have four triad axes equally inclined to one another; the point-group symbol always has a 3 in the second place.

(ii) Crystals of the tetragonal system have one 4 or $\bar{4}$ axis; the point-group symbol has this figure in the first place.

(iii) Crystals of the hexagonal system have one 3, $\bar{3}$, 6, or $\bar{6}$ axis; the point-group symbol has this figure in the first place. The hexagonal division comprises those with 6 or $\bar{6}$ axes, and the trigonal division those with 3 or $\bar{3}$ which cannot alternatively be written with 6 or $\bar{6}$ (e.g., $3/m$ must be written as $\bar{6}$ for purposes of classification).

(iv) Crystals of the orthorhombic system have three diad symmetry directions at right angles to each other, and no axes of higher symmetry. (Note that "symmetry directions" include mirror-plane normals, which are $\bar{2}$ axes.) The point-group symbol has three figures (or in abbreviated form sometimes two figures), which refer to the x, y, z axes, in that order.

(v) Crystals of the monoclinic system have one diad symmetry direction only. The point-group symbol has one figure (counting an axis and a mirror-plane normal sharing the same direction as one, e.g., $2/m$).

(vi) Crystals of the triclinic system have no symmetry direction. The point-group symbol is 1 or $\bar{1}$.

PROBLEMS AND EXERCISES

1. Draw sketch stereograms of the effects of inversion axes $\bar{5}$, $\bar{7}$, $\bar{8}$, $\bar{9}$, $\overline{10}$, and $\overline{12}$. Which of these are also the effect of a combination of simpler point operators?

2. Draw sketch stereograms of the effects of the following combinations of point operators. In each case, notice what other operators, if any, have been generated. Notice also whether any of those listed separately are in fact identical.
    $3/m$, $4/m$, $6/m$; $\bar{3}/m$, $\bar{4}/m$, $\bar{6}/m$; 32, 42, 62; $\bar{3}2$, $\bar{4}2$, $\bar{6}2$; $3m$, $4m$, $6m$; $\bar{3}m$, $\bar{4}m$, $\bar{6}m$; $3/m2$, $4/m2$, $6/m2$.
    For each, write the symbol of the point group (a) in its full form, (b) in its standard form.

3. Show that in each of the five point groups of the cubic system the form {101} consists of the same set of twelve faces.

4. In which of the cubic point groups does the form {111} constitute a tetrahedron?

5. Draw a sketch stereogram of the form {210} in point group 23. Is the form the same in any higher point group?

6. For each of the cubic point groups, write down the indices of all faces of the form {321}.

7. For each of the tetragonal point groups, write down the indices of all faces of the form {321}.

8. For each of the orthorhombic, monoclinic, and triclinic point groups, write down the indices of all faces of the form {321}.

9. In the quartz crystal of Chapter 5, Question 2, the angles between the normals to the prism faces are exactly 60°, and the faces belong to the same form; the angular measurements of the pyramid faces are all alike. The crystal is broken so that the lower end does not show. What, without further information, can you deduce about the symmetry?

10. In other crystals of quartz, there are small faces, referred to as *s*, bevelling the corners between prism and pyramid faces, as described in Chapter 5, Question 10, but these occur only on alternate corners. The lower end of the crystal is again broken off. What can now be said about the symmetry? Insert this set of faces in a sketch stereogram, showing the zones in which they lie, and marking their indices.

11. In yet other crystals of quartz, there are systematic differences in the size and appearance of alternate pyramid faces which show that they belong to different forms. Assuming that all the crystals of quartz described here and in Questions 9 and 10 are of the same phase (i.e., have the same structure and symmetry) deduce the point group.

12. Draw a sketch stereogram of a regular octahedron with its triad axis normal to the plane of projection. Choosing hexagonal axes of reference, index all faces of the form.

13. The unit cell of a crystal is a rhombohedron differing only slightly from a cube; as compared with a cube, it is elongated by a small fraction x along one body diagonal and compressed by $x/2$ in the directions at right angles to this. If hexagonal axes of reference are used to describe the structure, calculate the axial ratio.

14. A crystal of cubic symmetry, with the external shape of a regular octahedron, undergoes a distortion in which its unit cell changes to that described in Question 13. Which faces shown in the answer to Question 12 now belong to the same form?

Calculate (in terms of x) the change of inclination of the original octahedron faces to the direction of elongation.

Chapter 7 SYMMETRY OF LATTICES

7.1 INTRODUCTION

A lattice is a symmetry operator because, by using any of its translation vectors, the infinite crystal can be moved in a way which leaves it just as it was originally. We therefore expect that there will be combinations of lattices with other symmetry operators. It is very easy to show that there are only 5 possible combinations in two dimensions, and 14 possible combinations in three dimensions—the latter known as the 14 Bravais lattices. We shall derive them in this chapter.

7.2 ROTATION AXES AND TWO-DIMENSIONAL LATTICES

We begin by noticing that not all n-fold axes can be combined with a lattice. Let us take the 5-fold axis as illustration, and try to generate a lattice with it.

In Figure 7-1, let A and B be lattice points separated by the shortest existing translation vector, and let there be 5-fold axes perpendicular to the paper at A and B. Operation of the axis at A on B gives C, and operation of the axis at B on A gives D, C and D being new lattice points. But CD is shorter than AB, postulated as the shortest lattice vector. If, ignoring this contradiction, we went on operating with the new axes at C and D, and at each further point as it was generated, we should eventually cover the whole plane with a continuum of 5-fold axes perpendicular to it.

Figure 7-1 Attempt to generate a two-dimensional lattice using pentad axes ($2\pi/n = 72°$).

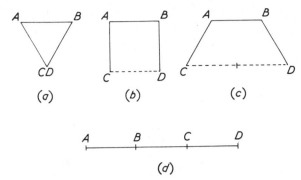

Figure 7-2 Initial stages of generating two-dimensional lattices with symmetry axes perpendicular to the paper. Axes initially given are at A and B; operation of axes gives C and D; angles CAB and $ABD = 2\pi/n$. (a) Hexad axis; (b) tetrad axis; (c) triad axis; (d) diad axis.

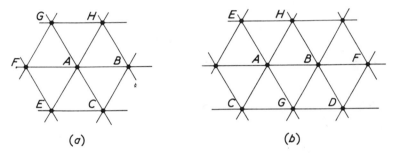

Figure 7-3 Hexagonal net, generated by (a) hexad axes, (b) triad axes, with axes originally given at A and B. Though the points are generated in a different order, the final results are the same.

Suppose now we try the same procedure with other n-fold axes. It can be seen that the condition for generating a lattice is that CD must be an integral multiple of AB. The possibilities are as shown in Figure 7-2, with $CD/AB = 0, 1, 2,$ or 3, for $n = 6, 4, 3,$ or 2, respectively.

We may compare the operation of a 6-fold axis and a 3-fold axis. In Figure 7-3(a), a 6-fold axis at A, operating on B, generates new points round it at C E F G H; B, operating on A, repeats C and H as well as generating three others not shown. In Figure 7-3(b), a 3-fold axis at A, operating on B, generates only C and E; but the lattice property requires B to have the same environment as A, and therefore generates F, G, and H. In both cases, the final result is the same.

7.3   DERIVATION OF THE FIVE TWO-DIMENSIONAL LATTICES

We can now enumerate all the possible types of two-dimensional lattices. This will illustrate some principles we shall need for dealing with three dimensions.

(i) The first type is characterised by a hexad axis perpendicular to its plane. Conventional axes of reference lie along two equal shortest vectors enclosing an angle of 120°. The unit cell is thus a rhombus with angles of 120° and 60°. This is the *hexagonal net*. A triad axis would, as we have seen, give us the same net.

(ii) A tetrad axis perpendicular to the plane gives rise to a *square net*. The unit cell is conventionally the square outlined by shortest vectors.

(iii) A diad axis perpendicular to the plane generates only a line of points; it therefore adds nothing to the symmetry inherent in the lattice translations, and is compatible with the most general kind of net, the *oblique net*. Conventionally, axes of reference are taken along the two shortest vectors, thus defining a primitive unit cell of which the sides are not required to be equal nor the angles right angles.

(iv) The only other symmetry element at our disposal is a mirror line (the trace of a mirror plane whose normal is in the plane of the paper). Translation vectors must then lie parallel and perpendicular to the mirror line, which means that the net must be rectangular, and that a second mirror line has been generated at right angles to the first. There need, however, be no relationship between the lengths of the vectors in the two directions at right angles.

Conventional axes are taken parallel to the two mirror lines. If the translation vectors in *both* these directions are shorter than any in an inclined direction, the unit cell is primitive and we have a *primitive rectangular net*, as shown in Figure 7-4(a).

(v) But suppose, with two mirror lines, there are translation vectors inclined to the mirror lines which are shorter than one of the vectors parallel or perpendicular to the mirror lines, as shown in Figure 7-4(b). Keeping the

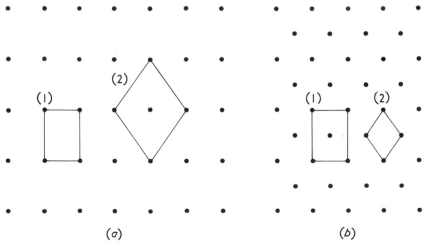

Figure 7-4 (a) Primitive rectangular net; (b) centred rectangular net. In each case (*1*) represents the conventional choice of unit cell, (*2*) a possible alternative choice.

same axes of reference, the unit cell is now a centred rectangle, giving a *centred rectangular net*.

These five kinds of net cover all the possibilities.

7.4 COMPARISON OF PRIMITIVE AND CENTRED NETS

The contrast between primitive and centred nets deserves further emphasis. We saw in §3.2 that any lattice (or net) can be described as primitive by a suitable choice of axes. What then is the fundamental difference between the nets we have called "primitive rectangular" and "centred rectangular"?

In Figure 7-4(a) and (b) we see an extended area of each lattice with the conventional unit cells outlined at (1). If, instead, we chose axes equally inclined to the symmetry directions, each unit cell would be a rhombus as shown at (2). In (a), the primitive rectangular net, the new unit cell would be centred, with double the area of (1); but in (b), the centred rectangular net, it would be primitive, with half the area of (1). For (b) it can be seen that this second choice of axes is likely sometimes to be more convenient than the conventional choice.

Now since centring the unit cell of the primitive rectangular net gives us a new kind of net, we ought to ask whether centring any of the other kinds of net would give us something new. Consider first the oblique net. If we centre the original lattice, Figure 7-5(a), giving a new lattice as in Figure 7-5(b), we can choose new axes to give us a primitive unit cell *EFGH*. This cell is oblique, like the original, with no special relation to symmetry directions, since there are none in its plane. Thus the centring has given us nothing essentially new in kind.

Similarly, a centred square net, like that in Figure 7-6(1), is simply a primitive square net of $1/\sqrt{2}$ times the size, with its axes of reference turned through 45°, as shown in Figure 7-6(2).

For the hexagonal net, the only kind of centring which results in a lattice is the insertion of a pair of points dividing the long diagonal of the 60°

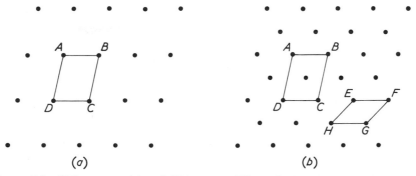

(a) (b)

Figure 7-5 Oblique net: (a) and (b) are two different lattices, but of the same type, though (b) with primitive unit cell *EFGH* is obtained by centring (a).

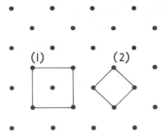

Figure 7-6 Square net: (1) is a centred unit cell; (2) is primitive, and is the conventional choice.

rhombus (Figure 7-7(1)) into equal thirds. But the result can more conveniently be described using a primitive unit cell of the same shape but with sides reduced by a factor $1/\sqrt{3}$, rotated by 30° (or 90°) relative to the original, as in Figure 7-7(2). In fact, the use of axes of reference as in (1) is forbidden by convention, because the notation for the trigonal and hexagonal system can only be made unambiguous if we may assume that the x and y axes coincide in direction with the shortest vectors (a difficulty which does not apply in the tetragonal case).

To sum up, though centred tetragonal, hexagonal, and oblique nets can be envisaged, they are not new types but merely unconventional descriptions of one of the five types listed in §7.3.

7.5 ENUMERATION OF THE 14 BRAVAIS LATTICES

We consider three-dimensional lattices in the same way.

The most general lattice has a centre of symmetry but no symmetry directions; hence any choice of axes to give a primitive unit cell is conventionally permissible.

Four other lattices with primitive unit cells may be at once derived from the four primitive two-dimensional nets: a diad axis perpendicular to an oblique net gives a monoclinic lattice; a diad axis perpendicular to a rectangular net gives an orthorhombic lattice; a tetrad axis perpendicular to

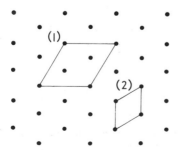

Figure 7-7 Hexagonal net: (1) is a centred unit cell; (2) is primitive, and is the conventional choice.

a square net gives a tetragonal lattice; a hexad (or triad) axis perpendicular to a hexagonal net gives a hexagonal lattice.

The sixth primitive lattice is cubic, with triad axes along the body diagonals of the cubic unit cell.

The remaining lattices are most easily derived by centring the primitive lattices and examining the results to see which are new types.

For an orthorhombic lattice, in which all three principal sections constitute rectangular nets, any one of these may be centred, or all three may be; or body centring is also possible. (Two-face-centring is of course impossible, as shown in §3.3.)

For a monoclinic lattice, in which the symmetry direction is conventionally chosen as the y axis, centring on the face perpendicular to this, B, gives nothing new (any more than did centring of the oblique net in two dimensions); but faces A and C, parallel to the symmetry axis, have rectangular nets, and centring on one of them (it does not matter which) gives a new type of lattice. Just as B-face-centring gives no change of type from primitive, so all-face-centring and body-centring give no change of type from A-face-centring or C-face-centring. There are thus only two types of monoclinic lattice.

For a triclinic lattice, as for an oblique net, no centring gives any new type.

For a cubic lattice, body-centring and all-face-centring are possible, and give new types; one-face-centring is incompatible with the presence of the triad axes, which make all faces of the cube alike.

For a tetragonal lattice, centring on C (perpendicular to the tetrad) is possible but gives nothing new. Centring on A or B alone is forbidden by symmetry (and two-face-centring is, as always, forbidden by the lattice property). Body-centring is possible and gives a new type; all-face-centring is possible but gives the same type as body-centring.

For a hexagonal lattice, centring of the equilateral triangles of the net, as in Figure 7-7, gives a lattice with smaller parameters but no new features. As in the two-dimensional case, convention strictly requires us to use the shortest vectors of this net as our axes of reference.

There is, however, a new lattice which can be derived by centring the primitive hexagonal lattice in a special way, namely, by adding equivalent points at $\pm(2/3, 1/3, 1/3)$. In projection down the symmetry axis, these lie at the centres of equilateral triangles of the net. The fact that this set of points *does* constitute a lattice can be seen from Figure 7-8, where the environments of E and F are exactly like those of A. (In checking this, it must be remembered that integers may be added to any atomic parameter: thus, if there is an atom at height $1/3$, there are others vertically above and below it at heights $4/3$ and $-2/3$.) The hexad axis has been destroyed by the centring, but a triad axis remains. The lattice so derived may be called *rhombohedrally-centred*; we shall show in §7.6 that it is the same as the *rhombohedral lattice*.

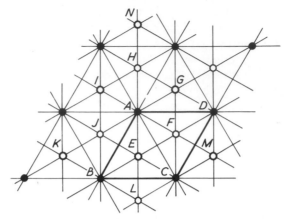

Figure 7-8 Rhombohedrally-centred lattice, projected down the triad axis. Solid black circles are at height zero, thick circles at height 1/3, and thin circles at height 2/3. *ABCD* is the projection of the hexagonal unit cell.

This completes the list of possible and distinct three-dimensional lattices—*the Bravais lattices*. They are summarised in Table 7-1. (Consideration of inversion axes would not add anything new, because all lattices have centres of symmetry at their nodes.)

7.6 THE RHOMBOHEDRAL LATTICE

A rhombohedral lattice is usually defined as one whose primitive unit cell is described by three lattice vectors related to one another by a triad axis of

Table 7-1 The Fourteen Bravais Lattices

SYMMETRY ↘ CENTRING	TRICLINIC	MONO-CLINIC	ORTHO-RHOMBIC	TETRA-GONAL	HEXAG-ONAL OR TRIGONAL	CUBIC
Primitive, *P*	*	*	*	*	*	*
One-face-centred, *A*, *B*, or *C*	0	*A* * *B* 0 *C* 0	*	*A* — *B* — *C* 0	*A* — *B* — *C* †	—
All-face-centred, *F*	0	0	*	0	—	*
Body-centred, *I*	0	0	*	*	—	*
Rhombohedrally-centred, *R*	—	—	—	—	*	—

Key:

* Independent Bravais type, conventional description.

0 Exists, but is the same type as one of the others marked with a star.

— Does not exist.

† For hexagonal and trigonal lattices centring of the *C* face is impossible; centring of the triangle which forms half of the *C* face is possible but gives no new type.

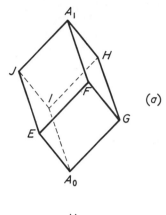

(a)

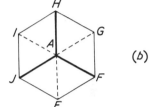

(b)

Figure 7-9 Rhombohedral unit cell, lettered to correspond with Figure 7-8: (a) perspective drawing, (b) projection down the triad axis (heavy lines are edges in upper half, broken lines edges in lower half).

symmetry. It can be seen that Figure 7-8, constructed to represent a rhombohedrally-centred hexagonal lattice, satisfies this condition. The vectors AE, AG, AI constitute the edges of a primitive rhombohedron. The complete rhombohedron is shown in perspective in Figure 7-9(a) and in projection down the triad axis in Figure 7-9(b). The directions of the vectors AE, AG, AI are the *rhombohedral axes of reference*; the axial lengths are equal, and the interaxial angles are equal to each other but of arbitrary value.

We thus have two useful ways of describing the rhombohedral lattice: using rhombohedral axes of reference, AE, AG, AI, which give a primitive unit cell; or using hexagonal axes of reference, AB, AD, and the direction of the triad axis, which give a unit cell of three times the volume, containing three lattice points. For many purposes, the latter description is more convenient, because it is much easier to visualise and discuss a structure which has only one oblique interaxial angle.

It would have been possible to have chosen other sets of symmetry-related vectors for our axes of reference. For example, the choice of AK, AM, AN (Figure 7-8) would give an all-face-centred rhombohedron, whose volume is four times that of the primitive rhombohedron (since the projected length of AK is twice that of AE and the height of the cell is unaltered). Alternatively, the choice of AF, AH, AJ, whose directions are nearer to the hexad axis than are those of AE, AG, AI, would give a body-centred rhombohedron with volume double that of the primitive rhombohedron (since it has the same area in projection, but twice the height). Such choices, though unconventional, may occasionally be useful (cf. Chapter 10).

It must be emphasised that the distinction between rhombohedral and hexagonal lattices is a real one, and does not depend on the axes of reference we choose. The relationship is in some ways like that between the centred and primitive rectangular nets, discussed in §7.3(iv) and (v).

The choice of rhombohedral axes of reference for a structure with a primitive hexagonal lattice, giving three lattice points per rhombohedral unit cell, though formally possible, has rarely any advantages.

The cube is a special case of a rhombohedron with 90° interaxial angles. The primitive unit cell of a face-centred cubic lattice is also a rhombohedron, but with interaxial angles of 60°.

Two additional points may be mentioned which are illustrated in Figure 7–8.

(i) The volume of the rhombohedron whose projection is *EFGHIJ* is one-third the volume of the prism of the same height whose base is *ABCD*, and is equal to the volume of the prism of the same height whose base is the projection of *AEFG*.

(ii) We have chosen our centring points as $\pm(2/3, 1/3, 1/3)$; if, instead, we had chosen $\pm(1/3, 2/3, 1/3)$, it would have been equivalent to turning the rhombohedron through 180° about its triad axis, i.e., taking the *reverse* instead of the *obverse* rhombohedron. To avoid confusion, it is desirable to keep to the conventional (obverse) setting, unless there are special reasons for doing otherwise (in which case the reader's attention must be called to the fact). For the conventional setting, the horizontal component of the rhombohedral *x* axis lies 30° anticlockwise from the hexagonal *x* axis, and the hexagonal *y* axis and the horizontal components of the rhombohedral *y* and *z* axes follow in anticlockwise sequence—the hexagonal *z* axis and the vertical components of the rhombohedral axes pointing upwards from the plane of the paper (cf. Figure 10-3).

We may mention here a further point sometimes causing difficulty. It has been seen earlier that a hexagonal lattice does not need a hexad axis to generate it; a triad axis will suffice. Hence, when we consider the symmetry of crystals, we cannot assume that all crystals whose point groups have triad axes have also rhombohedral lattices; they may have *either* hexagonal *or* rhombohedral lattices. On the other hand, all crystals whose point groups have hexad (or inversion hexad) axes *must* have hexagonal lattices. In making subdivisions of the hexagonal system, *classification by point-group symmetry does not coincide with classification by lattice type*.

PROBLEMS AND EXERCISES

1. Draw diagrams illustrating the derivation, from the six primitive Bravais lattices listed at the beginning of §7.5, of the remaining eight Bravais lattices, and justifying the statement that no new types other than these can be produced.

2. Show that, while the array of points occupied by atoms in cubic close packing constitutes a lattice, that occupied by atoms in hexagonal close packing does not.

3. Show that the array of occupied sites in diamond does not constitute a lattice. How is the primitive unit cell of diamond related to (a) the atomic positions, (b) the cubic unit cell? What is the Bravais lattice type?

4. Construct a section of the rhombohedral lattice parallel to $(11\bar{2}0)$, showing the arrangement of lattice points.

5. Show that if, for a rhombohedral lattice, the triad axis is replaced by a hexad axis without destroying any translation vectors of the lattice, the result is a smaller primitive hexagonal lattice. Find the lengths and orientations of its unit-cell edges relative to those of the original.

Chapter 8 SPACE GROUPS

8.1 INTRODUCTION

A periodic structure, we saw, is described by a lattice operating on a lattice complex. We have considered separately the symmetry of lattices (Chapter 7) and the symmetry of the objects on which they operate (Chapter 6); we now consider the symmetry of the resulting structure.

A three-dimensionally periodic structure can have any of the point-group operators—rotation or inversion axes, or mirror planes—and also two new kinds of operators, namely, *glide planes* and *screw axes*. The possible combinations of the various operators are 230 in number, and are known as the *space groups*.

A complete mathematical derivation of the space groups was achieved at the end of the nineteenth century independently by Federov in Russia and Schoenflies in Germany, using different methods. (The term *Fedorov group* is sometimes used instead of space group in commemoration of this.) The mathematical tools for the systematic description of crystal structures were thus made available long before there was any hope of knowing the structures by experiment, which became possible only after the discovery of X-ray diffraction in 1912.

We are not concerned with the mathematical theory of space groups, but rather with their practical application. This involves an understanding of the notation, so that when we are given a space-group symbol we know the operators implied by it, and if we are given also the coordinates of any one atom we can construct the coordinates of all others related to it by symmetry. Any atom of the set may be taken as the representative or *prototype atom* from which all the others can be derived.

Knowing the space group does not, of course, tell us the structure. In general, a structure is built up from more than one set of symmetry-related atoms, and a prototype atom of each set must be known; together, the whole group of prototype atoms constitutes the *asymmetric unit*. To build the structure, we can either first pick out the asymmetric unit and allow the space group to operate on it, or allow the space group to operate on one prototype atom at a time, and superpose the results.

In this chapter, we are interested in the space-group operation, rather than in what is operated on. We therefore consider the object to be a point (or a point atom). The set of symmetry-related points (or point atoms) which results is known as an *equipoint*.

As mentioned in §6.2, the original notation of Schoenflies proved inconvenient in practice. Originally devised for point groups, it was extended to space groups by adding an arbitrary serial number, which meant that the symbol did not carry all the requisite information. It has been superseded by the internationally adopted Hermann-Mauguin notation, which does convey the full information with very few arbitrary points to be remembered, and which will be used throughout this book. This is the notation used in the *International Tables for X-ray Crystallography* (Volume 1)— the source book for all the material in this chapter.

8.2 GLIDE PLANES AND SCREW AXES

A *glide plane** has a particular direction in the structure, like a mirror plane, and also a direction of glide, parallel to its own plane. Its operation combines a mirror reflection in the plane with a displacement equal to a half (or in certain less usual cases a quarter) of the lattice vector in the glide direction.

In Figure 8-1(a) let the broken line represent the trace of a glide plane parallel to (100), with glide direction [010]. A single operation reflects A in the plane and displaces it by a distance $b/2$ to B; repetition of the operation

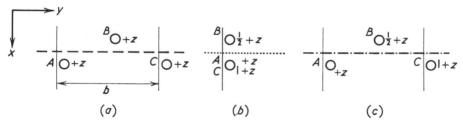

(a) (b) (c)

Figure 8-1 Glide planes. (a) Axial glide: the glide direction is horizontal across the page. (b) Axial glide: the glide direction is perpendicular to the page. (c) Diagonal glide, with components in both horizontal and perpendicular directions. In each case the normal to the plane is up and down the page, and the trace of the plane is shown by a heavy line— broken for (a), dotted for (b), and dot-dash for (c).

---

* The terms *glide plane* and *glide direction* have sometimes been used in connection with the plastic properties of crystals, where they refer to planes of atoms slipping over one another. This is now regarded as incorrect; the correct terms for describing the mechanical movements are *slip plane* and *slip direction*. In contexts in which there is danger of confusion, we might, however, speak of the symmetry operators as *symmetry glide planes* associated with *symmetry glide directions*.

brings it back to C, where it is a lattice repeat of A. Its height remains at $+z$ throughout.

If the glide is in the direction [001] (conventionally indicated by making its trace a heavy dotted line instead of a heavy broken line) the effect of the operation is as shown in Figure 8-1(b); B is now at height $1/2 + z$, and the repetition of the glide superposes C on top of A at height $1 + z$.

If the glide has components $b/2$ and $c/2$ (indicated by making the trace of the plane a heavy dot-dash line) the effect is as shown in Figure 8-1(c).

Glide components perpendicular to the plane are impossible; thus, a glide plane parallel to (100) can only have components along [010] and [001].

The symbol for a glide plane depends on the direction of its glide; those gliding by half the lattice vector along [100], [010], or [001] are called a, b, or

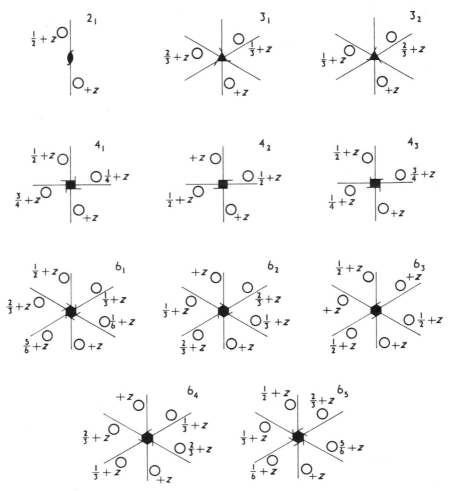

Figure 8-2 Screw axes. Axes are perpendicular to the paper. The symbols used for the axes are the conventional ones.

c respectively, while those with glide components in both permissible directions are called *n* ("diagonal glide plane"). The less common type of glide plane that glides by one quarter of the lattice vector in both permissible directions is called *d* ("diamond glide plane" because of its occurrence in the diamond structure).

The direction of the glide plane cannot be deduced from the letter given to it, but from the position occupied by that letter in the completed symbol, in exactly the same way as the direction of a mirror plane was deduced from the position of its letter *m* in the point-group symbol. Just as, in the point group, the direction of the mirror plane is defined by its *normal*, so is the direction of the mirror plane or glide plane in the space group.

Examples will be given in §8.5.

A *screw axis*, like a glide plane, is a translation operator. A screw axis described by the symbol n_m is one which combines a rotation of $2\pi/n$ about a line in the structure with a displacement equal to m/n times the lattice vector in the direction of the line.

Figure 8-2 illustrates the effects of the screw axes, in projection down the axes. It can be seen that n_m and n_{n-m} are *enantiomorphous*, being related like mirror images to one another. It can also be seen that, if *m* is a submultiple of *n*, the n_m screw axis *includes* an *m*-fold rotation axis.

8.3 THE SPACE-GROUP SYMBOL

The first letter in the symbol is the *lattice symbol*. This refers only to the character of the lattice centring, and does not tell us the lattice type, or the crystal system. For example, *Pban* and *P6₃mc* are space groups of different systems with primitive lattices, *C2/c* and *Ccca* are face-centred on (001), *I4̄2d* and *I4₁32* are body-centred, *Fdd2* and *Fm3c* are all-face-centred. The one case in which the symmetry is deducible from the lattice symbol is the rhombohedral lattice *R*, which can only occur in the trigonal subdivision of the hexagonal system.

The remaining part of the space-group symbol has a very close relation to the point-group symbol. To derive the point group, we replace all translation operators by point-group operators; i.e., we write *m* instead of *a*, *b*, *c*, *n*, or *d*, and drop the subscripts of all screw axes. Thus we see that the space groups mentioned previously belong respectively to point groups *mmm*, *6mm*, *2/m*, *mmm*, *4̄2m*, *432*, *mm2*, and *m3m*. Applying the rules of §6.12, we recognise that *I4₁32* and *Fm3c* are cubic, *P6₃mc* hexagonal, *I4̄2d* tetragonal, *Pban*, *Ccca*, and *Fdd2* orthorhombic, and *C2/c* monoclinic.

This close relationship to the point-group symbol means that the positional rules developed in Chapter 6—notably in §6.4, 6.7, and 6.12—can all be carried over and applied directly to the space-group symbol. There is, however, one thing needing care: whereas for point groups of the tetragonal and hexagonal systems two alternative choices of *x* and *y* axes were permissible, giving rise to alternative names for the same group (e.g., 321 and 312), for space groups the x and y axes are fixed by the decision that the unit cell must be primitive,* and therefore $P321$ and $P312$ are truly different (cf. §8.11).

Just as in point groups, so in space groups, some combinations of symmetry operators generate others. We shall not attempt to work through these systematically, but the close analogy with point groups is a help in recognising and remembering some useful generalisations which emerge from the examples that follow.

One obvious rule is as follows. If, in a point group, two symmetry operators in specified directions combine to give a third in a related direction, the same holds good in a space group, with the proviso that where there were mirror planes in the point group there may be mirror planes or glide planes in the space group, and where there were rotation axes in the point group there may be rotation axes or screw axes in the space group.

This rule is of particular importance in allowing us to deduce, from the abbreviated space-group symbols commonly used, the presence of symmetry operators other than those named. For example, the orthorhombic space group $Pbam$, like the point group mmm, must contain diad axes in all three axial directions, though we cannot tell without further inquiry whether they are rotation diads 2 or screw diads 2_1. Similarly, the trigonal space group $P\bar{3}c$ must, like its point group $\bar{3}m$ (which, as we saw, may be written in full as $\bar{3}2/m1$), contain diad axes parallel to the x axis, i.e., (since the lattice is primitive) to the direction of the shortest lattice vector perpendicular to the unique axis z.

8.4 ALTERNATIVE FORMS OF THE SPACE-GROUP SYMBOL

Just as a point group may have different symbols according to the choice of axes of reference (see §6.7), so may a space group. For each space group there is a *standard form*, recorded in the *International Tables for X-ray Crystallography*, but this is purely for convenience of reference, and it is explicitly stated that "*no official importance is attached to the particular setting of the space group adopted as standard.*" In practice, we must always be prepared for nonstandard forms, because choices of axes are fixed by other reasons than these conventions. Methods of conversion from one form to another will be given in Chapter 10; in the present chapter, we shall merely consider particular examples as they arise.

---

* This does not mean that we may not, for working purposes, choose axes of reference giving a larger unit cell. It *does* mean that for hexagonal and trigonal space groups, and some tetragonal space groups, the geometry of the translational symmetry operators in the larger cell cannot be described by existing symbols of the Hermann-Mauguin notation.

8.5 CONSTRUCTION OF SPACE-GROUP DIAGRAMS

We shall illustrate this process for a number of representative space groups. drawing them in the same way às in the *International Tables for X-ray Crystallography*. One unit cell is shown in projection on the *xy* plane, together with parts of adjacent cells. The object to be operated on is shown as a circle, placed in a general position with respect to any symmetry elements; its height is marked. For clarity, symmetry operators and the objects operated on are shown in separate diagrams.

In constructing these diagrams it is helpful (at least, until one is fairly experienced) to distinguish between symmetry operators put in deliberately and the others which are generated from them. In drawing one's own diagrams, colour can be used to distinguish them; in the examples in this book, separate diagrams on the left side have been drawn showing the only *necessary* operators in each case, while those on the right show the complete set generated by them.

The conventions for showing symmetry operators in the diagrams are those of the *International Tables for X-ray Crystallography*.

After the Hermann-Mauguin symbol of each space group used as an example, we give in round brackets (parentheses) its serial number in the *International Tables* and its Schoenflies symbol.

8.6 SOME SPACE GROUPS WITH GLIDE PLANES

Pc (No. 7, C_s^2) (Figure 8-3)

Applying the rule of §8.3, we see that the point group is *m*; it is therefore monoclinic, with the symmetry direction chosen as the *y* axis. Thus the space group has a glide plane normal to *y*, gliding by a vector distance $(1/2)c$. It is more convenient to use coordinates which are fractions of the unit-cell edge, and say that it glides by $1/2$ in the *z* direction (or the *c* direction, which means the same thing).

We are free to take our origin on the glide plane, making it $y = 0$; there is a translation-repeat at $y = 1$. These are marked on the left-hand diagram.

In the middle diagram, we take an atom *D* in a general position. (For clarity, we take it fairly near the origin.) We add the translation-repeats of *D* near the other corners of the unit cell, namely, *E, F, G*. Operating with the mirror plane, we get *H* at height $1/2 + z$; we mark this with a comma inside a circle to show that it has been derived by an inversion operation. There are corresponding partners for *E, F*, and *G*.

Inspection shows that an additional *c*-glide plane has developed at $y = 1/2$; we insert this (in addition to the original planes at $y = 0$ and $y = 1$) in the right-hand diagram.

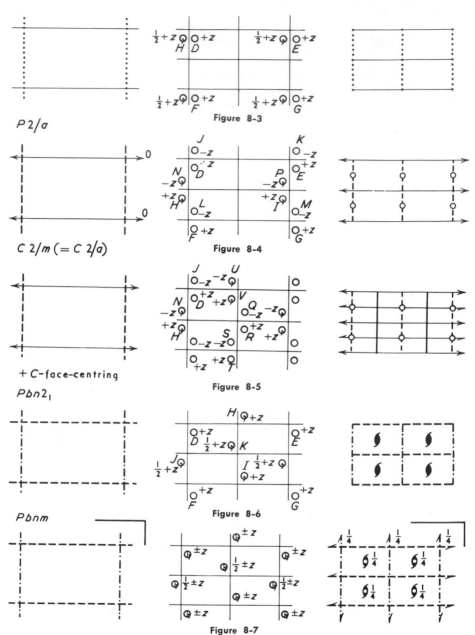

Figures 8-3 to 8-7 Space-group diagrams. The left-hand diagram in each figure shows the symmetry used to generate the rest; the middle diagram shows the result of its operation (including the lattice operation) on the prototype atom D; the right-hand diagram shows the complete symmetry.

Direction of axes of reference: OX from top to bottom of the page; OY from left to right, horizontally on the page; OZ upwards from the page at the top left-hand corner of the parallelogram (sloping backwards in Figures 8-3 to 8-5, normal to the paper in the rest). Heights of points or of lines or planes parallel to the xy plane are marked if they are not zero. The parallelogram in the left-hand diagram represents the base of the unit cell.

We can state a general rule, which holds if there is no symmetry greater than 2-fold: *For every symmetry plane or 2-fold axis, there is another of the same type, in the same direction, midway between the first and its translation-repeat.*

The same is true for centres of symmetry, except that they have no direction. When axes of higher symmetry are present, additional symmetry is again generated, but with less simple rules.

P2/a (No. 13, C_{2h}^4) (Figure 8-4)

This is again monoclinic (point group 2/m). We take the a-glide plane at $y = 0$, operate on D and E to obtain H and I, and find ourselves with another a-glide plane, as expected by the rule.

Now we add the rotation diad, passing through the origin. This, operating on D, E, F, G, gives J, K, L, M; operating on the translation-repeats of H and I, it gives N and P. We note that additional diads have been generated at $x = 1/2$, $z = 0$; $x = 0$, $z = 1/2$; $x = 1/2$, $z = 1/2$—the last named, for example, relating I at $+z$ to P at $1 - z$.

Since we know that the point group has a centre of symmetry, we look for centres of symmetry in the diagram, and find them midway between N and D, P and E, and other corresponding points. The complete set is inserted in the right-hand diagram. Since I is derived from E by a glide plane, which produces inversion, and P from I by a rotation axis, which does not invert, P and E are related by inversion—as they should be for a centre of symmetry.

The centre of symmetry is not at the origin, and this could be inconvenient. It would have been better to choose the origin differently, putting the first diad at $x = 1/4$, $z = 0$ instead of $x = 0$, $z = 0$, with the others to correspond. This would have been the conventional choice. With experience, one can learn to make the best choice at the outset; but in any case it is easy to redraw the diagram after the better choice has been recognised. Methods of handling a change of origin will be discussed in §10.2.

We cannot check this diagram directly against that in the *International Tables for X-ray Crystallography*,* because P2/a is not the standard form; the standard form is P2/c. On the basis of symmetry, we might just as well have called the glide direction z (or c), which would have given the standard form to the space-group symbol; but in an actual structure we might have had very good reasons for preferring to call it x (or a). If checking against the Tables were important, we could easily redraw the diagram with the z axis rather than the x axis in the plane of the paper.

---

* When the *International Tables for X-ray Crystallography* were compiled, a change of convention to take the symmetry direction in the monoclinic system as the z axis was being considered. Diagrams labelled "1st setting" were drawn with this convention, which was later rejected. Diagrams labelled "2nd setting" use the accepted convention.

C2/a (No. 12, C_{2h}^3) (Figure 8-5)

This has the same point group as $P2/a$. In fact, it contains all the symmetry operators of $P2/a$ plus a face-centring operation. We may begin by inserting all the atoms of Figure 8-5, and then move each of them by a vector $\frac{1}{2}\mathbf{a} + \frac{1}{2}\mathbf{b}$; this relates J, D, to Q, R; N, H, to U, V, and to S, T.

In addition to the symmetry elements of $P2/a$ we now find others. There are mirror planes at $y = 1/4$ and $y = 3/4$ midway between the glide planes; and there are screw diads at $x = 1/4$, $z = 0$ and corresponding positions, midway between the rotation diads.

This generating of new planes and axes, interleaving those originally put in and parallel to them, but not necessarily of the same type, is characteristic of lattice centring. If we had begun with the mirror planes instead of the glide planes, or the screw diads instead of the rotation diads, we should have achieved exactly the same set of points.

There is no accepted convention for naming all the symmetry elements in space groups with centred lattices, but for the moment we may write this space group as $C(2, 2_1)/(a, m)$. Only one of the figures in each pair of round brackets (parentheses) need be named, and it is conventional to name a rotation axis rather than a screw axis, a mirror plane rather than a glide plane; hence, conventionally, the space group is called $C2/m$. Here there is no question of change of axes, and to call it $C2/a$ is not incorrect but merely makes it harder to recognise.

Pbn2₁ (No. 33, C_{2v}^9) (Figure 8-6)

This is point group $mm2$, and is therefore orthorhombic.

We begin by using the two glide planes b and n. The order in which operations are done does not, of course, affect the result, but there is less danger of mistakes (as we shall see later) in using planes rather than axes to begin with.

Operation of the b-glide plane normal to the x axis on D and F gives H and I; operation of the n-glide plane normal to the y axis on D and the translation-repeat of H gives J and K.

We now recall that the point group $mm2$ could be written as mm because the diad is generated by the two planes; hence, by the rule of §8.3, the 2_1 axis of the space group should already be present. We find one perpendicular to the paper at $x = 1/4$, $y = 1/4$, and others at corresponding points, shown in the right-hand diagram. No other symmetry is predicted by the point group or found in the diagram.

If we had put in a screw diad unthinkingly, we might have put it at $x = 0$, $y = 0$, and thereby produced quite a different space group; by completing the diagram, it can be shown to be $Cmc2_1$. The position of the screw diad relative to the glide plane is *not* arbitrary, and cannot be altered to suit our convenience.

On the other hand, there is nothing to fix the height of the origin in the z direction, and we can take it where we choose.

This space group is again not in the standard form. For that, we should have to choose our axes to make it $Pna2_1$.

Pbnm (No. 62, D_{2h}^{16}) (Figure 8-7)

We begin exactly as in the last example, but must add the mirror plane m. Let us put it at $z = 0$; this fixes the origin in the z direction.

Additional symmetry appears. From the point group, mmm, we expect to find diads of some kind parallel to x and y, and also centres of symmetry. The diads turn out to be all screw axes, at $y = 0$, $z = 1/4$ and corresponding points for one set, $x = 0$, $z = 0$ and corresponding points for the other; the centres are at $1/4$, $1/4$, $1/4$, and corresponding points. Here, as for $P2/a$, it would have been better to choose the origin at a symmetry centre. $Pbnm$ is not the standard form, which is $Pnma$ (cf. §10.8).

8.7 NUMBER OF ATOMS IN THE EQUIPOINT

It is desirable to be able to predict from the space group the number of symmetry-related atoms per unit cell. Except for the cubic system, it is very easily done. We first write down the point - group symbol in its shortest possible form. Each plane m contributes a factor 2, and each n-fold axis a factor n. The product of these factors is multiplied by a factor for lattice centring—1 for P, 2 for A, B, C, or I, 4 for F—to give the total number of atoms per equipoint. For example, for $Pbn2_1$ the abbreviated form of the point group is mm; hence the number of atoms is $1 \times 2 \times 2$, in agreement with Figure 8-6. Unit - cell contents of all other examples illustrated can be checked in the same way.

8.8 THE DIAMOND-GLIDE PLANE

An example of the diamond-glide plane is found in $Fdd2$ (No. 43, C_{2v}^{19}) illustrated in Figure 8-8. The d-glide plane at $x = 1/8$ operates on D to give H, I, J; that at $y = 1/8$ operates on D to give K, L, M, and on H to give P, Q, N. The face-centring of the lattice operates on the pair ND to give RI, PL, ST, and on the pair HK to give JU, QM, VW. Diad axes in the z direction have been generated, as expected: rotation diads through $x = 0$, $y = 0$, and $x = 1/4$, $y = 1/4$, and the points obtained from these by face-centring operations; screw diads at $x = 1/4$, $y = 0$, and $x = 0$, $y = 1/4$, and corresponding points.

The d-glide plane is found in only a few space groups—two ortho-rhombic ($Fdd2$ and $Fddd$), five tetragonal, and five cubic, all either

Fdd 2

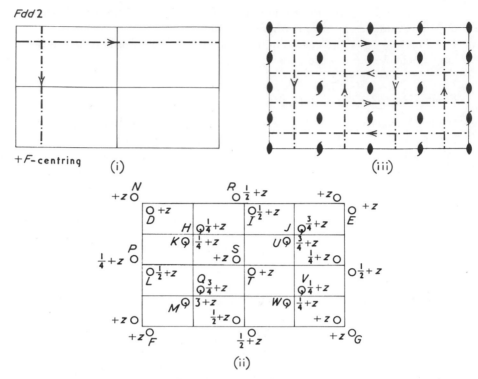

+*F*-centring   (i)   (iii)

(ii)

Figure 8-8 Orthorhombic space group, illustrating the operation of a *d*-glide plane. Conventions as for Figures 8.3 to 8.7.

body-centred or all-face-centred—but some structures with these space groups are extremely important, notably that of diamond. Other examples are found in Chapters 11, 12, 13.

8.9 INTERSECTING AND NON-INTERSECTING AXES

*P*222 (No. 16, D_2^1) and *P*222$_1$ (No. 17, D_2^2)

When it is necessary, as in this point group, to operate with two sets of diad axes at right angles, the question arises whether they lie in the same plane and intersect one another or are non-intersecting.

In Figure 8-9 we make rotation axes parallel to x and y intersect by putting them at height $z = 0$ (and of course at $z = 1/2$ as well). They generate a new rotation diad through the point of intersection of the first two, as shown in the right-hand diagram. This is therefore the space group *P*222.

In Figure 8-10 we take the rotation axes along the lines x, 0, 0 and y, 0, 1/4, which are non-intersecting. The new axis is along the line 0, 0, z,

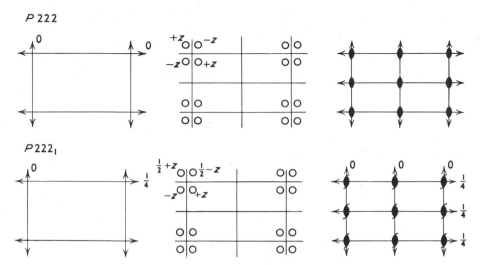

Figures 8-9 and 8-10 Orthorhombic space groups of point group 222. Conventions as for Figures 8-3 to 8-7.

intersecting both the others, but it is now a screw diad. The space group is $P222_1$.

Obviously in this case we cannot abbreviate the space-group symbol (as we did the point-group symbol) without losing information.

Neither of these space groups has a centre of symmetry, but (in contrast to $Pbn2_1$) the choice of origin along the z axis is not arbitrary, since it can be defined in relation to the position of the x or y symmetry axis.

It is possible to derive certain simple generalisations about intersection of axes. If diad axes in the x and y directions intersect, there will be rotation diads in the z direction; if not, there will be screw diads. If axes in the x and y directions are both rotation diads, the axis in the z direction will intersect both; if they are both screw diads, it will intersect neither; if one is a screw axis and the other a rotation axis, it will intersect the screw axis. It may however be simpler to work out each example from first principles than to remember such rules.

$I222$ (No. 23, D_2^8) and $I2_12_12_1$ (No. 24, D_2^9)

We begin by constructing $P222$ and then operate with the body-centring to get $I222$. We find (as always with centred lattices) additional symmetry elements parallel to those named in the symbols; in this case, they constitute an intersecting set of screw diads.

How can we get anything different for $I2_12_12_1$? We try taking non-intersecting screw diads as in $P2_12_12_1$ and operating with the body-centring. We find we have generated rotation diads, but a non-intersecting set.

The second space group is certainly different from the first, but, since both have rotation diads and screw diads in all three axial directions, the decision about their names is arbitrary. The group with intersecting axes of the same type is called *I*222, and that with non-intersecting axes $I2_12_12_1$. These two space groups, and two in the cubic system where the same problem arises (*I*23 and $I2_13$), are the only space groups for which the Hermann-Mauguin symbol is ambiguous without the addition of an arbitrary convention.

8.10 SPACE GROUPS WITH TETRAD AXES

$P4_1$ (No. 76, C_4^2) (Figure 8-11)

Putting in the screw tetrad axis at the origin, perpendicular to the paper, and its translation-repeats, we get the set of points in the second diagram. We find additional symmetry has developed: screw tetrads at the centre of the square, screw diads at the midpoints of the edges. These are shown in the third diagram.

In general, whatever kind of tetrad axis is placed at the corners, the same kind is developed at the centre of the square, and some kind of axis (tetrad or diad) at the midpoint of the edge.

$P4_2/n$ (No. 86, C_{4h}^4) (Figure 8-12)

The 4_2 axes are put in first. The glide is parallel to the paper, and since there is nothing as yet to fix the origin in the z direction, we may put the glide plane at height 0. (Note the symbol for this glide plane at the corner of the first and third diagrams.) Operating on the groups at F and G, this gives us the groups at H and I. We notice that the 4_2 axis recurs at the centre of the group represented by H.

Since the point group $4/m$ is centrosymmetric, we look for centres of symmetry in the space group, and find them at height 0 (and 1/2) at the centres of the small squares. Moreover, we notice tetrad symmetry around the midpoint J of an edge; in detail, we see that there is an inversion tetrad axis here, relating, for example, the four atoms nearest to J in projection lying at heights $+$ and $\frac{1}{2}-$. Extension of the drawing in other directions would show similar inversion tetrads, again with their centres of inversion at height 1/4. (Notice that though the $\bar{4}$ axes are present, they are not mentioned in the symbol.)

Our choice of origin is not the conventional one for this space group.

$P\bar{4}2_1m$ (No. 113, D_{2d}^3) (Figure 8-13)

We begin by putting in the $\bar{4}$ axis at the corners of the unit cell, with inversion centre at height 0, and we notice the $\bar{4}$ axis at the centre of the square, and the diad axes at the midpoints of the sides.

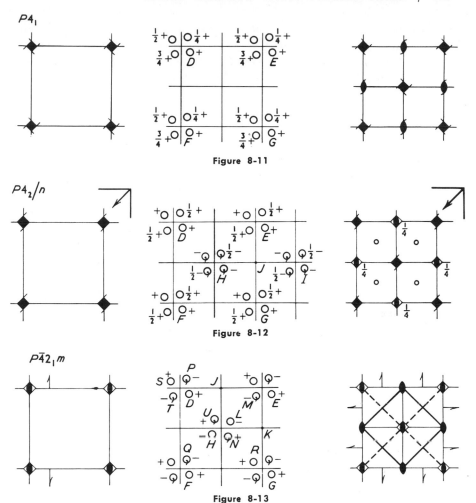

Figure 8-11

Figure 8-12

Figure 8-13

Figures 8-11 to 8-13 Space groups of the tetragonal system. Conventions as for Figures 8-3 to 8-7.

Where do we put in the screw diad 2_1, which must be parallel to the x axis? Let us jump at once to the right answer and make it the line x, 1/4, 0, leaving the consequences of other choices to be examined later. It acts on D and its three neighbours to give H and its three neighbours. Knowing from the point group that the symmetry plane at 45° must follow automatically, we look for it and find the trace of a mirror plane along the line JK and other symmetry-related lines. There is thus no need to insert any other symmetry elements.

We have not, however, yet noticed all the symmetry in the second diagram. There are glide planes parallel to JK passing through the tetrad inversion axes, relating D to N, P to H, and other corresponding planes at 90°. These are not diagonal glide planes, because there is no component along the

z axis, but they are not axial glide planes, because the plane itself is not parallel to two axes. The notation includes no symbol to describe such glide planes, because it is not necessary—they follow as a consequence of other symmetry operators which do have symbols.

$P\bar{4}b2$ (No. 117, D_{2d}^7) (Figure 8-14)

Suppose in trying to construct $P\bar{4}2_1m$ we had unthinkingly put the diad axis along x, 0, 0, to intersect the tetrad. Operation of this on D gives H, from which the tetrad at 0, 0, z gives I; the diad along x, 1, 0 operates on this to give J, which is the translation-repeat of K. Now it can be seen that the whole group near K is a translation-repeat of that near D; the same is true for H and I. Hence the lattice, on these axes, is not primitive but C-centred. The set of points *is* consistent with a space group, but the choice of axes of reference is wrong if we insist on a primitive unit cell. Taking x and y at 45° to the original, as in the third diagram, 2_1 must occupy the third place in the symbol, the second being occupied by a plane normal. There is the trace of a glide plane parallel to the diagonal of the original square, midway between the nearest tetrad axes, turning for example L into I and K into M. With the new axes, this is an a glide. Correspondingly, the plane normal to the x axis has a b glide. Hence the space-group symbol can be written $P\bar{4}b2_1$.

We notice that additional rotation diads are found parallel to the screw diads but midway between them. Mention of these could have been included in the space-group symbol, and indeed it is conventional to name them in preference to the screw diads when both are present. The conventional space-group symbol is thus $P\bar{4}b2$.

$P\bar{4}n2$ (No. 118, D_{2d}^8) and $P\bar{4}2_1c$ (No. 114, D_{2d}^4)

By starting with an inversion tetrad, as in the last two examples, diad screw axes could have been added along two other lines, x, 0, 1/4, or x, 1/4, 1/4.

$P\bar{4}b2$

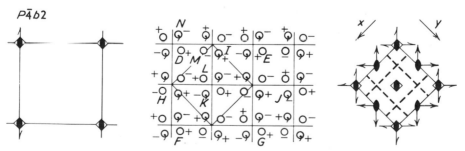

Figure 8-14 Space-group $P\bar{4}b2$. With axes of reference chosen as for the previous figures (i.e., with the unit cell shown in the first diagram), the lattice is C-face-centred. The correct choice of axes to give the symbol $P\bar{4}b2$ is shown in the third diagram, and the base of the corresponding unit cell (which is primitive) is outlined in the second.

The former gives $P\bar{4}n2$, which could also be $P\bar{4}n2_1$; the latter gives $P\bar{4}2_1c$, which is *not* the same as $P\bar{4}2c$ (No. 112, D_{2d}^2) or $P\bar{4}c2$ (No. 116, D_{2d}^6) ($= P\bar{4}c2_1$).

These examples illustrate what was said in §8.3, that, in space groups, as distinct from point groups, the order of the two last symbols indicates a real difference of arrangement, not merely our choice of axes of reference.

It should be said that the space groups $P\bar{4}2m$, $P\bar{4}b2$, and $P\bar{4}n2$ are among the more difficult ones, chosen as examples here in order to show some of the problems that occasionally arise, and that could cause confusion to anyone unaware of them.

8.11 SPACE GROUPS WITH A PRIMITIVE HEXAGONAL LATTICE

P6 (No. 168, C_6^1) (Figure 8-15)

Operating only with the rotation hexad axes at the corners of the unit cell, we find we have generated rotation triads at the centre of each equilateral triangle between them, and rotation diads midway between the hexads.

In general, whether the first set of axes are hexads or triads, and whatever their type, there is a triad axis of some kind at the centre of the triangle; the diads are only a necessary consequence if the first set of axes are hexads.

P3m1 (No. 156, C_{3v}^1) (Figure 8-16)

The *normal* to the mirror plane lies along the unit cell edge; i.e., the trace of the plane bisects the 60° angle. It is perhaps more convenient to operate first with it on the prototype atom D, giving H, and then to operate with the triad axis on the pair DH.

The three sets of mirror planes related by the triad axis are shown in the third diagram. Additional glide planes have been generated, lying midway between them. To test this in the second diagram, consider the plane whose trace lies along KL. The pairs DH, MN, are related by reflection in this plus a translation equal to half the long diagonal of the unit cell base, i.e., half the repeat distance in this direction.

P31c (No. 159, C_{3v}^4) (Figure 8-17)

Here the *normal* to the c-glide plane lies along the bisector of the 60° angle; hence the trace of the plane lies along the unit cell edge. The glide is in the vertical direction. Operating first with the glide plane (which gives H from D) and then with the triad, we get the set of points in the second diagram. Again there are glide planes midway between the first set—this time with a vertical as well as a horizontal glide component. In the second diagram, for example, the plane along KL relates the pair MN to the pair DH, with a vertical translation of $\frac{1}{2}c$ and a horizontal translation of $\frac{1}{2}a$.

Space groups of the trigonal system can be very confusing to the inexperienced, and there have been unfortunate mistakes in the literature. It is essential to remember that *the direction of any plane is specified by its normal and not its trace*. If this principle is firmly grasped, most of the difficulties disappear.

8.12 SPACE GROUPS WITH A RHOMBOHEDRAL LATTICE

Rhombohedral lattices can occur in any point group of the trigonal subdivision of the hexagonal system, i.e., those with one 3 or $\bar{3}$ axis but no 6 or $\bar{6}$ axis, namely 3, $\bar{3}$, 32, 3*m*, $\bar{3}2/m$. In some of the older crystallographic literature these were called the rhombohedral classes—sometimes even the rhombohedral system—and this may still cause confusion. In fact, while rhombohedral lattices are *permitted* in these classes, they are not *required*, and cannot be deduced from the point group symmetry (any more than the character of the lattice centring can be deduced for other point groups). In §8.11 we had examples of space groups of the point group $\bar{3}2/m$ which were not rhombohedral.

For space groups of the point groups 3*m* and $\bar{3}2/m$, with rhombohedral lattices, as for those with primitive hexagonal lattices, it is essential to remember that *the letter for the symmetry plane specifies the direction of its normal* and not its trace.

Figure 8-15

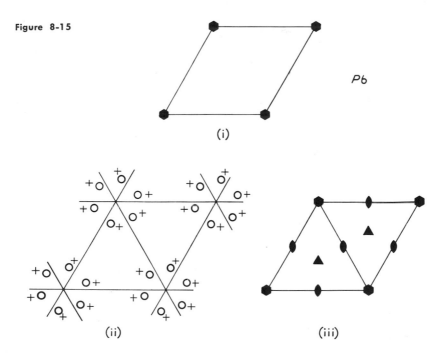

$P6$

(i)

(ii)   (iii)

Figure 8-16

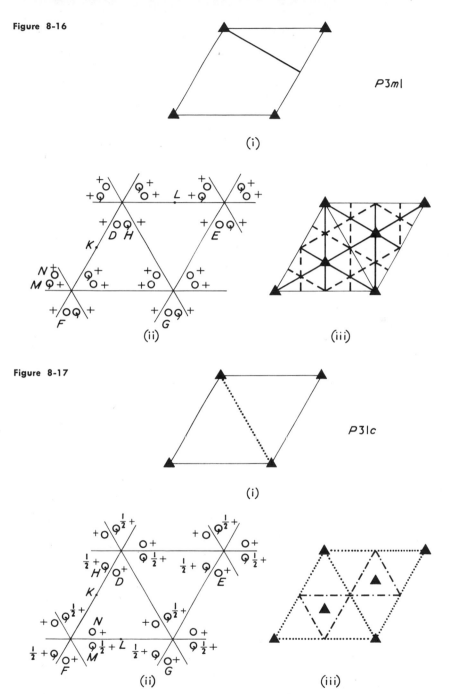

P3m1

(i)

(ii)

(iii)

Figure 8-17

P31c

(i)

(ii)

(iii)

Figures 8-15 to 8-17 Space groups of the hexagonal system. These are drawn with the same conventions as for Figures 8-3 to 8-7, except that the *OX* axis slopes down the page towards the lower left-hand corner, while *OY* still extends from left to right horizontally.

R3 (No. 146, C_3^4) (Figure 8-18)

We continue to work with hexagonal axes of reference, which are generally far more convenient than rhombohedral axes. The lattice is then a centred one, with the centring operator $(0, 0, 0; 2/3, 1/3, 1/3; 1/3, 2/3, 2/3) +$.

We first use the rotation triad axis, and then the centring operator, which repeats the prototype atom D at H and I, at heights $1/3 + z$, $2/3 + z$, respectively. The triad axis at the origin is of course repeated at the centres of the basal triangles. Moreover, two screw axes 3_1 and 3_2 have been generated between each pair of rotation triads, at equally spaced distances.

As a check, it is worth calculating the number of atoms in the unit cell (cf. §8.7). There is a factor of 3 for the rhombohedral centring, and 3 for the triad axis, giving 9—in agreement with the diagram.

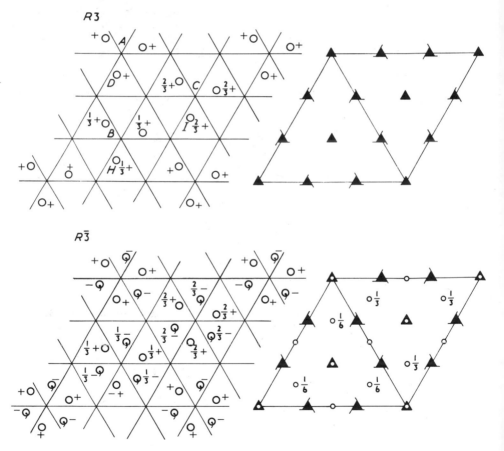

Figures 8-18 and 8-19 Space groups of the hexagonal system with rhombohedral lattices. These are drawn with the same conventions as for Figures 8-15 to 8-17, except that the first (preliminary) diagram of the set of three has been omitted, because it would show only the 3 or $\bar{3}$ axes at the corners of the parallelogram.

$R\bar{3}$ (No. 148, C_{3i}^2) (Figure 8-19)

Beginning with $R3$, we add a centre of symmetry at $(0,0,0)$, thereby converting the rotation triad into an inversion triad. Centres of symmetry occur at appropriate heights on the other triads, and a second set of centres is found midway between pairs of adjacent centres of the first set.

8.13 CUBIC SPACE GROUPS

These groups are generally fairly complicated to handle, and always difficult to draw, because the very important triad axes are not orthogonal. We shall not attempt any general discussion, but shall deal with examples as they occur in later chapters.

8.14 DERIVATION OF ATOMIC COORDINATES

If the choice of origin relative to the symmetry elements has been made, it is possible from the space-group symbol and the coordinates x, y, z of one prototype atom to derive the coordinates of all atoms of the equipoint. For one operator at a time, examples of results are as follows:

Centre at $(0, 0, 0)$:

$$x, y, z; \quad \bar{x}, \bar{y}, \bar{z};$$

$$\text{or} \quad \pm (x, y, z) \tag{8.1}$$

Centre at $(\frac{1}{4}, 0, 0)$:

$$x, y, z; \quad \tfrac{1}{2} - x, \bar{y}, \bar{z} \tag{8.2}$$

Mirror plane at $y = 0$:

$$x, y, z; \quad x, \bar{y}, z \tag{8.3}$$

Mirror plane at $y = \frac{1}{4}$:

$$x, y, z; \quad x, \tfrac{1}{2} - y, z \tag{8.4}$$

c-glide plane at $y = 0$:

$$x, y, z; \quad x, \bar{y}, \tfrac{1}{2} + z \tag{8.5}$$

c-glide plane at $y = \frac{1}{4}$:

$$x, y, z; \quad x, \tfrac{1}{2} - y, \tfrac{1}{2} + z \tag{8.6}$$

n-glide plane at $y = 0$:

$$x, y, z; \quad \tfrac{1}{2} + x, \bar{y}, \tfrac{1}{2} + z \tag{8.7}$$

n-glide plane at $y = \frac{1}{4}$:

$$x, y, z; \quad \tfrac{1}{2} + x, \tfrac{1}{2} - y, \tfrac{1}{2} + z \qquad (8.8)$$

Rotation diad along line 0, 0, z:

$$x, y, z; \quad \bar{x}, \bar{y}, z \qquad (8.9)$$

Rotation diad along line 0, $\frac{1}{4}$, z:

$$x, y, z; \quad \bar{x}, \tfrac{1}{2} - y, z \qquad (8.10)$$

Screw diad along line 0, 0, z:

$$x, y, z; \quad \bar{x}, \bar{y}, \tfrac{1}{2} + z \qquad (8.11)$$

Screw diad along line $\frac{1}{4}$, 0, z:

$$x, y, z; \quad \tfrac{1}{2} - x, \bar{y}, \tfrac{1}{2} + z \qquad (8.12)$$

Rotation tetrad along line 0, 0, z (Figure 8-20 a):

$$x, y, z; \quad y, \bar{x}, z; \quad \bar{x}, \bar{y}, z; \quad \bar{y}, x, z \qquad (8.13)$$

Inversion tetrad along line 0, 0, z, with centre of inversion at $(0, 0, 0)$:

$$x, y, z; \quad y, \bar{x}, \bar{z}; \quad \bar{x}, \bar{y}, z; \quad \bar{y}, x, \bar{z} \qquad (8.14)$$

Inversion tetrad along line 0, 0, z, with centre of inversion at $(0, 0, 1/4)$:

$$x, y, z; \quad y, \bar{x}, \tfrac{1}{2} - z; \quad \bar{x}, \bar{y}, z; \quad \bar{y}, x, \tfrac{1}{2} - z \qquad (8.15)$$

Rotation triad along line 0, 0, z (Figure 8-20 b):

$$x, y, z; \quad \bar{y}, x - y, z; \quad y - x, \bar{x}, z \qquad (8.16)$$

Inversion triad along line 0, 0, z, with centre of inversion at $(0, 0, 0)$:

$$\pm (x, y, z; \quad \bar{y}, x - y, z; \quad y - x, \bar{x}, z) \qquad (8.17)$$

Rotation hexad along line 0, 0, z:

$$x, y, z; \quad x - y, x, z; \quad \bar{y}, x - y, z; \quad \bar{x}, \bar{y}, z; \quad y - x, \bar{x}, z; \quad y, y - x, z \qquad (8.18)$$

Lattice-centring on (100) (A-face-centring):

$$x, y, z; \quad x, \tfrac{1}{2} + y, \tfrac{1}{2} + z; \quad \text{or} \quad (0, 0, 0; \ 0, \tfrac{1}{2}, \tfrac{1}{2}) + \qquad (8.19)$$

Body-centring of lattice:

$$x, y, z; \quad \tfrac{1}{2} + x, \tfrac{1}{2} + y, \tfrac{1}{2} + z; \quad \text{or} \quad (0, 0, 0; \ \tfrac{1}{2}, \tfrac{1}{2}, \tfrac{1}{2}) +$$

$$(8.20)$$

All-face-centring of lattice:

$$x, y, z; \quad \tfrac{1}{2} + x, \tfrac{1}{2} + y, z; \quad x, \tfrac{1}{2} + y, \tfrac{1}{2} + z; \quad \tfrac{1}{2} + x, y, \tfrac{1}{2} + z;$$

$$\text{or} \quad (0, 0, 0; \ \tfrac{1}{2}, \tfrac{1}{2}, 0; \ 0, \tfrac{1}{2}, \tfrac{1}{2}; \ \tfrac{1}{2}, 0, \tfrac{1}{2}) + \qquad (8.21)$$

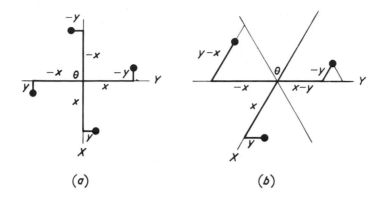

(a)   (b)

Figure 8-20 Derivation of coordinates of points resulting from the operation of (a) a tetrad axis at 0, 0, z, (b) a triad axis at 0, 0, z.

Rhombohedral centring:

$$x, y, z; \quad \tfrac{2}{3} + x, \tfrac{1}{3} + y, \tfrac{1}{3} + z; \quad \tfrac{1}{3} + x, \tfrac{2}{3} + y, \tfrac{2}{3} + z;$$

$$\text{or} \quad (0, 0, 0; \ \tfrac{2}{3}, \tfrac{1}{3}, \tfrac{1}{3}; \ \tfrac{1}{3}, \tfrac{2}{3}, \tfrac{2}{3}) + \tag{8.22}$$

We now consider, as an illustration, the space group $C2/a$ (Figure 8-5). Taking the origin at a centre of symmetry, there is an a-glide plane at $y = 0$ and a rotation diad along the line $1/4, y, 0$. The glide plane gives

$$x, y, z; \quad \tfrac{1}{2} + x, \bar{y}, z \tag{8.23}$$

Operation of the rotation diad on x, y, z gives $1/2 - x, y, \bar{z}$; on $1/2 + x, \bar{y}, z$ it gives $\bar{x}, \bar{y}, \bar{z}$. Operation of the glide plane on $1/2 - x, y, \bar{z}$ also gives $\bar{x}, \bar{y}, \bar{z}$. Moreover, we see that the set of 4 points is centrosymmetric about the origin, and can be written

$$\pm (x, y, z; \ \tfrac{1}{2} + x, \bar{y}, z) \tag{8.24}$$

Indeed, it would have been much easier to use the glide plane plus the symmetry centre to derive the same result. The result is still incomplete—it represents the equipoint of the space group $P2/a$. Adding the lattice centring, we get altogether 8 points:

$$\left. \begin{array}{ll} x, y, z & \tfrac{1}{2} + x, \tfrac{1}{2} + y, z \\[4pt] \tfrac{1}{2} + x, \bar{y}, z & x, \tfrac{1}{2} - y, z \\[4pt] \tfrac{1}{2} - x, y, \bar{z} & \bar{x}, \tfrac{1}{2} + y, \bar{z} \\[4pt] \bar{x}, \bar{y}, \bar{z} & \tfrac{1}{2} - x, \tfrac{1}{2} - y, \bar{z} \end{array} \right\} \tag{8.25}$$

Inspection shows pairs such as x, y, z; $x, 1/2 - y, z$; characteristic of a mirror plane at $y = 1/4$. We had earlier (§8.6) recognised the existence of such a plane from the diagram.

As a second example, consider $Pbn2_1$ (No. 33), with the origin as shown in Figure 8-6. The b-glide plane at $x = 0$ gives

$$x, y, z; \qquad \bar{x}, \tfrac{1}{2} + y, z \qquad\qquad (8.26)$$

The n-glide plane at $y = 0$, operating on these two, gives

$$\tfrac{1}{2} + x, \bar{y}, \tfrac{1}{2} + z; \qquad \tfrac{1}{2} - x, \tfrac{1}{2} - y, \tfrac{1}{2} + z \qquad\qquad (8.27)$$

As a check, we see that the pair x, y, z; $1/2 - x, 1/2 - y, 1/2 + z$; is related by a screw diad along $1/4, 1/4, z$, and similarly for the other pair.

As with the space-group diagrams, it is always wise to check one's atomic coordinates against those listed in the *International Tables for X-ray Crystallography*. It must be remembered, however, that different choices of origin will give different expressions for the set of coordinates, and while alternative choices are sometimes given in the *International Tables for X-ray Crystallography*, they do not cover all the possibilities that one may find useful.

8.15 SPECIAL POSITIONS

Hitherto in this chapter we have taken care to place our point atom away from the symmetry operators, to get the most general case.

Suppose, instead, we had placed a point atom on a mirror plane. Operation of the mirror plane would have left the original atom unchanged, and created nothing new. Positions where this happens are called *special positions*. They comprise points lying on rotation axes or mirror planes, and also centres of inversion (including centres of symmetry).

As an illustration, consider the space group $Bmm2$ (Figure 8-21). General positions of the equipoint, and the symmetry operators, are shown in (a) and (b). In (c) the prototype atom is placed on the mirror plane at $x = 0$, and in (d) on the mirror plane at $y = 0$; in each case the number of atoms per unit cell is reduced from 8 to 4. In (e), the atom lies on both planes (and therefore also on the rotation axis), giving 2 atoms per cell. We see that each independent point-symmetry operator passing through the point reduces the number per cell by a factor of 2. (If the rotation axis had been n-fold, the factor would have been n.)

There is a distinction between the situations in (c) and (d). In (c), all the mirror planes parallel to (100)—those at $x = 1/2$ as well as at $x = 0$—carry atoms of the set; in (d), of those parallel to (010), only half—those at $y = 0$—carry atoms. Thus, if we are listing all the special positions of this space group, we must record prototype $x, 1/2, z$ as well as $x, 0, z$, but not $1/2, y, z$ if $0, y, z$ is already mentioned.

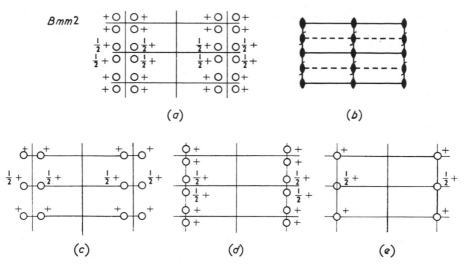

Figure 8-21 Space group *Bmm*2: (a) atoms in general position *x, y, z*; (b) symmetry elements; (c) atoms in special position 0, *y, z*; (d) atoms in special position *x*, 0, *z*; (e) atoms in special position 0, 0, *z*. (Axes of reference as in Figures 8-3 to 8-14.)

Consider case (d) a little further. Whether we put our first prototype atom at $y = 0$ or $y = 1/2$ is merely a matter of choice of origin; the resulting structure is the same in either case. In most actual structures, however, the asymmetric unit is more than a point atom, and when we have to insert a second independent atom we no longer have a free choice of origin. We shall get totally different structures according to whether we put the second atom on the same mirror plane parallel to (010) as the first, or on a plane midway between those occupied by the first. This situation does not arise in (c), where both sets of atoms lie on both sets of mirror planes anyway.

The complete set of atomic coordinates for general and special positions in *Bmm*2 is given in Table 8-1. Tables of this kind for each space group are given in the *International Tables for X-ray Crystallography*. Apart from the names given to each kind of special position, which are conventionally

Table 8-1 Equivalent Positions in Space Group *Bmm*2 (No. 38, C_{2v}^{14})

NUMBER OF POSITIONS	WYCKOFF SYMBOL	POINT SYMMETRY	COORDINATES OF EQUIVALENT POSITIONS
			$(0, 0, 0; \frac{1}{2}, 0, \frac{1}{2}) +$
8	*f*	1	$x, y, z;\ \bar{x}, \bar{y}, z;\ \bar{x}, y, z;\ x, \bar{y}, z$
4	*e*	*m*	$x, \frac{1}{2}, z;\ \bar{x}, \frac{1}{2}, z$
4	*d*	*m*	$x, 0, z;\ \bar{x}, 0, z$
4	*c*	*m*	$0, y, z;\ 0, \bar{y}, z$
2	*b*	*mm*	$0, \frac{1}{2}, z$
2	*a*	*mm*	$0, 0, z$

accepted letters based on Wyckoff's original systematisation of the subject in 1929, all this information can easily be deduced from the space-group symbol with the help of space-group diagrams such as those in earlier parts of this chapter.

Though we are not, in this book, concerned with methods of structure determination, it is important to understand the kind of argument by which atoms are assigned to special positions. Suppose that the number of chemically identical atoms per unit cell, p, is less than the number of atoms in the equipoint for the general position, q. Unless there is disorder (which we are not considering) p/q will be a simple fraction, and p will be the number of atoms in the equipoint of a special position. If, in a structure with space-group $Bmm2$, there were 4 atoms of one kind per unit cell, the only possible ordered positions for them would be $0, y, z$; or $x, 0, z$; or $x, 1/2, z$; or two independent pairs of positions on rotation diads. To decide among the possibilities (and to evaluate y, z or x, z), other arguments would be needed. Again, if there were 12 atoms of one kind, it is possible that 8 would be in general positions, and 4 in special positions; on the other hand, they might fall into three sets of 4, each in special positions. Even if there were 8 atoms of one kind, they might occupy special positions. There is nothing to require atoms of the same chemical species to occupy symmetry-related positions, and hence one cannot assume that, because the numbers would allow general positions, these are in fact chosen. Again, there is nothing in the symmetry requirements to prevent two unrelated atoms from occupying the same kind of special position with different parameters, e.g., $0, y_1, z_1$, and $0, y_2, z_2$; in such cases, however, packing considerations may limit the possibilities considerably.

Table 8-1 also specifies the point-group symmetry of the special positions. This may be useful if, as often happens, a group of atoms of known configuration has to be inserted in the structure. For example, a tetrahedral group could not be placed at a special position with centrosymmetric point-group symmetry, and if it is at a special position with mirror symmetry its orientation is thereby fixed.

8.16 GENERAL POSITIONS AND EUTYCHIC POSITIONS

A *general position*, as opposed to a special position, is one which allows the development of the full number of atoms per unit cell. This means that atoms on glide planes or screw axes are in general positions (contrary to what is sometimes mistakenly assumed); they are *not* in special positions, even though their coordinates happen to be 0 or 1/2, because operation of the glide plane or screw axis continues to generate as many atoms as if the prototype were not on the axis.

A somewhat similar situation occurs when atomic parameters not fixed by symmetry happen to have simple fractional values such as 1/4 and 1/8. In such circumstances, arrays of atoms may have partial or local symmetry greater than that indicated by the space group.

Positions with accidentally simple parameters—whether they are on glide planes or screw axes, or have other local symmetry—may be called *eutychic* (from the Greek τυχη—"good by chance"). Structures in which they occur need to be considered carefully to avoid misunderstandings.

8.17 SUBGROUPS AND BRAVAIS ARRAYS

A *subgroup* is a set of operators, itself forming a group, whose operations are all included among those of a higher group. We had examples among the point groups discussed. For example, the tetrahedral point group $\bar{4}3m$ (§6.11) is a subgroup of the octahedral point group $4/m\ 3\ 2/m$ (also §6.11), though this is less easily noticed when the latter is written in its shortened form $m3m$. A diad axis is a subgroup of either a rotation tetrad or an inversion tetrad (§6.3), but an inversion tetrad is not a subgroup of a rotation tetrad, or vice versa; a rotation triad is a subgroup of an inversion triad, but a rotation hexad is not a subgroup of an inversion hexad. The point groups of the figures obtained by stretching or compressing a cube along any of its symmetry axes (§6.11) are all subgroups of $m3m$, because such an operation destroys symmetry without introducing anything new.

Turning to the space groups, we saw in §8.6 that $C2/m$ resulted from adding the lattice centring C to $P2/a$; hence $P2/a$ is a subgroup of $C2/m$. Again $Pbn2_1$ is a subgroup of $Pbnm$ (§8.6), and $P222$ a subgroup of $I222$ (§8.9), which is itself a subgroup of $I23$.

In some cases the subgroup can be recognised at sight from the symbol; in others, recognition needs a little experience regarding the symmetry implied, but not named, by a particular symbol, or alternatively the ability to work this out.

A lattice, which is itself a symmetry group, is always a subgroup of the space group.* For example, in space group Pc (§8.6), the operation of the lattice alone gave us, from prototype atom D, its translation repeats $E\ F\ G$ (Figure 8-3) together with repeats outside the area illustrated.

An array of atoms produced from a prototype atom by the operation of a lattice, without explicit use of any other symmetry elements, is known as a *Bravais array*. (The symmetry does of course enter implicitly, in controlling

---

* This means that we cannot determine the symmetry of the lattice from the lattice vectors alone when the lattice is associated with a lattice complex. For example, in Figure 3-1 there are lattice vectors which are equal in two directions at right angles; i.e. the unit cell is geometrically square, but, because the lattice can tolerate a lattice complex without either a tetrad axis or mirror lines, it must be classed as oblique—the shape of the unit cell is accidental. Fortunately this kind of situation, where the lattice parameters suggest a higher symmetry than the structure as a whole possesses, is rare in actual crystals, but it does sometimes occur. For example, a unit cell like that of a hexagonal lattice occurs in coesite (a polymorph of SiO_2), though the true symmetry is monoclinic and the arrangement of the atoms is not even approximately hexagonal.

the relative lengths and directions of the lattice vectors.) The crystallographer is always aware that every atom in the unit cell is a prototype member of a Bravais array in the complete structure, and may think it sufficient to name one atom and take its repetitions for granted. In Figure 8-13, for example, there are 8 Bravais arrays to consider, with prototype atoms which may be chosen, according to taste, as *DQRMUHNL* or *DPSTUHNL* or otherwise. Physicists who are less used to describing any but very simple periodic structures, and who wish to emphasise the repetition, use the very misleading term "sublattice" for a Bravais array, and in this example would say there were 8 sublattices. The term is misleading because in fact the lattice operator is used in its entirety and is not broken up into parts. What *may* usefully be broken up into parts for separate treatment is the lattice complex, a physical or chemical entity of which one atom at a time can—if it is helpful—be picked out as a prototype on which the lattice translations operate.

8.18 SUMMARY AND COMMENTS

Many space groups are so simple that it is well worthwhile being able to use the rules to construct them from first principles. The more difficult will always need to be checked from the *International Tables for X-ray Crystallography*, and it is perhaps not worthwhile trying to remember many details about them. However, one is always liable to meet examples described in unconventional and unfamiliar form (often for very good reasons, though sometimes simply because the original authors were not familiar with the conventions), and it is important to know enough to be able either to work with them as they stand or to re-express them in a way which allows the *International Tables for X-ray Crystallography* to be consulted. Practice in deriving space groups and familiarity with the principles underlying the notation is a great help in such cases.

(a) In deriving space groups, some useful things to remember are as follows.

(i) Note the minimum symmetry needed to define the point group, and operate, in the first instance, with the corresponding symmetry elements of the space group; then add the lattice centring.

(ii) Note the additional symmetry elements generated in the point group; look for corresponding elements in the space group.

(iii) Expect to find symmetry elements of the same type midway between those which are translation-repeats of each other—except for triad and hexad axes, where one expects new triads at the centre of a triangle of the original ones; and tetrad axes, where one expects new tetrads at the centre of a square of the original ones.

(iv) With *n*-fold axes where *n* is greater than 2, and with centred lattices, look for additional symmetry elements parallel to those named in the symbol and interleaving them.

(b) The space-group symbol is designed to give directions, not positions, of symmetry elements. The answer to the question whether axes in a given space group are intersecting or non-intersecting is therefore not obvious; but (if the *International Tables for X-ray Crystallography* are not at hand and the rules are not remembered) it can always be found by constructing diagrams to test the possibilities and see which leads to the desired space group.

(c) Similarly, the position of a centre of symmetry relative to any planes or axes is not indicated directly by the symbol but can always be found by constructing the diagram.

(d) Atomic coordinates of the equipoint can always be constructed from that of a prototype atom if the positions of the minimum essential set of symmetry elements relative to one another are known.

(e) Special positions are those on mirror planes, rotation axes, or centres of symmetry. They can easily be derived from general positions, using either the diagram or the set of atomic coordinates. The reverse process, derivation of general positions from special positions, is of course not possible without reference to the space-group symbol.

PROBLEMS AND EXERCISES

1. (a) By trying all possible combinations, develop all the space groups with point group 222 and a primitive lattice.

   (b) By adding various kinds of lattice centring, develop all the different space groups with point group 222.

2. Draw diagrams of the symmetry elements and atoms in general and special positions in the following space groups:

   (a) $C2/m$, $Pmm2$, $Imm2$, $Pnn2$, $Pmc2$, $Cmc2$
   (b) $Pbca$, $Pnma$, $Cmca$
   (c) $P3_121$, $P3_212$, $P\bar{3}m1$, $R3m$, $R3c$, $R\bar{3}m$, $R\bar{3}c$
   (d) $P6_222$, $P6_422$, $P6_3$, $P6_322$, $P6_3mc$, $P6_3/mmc$, $P6/mmc$
   (e) $P4mm$, $P4_12_12$, $P4/mnm$, $P\bar{4}2_1m$
   (f) $I4_1/a$, $I4_1/amd$, $I4/mcm$, $I\bar{4}2d$

   Notes: It is helpful to arrange the diagrams as in Figures 8-3 to 8-17, with a left-hand diagram showing only the limited selection of symmetry elements actually needed to construct the set of general positions, and a right-hand diagram showing the full symmetry generated. Alternatively, the symmetry elements actually used can be shown on the right-hand diagram in a different colour.

   To avoid confusion, it is best to have separate diagrams for atoms in special positions.

   For centred space groups, it is generally easiest to leave the lattice-centring operation to the last, after all the atoms of the corresponding primitive space group have been inserted.

Always check whether the number of atoms of the general form within the boundaries of the unit cell agrees with that expected from the symmetry group. If they are too few, look for what you have omitted. If they are too many, consider whether you have actually constructed a different space group with a smaller lattice.

Check your results against the *International Tables for X-ray Crystallography*. If they differ, is it merely by a different choice of origin? If so, redraw with the conventional origin.

3. Show that the first member of each of the following pairs of space groups contains only symmetry elements which are also present in the second member of the pair, and hence that the first is a subgroup of the second. (Examples of materials characterised by these space groups are given.)

   (a) *Im2m*, *Immm* (sodium nitrate)
   (b) *P4mm*, *Pm3m* (barium titanate)
   (c) *Pmm2*, *Pm3m* (barium titanate)
   (d) *P2$_1$ma*, *Pnma* (thiourea)
   (e) *P6$_3$22*, *P6$_3$/mmc* (barium aluminate, tridymite)
   (f) *P6$_2$22*, *P3$_2$21* (quartz)
   (g) *Pnnm*, *P4$_2$/mnm* (calcium chloride, rutile)
   (h) *Cccm*, *P6/mcc* (low cordierite, beryl)

4. Write down all the subgroups of the space group *P6$_3$/mmc*. In which of them are there general or special positions with 6 atomic sites in the equipoint? In each case, write down the coordinates and the point symmetry of the site.

5. Draw diagrams of the space group *P6/mcc* (No. 192). (Note that the hexad axis is the line of intersection of two *c*-glide planes, and its intersection with the mirror plane is taken as origin.)

   List all the special positions and their point symmetry.

6. In a certain cubic space group, the points of the general position are u, v, w; $1/4 + v, 1/4 + u, 1/4 + w$; together with others derived from them by the operation of the following: an *a*-glide plane at $y = 1/4$, a *b*-glide plane at $z = 1/4$, and a *c*-glide plane at $x = 1/4$; a triad axis through the origin; a centre of symmetry at the origin; body-centring.

   (a) Show how to derive the complete set of symmetry-related points.

   (b) Show that there are four kinds of special position which have no variable parameter. Note their point symmetries, and the number of atoms in each set, and list all their coordinates.

   (c) Assuming u, v, w to be small but not negligible, write down the coordinates of points of a general position grouped respectively round $0, 0, 0$; $0, 0, 1/2$; $1/4, 1/4, 1/4$; and $1/4, 3/4, 1/4$.

   (d) Show that the space group is *Ia3d* (No. 230), and find the three kinds of special position not considered under (b).

Chapter 9 PARTLY PERIODIC GROUPS AND COLOUR GROUPS

9.1 INTRODUCTION

This chapter is to some extent a digression: it is an extension of the ideas of Chapters 6, 7, and 8 in new directions which are important for crystal physics and crystal chemistry, but which (unlike Chapters 6 to 8) we shall not need for the descriptions and discussion of crystal structures in the second half of the book.

In Chapter 8 we considered the three-dimensional space groups, sometimes called the Federov groups. In Chapter 7 we not only considered the three-dimensional lattices, the Bravais lattices, but led up to them by a discussion of the possible two-dimensional lattices; we could, had we wished, have used the methods of Chapter 8 to derive the corresponding two-dimensional space groups. In doing so we would not have allowed atoms or points to exist anywhere except in the plane of the lattice. Such structures exist in a two-dimensional world and are fully periodic in that world. The numbers of point groups, lattices, and space groups are shown in Table 9-1.

We now proceed to consider what happens in less restricted cases—if we allow ourselves a different use of dimensions in space, or attribute to the atoms some physical property which is not directly shown by their position in space.

9.2 PARTLY PERIODIC GROUPS

Structures existing in three-dimensional space need not be three-dimensionally periodic. A piece of basketwork or a woven cloth has two-dimensional periodicity, but uses the third dimension for its over-and-under effects; a chain has one-dimensional periodicity, though each link has its own three-dimensional symmetry. Again, on a two-dimensional surface, a piece of paper or a wall, we may have a one-dimensionally repeating border. In each case we can develop the complete set of symmetry groups which are mathematically possible.

The first step is to see what symmetry operators are allowed. Glide planes, for example, can only exist if their glides are in a direction with periodicity; screw axes can only exist parallel to such directions. If there is only one-dimensional periodicity, there can be no n-fold axes perpendicular to the periodic direction with n greater than 2; if there is only two-dimensional periodicity, there can be no such axes lying in the periodic plane. These restrictions arise because if we used the forbidden operators we would generate periodic repetition in the directions in which we have postulated it does not occur. Having, in the light of these restrictions, obtained a set of allowable operators, we can combine them in various ways, just as we did to derive the ordinary three-dimensionally periodic space groups.

Consider, for example, a one-dimensional border in two-dimensional space. We cannot have axes of more than 2-fold symmetry in the plane of the paper, because they would generate points out of the plane; neither can we have such axes perpendicular to the paper, because they would generate a two-dimensional net.* A 2-fold axis perpendicular to the plane is equivalent to a centre; along the periodic direction a 2-fold rotation axis is equivalent to a mirror line in the same direction, a 2_1 axis to a glide line in the same direction, and a $\bar{2}$ axis to a mirror line in the perpendicular direction. These are the only allowable operators. Combinations of them give us the 7 groups shown in Figure 9-1. The group in Figure 9-1(g) is drawn in two different ways to illustrate different ways of deriving it: if to a point group with a centre of symmetry we add a glide line, we find we have generated a mirror line perpendicular to the periodic direction.

Similar methods could be used to derive all the groups with two-dimensional periodicity in three dimensions—the *diperiodic groups*. There are 80 of them. Alternatively, they could be picked out from the 230 space groups by taking account of the restrictions previously noted—that n-fold axes with $n > 2$ can only be perpendicular to the periodic plane, and that only those lattice centrings, glide planes, and screw axes are allowable which give translations *in* the periodic plane—and noticing in addition that, for some orthorhombic and monoclinic space groups where different choices of periodic plane are possible, more than one diperiodic group can be derived from the same space group. For example, of the space-group diagrams in Chapter 8, Figures 8-4, 8-5, 8-9, 8-13, 8-14, and 8-16 (showing $P2/a$, $C2/m$, $P222$, $P\bar{4}2_1m$, $P\bar{4}b2$, and $P3m1$ respectively) all could be taken to represent

---

* This does not mean, of course, that the groups operated on—the lattice complexes—may not have their own local symmetry. They may for example be squares, pentagons, or hexagons, but this local symmetry does not extend to the pattern as a whole, as would be necessary if it were to be part of the symmetry group.

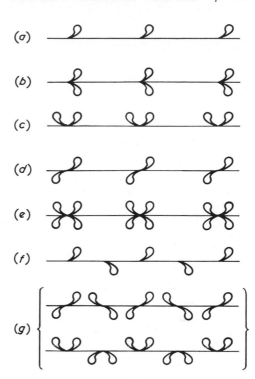

Figure 9-1 Space groups with spatial extension in two dimensions and one-dimensional periodicity: (a) periodic translation only; (b) mirror line parallel to translation; (c) mirror line perpendicular to translation; (d) centre; (e) mirror lines parallel and perpendicular to translation, and centre; (f) glide line; (g) glide line, centre, and mirror line perpendicular to translation (two examples).

diperiodic groups with the plane of the paper as their periodic plane, but Figures 8-6, 8-7, 8-12, and 8-17 ($Pbn2_1$, $Pbnm$, $P4_2/n$, and $P31c$) could not. Figure 8-3 (Pc) does not represent a diperiodic group as drawn, projected on (001); but if either (100) or (010) had been chosen as the periodic plane a diperiodic group would have been obtained, and the two choices give different groups (Figure 9-2).

Partly periodic groups can be of use in the study of particular kinds of materials which have only developed regularly in a two-dimensional sheet. These include the so-called *liquid crystals*, thin crystals grown as surface layers, and, in some twinned crystals, the parts of the crystal near the plane common to the two separate components.

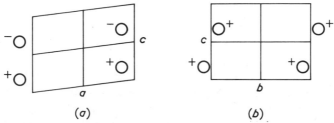

Figure 9-2 Space-group diagram of Pc, projected (a) on (010), (b) on (100). Both (a) and (b) could represent diperiodic groups, with the plane of the paper as the periodic plane; but they would be *different* groups.

9.3 BLACK AND WHITE SYMMETRY

Study of partly periodic symmetry is useful not only in its application to the kinds of material mentioned in the last paragraph but also for the help it gives us towards understanding a further development, the subject of *black-and-white symmetry*.

Suppose we are concerned with something more than the positions of atoms—suppose they have a property which may have two values, which we can represent by + and −. It might, for example, be the *sense* of a vector representing magnetic moment (the direction of the moment but not its sense being fixed by the position of the atom and its neighbours). Suppose that the + and − values do not occur at random, but in a way dependent on the spatial symmetry, though not such as to make all spatially equivalent atoms have the same sign. Then, if the atoms of interest to us are limited spatially to a two-dimensional sheet, we can take over the formal treatment of diperiodic groups, but now the + and − signs no longer refer to spatial positions above or below, but represent the physical property.

Instead of using + and −, however, it is often more convenient—since all we want is to indicate an "either/or" distinction—to use the colours black and white. Then we may say that the diperiodic groups in three dimensions, where the atoms are given no property but spatial positions, are completely equivalent to two-dimensional black-and-white groups, where black and white represent a two-valued property of the atom.

The relevance of these groups arises when, as in magnetism, the properties labelled + and −, or black and white, are related in the symmetry of their distribution to the symmetry of the atomic positions, without necessarily being identical. For example, two atoms which are translation-equivalent in position and environment may have antiparallel magnetic moments. A symmetry operator which, in a diperiodic group, changed the height from + to − or vice versa, in a black-and-white group becomes an *antisymmetry operator*. We define an *antisymmetry operator* as one which changes the spatial position of an atom in the same way as a symmetry operator but also reverses its "colour" (or rather, the property represented by colour).

For example, consider the singly-periodic groups of Figure 9-1. There we had three kinds of point operators—mirror lines parallel and perpendicular to the periodic direction x, and symmetry centres—which, combined separately with the lattice, gave the groups of Figure 9-1(b), (c), and (d). Now we have the corresponding antisymmetry operators, which, with the same lattice, give us those in Figure 9-3(a), (b), and (c). Figure 9-1(f) showed a glide line; Figure 9-3(d) shows the corresponding antisymmetry glide line.

We must not forget, however, that lattice translation is itself a symmetry operation, and we can have the corresponding anti-translation shown in Figure 9-3(e). Combinations of the operators are not all independent: for example, an anti-translation combined with an anti-centre generates a centre

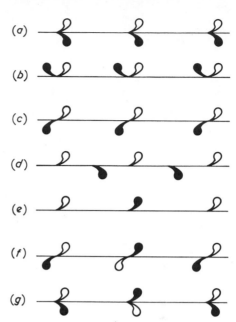

Figure 9-3 Examples of black-and-white space groups with spatial extension in two dimensions and one-dimensional periodicity: (a) anti-mirror line parallel to translation; (b) anti-mirror line perpendicular to translation; (c) anti-centre; (d) anti-glide line; (e) anti-translation; (f) anti-translation plus anti-centres, with centres generated midway between anti-centres; (g) anti-translation plus anti-mirror line parallel to translation, with glide line generated.

midway between the anti-centres, as shown in Figure 9-3(f); an anti-translation with an anti-mirror line along it generates a glide line in the same direction, as shown in Figure 9-3(g). Altogether there are 17 different black-and-white combinations.

Figure 3-1 is really an example of colour symmetry. Suppose in the first instance, ignoring the colour of the horses and considering only their outline, we pick out a primitive unit cell with one edge b lying horizontal on the page and one edge a sloping. But then we decide to take colour into account: we have to double the length of the unit cell in the horizontal direction. The anti-translation changes the colour at each operation; two operations are needed for a true repeat. The space-group symbol is written $p_b'1$, the subscript b showing the direction of the anti-translation. (An alternative choice of primitive unit cell would have been a square placed diagonally on the page, centred by a point of opposite colour; but this, though permissible, is inconvenient for the formal notation.)

In counting up the total numbers of colour groups, there is another feature to be borne in mind. A diperiodic group in three dimensions may have a mirror plane or diad axis lying in its periodic plane, so that, in projection, + and − atoms are superposed. In the two-dimensional black-and-white group they are replaced by a single atom, which must at the same time be black *and* white—this means it must·be neutral, with its property zero or its colour grey. There are, similarly, grey groups in our one-dimensional black-and-white example. They are produced by the anti-identity operation.

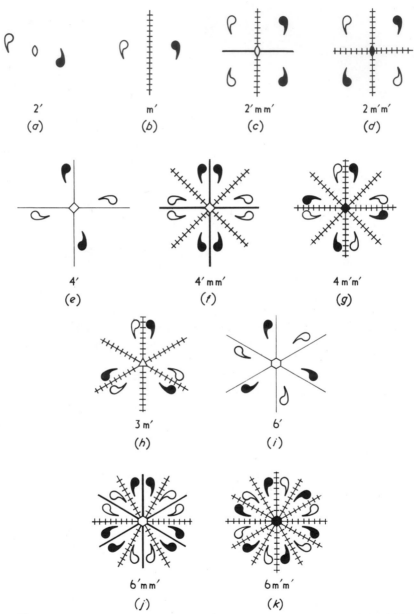

Figure 9-4 The two-dimensional two-colour point groups: (a) 2′; (b) m′; (c) 2′mm′; (d) 2m′m′; (e) 4′; (f) 4′mm′; (g) 4m′m′; (h) 3m′; (i) 6′; (j) 6′mm′; (k) 6m′m′.

Altogether, then, in our one-dimensional example, we have 7 one-colour groups, those shown in Figure 9-1 (it does not matter whether we call them black or white), 7 grey groups with exactly the same spatial configurations as the one-colour, and 17 black-and-white groups.

Next, we may consider fully-periodic two-dimensional black-and-white symmetry. Here we may have triad, tetrad, or hexad axes perpendicular to the plane. Anti-triad axes are impossible, because they are of odd order and would always give grey points, but anti-tetrad axes and anti-hexad axes are possible, as shown in Figure 9-4(e) and (i). Using these, we find that there are 5 new two-colour lattices in addition to the 5 one-colour discussed in §7.3; the triangular (or hexagonal) lattice does not give one, but the primitive rectangular lattice gives two, and the others give one each, by adding black points either at the centre of a white-to-white translation vector or at the centre of a rectangle (or square) formed by four white-to-white translation vectors.

There are 11 two-colour point groups in two dimensions, shown in Figure 9-4 with their point group symbols. In the diagrams, an anti-plane is shown by a cross-barred line, an anti-axis by a symbol of the same shape as that for the corresponding rotation axis, but open instead of solid. To obtain the complete number of point groups, we must add 10 single-colour groups and 10 grey groups, giving a total of 31. In physical terms, the single-colour groups are those which apply when the property is always of the same kind for every atom, and the grey groups are those in which it is always zero.

Combining the lattice with the point groups, we find 46 two-colour space groups, which with 17 one-colour and 17 grey groups makes a total of 80—the same as for the diperiodic groups in three dimensions. (The one-colour groups correspond to those of the diperiodic groups in which the heights of all atoms in the equipoint are the same.)

Two-dimensional black-and-white groups have been a help in certain methods of Fourier analysis ("generalised projections"), where, at a particular stage of the work, positive or negative densities are associated with symmetry-related atoms. They are also sometimes useful for discussing arrangements of magnetic moments—but here, of course, the limitation to two directions in space may be a difficulty. We really want to be able to use the fully three-dimensional black-and-white symmetry.

In recent years, much attention has been given to this subject, notably by Russian workers, and in particular by A. V. Shubnikov. The three-dimensional black-and-white space groups are commonly called the Shubnikov groups in recognition of his contribution. Full tables of them have been published with diagrams. The numbers of lattices, point groups, and space groups are given in Table 9-1.

It is interesting to pursue the train of thought suggested by the relationships of the two-dimensional black-and-white groups to the diperiodic groups in three dimensions. We should expect a similar relationship between the

Table 9-1 Comparison of Ordinary (Colourless) Symmetry and Black-and-White Symmetry For Fully Periodic Structures

KIND OF SYMMETRY GROUP	NUMBER OF DIMENSIONS	NUMBER OF GROUPS				
			BLACK-AND-WHITE			
		ORDINARY	1-COLOUR	GREY	2-COLOUR	TOTAL
Lattices	1	1	1		1	2
	2	5	5		5	10
	3	14	14		22	36
Point groups	1	2	2	2	1	5
	2	10	10	10	11	31
	3	32	32	32	58	122
Space groups	1	2	2	2	2	6
	2	17	17	17	46	80
	3	230	230	230	1191	1651

Shubnikov groups and triperiodic groups in four dimensions. This in fact exists. By endowing the atom with a property whose symmetry is not necessarily that of the atomic positions, we have in effect called in a new dimension.

In deriving all the possibilities in such a complicated set, more rigorous mathematical methods are of course necessary than the step-by-step building-up arguments we have used here. The use of the step-by-step approach is to allow us to understand and use the groups after they have all been listed and displayed in tables, and to equip us with ideas from which we may work out some of the simple cases that often occur in practice.

9.4 MANY-COLOUR SYMMETRY

There is, of course, no reason why we should limit ourselves to two colours. We could, for example, postulate a three-colour triad axis, with the spatial properties of an ordinary triad but demanding the colour sequence red, green, blue for successive repetitions of the object. Such ideas have been used two-dimensionally by the Dutch artist M. C. Escher for constructing beautiful designs, and some of them have been used as illustrations for teaching.* Three-dimensionally, the complexity of the possible arrangements increases further. Only a limited amount of exploration has been done, but it has not so far produced new working tools for the crystallographer.

* *Symmetry Aspects of M. C. Escher's Periodic Drawings*, by C. H. MacGillavry, produced for the International Union of Crystallography by a collaboration between the artist and the author, a crystallographer.

PROBLEMS AND EXERCISES

1. By drawing diagrams of symmetry elements and atomic positions like those in Chapter 8, derive diperiodic groups from the space groups associated with point group 2. Show that one of the three space groups gives rise to two different diperiodic groups, one rectangular and one oblique, and each of the other two gives rise to one rectangular group.

2. Repeat the procedure for point groups m and $2/m$ (with four and six space groups respectively). Show that in each there is one space group which does not give rise to a diperiodic group, and two that each give rise to one rectangular and one oblique group.

3. Which of the following orthorhombic space groups give rise to two different diperiodic groups, which to one only, and which to none? In each case, write the space group symbol in the form appropriate to the choice of the z axis as the axis without periodicity. (See Chapter 10 for rules for transformation of space-group symbols.)

    $P222$ (No. 16), $P2_12_12_1$ (No. 19), $C222$ (No. 21), $F222$ (No. 22), $Pmm2$ (No. 25), $Pmc2_1$ (No. 26), $Pnc2$ (No. 30), $Pna2_1$ (No. 33), $Cmm2$ (No. 35), $Ccc2$ (No. 37), $Pnnn$ (No. 48), $Pccm$ (No. 49), $Pmma$ (No. 51), $Cmcm$ (No. 63), $Cmma$ (No. 67), $Fmmm$ (No. 69)

4. Derive the five two-colour fully two-dimensional lattices, and show that there are no others.

5. Show, by attempting other combinations of symmetry operators, that there are no two-colour two-dimensional point groups other than those shown in Figure 9-4.

6. Draw diagrams like those of Chapter 8, but including the symmetry operators of Figure 9-4, to illustrate all the singly periodic space groups in two dimensions obtainable with two-colour symmetry operators. Indicate which groups are one-colour and which are grey.

7. Of the seven oblique diperiodic groups (including two derived from triclinic space groups as well as the five discussed in Questions 1 and 2), which correspond to one-colour groups and which to grey groups?

8. Which of the rectangular diperiodic groups discussed in Question 3 are one-colour and which are grey?

Chapter 10 GEOMETRICAL TRANSFORMATIONS

10.1 INTRODUCTION

We have seen in earlier chapters that, to describe a crystal structure, we must choose three nonparallel lattice vectors to define axes of reference and a unit cell, and also choose a point for an origin. Though there are conventions directing the choice (summarised in the appendix to this chapter), based on what is generally most convenient, they do not always give an unambiguous decision; even when they do, sometimes an unconventional choice may be much more convenient. It is therefore important to be able to transform from one set of axes to another at will, and to give a clear statement of what has been done.

Changes in axes of reference may lead to changes in any or all of the quantities dependent on them, notably:

> lattice parameters;
> atomic parameters;
> Miller indices, Bragg indices, zone indices;
> the point-group symbol;
> the type of lattice centring;
> the space-group symbol.

These will all be dealt with in this chapter.

It must be emphasised that the transformations considered in this chapter are changes in the *description* of the structure and not changes in the structure itself. Actual changes of structure will be left for consideration in Chapter 15.

10.2 CHANGES OF ORIGIN ONLY

Changes of origin, keeping the unit cell unchanged, affect only the atomic coordinates. The rule is simple: if, for the original choice of origin, the coordinates of any point are x_0, y_0, z_0, and the coordinates of the point to be taken as the new origin are u_0, v_0, w_0, then the new coordinates are

$$x_N = x_0 - u_0$$

$$y_N = y_0 - v_0 \qquad\qquad (10.1)$$

$$z_N = z_0 - w_0$$

Simple examples have already been given in §4.4 and 4.6, where the origin was originally taken on one of the atoms and then moved to a centre of symmetry, as follows:

	OLD COORDINATES	NEW ORIGIN WITH RESPECT TO OLD	NEW COORDINATES
Diamond	$0, 0, 0;\ \frac{1}{4}, \frac{1}{4}, \frac{1}{4}$	$\frac{1}{8}, \frac{1}{8}, \frac{1}{8}$	$\pm\ (\frac{1}{8}, \frac{1}{8}, \frac{1}{8})$
Magnesium	$0, 0, 0;\ \frac{2}{3}, \frac{1}{3}, \frac{1}{2}$	$\frac{1}{3}, \frac{2}{3}, \frac{3}{4}$	$\pm\ (\frac{2}{3}, \frac{1}{3}, \frac{1}{4})$

Again, in §8.6, our original choice of origin for space group $P2/a$ (No. 13) gave a centre of symmetry at $1/4, 0, 0$. The original coordinates of the equipoint were

$$x, y, z; \qquad \tfrac{1}{2} + x, \bar{y}, z; \qquad \bar{x}, y, \bar{z}; \qquad \tfrac{1}{2} - x, \bar{y}, \bar{z}$$

Referred to an origin at $1/4, 0, 0$, they become

$$-\tfrac{1}{4} + x, y, z; \qquad \tfrac{1}{4} + x, \bar{y}, z; \qquad -\tfrac{1}{4} - x, y, \bar{z}; \qquad \tfrac{1}{4} - x, \bar{y}, \bar{z}$$

or

$$\pm\ (-\tfrac{1}{4} + x, y, z; \ \tfrac{1}{4} + x, \bar{y}, z)$$

It is convenient to choose a new parameter $x' = -1/4 + x$, giving

$$\pm\ (x', y, z; \ \tfrac{1}{2} + x', \bar{y}, z)$$

Unless there is any need to continue to refer to the original parameters, the prime can subsequently be dropped.

In transferring the origin to a centre of symmetry, it is important to remember that all centres of symmetry are not alike, once atoms have been inserted. For example, in diamond, with the original choice of origin and axial directions, the centre of symmetry at $1/8, 1/8, 1/8$ lies at the middle of a short C—C bond, while that at $1/8, 5/8, 1/8$ is rather far away from all atoms. Geometrically, either point can be chosen for the new origin; physically, they are very different, and one should take care to choose that which will be most convenient.

10.3 CHANGES OF UNIT CELL

When a new unit cell is to be chosen, the easiest way to display its relationship to the old is to draw a projection of the old cell down one of its axes, with the new cell superposed on it. By inspection, we can then write down vector equations which will give us the transformation matrix. If the original unit-cell vectors are $\mathbf{a}_0$, $\mathbf{b}_0$, $\mathbf{c}_0$, and the new ones $\mathbf{a}_N$, $\mathbf{b}_N$, $\mathbf{c}_N$, we have

$$\mathbf{a}_N = u_1\mathbf{a}_0 + v_1\mathbf{b}_0 + w_1\mathbf{c}_0$$

$$\mathbf{b}_N = u_2\mathbf{a}_0 + v_2\mathbf{b}_0 + w_2\mathbf{c}_0 \qquad (10.2)$$

$$\mathbf{c}_N = u_3\mathbf{a}_0 + v_3\mathbf{b}_0 + w_3\mathbf{c}_0$$

where the u, v, w are all integers if the original unit cell was primitive, and simple fractions or integers if it was centred. The set of u, v, w constitutes the matrix for expressing *the new unit-cell vectors in terms of the original*.

(i) Suppose, for example, we have a monoclinic crystal with a primitive lattice, as shown in Figure 10-1. Suppose that two authors have chosen different unit cells as indicated by the subscripts P and Q, the $\mathbf{b}$ vector being the same for both and pointing up from the paper. To get right-handed axes with obtuse β, we must take the origins at O_P and O_Q as shown.

From the diagram, we can describe the Q cell in terms of the P cell as follows:

$$\mathbf{a}_Q = -\mathbf{a}_P - 3\mathbf{c}_P$$

$$\mathbf{b}_Q = \mathbf{b}_P \qquad (10.3)$$

$$\mathbf{c}_Q = \mathbf{a}_P + \mathbf{c}_P$$

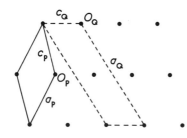

Figure 10-1 Section of a monoclinic (or triclinic) lattice, showing two choices of unit cell.

Then the transformation matrix to get Q in terms of P is formed from the coefficients, giving

$$\begin{pmatrix} -1 & 0 & -3 \\ 0 & 1 & 0 \\ 1 & 0 & 1 \end{pmatrix} \qquad (10.4)$$

To describe the P cell in terms of the Q cell, we could simply take the inverse of the matrix 10.4. However, in practice it is often easier and safer to go back to the diagram and construct the equations for $\mathbf{a}_P$, $\mathbf{b}_P$, $\mathbf{c}_P$ from it. Here we have

$$\mathbf{a}_P = \tfrac{1}{2}(\mathbf{a}_Q + 3\mathbf{c}_Q) = \tfrac{1}{2}\mathbf{a}_Q + \tfrac{3}{2}\mathbf{c}_Q$$

$$\mathbf{b}_P = \mathbf{b}_Q \qquad (10.5)$$

$$\mathbf{c}_P = -\tfrac{1}{2}(\mathbf{a}_Q + \mathbf{c}_Q) = -\tfrac{1}{2}\mathbf{a}_Q - \tfrac{1}{2}\mathbf{c}_Q$$

giving the matrix

$$\begin{pmatrix} \tfrac{1}{2} & 0 & \tfrac{3}{2} \\ 0 & 1 & 0 \\ -\tfrac{1}{2} & 0 & -\tfrac{1}{2} \end{pmatrix} \qquad (10.6)$$

which is the inverse of the matrix 10.4.

(ii) Another example, which occurs commonly in practice, is given in Figure 10-2. Here both unit cells have been drawn with the same origin; as

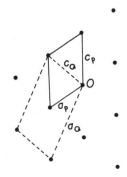

Figure 10-2 Second example of a monoclinic (or triclinic) lattice, with two choices of unit cell.

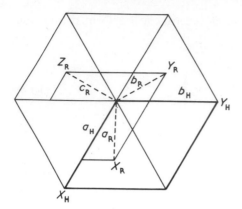

Figure 10-3 Relation of hexagonal and rhombohedral axes of reference, in the standard setting.

in the first example, **b** is the same for both. Applying the same procedure as in the first example, we find matrices as follows:

$$Q \text{ in terms of } P \qquad\qquad P \text{ in terms of } Q$$

$$\begin{pmatrix} 1 & 0 & -1 \\ 0 & 1 & 0 \\ 1 & 0 & 1 \end{pmatrix} \qquad \begin{pmatrix} \tfrac{1}{2} & 0 & \tfrac{1}{2} \\ 0 & 1 & 0 \\ -\tfrac{1}{2} & 0 & \tfrac{1}{2} \end{pmatrix} \qquad (10.7)$$

(iii) The same method applies when all three unit-cell vectors are to be changed, though it is a little harder to draw and to visualise. One of the most important examples is the transformation from rhombohedral to hexagonal axes and vice versa. Figure 10-3 shows the relative positions of the two sets of axes according to the usual convention in which the rhombohedron is in its *obverse* position (cf. §7.6). The rhombohedral cell edges, $\mathbf{a}_R$, $\mathbf{b}_R$, $\mathbf{c}_R$, are all inclined equally to the plane of the diagram, and are of course equal in magnitude, though different in direction. The hexagonal edges $\mathbf{a}_H$ and $\mathbf{b}_H$, in the plane of the diagram, are also equal in magnitude to each other; $\mathbf{c}_H$ is different, and is perpendicular to the plane. By inspection, we have

$$\mathbf{a}_R = \tfrac{2}{3}\mathbf{a}_H + \tfrac{1}{3}\mathbf{b}_H + \tfrac{1}{3}\mathbf{c}_H$$

$$\mathbf{b}_R = -\tfrac{1}{3}\mathbf{a}_H + \tfrac{1}{3}\mathbf{b}_H + \tfrac{1}{3}\mathbf{c}_H \qquad (10.8)$$

$$\mathbf{c}_R = -\tfrac{1}{3}\mathbf{a}_H - \tfrac{2}{3}\mathbf{b}_H + \tfrac{1}{3}\mathbf{c}_H$$

$$\mathbf{a}_H = \mathbf{a}_R - \mathbf{b}_R$$

$$\mathbf{b}_H = \mathbf{b}_R - \mathbf{c}_R \qquad (10.9)$$

$$\mathbf{c}_H = \mathbf{a}_R + \mathbf{b}_R + \mathbf{c}_R$$

and hence transformation matrices as follows:

$$R \text{ IN TERMS OF } H \qquad\qquad H \text{ IN TERMS OF } R$$

$$\begin{pmatrix} \frac{2}{3} & \frac{1}{3} & \frac{1}{3} \\ -\frac{1}{3} & \frac{1}{3} & \frac{1}{3} \\ -\frac{1}{3} & -\frac{2}{3} & \frac{1}{3} \end{pmatrix} \qquad\qquad \begin{pmatrix} 1 & -1 & 0 \\ 0 & 1 & -1 \\ 1 & 1 & 1 \end{pmatrix} \qquad (10.10)$$

(Other examples of transformations involving hexagonal or rhombohedral lattices will be given in §10.10.)

10.4 CHANGES OF COORDINATES RESULTING FROM CHANGES IN UNIT CELL

It can be shown that, if the matrix giving the edges of the new unit cell in terms of the old is A, and the new and old cells have the same origin, then *the matrix giving the new coordinates in terms of the old is the inverse of A with rows and columns interchanged*. We may call this matrix B. If the element of matrix A is s_{ij}, and that of its inverse is t_{ij}, then the element of matrix B is t_{ji}.

We may illustrate this by deriving matrix B from first principles for the example in Figure 10-2. Figure 10-4 is an enlargement of part of this, showing an arbitrary point A with coordinates x_P, z_P relative to the first set of axes, and x_Q, y_Q relative to the second. From the diagram, considering

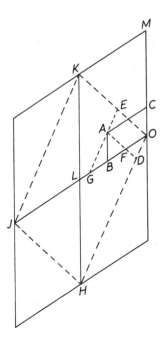

Figure 10-4 Enlargement of the same lattice as in Figure 10-2, with construction lines to show the transformation of the coordinates of an arbitrary point A. (The primitive unit cell is $OLKM$.)

similar triangles, we have

$$x_Q = \frac{AE}{a_Q} = \frac{OD}{OH} \equiv \frac{OF}{OJ} = \frac{1}{2}\left(\frac{OB}{OL} - \frac{FB}{OL}\right) = \frac{1}{2}\left(x_P - \frac{AB}{KL}\right) = \frac{1}{2}(x_P - z_P)$$

(10.11)

$$z_Q = \frac{AD}{c_Q} = \frac{OE}{OK} = \frac{OG}{OJ} = \frac{1}{2}\left(\frac{OB}{OL} + \frac{BG}{OL}\right) = \frac{1}{2}\left(x_P + \frac{AB}{KL}\right) = \frac{1}{2}(x_P + z_P)$$

Hence the matrix giving the atomic coordinates of Q in terms of P is

$$\begin{pmatrix} \frac{1}{2} & 0 & -\frac{1}{2} \\ 0 & 1 & 0 \\ \frac{1}{2} & 0 & \frac{1}{2} \end{pmatrix}$$

(10.12)

which is the same as the right-hand matrix of equation 10.7 with rows and columns interchanged.

Exactly the same argument can be used for the first example of §10.3. If the diagram is used as it stands, we obtain

$$x_Q = \tfrac{1}{2}(1 + x_P) - \tfrac{1}{2}z_P$$

$$z_Q = \tfrac{3}{2}(1 + x_P) - \tfrac{1}{2}z_P$$

(10.13)

The integers have come in because the origins of the two unit cells are separated by $\mathbf{a}_P$. Since the matrix is concerned with transformations of atomic coordinates relative to the origin of the unit cell, the integers may be ignored; the matrix for the transformation of atomic coordinates is then seen to be

$$\begin{pmatrix} \frac{1}{2} & 0 & -\frac{1}{2} \\ 0 & 1 & 0 \\ \frac{3}{2} & 0 & -\frac{1}{2} \end{pmatrix}$$

(10.14)

It is easiest in general to keep the origin constant during any change of lattice vectors, and make any necessary change of origin afterwards, by the method of §10.2.

10.5 CHANGES IN RECIPROCAL LATTICE VECTORS AND IN INDICES

If the new lattice vectors in direct space are given in terms of the old by matrix A, it can be shown that the new *reciprocal* lattice vectors are given in terms of the old by matrix B (the inverse of A with rows and columns interchanged).

This matrix B is the same as that used for transformation of *coordinates in direct space*. Because of the reciprocity of the two spaces, it follows that *matrix A applies to the transformation of coordinates in reciprocal space.*

We can now draw conclusions about the matrices needed for transformation of indices. Since the Bragg indices are coordinates in reciprocal space, they transform according to matrix A. This rule does not depend on the fact that they are integers; if, for any reason, we want to work with a set of axes that makes them nonintegral, or if we want to describe effects at points in reciprocal space where Bragg maxima do not occur, the rule remains applicable.

To obtain Miller indices, we use the same matrix A but divide by any common factor or multiply by a common denominator to make them all integers.

Zone indices represent coordinates in direct space, and therefore transform according to matrix B; again we divide the result by any common factor or multiply by a common denominator.

Thus all transformations to obtain quantities pertaining to the new lattice in terms of those pertaining to the old can be carried out using one of two matrices as follows:

MATRIX A	MATRIX B
	(Inverse of A with rows and columns interchanged)
Edges of unit cell (lattice parameters in direct space)	Bragg spacings of parametral planes (lattice parameters in reciprocal space; edges of reciprocal unit cell)
Coordinates in reciprocal space	Coordinates in direct space
Bragg indices, Miller indices	Zone indices

For the reverse transformation—to obtain the old in terms of the new—we use the corresponding inverse matrices. Thus, for example, the complete set of matrices needed for transformation between hexagonal and rhombohedral axes, in the setting of Figure 10-3, is as follows:

	H IN TERMS OF R		R IN TERMS OF H	
Matrix A: lattice parameters in direct space, etc	$\begin{pmatrix} 1 & -1 & 0 \\ 0 & 1 & -1 \\ 1 & 1 & 1 \end{pmatrix}$	(10.15)	$\begin{pmatrix} \frac{2}{3} & \frac{1}{3} & \frac{1}{3} \\ -\frac{1}{3} & \frac{1}{3} & \frac{1}{3} \\ -\frac{1}{3} & -\frac{2}{3} & \frac{1}{3} \end{pmatrix}$	(10.16)
Matrix B: coordinates in direct space, etc	$\begin{pmatrix} \frac{2}{3} & -\frac{1}{3} & -\frac{1}{3} \\ \frac{1}{3} & \frac{1}{3} & -\frac{2}{3} \\ \frac{1}{3} & \frac{1}{3} & \frac{1}{3} \end{pmatrix}$	(10.17)	$\begin{pmatrix} 1 & 0 & 1 \\ -1 & 1 & 1 \\ 0 & -1 & 1 \end{pmatrix}$	(10.18)

All these are of course derivable mathematically from any one of them. For working purposes, however, it is probably simplest to write down the two matrices for the unit-cell edges from inspection of a diagram showing their relation in direct space, and derive the other two from them by interchanging rows and columns.

10.6 CHANGES IN LATTICE SYMBOLS, POINT-GROUP SYMBOLS, AND SPACE-GROUP SYMBOLS

As we saw in earlier chapters, the symbols of the Hermann-Mauguin notation presuppose a conventional choice of axes of reference with respect to symmetry directions. The conventions are summarised in the appendix to this chapter. If we disregard them, we may find ourselves unable to make a formal statement of the symmetry in terms of our chosen axes. In that case, the only course open to us, within the Hermann-Mauguin system, is to use the conventional symbol enclosed in a curly bracket (brace) to dissociate it from any particular choice of axes.

Some conventions, however, are not essential to the use of the notation, and disregard of these is only important in so far as it may make the working clumsier or comparison with the *International Tables for X-ray Crystallography* harder. For example, in the triclinic system, with no symmetry directions, and in the orthorhombic system, with three directions all of twofold symmetry, there are no important rules to decide which of the unit cell edges should be called **a**, which **b**, and which **c**; and the same freedom applies to the choice of **a** and **c** in the monoclinic system. Many of the conventions preferring a primitive unit cell to a centred one, or one kind of centring (from among those to which symbols have been given) to another, are equally unimportant. In general, if any choice of axes of reference leads to an unambiguous Hermann-Mauguin symbol for the space group, it is a reasonable choice on which to base a description of the structure, at least so far as the symmetry is concerned.

We begin by considering the *lattice symbol* in some simple examples.

Figure 10.1 can be taken to represent the projection down [010] of either a monoclinic or a triclinic lattice, according to whether [010] is perpendicular or inclined to the plane containing **a** and **c**. Obviously P represents a primitive unit cell, Q a B-face-centred unit cell. The lattice symbol is P for the former choice of unit cell, B for the latter.

Similar considerations apply to the unit cells P and Q in Figure 10-2. Suppose, however, that the B-face-centred unit cell Q is a rectangular parallelepiped, and its edges $\mathbf{a}_Q$ and $\mathbf{c}_Q$ are symmetry directions, making the structure orthorhombic (Figure 10-5). It is still possible to work with the primitive unit cell P, and the transformation matrix derived from Figure 10-2 (matrix 10.7) is valid irrespective of the symmetry; but it is not possible to formulate a point-group symbol or space-group symbol appropriate to

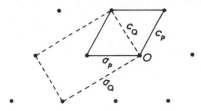

Figure 10-5 Relation between the primitive unit cell P and the conventional unit cell Q of a B-face-centred orthorhombic lattice.

the P axes of reference, because the symmetry directions do not coincide with them.

This example is important in practice. The base of the primitive unit cell P is a rhombus, with sides $\mathbf{a}_P$ and $\mathbf{c}_P$ equal. (It is in fact the geometrical relation of this rhombus to the rectangle that gave the name orthorhombic to the whole system.) Another related example is one in which the base of the primitive unit cell P is a rectangle, and that of the larger B-face-centred cell Q is a rhombus.

Another important case is illustrated in Figure 10-6. It represents a tetragonal lattice, with two choices of axes of reference. The conventional choice is P, although Q may be more convenient in some cases. If P is primitive, Q is C-face-centred; if P is body-centred (lattice symbol I), Q is all-face-centred (lattice symbol F).

We may note here that there are no 'permissible' alternatives comparable with this in the hexagonal (including trigonal) system, because we have no symbol for the kind of lattice centring which occurs if axes OX and OY are rotated by $30°$ from the direction of the shortest lattice vector, which is their conventional direction (Figure 10-7). Referred to the new axes, lattice points occur at $(2/3, 1/3, 0)$ and $(1/3, 2/3, 0)$ as well as at $(0, 0, 0)$. In an earlier system of notation, this kind of centring was called H, but its use caused confusion and it was dropped. If for any reason a choice of axes corresponding to the large unit cell is temporarily necessary, the results can

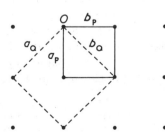

Figure 10-6 Relation between the primitive unit cell P and the C-face-centred unit cell Q of a tetragonal lattice.

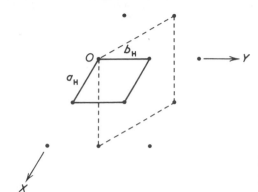

Figure 10-7 Projection of a hexagonal lattice, showing the relation between the primitive unit cell (solid lines) and a large unit cell of three times the volume (broken lines).

later be transformed back into terms of the conventional·axes for purposes of record.

Though inspection of a diagram is often the easiest way of finding the new lattice symbol, it is also possible to derive it from the matrix expressing *coordinates* of the new lattice in terms of the old. When the transformation is carried out, all sets of new coordinates of lattice points which involve fractions are listed, and the lattice symbol is deduced from them. For example, if there are lattice points with new coordinates (1/2, 1/2, 1/2) plus or minus integers, and nothing else nonintegral, the new lattice description is body-centred, with symbol I. Further examples will be found in §10.7.

There are not many *point groups* for which the symbol depends on the choice of axes of reference, but for those where it does the matter is important. They are those in which the directions referred to in the symbol correspond to different kinds of diad symmetry (or absence of symmetry). For example, if we say that the point group of a particular crystal is $mm2$, we mean that the diad axis lies along OZ; if we re-name this direction OX, the point group must be written as $2mm$. (If we do not know, or do not wish to mention, the relation of the diad axis to the axes of reference, we write it as $\{mm2\}$, $\{2mm\}$, or $\{m2m\}$ indifferently.) On the other hand, point group 222 retains the same name for any choice of axes of reference along symmetry directions. Again, in the tetragonal system, rotation of OX and OY through 45° requires an interchange of the second-position and third-position figures in the point-group symbol, since they refer to the OX axis and the direction at 45° to it; thus $\bar{4}2m$ must be re-named $\bar{4}m2$ and vice versa (cf. §6.10). A similar relation holds for a 30° rotation of OX and OY in the hexagonal (and trigonal) system, for example in the re-naming of $\bar{3}\ 2/m\ 1$ as $\bar{3}\ 1\ 2/m$.*

---

* Although this re-naming for changed axes is possible for *point* groups, we should not think it holds equally for *space* groups of the hexagonal system. The fact that there is no conventional symbol to describe the kind of centring it involves prevents it from doing so. The symbols $P\bar{3}\ 2/m\ 1$ and $P\bar{3}\ 1\ 2/m$ actually name different space groups ($\overline{No}$s. 164 and 162 respectively)—cf. §8.11.

The *space-group symbol* is additionally dependent on the choice of axes when there are glide planes present, or (in class 222 of the orthorhombic system) when there are rotation diads in some directions and screw diads in others. In transforming a space-group symbol, two questions have to be asked about any glide plane: how, if at all, has the name of the axis normal to the plane been changed, and how, if at all, has the name of the axis parallel to the glide direction been changed? The first question only arises in the orthorhombic and tetragonal systems (since in the monoclinic system the normal to the symmetry plane is always OY); the answer to it determines the position in the space-group symbol carrying information about that plane. To answer the second question, we need the matrix giving the original unit cell in terms of the new unit cell. The actual glide, given by the symbol as a fraction of one of the original unit-cell edges, is thereby re-expressed in terms of the new unit-cell edge, and it can be seen at once whether its form corresponds to a conventional glide symbol. Examples illustrating these points in detail will be given in §10.7 and 10.8.

In deriving the names of space groups, it must be remembered that they do not necessarily mention all the symmetry elements present, and that the conventions as to which should be mentioned and which omitted in the "standard form" (cf. §8.4) (though always such as to make possible the derivation of any that are omitted from those that are mentioned) can sometimes be rather arbitrary. For example, in §8.6 we saw that space group $C2/m$ possessed a-glide planes parallel to and midway between the mirror planes, and could legitimately have been written $C2/a$, but it would not have been found as such in the main table of space groups. This doubling of the sets of symmetry elements in a given direction is a characteristic of centred space groups (cf. §8.6). Unless one is aware of its occurrence, there may sometimes be difficulties in recognising that two different forms of the space-group symbol in fact refer to the same space group in the same orientation. We shall have examples later in the chapter.

As an alternative to working out the space-group symbols for new settings, one can look them up in the *International Tables for X-ray Crystallography*, Volume 1, Table. 6.2.1, where most of them are listed. The additional symmetry elements in centred space groups are included in the same table; or they can be worked out by the methods of Chapter 8 for each particular case.

10.7   EXAMPLES IN THE MONOCLINIC SYSTEM

In this system, the symmetry plane must be perpendicular to **b**, but the choice of **a** and **c** in the plane is free.

Since the roles played by **a** and **c** are the same, the following pairs of symbols obviously represent the same space groups: *Pa, Pc* (No. 7); *Cm, Am* (No. 8); *Cc, Aa* (No. 9). In each case, the transformation matrix giving the unit cell of either member of the pair in terms of the other is

$$\begin{pmatrix} 0 & 0 & 1 \\ 0 & -1 & 0 \\ 1 & 0 & 0 \end{pmatrix}$$ (10.19)

The reversal of sense of **b** is necessary to keep the axes right-handed, but makes no difference to the space-group symbol.

Figure 10-8(a) illustrates another possibility. The matrix giving the unit cell indicated by the broken line in terms of the unit cell indicated by the

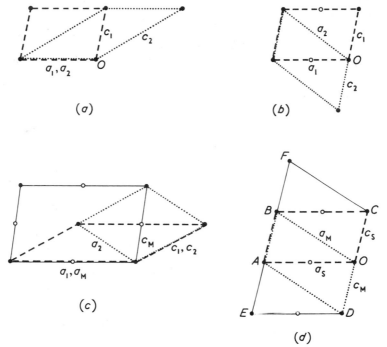

(a)

(b)

(c)

(d)

Figure 10-8 Projection on (010) of various monoclinic lattices, showing different choices of unit cell. Black circles and open circles are lattice points at heights 0 and 1/2 respectively. The origin is taken at O; the [010] axis points upward from the paper unless otherwise stated. (a) Primitive lattice. (This diagram serves equally well as an oblique projection down [010] of a triclinic lattice.) (b) Centred lattice. For the dotted unit cell (subscript 2), [010] points downward. (c) Lattice of gypsum. Solid lines show the unit cell corresponding to the morphological axes; broken lines (subscript 1) and dotted lines (subscript 2) show unit cells used by different authors. (d) Lattice of amphibole. Dotted lines (subscript M) show the unit cell corresponding to the morphological axes; broken lines (subscript S) show a unit cell used by some authors for describing the structure. For the dotted (M) unit cell, [010] points downward.

dotted line is

$$
\begin{pmatrix}
1 & 0 & 0 \\
0 & 1 & 0 \\
1 & 0 & 1
\end{pmatrix}
\tag{10.20}
$$

If the space-group symbol referred to the broken-line axes (subscript 1) is Pc, the same glide referred to the dotted-line axes (subscript 2) is

$$
\tfrac{1}{2}\mathbf{c}_1 = \tfrac{1}{2}\mathbf{a}_2 + \tfrac{1}{2}\mathbf{c}_2
\tag{10.21}
$$

and hence the glide symbol is n and the space-group symbol Pn.

Figure 10-8(b) shows a centred lattice. The matrix giving the unit cell indicated by the broken line in terms of the unit cell indicated by the dotted line is

$$
\begin{pmatrix}
1 & 0 & 1 \\
0 & -1 & 0 \\
0 & 0 & -1
\end{pmatrix}
\tag{10.22}
$$

the reversal of $\mathbf{c}$ being needed to keep the β-angle obtuse, and of $\mathbf{b}$ to keep the axes right-handed. If the space-group symbol referred to the broken-line axes is Cc, the glide with respect to the new axes will remain in the $\mathbf{c}$ direction, since the only change in this is in its sense, which is irrelevant here. On the other hand, it can be seen from the diagram that the new lattice symbol is I. This could have been deduced as follows from the matrix giving coordinates for the dotted-line axes in terms of the broken-line axes, which is matrix 10.22 with rows and columns interchanged, namely,

$$
\begin{pmatrix}
1 & 0 & 0 \\
0 & -1 & 0 \\
1 & 0 & -1
\end{pmatrix}
\tag{10.23}
$$

On the new (dotted-line) axes, the lattice centring point $(1/2, 1/2, 0)$ becomes

$$
\tfrac{1}{2} + 0 + 0, \qquad 0 - \tfrac{1}{2} + 0, \qquad \tfrac{1}{2} + 0 + 0
$$

i.e., $(1/2, -1/2, 1/2)$, equivalent by the lattice translation to $(1/2, 1/2, 1/2)$. Thus the new unit cell is body-centred. The new space-group symbol is accordingly Ic. This, however, is not its most usual form. It is easy to show, by the methods of Chapter 8, that the combination of body-centring with c-glide planes implies the presence of a-glide planes midway between the c-glide planes (cf. §10.6); in fact, either set of glide planes, in combination with the body-centring, implies the presence of the other. Conventionally it is usual to mention a-glide planes in preference to c-glide planes where both occur in the same direction; hence the new space-group symbol is more conventionally written Ia.

An example of the use of this transformation occurred in the structure determination of *afwillite*, $Ca_3(SiO_3OH)_2 \cdot 2H_2O$. The morphologically-chosen axes corresponded to the dotted-line axes of Figure 10-8(b), with space-group symbol Ia, and the axes chosen for the structure analysis to the broken-line axes, with space-group symbol Cc. The latter is the "standard setting," allowing easy comparison with the *International Tables for X-ray Crystallography*, but the choice was nevertheless unfortunate. Not only is the morphological unit cell a more convenient shape, with a β-angle of 98° as compared with 135° for the new one, but it also shows up the physical features of the crystal, such as its cleavage, in a more natural way.

An example where long-established morphological axes proved unsatisfactory for the description of the structure occurred for *gypsum*, $Ca_2SO_4 \cdot 2H_2O$. The shortest vectors along these axes defined the unit cell shown by the solid lines in Figure 10-8(c). Two different structural choices were made by different workers and are shown in Figure 10-8(c) by broken lines, denoted (1), and dotted lines, denoted (2). From inspection of the diagram, we have the following matrices for the unit cells:

M IN TERMS OF (1)	M IN TERMS OF (2)	(2) IN TERMS OF (1)
$\begin{pmatrix} 1 & 0 & 0 \\ 0 & 1 & 0 \\ 1 & 0 & 2 \end{pmatrix}$ (10.24)	$\begin{pmatrix} 1 & 0 & -1 \\ 0 & 1 & 0 \\ 1 & 0 & 1 \end{pmatrix}$ (10.25)	$\begin{pmatrix} 1 & 0 & 1 \\ 0 & 1 & 0 \\ 0 & 0 & 1 \end{pmatrix}$ (10.26)

Referred to unit cell (1), the space-group symbol is $C2/c$ (No. 15); this corresponds to the "standard setting" of the *International Tables for X-ray Crystallography*. Its disadvantage is its very inconvenient shape, the β-angle being about 150°. The dotted unit cell (2) has the same volume and a better shape. The matrix for (2) in terms of (1), matrix 10.26, is the same as matrix 10.22, except for the senses of OY and OZ, which, while important in dealing with atomic coordinates, are irrelevant to discussion of the space group. Hence, by the arguments used earlier in considering the transformation of space group Cc by the matrix given in 10.22, the space-group symbol referred to axes (2) is $I2/a$. This choice of unit cell is therefore convenient with respect to the space-group symbol as well as the unit-cell shape.

The morphological unit cell is not automatically condemned because it has double the volume of the others. We must, however, find whether it can be given a satisfactory space-group symbol. The lattice symbol can be found either by inspection of Figure 10-8(c), or, more formally, from the matrix giving coordinates for M in terms of (1), which is

$$\begin{pmatrix} 1 & 0 & -\frac{1}{2} \\ 0 & 1 & 0 \\ 0 & 0 & \frac{1}{2} \end{pmatrix} \tag{10.27}$$

Because of the occurrence of halves in this, we have to work out the new coordinates of lattice points with old coordinates $(0, 0, 1)$ and $(1/2, 1/2, 1)$ as well as $(1/2, 1/2, 0)$. They come out as $(-1/2, 0, 1/2)$, $(0, 1/2, 1/2)$, and $(1/2, 1/2, 0)$ respectively. Thus the lattice referred to the morphological axes is all-face-centred.

Next, we consider how a glide of $(1/2)\mathbf{c}_1$ can be expressed in terms of the morphological axes. We use the matrix giving unit-cell edges of (1) in terms of M, which is matrix 10.27 with rows and columns interchanged, namely,

$$\begin{pmatrix} 1 & 0 & 0 \\ 0 & 1 & 0 \\ -\frac{1}{2} & 0 & \frac{1}{2} \end{pmatrix} \tag{10.28}$$

From this,

$$\tfrac{1}{2}\mathbf{c}_1 = \tfrac{1}{2}(-\tfrac{1}{2}\mathbf{a}_M + \tfrac{1}{2}\mathbf{c}_M) = \tfrac{1}{4}(-\mathbf{a}_M + \mathbf{c}_M) \tag{10.29}$$

This has the form of a d-glide with respect to the new axes, but, unlike the d-glide plane in systems with orthogonal axes, the direction of the glide is not uniquely specified by the symbol. There is in fact no satisfactory Hermann-Mauguin symbol for this space group referred to the morphological axes, and therefore these axes are not satisfactory for describing the structure.

Another example of different choices of axes, which has caused some confusion in the literature, concerns the *amphiboles* (silicates with double chains of Si tetrahedra, linked laterally by other cations). Using the original morphological axes, the projection of the unit cell is shown in Figure 10-8(d) as $OBAD$ (the axis OY pointing downward into the paper); this unit cell is body-centred. An alternative choice is $OABC$ (with OY pointing upward); this is C-face-centred. Unfortunately, in the early structural descriptions, the necessity for retaining an obtuse angle β was not appreciated, and it was not realised that the morphological axes themselves imply an obtuse interaxial angle (the symbol β being used, in the old mineralogical convention, for the interfacial angle between the normals to (100) and (001) rather than for the interaxial angle—i.e., for $180° - XOZ$ rather than for XOZ itself).

Hence the unit cell $OBFC$ was sometimes used instead of $OBAD$, and $OAED$ instead of $OABC$. The fact that some of the lattice vectors and angles of different cells are accidentally similar added to the confusion.

Referred to cell S, indicated by the broken line, the space-group symbol is $C2/m$; referred to cell M, indicated by the dotted line, it is $I2/m$. (In either case the space group is No. 12.) The matrix giving the unit cell of

S in terms of M is

$$\begin{pmatrix} 1 & 0 & 1 \\ 0 & -1 & 0 \\ 0 & 0 & -1 \end{pmatrix} \qquad (10.30)$$

In this case, the reversal of the sense of **b** is not important even when atomic coordinates have to be dealt with, because of the existence of the mirror plane of symmetry; on the other hand, to ignore the reversal of **c** gives a wrong sign to all z coordinates.

The S unit cell has a slightly smaller β-angle than does the M unit cell, but the difference is too small to affect their relative convenience in use. On the other hand, using the M cell allows easier comparison with morphological features, and also with an important group of related silicates, the *pyroxenes*.

10.8 EXAMPLES IN THE ORTHORHOMBIC SYSTEM

In this system, while the set of three axial directions as a whole is fixed, there is no overriding reason to decide which shall be given which name.

Because the set of directions does not change, there can be no change in the *kind* of lattice centring or the *kind* of glide in a glide plane: I and F centring remains so for any choice of axes, and so do n-glide and d-glide planes, where the glide has equal components in both axial directions parallel to the plane. The one-face-centred lattices A, B, and C, and the glides parallel to one axis, a, b, or c, will however change their symbol. Moreover, since the three positions following the lattice symbol refer to the OX, OY, and OZ directions respectively, interchange of these directions will interchange the positions of the symbols referring to the different kinds of symmetry elements.

If we had to take into account the two different senses of each of three orthogonal directions, there would be 24 ways of choosing a set of right-handed axes. As we have seen, though the sense matters for dealing with coordinates (except when there is a mirror plane), it does not matter for dealing with space groups, and the possibilities are therefore reduced to 6, given by the following matrices, of which the first represents the original choice:

$$\begin{pmatrix} 1 & 0 & 0 \\ 0 & 1 & 0 \\ 0 & 0 & 1 \end{pmatrix} \text{(a)} \quad \begin{pmatrix} 0 & 1 & 0 \\ 0 & 0 & 1 \\ 1 & 0 & 0 \end{pmatrix} \text{(b)} \quad \begin{pmatrix} 0 & 0 & 1 \\ 1 & 0 & 0 \\ 0 & 1 & 0 \end{pmatrix} \text{(c)}$$

$$\begin{pmatrix} 1 & 0 & 0 \\ 0 & 0 & 1 \\ 0 & -1 & 0 \end{pmatrix} \text{(d)} \quad \begin{pmatrix} 0 & 1 & 0 \\ 1 & 0 & 0 \\ 0 & 0 & -1 \end{pmatrix} \text{(e)} \quad \begin{pmatrix} 0 & 0 & 1 \\ 0 & 1 & 0 \\ -1 & 0 & 0 \end{pmatrix} \text{(f)}$$

$$(10.31)$$

These six matrices allow the symbols for all 'permissible' alternative settings in the orthorhombic system to be written down.

Suppose our space group is *Pbnm* (No. 62), and we wish to change to a new setting given in terms of the old by the matrix 10.31(b). The old unit cell can be seen to be given in terms of the new by

$$\begin{pmatrix} 0 & 0 & 1 \\ 1 & 0 & 0 \\ 0 & 1 & 0 \end{pmatrix} \tag{10.32}$$

Hence the glide $(1/2)\mathbf{b}$ on the old axes becomes $(1/2)\mathbf{a}$ on the new. The old axial directions OX, OY, OZ, with which are associated symmetry planes b, n, m, respectively, become OZ, OX, OY, with which are associated symmetry planes a, n, m, respectively. Rearranging the symbols in the conventional order, and retaining the lattice symbol P unchanged, we obtain the new space-group symbol, *Pnma*.

These two different sets of axes have been used by different authors for *olivine* (§11.9) and compounds isomorphous with it. The space-group symbol *Pbnm* refers to the morphological choice, which is preferred, because it is well established in the older literature, even though *Pnma* is the "standard setting."

By the same kind of argument, it can be shown that the choices of axes represented by the four remaining matrices given in 10.31(c) to (f) give the symbols *Pmcn*, *Pcmn*, *Pnam*, *Pmnb*, for the same space group.

Again, suppose our space group is *Abm2* (No. 39), and our new axes are again related to the old by the matrix 10.31(b). To transform the lattice symbol, we note that A refers to the face of the unit cell defined by $\mathbf{b}$ and $\mathbf{c}$; on the new axes these have become $\mathbf{a}$ and $\mathbf{b}$ respectively, and the lattice symbol is therefore C. The rest of the symbol is found in the same way as for the last example; the complete symbol is *Cm2a*. The other four matrices of the set give *B2cm*, *Ac2m*, *Bma2*, and *C2mb*, respectively.

As a third example, suppose, having chosen convenient axes of reference for a new compound of orthorhombic symmetry, we find lattice centring and glide planes from which we deduce the space-group symbol *Ib2a*, and we want to know the "standard setting" for comparison with the *International Tables for X-ray Crystallography*. We could use all the matrices given in 10.31 in turn to construct new symbols, and inspect the results. We can, however, save labour by noticing that standard settings for this point group choose OZ as the direction of the diad axis; hence OY in our original setting corresponds to OZ in the standard setting, and the two possible matrices giving our unit cell in terms of the standard are

$$\begin{pmatrix} 1 & 0 & 0 \\ 0 & 0 & 1 \\ 0 & 1 & 0 \end{pmatrix} \text{(a)} \qquad \begin{pmatrix} 0 & 1 & 0 \\ 0 & 0 & 1 \\ 1 & 0 & 0 \end{pmatrix} \text{(b)} \tag{10.33}$$

Applying these, we get the following results for the name of each symmetry direction and the kind of symmetry associated with it:

Original: OX, b-glide; OY, diad axis; OZ, a-glide; $Ib2a$
Matrix (a): OX, c-glide; OZ, diad axis; OY, a-glide; $Ica2$
Matrix (b): OY, c-glide; OZ, diad axis; OX, b-glide; $Ibc2$

Neither of our new forms appears among the standard settings. However, we remember that, for a body-centred lattice, the presence of b-glide planes parallel to (100) implies also c-glide planes parallel to them and midway between them—in fact, the original symbol might have been written $I\,b/c\,2\,a/b$. Applying the two matrices to this, we obtain $I\,c/b\,a/c\,2$ and $I\,b/c\,a/c\,2$, which may both be written $Iba2$ (following the convention of using for preference the letter appearing earlier in the alphabet). This is now recognisable, on consulting the *International Tables for X-ray Crystallography*, as the standard setting of the space group, No. 45.

10.9 EXAMPLES IN THE TETRAGONAL SYSTEM

The most generally useful transformation in the tetragonal system is that corresponding to Figure 10-6, relating the smallest unit cell to that of a unit cell of twice the volume, with axes **a** and **b** at 45° to those of the first. We may notice two reasons why it is so often useful.

In the first place, from the external appearance of a tetragonal crystal it is not possible to predict with certainty (though one may sometimes guess) which direction normal to the tetrad axis will be that of the shortest vector. It is therefore not unusual for a tetragonal crystal, referred to axes chosen morphologically, to have a C-face-centred or all-face-centred unit cell. If these axes are well established in the literature, or obviously related to the external form of the crystal, it may be preferable to keep them for the description of the structure, rather than referring it to the conventional smaller unit cell; but in any case transformation between the two sets of axes is likely to be needed.

Secondly, it sometimes happens that the larger unit cell may show up important relationships. Suppose, for example, we distorted a face-centred cubic structure by elongating or compressing it in the direction of a cell edge, thereby making it tetragonal. If we keep the same axes of reference, the unit cell remains all-face-centred. But the shortest lattice vectors in the plane at right angles to the tetrad do not lie along these axes, but at 45° to them. Conventionally we should choose our axes along these shortest vectors, and so get a body-centred unit cell. This, however, obscures the resemblance to the original cube. It may therefore be better to keep the larger all-face-centred cell. There are many examples in later chapters, and we therefore give only one here.

Hausmannite, Mn_3O_4 (§12.13), with a conventional choice of axes, has space group $I\,4_1/amd$ (No. 141), which may alternatively be written $I\bar{4}/amd$.

Taking **a** and **b** axes at 45° to the original, we obtain an all-face-centred unit cell which is almost a cube, and all the atoms are close to positions of cubic symmetry. To find the new space-group symbol, we begin by noting that, on the old axes, the mirror plane is perpendicular to OX and the d-glide plane at 45° to it. Relative to the new axes, these are interchanged, but the character of the glide is not altered; the space-group symbol therefore ends in dm. The $\bar{4}$ axis is unchanged, but the original a-glide perpendicular to it now has equal components along the new **a** and **b** axes, and each represents only one quarter of the repeat distance in these directions; hence the a-glide plane has become a d-glide plane. Putting all these together, we obtain the new space-group symbol $F\bar{4}/ddm$. The structure is closely related to the cubic *spinel* structure (§11.3) with space group $Fd3m$ (No. 227).

10.10 EXAMPLES IN THE HEXAGONAL AND CUBIC SYSTEMS

For hexagonal (including trigonal) crystals as for tetragonal crystals, it is not possible to predict the direction of the shortest vector perpendicular to the axis from the external symmetry. If the axes of reference are chosen before the structure is known, we may find that OX and OY lie at 30° to the direction of the shortest lattice vector in their plane, giving the large unit cell shown in Figure 10-7. This may happen whether or not the small cell is rhombohedrally centred. There is no space-group symbol corresponding to the large unit cell. Unlike what happens in the tetragonal system, it is therefore necessary to transform to the small unit cell and adopt its axes for the description of the structure for permanent record.

EXAMPLE (i)

This does not mean that we are not allowed, for particular purposes when it happens to be convenient, to work in terms of the large unit cell. The matrices for transforming from one to the other are valid whether or not they lead to a permissible space-group symbol. From Figure 10-7 we may construct those relating cell edges, and then deduce those relating the coordinates, as follows:

	L IN TERMS OF S	S IN TERMS OF L
Cell edges	$\begin{pmatrix} 2 & 1 & 0 \\ -1 & 1 & 0 \\ 0 & 0 & 1 \end{pmatrix}$ (a)	$\begin{pmatrix} \frac{1}{3} & -\frac{1}{3} & 0 \\ \frac{1}{3} & \frac{2}{3} & 0 \\ 0 & 0 & 1 \end{pmatrix}$ (b)
Coordinates	$\begin{pmatrix} \frac{1}{3} & \frac{1}{3} & 0 \\ -\frac{1}{3} & \frac{2}{3} & 0 \\ 0 & 0 & 1 \end{pmatrix}$ (c)	$\begin{pmatrix} 2 & -1 & 0 \\ 1 & 1 & 0 \\ 0 & 0 & 1 \end{pmatrix}$ (d)

$$(10.34)$$

If we wish to use the large unit cell, we can find the atomic coordinates relative to the small cell from the space group, and transform them into terms of the large cell. Suppose, for example, that the only symmetry is a triad axis, giving a set of three points, at

$$x, y, 0; \quad -y, x - y, 0; \quad -x + y, -x, 0$$

relative to the small cell. Relative to the large cell, using matrix 10.34(c), we obtain

$$\tfrac{1}{3}x + \tfrac{1}{3}y, -\tfrac{1}{3}x + \tfrac{2}{3}y, 0; \quad \tfrac{1}{3}(-y) + \tfrac{1}{3}(x - y), (-\tfrac{1}{3})(-y) + \tfrac{2}{3}(x - y), 0;$$

$$\tfrac{1}{3}(-x + y) + \tfrac{1}{3}(-x), (-\tfrac{1}{3})(-x + y) + \tfrac{2}{3}(-x), 0$$

i.e.,

$$\tfrac{1}{3}(x + y), \tfrac{1}{3}(-x + 2y), 0; \quad \tfrac{1}{3}(x - 2y), \tfrac{1}{3}(2x - y), 0;$$

$$\tfrac{1}{3}(-2x + y), \quad \tfrac{1}{3}(-x - y), 0 \tag{10.35}$$

There are, however, three times as many atoms in the large cell as in the small one; to get the complete set, we must apply to those we have already found the lattice operator (0, 0, 0; 2/3, 1/3, 0; 1/3, 2/3, 0) +.

EXAMPLE (ii)

The primitive unit cell of a face-centred cubic lattice is a rhombohedron. Figure 10-9(a) shows the projection of the cubic unit cell down its triad

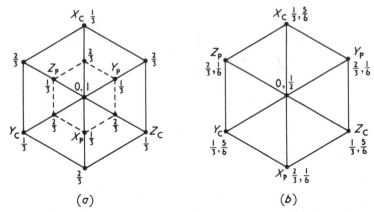

(a)   (b)

Figure 10-9 (a) Projection of a face-centred cube down a triad axis. Heights marked are fractions of the length of the body diagonal, measured upwards from the lowest point of the cube. Solid lines are edges of the cube; broken lines outline the projection of the primitive rhombohedron. Points X_C, Y_C, and Z_C, at height 1/3, are corners of the cube; X_P, Y_P, and Z_P, also at height 1/3, are corners of the primitive rhombohedron. (b) Projection of part of a body-centred cubic lattice down a triad axis. Heights are marked in same way as in (a) and the cube is shown similarly. The primitive rhombohedron has the same outline as the cube, but its lower corners X_P, Y_P, and Z_P are at height 1/6, and its upper corners at 1/3 and 1/2.

axis (solid lines) and the projected outline of the primitive unit cell (broken lines). The matrices relating the unit cells are as follows:

$P(F)$ IN TERMS OF C $\qquad\qquad\qquad$ C IN TERMS OF $P(F)$

$$\begin{pmatrix} 0 & \frac{1}{2} & \frac{1}{2} \\ \frac{1}{2} & 0 & \frac{1}{2} \\ \frac{1}{2} & \frac{1}{2} & 0 \end{pmatrix} \qquad\qquad \begin{pmatrix} -1 & 1 & 1 \\ 1 & -1 & 1 \\ 1 & 1 & -1 \end{pmatrix} \quad (10.36)$$

On hexagonal axes of reference, we can see at a glance that

$$\mathbf{c}_{P(F)} = \mathbf{c}_C, \qquad \mathbf{a}_{P(F)} = -\tfrac{1}{2}\mathbf{a}_C \qquad (10.37)$$

EXAMPLE (iii)
The primitive unit cell of a body-centred cubic lattice is also a rhombohedron. In Figure 10-9(b) its projection coincides with that of the cube, but its corners are the lattice points at heights 0, 1/6, 1/3, and 1/2 (whereas those of the cube are at 0, 1/3, 2/3, and 1). The matrices relating the unit cells are as follows:

$P(I)$ IN TERMS OF C $\qquad\qquad\qquad$ C IN TERMS OF $P(I)$

$$\begin{pmatrix} -\frac{1}{2} & \frac{1}{2} & \frac{1}{2} \\ \frac{1}{2} & -\frac{1}{2} & \frac{1}{2} \\ \frac{1}{2} & \frac{1}{2} & -\frac{1}{2} \end{pmatrix} \qquad\qquad \begin{pmatrix} 0 & 1 & 1 \\ 1 & 0 & 1 \\ 1 & 1 & 0 \end{pmatrix} \quad (10.38)$$

On hexagonal axes of reference,

$$\mathbf{c}_{P(I)} = \tfrac{1}{2}\mathbf{c}_C, \qquad \mathbf{a}_{P(I)} = -\mathbf{a}_C \qquad (10.39)$$

EXAMPLE (iv)
Since a cube is a special case of a rhombohedron, with $\alpha = 90°$, and since the form of the matrices does not depend on the magnitude of the angle, the relations of matrices 10.36 to 10.39 are valid for any face-centred or body-centred rhombohedron and the primitive rhombohedron derived from it.

EXAMPLE (v)
To relate hexagonal and rhombohedral axes of reference for a rhombohedral lattice, matrices have already been given in 10.10 and 10.15 to 10.18. We may add here a selection of formulae usable for numerical calculations, all of which can be derived by simple geometry from Figure 10-3.

$$a_R = \tfrac{1}{3}a_H \left[3 + \left(\frac{c_H}{a_H} \right)^2 \right]^{\frac{1}{2}} \tag{10.40}$$

$$\alpha = 2 \sin^{-1} \frac{3}{2 \left[3 + \left(\frac{c_H}{a_H} \right)^2 \right]^{\frac{1}{2}}} \tag{10.41}$$

$$a_H = 2a_R \sin \alpha/2 \tag{10.42}$$

$$\frac{c_H}{a_H} = \frac{3 \left[1 - \frac{4}{3} \sin^2 \frac{\alpha}{2} \right]^{\frac{1}{2}}}{2 \sin \frac{\alpha}{2}} \tag{10.43}$$

For a face-centred cubic lattice, the angle α of the primitive rhombohedron is $60°$; for a body-centred cubic lattice it is $109°28'$.

APPENDIX: CONVENTIONS GOVERNING CHOICE OF AXES OF REFERENCE

Axes of reference are a right-handed set OX, OY, OZ, lying along edges **a**, **b**, **c** of the unit cell.

The unit cell is a parallelepiped, primitive or centred, chosen in accordance with the symmetry as outlined in the following paragraphs.

[Note: there are some ambiguities not covered by convention, or covered only by subordinate and sometimes conflicting conventions; also there are alternative choices which, while unconventional, are still permissible in the sense that they still allow full use of the Hermann-Mauguin notation. Comments on these are enclosed in square brackets.]

Cubic Unit Cell. The smallest *cube* whose sides are lattice vectors. The lattice may be P, F, or I.

Tetragonal Unit Cell. A tetragonal prism defined by the shortest lattice vector parallel to the *tetrad* axis (called **c**) and the two shortest (equal) lattice vectors perpendicular to it (**a** and **b**). The lattice may be P or I.

[A permissible alternative is the tetragonal prism of the same height but twice the volume, whose base is a centred square with sides **a** and **b** at $45°$ to those of the first choice. Its lattice symbol is C or F, according to whether the first choice is P or I.]

Hexagonal Unit Cell. A $120°$ prism defined by the shortest lattice vector parallel to the *hexad* or *triad* axis (called **c**) and two shortest (equal) lattice vectors perpendicular to it. The lattice may be P or R. If it is R, directions of OX and OY must be chosen so that the lattice operator is $(0, 0, 0; \ 2/3, 1/3, 1/3; \ 1/3, 2/3, 2/3)+$, giving an *obverse* rhombohedron.

[If the symmetry is trigonal and not hexagonal, and the lattice is P, there are two equally good but different choices for the axes perpendicular to the triad, one rotated by 180° about it with respect to the other.

If the lattice is R, the symmetry must be trigonal and not hexagonal, but the two choices are not equally good according to convention; that giving a *reverse* rhombohedron, whose lattice operator is $(0, 0, 0; 1/3, 2/3, 1/3; 2/3, 1/3, 2/3)+$, is only used if some special reason requires it. Unless stated otherwise, the obverse setting is always assumed (cf. §7.6).]

Rhombohedral Unit Cell. A rhombohedron defined by the three shortest (equal) lattice vectors neither parallel nor perpendicular to the *triad* axis.

Orthorhombic Unit Cell. A rectangular parallelepiped whose sides are the shortest lattice vectors parallel to the *three diad* symmetry directions. The lattice may be P; $A, B,$ or C; I; or F.

[Three criteria, which are often conflicting, are used in allocating letters $a, b,$ and c to the three directions:

(i) to show up the relationship of the structure to other structures previously studied,

(ii) to give the space-group symbol its "standard form,"

(iii) to keep axial lengths in the sequence $c < a < b$.

Of these criteria, (i) is by far the most important. Transformation to a form satisfying (iii) may be necessary for determinative purposes, to look up a structure in tables of data; but it can be done temporarily when required. Much confusion has been introduced into descriptions of structures by blind application of rules (ii) and (iii), especially when it has meant changing from a previously accepted choice of axes.]

Monoclinic Unit Cell. A parallelepiped defined by the shortest vector parallel to the *diad* symmetry direction (called **b**) and two short lattice vectors perpendicular to it (**a** and **c**).

Whatever way the vectors **a** and **c** are chosen, their *senses* should always be such as to enclose an obtuse angle $XOY = \beta$. Any breach of this rule could lead to serious confusion.

[There are several criteria, often conflicting, to decide the choice of **a** and **c** vectors:

(i) to make the axes coincide with morphologically important directions,

(ii) to keep the interaxial angle β as near 90° as possible,

(iii) to show the relationship to other structures previously studied,

(iv) to give the space-group symbol, and in particular the lattice symbol, its "standard form,"

(v) to keep axial lengths in the sequence $c < a$.

Of these criteria, (v) (like the corresponding criterion in the orthorhombic system) is useful for determinative purposes, but should not be allowed to override other criteria because transformations can be made temporarily when needed. Rule (iv) should not be rigorously applied; there are no important general reasons for preferring a to c in the symbol, or vice versa, and choices giving B centring instead of P, or I or even F instead of A or C, or an n-glide plane instead of a c-glide, can be considered on their merits. It is important, however, that the choice should lead to some space-group symbol compatible with the notation. Rule (ii) is only a matter of working convenience, but often suggests the same choice as (i) and (iii).]

Triclinic Unit Cell. Any primitive parallelepiped.

[Different criteria for choice are as follows:

(i) to make the axes coincide with morphologically important directions,

(ii) to keep all interaxial angles near 90°,

(iii) to keep all interaxial angles greater than 90°,

(iv) to show the relationship to other structures previously studied,

(v) to keep axial lengths in the sequence $c < a < b$.

As before, (v) is useful for determinative purposes, but it is not preponderant. Rule (iii) is not always compatible with (i) and (ii), and in such cases it may be necessary to override it.

The condition that the unit cell must be primitive may be waived: choices making its centring A, B, C, I, or even F are acceptable, if they satisfy (i), (ii), or (iv) better than a primitive unit cell would.]

PROBLEMS AND EXERCISES

1. Strontium titanate, which is cubic with one formula unit per unit cell, is sometimes described with the origin taken at the Ti atom; the oxygen positions are then $1/2, 0, 0$; $0, 1/2, 0$; $0, 0, 1/2$; and the Sr position is $1/2, 1/2, 1/2$. Rewrite the coordinates of all atoms with the origin taken at the Sr atom.

2. In space group $P2_1/n$, if the origin is taken at a centre of symmetry, show that the atomic coordinates of a set of atoms in general positions will remain unchanged if the sense of the y axis is reversed and the origin is moved by a translation $(\mathbf{a} + \mathbf{b} + \mathbf{c})/2$.

3. In the *International Tables for X-ray Crystallography* the origin in the space group $P3_121$ is chosen so that the diad axis along the x-axis intersects the screw triad at height $z = 1/3$. For a particular purpose it is more convenient to take the origin at this intersection of axes. Write down the complete set of coordinates of atoms in a general position x, y, z, using the first choice; transform it to that for the second choice; and show that it is identical with that for the set derived from an atom at x', y', z' by direct use of the diad and screw triad axes. State the connection between x, y, z and x', y', z'.

4. (a) In his original work on aragonite (space group *Pmcn*), W. L. Bragg found atomic positions as follows:

Ca: $0, \frac{7}{12}, 0$; C: $0, \frac{1}{4}, \frac{1}{6}$; O(1): $0, 0.09, \frac{1}{6}$; O(2): $0.23, 0.32, \frac{1}{6}$

His origin, however, was not at a centre of symmetry. Move the origin to the centre of symmetry at $1/4, 0, -1/4$, and find the new atomic coordinates. By use of the space-group symmetry, show that they agree with those given in §11.8.

(b) Early work on potassium nitrate, isomorphous with aragonite, gave atomic positions as follows:

K: $0, 150°, 0$; N: $0, 270°, 60°$; O(1): $0, 318°, 60°$;
O(2): $70°, 247°, 60°$

The lattice parameters are $a = 5.42$ Å, $b = 9.17$ Å, $c = 6.45$ Å. Show what changes are needed in the choice of origin and sense of the axes to make the description correspond with that in (a). Rewrite the atomic positions accordingly.

5. For one form of sodium niobate (Phase *Q*), with space group $P2_1ma$, the origin is chosen so that the mirror plane is at $y = 0$ and the *a*-glide plane at $z = 0$; the coordinates of certain atoms are then as follows:

Nb: 0.260 0.253 0.243
O(3): 0.005 0.231 0.531
O(4): 0.451 0.277 0.969

(a) Derive the coordinates of all symmetry-related atoms.

(b) Move the origin so that one Nb is at $0, 0.003, -0.007$, and find the new coordinates of all the other atoms.

(c) Express the original coordinates in the form $n/4 + p$, where n is an integer and p a small parameter, and repeat (a) and (b) to obtain the coordinates of all atoms in terms of their p's.

(d) If you were originally given the answer to (b) as a description of the structure, show how you could deduce the presence and position of the symmetry elements. Check that the positions are the same as those derived by change of origin from the positions given at the beginning of the question.

(e) Check your results by drawing sketch diagrams.

6. In spinel (which is cubic face-centred), with the conventional choice of origin there are oxygen atoms at x, x, x; $x, \bar{x}, \bar{x}$; $\bar{x}, x, \bar{x}$; $\bar{x}, \bar{x}, x$, with $x = u$ and $x = 1/4 - u$, together with others related to them by the face-centred lattice. Show that this array of atoms is centrosymmetric.

Taking a centre of symmetry as origin, find new expressions for the positions of all the oxygen atoms, and rewrite them more conveniently in terms of a new parameter x'. If x is 0.380, what is x'?

7. (a) Lead zirconate has an orthorhombic unit cell with $a = 5.88$ Å, $b = 11.77$ Å, $c = 8.22$ Å. Show that with very small changes of size this could be derived from a simple cubic unit cell with $a \simeq 4.15$ Å, and find the matrices for transforming one to the other.

(b) Diffraction maxima h, k, l, which are systematically weak in the true structure, are those which would be forbidden in the cubic structure because their indices would be nonintegral. What are the systematic rules applying to h, k, l?

8. Write down all the matrices for transformations of the orthorhombic unit cell allowing different choices of a, b, c as names for its edges (irrespective of their sense). Use these to find the complete set of symbols corresponding to each of the following space groups:

$Pmn2$, $P2na$, $Pba2$, $B2mb$, $Pban$, $Pnma$, $Pmna$, $Cmcm$, $Ibam$, $Ibca$.

9. In describing aragonite, R. W. G. Wyckoff chose different axes from those of W. L. Bragg, which are given in part (a) of Question 4. With his origin at a centre of symmetry with coordinates $-1/4$, 0, $1/4$ on the Bragg axes, Wyckoff obtained for atom O(1) the coordinates -0.09, $11/12$, $1/4$. Find (a) the matrix for transformation of coordinates from those of Bragg, (b) the coordinates of the other atoms, (c) the matrix for the transformation of the unit cell, (d) the symbol for the space group in the Wyckoff setting.

10. High quartz, with space group $P6_2 22$ (No. 180), is described on hexagonal axes of reference with $a = 5.01$ Å, $c = 5.47$ Å. With the origin chosen at the intersection of the triad axis and the diad axis parallel to x, Si atoms are at $1/2$, 0, 0 and O atoms at x, $\bar{x}$, $5/6$, with $x = 0.20$. Using rhombohedral axes of reference, find (a) the lattice parameters, (b) the coordinates of the atoms.

Derive the coordinates of all atoms in the hexagonal unit cell, and hence obtain those of all atoms in the rhombohedral unit cell.

11. Corundum, Al_2O_3, is usually described using hexagonal axes of reference, as in §11.4. Transform to rhombohedral axes and find the lattice parameters and the positions of all atoms in the rhombohedral unit cell.

12. Sodium nitrate, described with rhombohedral axes of reference, has $a = 6.32$ Å, $\alpha = 47°14'$; Na in 0, 0, 0; N in $1/4$, $1/4$, $1/4$; O in x, $1/2 - x$, $1/4$; $1/2 - x$, $1/4$, x; $1/4$, x, $1/2 - x$, with $x = 0.495$. Find the lattice parameters and positions of atoms for the description using hexagonal axes.

13. Cobalt trifluoride (space group $R\bar{3}c$) has a primitive rhombohedral unit cell with $a = 5.28$ Å, $\alpha = 57.0°$, containing two formula units, $2CoF_3$. The F atoms are in positions 6(e), x, $1/2 - x$, $1/4$, with $x = -0.15$.

(a) Find the lattice parameters for hexagonal axes of reference.

(b) Show that one F atom is at 0.07, 1/3, 1/12.

(c) Find the positions of all other atoms.

(d) Find the lattice parameters and the atomic positions if the axes are chosen to give a face-centred rhombohedral unit cell of the same height containing eight formula units.

14. Boron phosphate, BPO_4, is tetragonal, with $a = 4.332$ Å, $c = 6.640$ Å, 2 molecules per unit cell, and space group $I\bar{4}$. The B atoms are at 0, 1/2, 1/4, P at 0, 0, 0; oxygen atoms are in general positions x, y, z, with $x = 0.14$, $y = 0.26$, $z = 0.13$. An alternative choice of unit cell takes the x and y axes at 45° to those in the first choice.

(a) Write down the matrices for transforming lattice parameters and atomic coordinates from each setting to the other.

(b) Find the lattice parameters and positions of all B and P atoms in the new setting. Find the axial ratio, and show that, with a fairly small change in this, the structure would be one of those described in Chapter 4.

(c) Write down the positions of all oxygen atoms in the small unit cell and in the large unit cell.

15. Coesite (one form of SiO_2) is monoclinic, with space group $\{C2/c\}$ (No. 15). With the original choice of unit cell, the face perpendicular to the symmetry direction was an exact 120° rhombus (within the accuracy of measurement); hence an unconventional setting was chosen, with the symmetry direction as the z axis. Then $a = b = 7.17$ Å, $c = 12.38$ Å, $\beta = 120.0°$, and the two independent Si atoms were at 0.14, 0.07, 0.11; and 0.51, 0.54, 0.16, respectively. For comparison with the felspar structure, however, it was more helpful to choose a unit cell derived from the original by the matrix $110/001/0\bar{1}0$, followed by a change of origin to 1/2, 0, 3/4.

(a) With the original axes, write down the space-group symbol and the coordinates of the complete set of atoms derived by symmetry from one atom at x, y, z.

(b) Draw a sketch diagram showing the relation of the new unit cell to the old.

(c) Find the space group symbol, lattice parameters, and position parameters referred to the new axes.

(d) Find the new indices of planes originally indexed as (100), (010), (110), ($1\bar{1}0$), and directions originally indexed as [110], [$1\bar{1}0$], [101], [$10\bar{1}$], [120], [210].

16. Sodium bicarbonate, $NaHCO_3$, is monoclinic, and was originally described with lattice parameters $a = 7.51$ Å, $b = 9.70$ Å, $c = 3.53$ Å, $\beta = 93.3°$, space group $P2_1/n$ (No. 14). With these axes, the atomic position parameters are those shown as set (1). Later workers, however, chose a unit cell with $a = 3.53$ Å, $b = 9.70$ Å, $c = 8.11$ Å, $\beta = 112°25'$, space group $P2_1/c$, and obtained the position parameters of set (2). Show that these results are compatible. Give the transformation matrix by which the original position parameters can be obtained from the later set.

Are the atoms whose position parameters are recorded in sets (1) and (2) the same, or how are they related by symmetry?

	SET (1)			SET (2)		
	x	y	z	x	y	z
Na	0.286	0.004	0.713	0.427	0.004	0.714
C	0.077	0.237	0.287	0.210	0.237	0.923
O(1)	0.071	0.367	0.261	0.190	0.367	0.929
O(2)	0.205	0.163	0.193	0.988	0.163	0.795
O(3)	0.940	0.171	0.436	0.496	0.171	0.060

17. (a) For oxalic acid dihydrate (monoclinic), the original choice of axes of reference gives $a = 6.12$ Å, $b = 3.61$ Å, $c = 12.03$ Å, $\beta = 106.2°$, and the space group symbol is $P2_1/n$. Certain later authors have preferred to use the setting for which the symbol is $P2_1/a$. What is the simplest change of axes which will allow this? Express your result as a matrix giving the new lattice parameters in terms of the old, and calculate their values.

(b) There are two $(COOH)_2$ molecules per unit cell, and one C atom has coordinates -0.04, 0.04, 0.05, referred to the original axes, with the origin at a centre of symmetry. In the new description, the origin is taken at a different centre of symmetry, midway between two molecules related by the shortest lattice vector. Write down the coordinates of all the carbon atoms according to each description.

(c) The projection of the C—C bond on the (010) plane lies very nearly along a simple rational direction of the lattice. Using the information in (b), determine the zone indices of this direction in the original description and in the new description.

(d) According to the original description, there are two oxygen atoms with coordinates as follows:

O(1): 0.09, 0.94, 0.15;   O(2): 0.78, 0.22, 0.04

In a closely related structure, whose unit cell is derived from the first-named choice for the original structure by the matrix 101/010/100, followed by displacement of the origin to a different centre of symmetry, two oxygen atoms were located as follows:

O(1): 0.04, 0.20, 0.25;   O(2): 0.15, 0.85, 0.44

Rewrite the coordinates of the atoms in the second structure, using the original axes of reference of the first, so as to show the resemblance. [Note: Remember that you may, if you find it helpful, quote the coordinates of a symmetry-equivalent atom rather than of one of those listed above, but, if so, you should indicate the symmetry operation you used to derive it.]

Chapter 11 STRUCTURES OF OXIDES

11.1 INTRODUCTION

The structures in this chapter are of materials of technological importance, or are closely related to them. For detailed studies of such materials, it is important to be provided with a complete formal description, and to be able to translate it into a qualitative description in which the feature of particular interest for one's purpose is given emphasis. As shown in earlier chapters, there can be many such descriptions—in terms of anion packing, or the pattern of cation sites, or coordination polyhedra and their linkage—and we must be able to move from one to another at will. Where not all atoms are in special positions, we want to know how interatomic distances will vary with the atomic position parameters. (There was an extreme example of the effect of such variation on the character of the nickel arsenide structure in Chapter 4.)

For each of the structures discussed, we shall therefore be first concerned with the derivation of complete sets of atomic positions from the formal description given in terms of the space group. We shall then see how the same structure may be qualitatively described in terms of packing, or linkage, or both. Often we shall begin with an idealised packing or coordination and then introduce the small deviations that are important for the true structure.

Each of the structures described in this chapter (with one or two exceptions) is the simplest representative of its kind, the *aristotype*. There are often closely related structures, *hettotypes*, with more complicated deviations from the ideal, and some of them are very important, but discussion of such *families of structures* (and definitions of the terms) will be left till §12.1. Structures are *strictly isomorphous* only if they have the same symmetry and the same sets of atomic positions, differing only in the species of atoms occupying the sites and the numerical values of lattice parameters and atomic position parameters.* In this chapter, besides the type materials chosen to illustrate the structures, we shall consider other materials only if they are strictly isomorphous with them.

It should be remembered that the names of structure types may not always have been taken from the simplest representative of the family, but from the first member of the family to have been carefully studied. In early work it often happened that the small differences of structure within a family could not be distinguished—and, indeed, there are still many purposes for which the distinction is unimportant, and for which an aristotype described in this chapter is an adequate approximation to the hettotypes considered in Chapter 12.

We shall be much concerned with coordination polyhedra and their dependence on cation radius. Reference to Chapter 2, Tables 2.2 to 2.6, may be useful. It should also be reiterated that in using the terms "cation," "anion," "ionic radii," we are *not* implying that atoms are necessarily fully ionised or bonds wholly ionic. "Cations" and "anions" are names for the more electropositive and more electronegative atoms, and their "ionic radii" or "atomic radii" are their contributions to the interatomic distance under undisturbed conditions.

11.2 THE IDEAL PEROVSKITE† STRUCTURE

This is probably the simplest example of a structure containing two different cations. It has no variable parameters except the unit-cell edge a. This means that it is easy to describe and visualise, but that, physically speaking, it has many special features.

General formula: ABO_3; example, $SrTiO_3$
Lattice: Primitive cubic, with $a = 3.905$ Å
Number of formula units per unit cell: 1
Space group: $Pm3m$ (No. 221)
Atomic coordinates (first choice of origin):
 Ti in $1(a)$: $0, 0, 0$
 Sr in $1(b)$: $\frac{1}{2}, \frac{1}{2}, \frac{1}{2}$
 O in $3(c)$: $\frac{1}{2}, 0, 0$

---

* It is arguable whether the geometrical resemblance is sufficient, or whether one ought only to call them isomorphous if there is also some chemical similarity, implying qualitative similarity of interatomic forces: should one say that ordered AgCd or β-brass is isomorphous with CsCl (§4.8)?

† Perovskite (sometimes spelled perewskite) is a mineral of composition $CaTiO_3$, in which the structure was first studied; later work showed that it actually had a more complicated variant of the structure described here (cf. Chapter 12), which we must therefore distinguish as the *ideal* (or *idealised*) perovskite structure

Though the above information* is sufficient, it is often helpful to record *all* the atomic coordinates, and not merely those of prototype atoms; in this case, the three O atoms are at 1/2, 0, 0; 0, 1/2, 0; 0, 0, 1/2. With an alternative choice of origin, the coordinates are:

Sr in $1(a)$: 0, 0, 0
Ti in $1(b)$: $\frac{1}{2}, \frac{1}{2}, \frac{1}{2}$
O in $3(d)$: $\frac{1}{2}, \frac{1}{2}, 0$

Both choices are equally good formally, but the first is more helpful for visualising the structure and understanding its physical relations. The first is used in Figure 11-1, and we shall keep to it throughout.

The unit cell is shown in perspective in Figure 11-1(a) and in projection in Figure 11-1(c). The *A* cations, Sr, and the *B* cations, Ti, both lie at special positions where they have the full cubic point-group symmetry. The O atoms do not; no triad axes go through them, and their point-group symmetry is only *mmm*.

Positions of atoms within the unit cell are shown in Figure 11-1(a) and (c). Each Ti is surrounded by 6 O atoms, as shown in Figure 11-1(d), and each Sr by 12 O atoms, as shown in Figure 11-1(e). The easiest way to picture the structure is as a framework of TiO_6 octahedra sharing corners, with the large Sr atom in the cavity between octahedra. This is illustrated in Figure 11-1(b). All octahedra are in parallel orientation, and the two Ti—O bonds to the same O are in a straight line, as indicated in Figure 11-1(f).

The dimensions of the framework are determined by the Ti—O bond length, which requires that $a = 2(r_{Ti} + r_O)$. This fixes the size of the cavity, and hence the permissible size of the *A* cation, assuming that Goldschmidt's rule is obeyed, i.e., that *A* just touches all its neighbours. In that case it can be seen from Figure 11-4(e) and (c) that

$$(r_A + r_O)_{str} = (\sqrt{2}/2)a = \sqrt{2}(r_B + r_O)_{str} \tag{11.1}$$

where by using the subscript "str(ucture)" we remind ourselves that the radii are those actually found in the structure. To compare with tabulated values we must multiply the left-hand side of this equation by 1.10 to correct for 12-coordination, and insert a tolerance factor t; then writing

$$1.1(r_A + r_O) = t\sqrt{2}(r_B + r_O) \tag{11.2}$$

---

* Inclusion of the conventional symbols for the different kinds of sites and the numbers of atoms in each—here $1(a)$, $1(b)$, $3(c)$—is not an essential part of the description of the structure. It is, however, very convenient, because it allows easy reference to the *International Tables for X-ray Crystallography*, where the coordinates of all the atoms in each kind of site in each space group are listed, together with their point symmetry. The symbols can also be helpful as an aid to conciseness in the discussion of structures. The alphabetical order within each space group is that of decreasing point symmetry, with the general position denoted by the last letter used. In some structures, especially those of high symmetry like perovskite, all atoms are in special positions, and the general position is not represented (cf. §8.15).

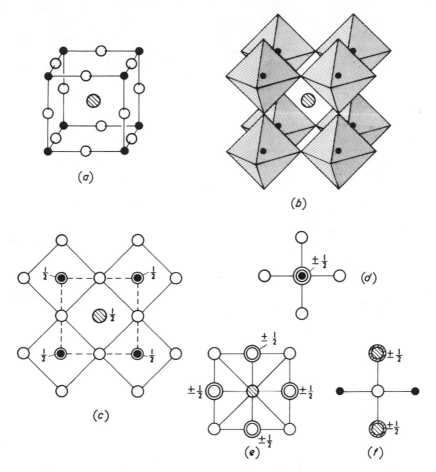

Figure 11-1 $SrTiO_3$ (ideal perovskite structure, ABO_3). (a) Perspective view of unit cell contents: black circles Ti, open circles O, hatched circle Sr. (b) Perspective view of framework, showing octahedra. (c) Projection of structure on (001). Octahedra are outlined by solid lines, and unit cell by broken line. (d), (e), (f) Parts of same projection, showing environment of Ti, Sr, and O respectively. Heights marked are relative to central atom; unmarked atoms are at same height as central atom. In (f), O has two Ti and four Sr neighbours.

we can substitute values for the radii from Table 2.2, and expect to find t very close to 1.

This allows us to predict, in a general way, the occurrence of perovskite structures in ABO_3 compounds: they should occur if t is nearly equal to 1, but not otherwise. With $SrTiO_3$, for example, (using Goldschmidt radii) we have $t = 1.02$. By contrast, $MgTiO_3$ would give $t = 0.83$, which is not satisfactory. One can see this in another way by noticing that Mg, with a radius of 0.78 Å as compared with 1.27 Å for Sr, is considerably too small for the cavity. In fact, $MgTiO_3$ adopts the ilmenite structure (§11.5), which allows Mg as well as Ti to be 6-coordinated.

When the misfit of size is less, we may get a modified structure which still belongs to the perovskite family. An example is $CdTiO_3$, for which $t = 0.92$, and which can occur with either the ilmenite structure or a modified perovskite structure. Discussion of such modified perovskite structures will be left to Chapter 12.

It is not possible to establish exact limits for t; not only does its numerical value depend on whether we use Goldschmidt or Pauling radii (or some other set), but the radii from which it is derived are not (as we saw in Chapter 2) strictly constant from one compound to another. Thus, although t serves as a useful guide, it cannot be expected to give very detailed discrimination.

The requirement of good fit for the A cation allows us to think of the structure in another way. Suppose the radius of B is at the lower limit for 6-coordination, so that the O atoms touch each other. (This is nearly true for TiO_6.) Then the O atoms as a whole fill three out of four positions in a cubic close packing, and the fourth is occupied by Sr. · If the SrO_3 array could be taken by itself, it would have exactly the same structure as $AuCu_3$ (§4.12), namely a cubic close packing of the majority atom, in which one out of four is systematically replaced by an atom of the minority species. Such a structure demands near equality in size of the two kinds of atoms, which is true for Sr (radius 1.27 Å) and O (radius 1.32 Å). But SrO_3 is of course not electrically stable by itself; the Ti atom must be added. It is inserted at the centre of those octahedral interstices of the close-packed array which have only O atoms surrounding them.

If the B atom were rather larger, the O atoms would be pushed farther apart, and the cavity could hold a larger A. For example, in $BaZrO_3$ the O—O distance ($\sqrt{2}a/2$) is 2.96 Å as compared with 2.79 Å in $SrTiO_3$; this makes a suitable space for the larger Ba. $BaZrO_3$ has again the ideal structure with $t = 1.02$.

Requirements with respect to electrical neutrality must be satisfied as well as size requirements. Applying Pauling's rules, we see that each bond from Ti has strength 4/6 and each bond from Sr has strength 2/12. Each O receives two bonds from Ti and four bonds from Sr, as shown in Figure 11-1(f); their sum is

$$2 \times \tfrac{4}{6} + 4 \times \tfrac{2}{12} = \tfrac{4}{3} + \tfrac{2}{3} = 2 \tag{11.3}$$

which satisfies the rule. But one can see that the rule would equally be satisfied for a more general perovskite of composition ABX_3 where X is any anion, provided that the structure is electrically neutral; i.e., that the valencies q_A, q_B, and q_X are connected by the relation

$$q_A + q_B = -3q_X \tag{11.4}$$

This sum then becomes

$$2 \times q_B/6 + 4 \times q_A/12 = (q_A + q_B)3 = -q_X \tag{11.5}$$

Examples of compounds of different valency belonging to the perovskite family are numerous. Those with the ideal structure include $KMgF_3$, $KNbO_3$ above 500° C, and $LaAlO_3$ above 535° C.

It is perhaps useful to note the characteristics of the ideal perovskite structure (the aristotype) which are possessed equally by the variants (the hettotypes) discussed in Chapter 12. They are: a framework of linked octahedra in nearly parallel orientation; a repeat or pseudorepeat distance of about 4 Å, the corner-to-corner length of an octahedron; the need for a fairly well-fitting A cation; and the continued satisfaction of the valency rules. Details of interatomic distances and atomic environments are of course different in the hettotypes, though the aristotype can still be used for many purposes as a rather coarse approximation.

11.3 THE SPINEL STRUCTURE

This is an example of a structure essentially dependent on a cubic close-packing of anions, bound together by suitably placed interstitial cations. It has one variable atomic position parameter, which allows it a little more freedom of adjustment than the ideal perovskite structure.

General formula:* B_2AO_4; example, Al_2MgO_4 (the mineral spinel)
Lattice: Face-centred cubic, with $a = 8.09$ Å
Number of formula units per unit cell: 8
Space group: $Fd3m$ (No. 227)
Atomic coordinates (using first choice of origin of the *International Tables for X-ray Crystallography*, which puts centre of symmetry at $\frac{1}{8}, \frac{1}{8}, \frac{1}{8}$):†
  Al in 16(c): $\frac{5}{8}, \frac{5}{8}, \frac{5}{8}$
  Mg in 8(a): 0, 0, 0
  O in 32(e): x, x, x, with $x = \frac{3}{8} + u$, where u is small ($\simeq 0.015$)

Our first task is to find how to derive the coordinates of all the atoms, construct diagrams, and describe the structure.

Since for any species of atom the number per unit cell is not a multiple of 3, atoms of each kind must lie in special positions on triad axes; this indeed has already been assumed in listing the coordinates of the prototype atoms set out above, all of which lie on the body diagonal of the cube. Since the space group has one diamond-glide plane parallel to cube face, it has two others at right angles; hence the orthorhombic space group $Fdd2$ considered in §8.8 is one of its subgroups, and can be used as a starting point for construction. We make the lattice cubic, move the atom S (Figure 8-8) to

---

* Often the formula is written AB_2O_4, e.g. $MgAl_2O_4$, and this has been used for the conventional naming of A and B sites.
  † Though it is usually more convenient to take the origin at a centre of symmetry, for this space group there are advantages in a different choice. We follow here the choice made by Wyckoff in *Crystal Structures*.

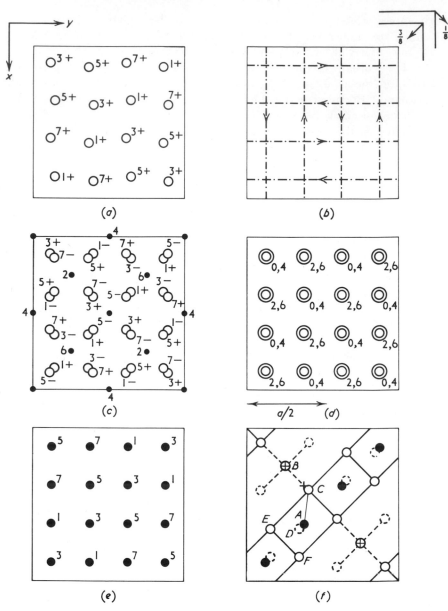

Figure 11-2 Spinel: projections of unit cell. (a) First stage of construction: operation of d-glide planes parallel to (100) and (010) (cf. Figure 8-8). (b) Position and direction of d-glide planes. (c) Array of O atoms (open circles) and Mg atoms (small black circles). Heights are marked in units of $a/8$ (except for Mg atoms at height zero); the symbol $+$ or $-$ is an abbreviation for $+u$ or $-u$, where $u \simeq 0.015$. (d) Array of O atoms if u were zero. Heights are marked in units of $a/8$ above a different base level from that in (c). Note that the repeat unit is a cube of side $a/2$. (e) Array of Al atoms: heights are marked in units of $a/8$ above the same base level as in (c). (f) Projection of part of structure, with u given double its value in (c). Black circles (Al) and open circles (O) are at or near height $3/8$; crossed circles (Mg) are at height $1/4$, and dotted circles (O) at or near height $1/8$. The separate cross marks what would be the position of atom C if u were zero.

$3/8 + u$, $3/8 + u$, $3/8 + u$, and the other atoms to match, and obtain the set shown in Figure 11-2(a). Adding the remaining d-glide plane parallel to the paper, as shown in Figure 11-2(b), we get the complete array of oxygens in Figure 11-2(c), where the development of the mirror plane m (with traces along the diagonals of the square) as a consequence of the symmetry operations already put in can be recognised; there are also four mirror planes of the same set inclined at 45° to the plane of projection.

The array of oxygens is very nearly close-packed; if u were zero it would be exactly so. This can be seen by subtracting 3 from all heights in Figure 11-2(c), and moving each pair of overlapping atoms to coincide in projection, giving Figure 11-2(d).

The Al atoms are midway between pairs of oxygen atoms, on special positions which are centres of symmetry. If u had been zero, it would not have mattered whether we put the prototype Al at 5/8, 5/8, 5/8, or 5/8, 5/8, 1/8, since they are both centres of symmetry; but because u is not zero the Al—O distances will be different in the two cases. As drawn, with $u > 0$, they will be shorter if Al is put at 5/8, 5/8, 5/8. The resultant array of Al atoms, with their heights, is shown in Figure 11-2(e).

The Mg atoms can easily be derived from the prototype at 0, 0, 0, using only the d-glide planes at height 1/8 and 3/8; they are shown in Figure 11-2(c). Taken by themselves, their array is identical with the structure of diamond, with the same unit-cell side a as that of the complete structure.

It is necessary to be careful about the choice of origin in this structure. We have taken it at the intersection of the {110} mirror planes, in such a way that along the line [111] through the origin we find Mg at 0, O at $3/8 + u$, Al at 5/8, and O at $7/8 + u$. Moving the origin to 1/4, 1/4, 1/4 would give O atoms at $1/8 + u$, $1/8 + u$, $1/8 + u$, and Al at 3/8, 3/8, 3/8; moving the origin to 5/8, 5/8, 5/8 (a centre of symmetry) would give O at $\pm(1/4 - u, 1/4 - u, 1/4 - u)$ and Mg at $\pm(3/8, 3/8, 3/8)$. Though such alternative choices might sometimes be useful, the danger of confusion is so great that it is better to avoid them if possible.

Having obtained a complete description of the atomic positions, we can now discuss the structure in terms of packing and linkage. Starting with an array of oxygens in cubic close packing, we insert Al in certain octahedral interstices and Mg in certain tetrahedral interstices, the selection of interstices being made in such a way that the repeat distance along each axis is double what it would be for the ideal close packing of oxygens alone but that this larger unit cell is face-centred. The unit-cell edge is $2\sqrt{2}$ times the O—O edge distance; since the usual O—O contact distance is about 2.8 Å, we expect a cell edge of not less than 8 Å.

The octahedra share edges to form ribbons extending parallel to the face diagonals of the cube, in two directions in the (001) plane (the plane of projection in Figure 11-3), and in four directions inclined at 45° to this plane. Three pairs of parallel edges of each octahedron are shared, those bounding one pair of parallel faces. The result is a three-dimensional framework of

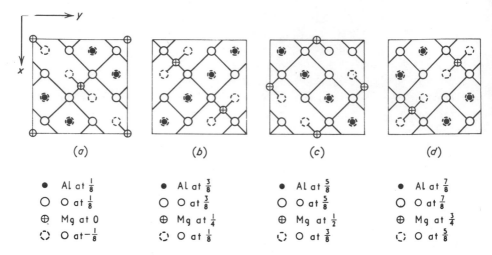

(a)	(b)	(c)	(d)

- Al at $\frac{1}{8}$
○ O at $\frac{1}{8}$
⊕ Mg at 0
◌ O at $-\frac{1}{8}$

- Al at $\frac{3}{8}$
○ O at $\frac{3}{8}$
⊕ Mg at $\frac{1}{4}$
◌ O at $\frac{1}{8}$

- Al at $\frac{5}{8}$
○ O at $\frac{5}{8}$
⊕ Mg at $\frac{1}{2}$
◌ O at $\frac{3}{8}$

- Al at $\frac{7}{8}$
○ O at $\frac{7}{8}$
⊕ Mg at $\frac{3}{4}$
◌ O at $\frac{5}{8}$

Figure 11-3 Spinel structure, projected on (001). The four diagrams show successive layers of the structure; to obtain the complete structure they should be superposed, with (a) at the bottom and (d) at the top. The oxygen atoms shown as dotted circles in (b), for example, are the same as those shown as solid circles in (a). Notice the ribbons of octahedra sharing edges, which run diagonally, linked by Mg atoms in tetrahedral environments. Similar ribbons of octahedra inclined to the plane of projection can be traced by superposing the diagrams.

composition AlO_2. The Mg atoms occupy interstices linking the ribbons as shown in Figures 11.2(f) and 11.3(a) to (d). All octahedra are symmetry-equivalent, and so are all tetrahedra; there are two octahedra to each tetrahedron.

We next consider the electrostatic balance. Each O atom has three Al neighbours, and one Mg; this environment is shown for the prototype O atom in Figure 11-4. The sum of the bond strengths to it is $3 \times (3/6) + 2/4 = 2$; thus the Pauling rule is exactly satisfied. Moreover, the distribution of the bonds in space is well-balanced; without the Mg, the three bonds to Al would have had a very one-sided effect.

Up to this point, we have not considered ionic sizes, except to assume that Al and Mg are capable of 6-coordination and 4-coordination respectively. This is true, but it is also true that Mg is larger than Al, and therefore we should have expected it to occupy for preference the octahedral site, which is larger than the tetrahedral site. The actual Goldschmidt radius sums for Al—O and Mg—O are 1.89 Å and 2.10 Å respectively for 6-coordination, 1.74 Å and 1.97 Å for 4-coordination. In solving the

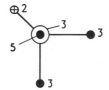

Figure 11-4 Environment of oxygen atom in spinel. The atoms projected all appear in Figure 11-3(b) (with the central O at 3/8, 3/8) except for the Al atom at height 5/8, which appears in Figure 11-3(c). (The scale is double that of Figure 11-3.)

structure, it would have been natural to put Mg into the octahedral sites, had it not been for the difficulty that there are twice as many octahedral sites as Mg atoms, and only half as many tetrahedral sites as Al atoms. This difficulty was faced by T. F. W. Barth and E. Posnjak when they investigated the structure about 1931, and they suggested a possible solution: that the Mg atoms did indeed occupy octahedral sites, but shared them at random with half the Al atoms, the other half of the Al atoms occupying tetrahedral sites. This was a novel idea, because up to that date it had been assumed that, when the chemical composition allowed it, crystallographically equivalent sites *must* be occupied by chemically identical atoms. To the new kind of arrangement they gave the name of "variate atom equipoints."

At that date, it was difficult to test the idea for spinel itself, because the X-ray scattering factors of Mg and Al are so nearly alike that they could not be distinguished. This is not true for their neutron scattering factors, and it is now known from neutron diffraction that the suggestion of a disordered structure for Al_2MgO_4 was wrong; it has a *normal spinel structure*, and the description given above is strictly correct, with Al in octahedral sites and Mg in tetrahedral sites. Nevertheless, the idea of Barth and Posnjak was a very fruitful one; it was found almost at once to apply to other closely related materials, such as Fe_2MgO_4. These are said to have the *inverse spinel structure*. Further discussion of such structures will be left to §12.12.

For spinel itself there is another answer to the difficulty. The relative sizes of octahedral and tetrahedral interstices depend quite sensitively on the parameter u; when u is positive, the tetrahedral interstice is increased and the octahedral one is decreased. This is illustrated in Figure 11-2(f), where AC is shorter and BC longer than if u were zero. The bond lengths can easily be calculated, if we neglect second-order terms in the small quantity u. We have

$$\text{Al—O} = a\{(\tfrac{1}{4} - u)^2 + u^2 + u^2\}^{\frac{1}{2}} \simeq \tfrac{1}{4}a(1 - 4u) \tag{11.6a}$$

$$\text{Mg—O} = a\{(\tfrac{1}{8} + u)^2 + (\tfrac{1}{8} + u)^2 + (\tfrac{1}{8} + u)^2\}^{\frac{1}{2}} \simeq \frac{\sqrt{3}}{8} a(1 + 8u) \tag{11.6b}$$

With the observed values $u = 0.012$ Å, $a = 8.09$ Å, these give Al—O = 1.93 Å, Mg—O = 1.92 Å, which are not far from the normal interatomic distances for 6-coordination and 4-coordination respectively.

At the same time, as can be seen from Figure 11-2(f), the octahedra are distorted; they are narrowed across the breadth of the ribbon and tilted relative to their neighbours. The short octahedron edges (DE, EF, DF) are given by $a\sqrt{2}(1/4 - 2u)$, and the long edges by $a\{2(1/4)^2 + (2u)^2\}^{\frac{1}{2}} \simeq \sqrt{2}a/4$, which for $u = 0.012$ are 2.57 Å, 2.86 Å respectively. The shortening of shared edges is a common feature of structures like this. It can be explained quite simply in electrostatic terms: each cation-oxygen bond has a component directed along the edge, giving an O—O attraction, and this attraction will be twice as great when two cations contribute to it. Thus the distortion of the octahedra needed to produce a stable AlO_2 framework itself requires a positive u parameter, and increases the size of the tetrahedral site available to the Mg atom.

Barth and Posnjak had not overlooked the possibility of adjusting the u parameter in this way, but they were not able to measure it with great accuracy, and their evidence suggested that it was less than 0.010, too small to be satisfactory for a normal structure.

A large number of materials can take up the spinel structure. As with the perovskite structure, we can have anions other than oxygen. The electrostatic requirement for neutrality in B_2AX_4 is that $2q_B + q_A = -4q_X$; this can be satisfied in many different ways. Examples of compounds with spinel structures are Mg_2TiO_4, Na_2MoO_4, Li_2NiF_4, Ti_2CuS_4, Fe_3O_4. By contrast with the perovskite structure, the existence of a variable parameter u means that we can have a much larger range of truly isomorphous structures, i.e., structures with the same symmetry and the same type of occupation of atomic sites. On the other hand, strictly speaking, normal and inverse spinels are not isomorphous; moreover, structures of the spinel family with lower symmetry do exist. Further discussion of the isomorphous examples will therefore be postponed till we consider the family as a whole, in §12.12 to 12.13.

11.4 THE CORUNDUM STRUCTURE

This, and the two following structures (ilmenite and lithium niobate), are based on a hexagonal close-packing of anions, with all cations in octahedral interstices. Dimensionally, therefore, they are very similar and can be described with similar axes of reference; but beyond that the differences are fundamental. In the corundum structure, both cations are alike.

General formula: A_2O_3; example, Al_2O_3 (the mineral corundum, which, with impurities added for colouring, is also sapphire and ruby)

Lattice: Rhombohedrally-centred hexagonal, with $a = 4.763$ Å, $c = 13.003$ Å

Number of formula units per unit cell: 6

Space group: $R\bar{3}c$ (No. 167)

Atomic coordinates:
   Al in $12(c)$: $0, 0, z$, with $z = 0.347$
   O in $18(c)$: $x, 0, \frac{1}{4}$, with $x = 0.306$

Both x and z are close to $\frac{1}{3}$; it is helpful to write them as

$$x = \tfrac{1}{3} + u, \quad \text{with} \quad u = -0.027$$
$$z = \tfrac{1}{3} + w, \quad \text{with} \quad w = 0.014.$$

Consider first the array of oxygens, which turns out to be very nearly hexagonal close-packing. Operation of the triad axis and the centre of

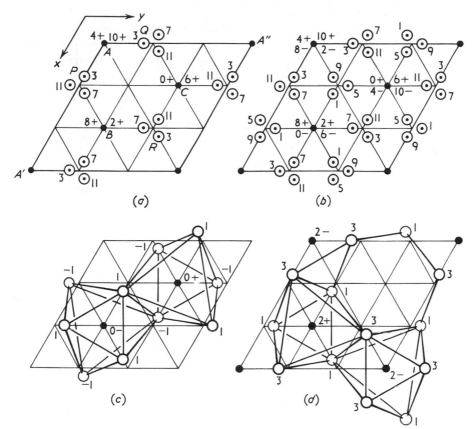

Figure 11-5 Al_2O_3: projection of parts of structure on (0001) (drawn with $u = 0.067$). Heights are marked in units of $c/12$. (a) Array of atoms produced by operation of $R3$ on prototype oxygen and $R3c$ on prototype Al. (b) Complete array of atoms, produced by adding to (a) a centre of symmetry at $(0, 0, 0)$; symmetry $R\bar{3}c$. (c) Octahedra surrounding cations at height near zero: black circles are Al; heavy and light open circles are O at $1/12$ and $-1/12$ respectively. (d) Sheet of octahedra immediately above those in (c): black circles are Al near $2/12$; heavy and light open circles are O at $3/12$ and $1/12$ respectively.

symmetry gives the set

$$\pm(x, 0, \tfrac{1}{4}; \ 0, x, \tfrac{1}{4}; \ \bar{x}, \bar{x}, \tfrac{1}{4})$$

It is clear therefore that the O atoms lie on layers at heights $1/4$, $3/4$. Applying the rhombohedral centring operator, $(0, 0, 0; \ 2/3, 1/3, 1/3; \ 1/3, 2/3, 2/3)+$, we find that there are identical layers, laterally displaced, at heights $1/3$ and $2/3$ above those first considered; altogether there are 6 layers in the unit cell, at heights $(2n + 1)/12$. If u had been zero, each layer would have formed a net of equilateral triangles, and the sequence of layers would have formed a hexagonally packed array.

The construction is illustrated in Figure 11-5. For the moment, consider only oxygen atoms. In Figure 11-5(a), the prototype atom at $1/3 + u, 0, 1/4$

has been inserted (with u exaggerated to show its effect), and has been operated on by the triad axis and the rhombohedral lattice-centring, i.e., by $R3$. The three atoms in each small triangular group would have projected on top of one another if u had been zero. In Figure 11-5(b) the centre of symmetry has been added. It can be seen that it has generated c-glide planes, whose traces extend vertically up and down the page and in directions at 60° to either side of this, all passing through the origin. (Since the traces of the glide planes are at 30° and 90° to the x and y axes, the normals to the glide planes are parallel to the axes, and the symbol for the glide planes is therefore correctly placed immediately after that for the triad axis—cf. §8.11.)

Even though we have used only the operator $R\bar{3}$, we have obtained an array of atoms with symmetry $R\bar{3}c$. Why should this be? The answer is that the prototype atom has been put in a special position on the plane defined by triad axes through adjacent lattice points, and therefore has been given symmetry to start with.

Now consider the Al atoms. These lie on the triad axes, and hence remain unchanged by the operation 3; the lattice centring gives the set

$$0, 0, z; \quad \tfrac{2}{3}, \tfrac{1}{3}, \tfrac{1}{3} + z; \quad \tfrac{1}{3}, \tfrac{2}{3}, \tfrac{2}{3} + z$$

i.e., one projecting at each of the positions A, B, C in Figure 11-5(a). Addition of the centre of symmetry gives, projecting at A, the pair $\pm(0, 0, z)$, with similar pairs at B and C. These do *not* automatically satisfy the space group $R\bar{3}c$. The glide plane must be added deliberately, giving, at A, the set of four:

$$\pm(0, 0, z; \ 0, 0, \tfrac{1}{2} + z)$$

with similar sets at B and C. In Figure 11-5(a) and (b), the glide plane operation has in fact been used; (a) has the symmetry $R3c$.

Inspection shows that the Al atoms lie in octahedral interstices of the hexagonal packing. There are six such interstices in the unit cell projecting at each of the positions A, B, C, and four of them are occupied, the two empty ones occurring at different heights in the columns at $A, B,$ and C. Figure 11-5(c) illustrates the sheet of octahedra having Al atoms at height near 0. Each shares three edges with its neighbours in the same sheet. Figure 11-5(d) shows the sheet of octahedra directly above that in (c), having its Al atoms nearly at 2/12, and sharing O atoms at 1/12 with the first sheet. By comparing the two diagrams it can be seen that a face is shared between the two octahedra in column B having Al atoms at $0 -$ and $(2/12) +$. Two things are noticeable: the edges of the shared face are shorter than those of the unshared face parallel to it (another illustration of the general principle mentioned in §11.3); and the Al atoms are farther apart than if they had remained midway between the oxygen layers (an illustration of the effect of cation-cation repulsion).

The next sheet of octahedra, which has Al atoms nearly at 4/12, shares a face in the A column with the 2/12 sheet, and a face in the C column with the

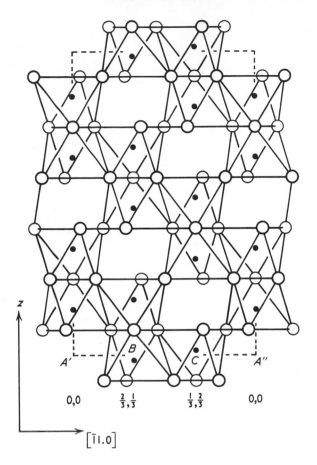

Figure 11-6 Al₂O₃: projection on (11$\bar{2}$0) of part of structure one octahedron thick (drawn with $u = -0.083$, $w_{Al} = 0.017$). Small black circles are Al in plane $x + y = 1$ whose trace in Figure 11-5 is the long diagonal $A'A''$; heavy and light open circles are O lying in front of and behind this plane, as viewed from the lower right-hand corner of Figure 11-5(b). Corners of the unit cell are indicated by broken lines.

6/12 sheet; and successive sheets are built on in the same way, each sharing faces above and below. We can think of the whole structure as consisting of pairs of octahedra sharing a face parallel to (0001), the pairs arranged on a rhombohedral lattice, and each octahedron linked to three others in its own plane by sharing three sloping edges. The arrangement can be seen in Figure 11-6 (where, however, only one of the three neighbours sharing a sloping edge is shown—the others lie above and below the paper). Notice that the pair of octahedra halfway up the cell is *not* a translation-repeat of the pair below it. In one pair, the triangle constituting the shared face has its vertex pointing upwards from the paper, in the other, pointing downwards below the paper.

If close-packing were ideal, and the octahedra therefore regular, the lattice parameters c and a could both be expressed in terms of the octahedron edge length l. Using the results of §6.11, we have

$$c = 6\sqrt{\frac{2}{3}}\, l, \qquad a = \sqrt{3}l, \qquad \frac{c}{a} = \sqrt{8} \qquad (11.7)$$

The actual axial ratio approximates to this. Writing $c/(\sqrt{8}a) = 1 + \eta$, we find that $\eta = -0.035$. The advantage of expressing the axial ratio in this form is its convenience for calculating the effect of changes of parameters either from one isomorphous material to another or for the same material under different conditions.

The Al—O bond lengths to the O atoms of the shared and unshared faces are different. From Figure 11-5(c) it can be seen that their projections are $(\frac{1}{3} + u)a$ and $\{\frac{1}{9} - \frac{1}{3}u + u^2\}^{\frac{1}{2}}a$ respectively. (To derive this latter expression, we use the relation for the sides a, b, c of a triangle with $ACB = 120°$, namely, $c^2 = a^2 + b^2 + ab$.) Their vertical components are $(\frac{1}{12} + w)c$ and $(\frac{1}{12} - w)c$. Hence

$$\text{Al—O}_{\text{shared}} = a\{(\tfrac{1}{3} + u)^2 + 8(1 + \eta)^2(\tfrac{1}{12} + w)^2\}^{\frac{1}{2}}$$

$$= 0.994a/\sqrt{6} = 1.93\ \text{Å} \quad (11.8a)$$

$$\text{Al—O}_{\text{unshared}} = a\{\tfrac{1}{9} - \tfrac{1}{3}u + u^2 + 8(1 + \eta)^2(\tfrac{1}{12} - w)^2\}^{\frac{1}{2}}$$

$$= 0.970a/\sqrt{6} = 1.89\ \text{Å} \quad (11.8b)$$

Neglect of second-order small quantities in u, w, and η, would have given us the expressions

$$\text{Al—O}_{\text{shared}} \simeq (a/\sqrt{6})(1 + 2u + 4w + \eta/3) \qquad (11.9a)$$

$$\text{Al—O}_{\text{unshared}} \simeq (a/\sqrt{6})(1 - u - 4w + \eta/3) \qquad (11.9b)$$

which, though leading to errors of 1 or 2 percent in the bond lengths, are very convenient when only *changes* in bond lengths due to changes in parameters are being considered. It can be seen at a glance from this expression—as also qualitatively from Figure 11-5(c) and (d)—that distortion of the octahedron shape (measured by $-u$) and displacement of Al within the octahedron (measured by w) act in opposite directions on the two kinds of Al—O bonds, with the result that they remain nearly equal. It is this kind of harmonious cooperation between deviations from the ideal demanded by different ingredients of a structure that tends to make the structure as a whole stable. (Compare the effect of the parameter u in spinel, which results in shortening of shared octahedron edges *and* enlargement of tetrahedral site.)

The octahedron edges can be calculated similarly. Those in the (0001) plane are

$$l_{\text{shared}} = \sqrt{3}\,(\tfrac{1}{3} + u)a = \frac{a}{\sqrt{3}}\,(1 + 3u)$$

$$= 0.919\,\frac{a}{\sqrt{3}} = 2.53\ \text{Å} \qquad\qquad (11.10a)$$

$$l_{\text{unshared}} = \sqrt{3}\,\{(\tfrac{1}{3})^2 - \tfrac{1}{3}u + u^2\}^{\frac{1}{2}}a = \frac{a}{\sqrt{3}}\,\{1 - 3u + 9u^2\}^{\frac{1}{2}}$$

$$= 1.043\,\frac{a}{\sqrt{3}} = 2.87\ \text{Å} \qquad\qquad (11.10b)$$

The two kinds of inclined edges have the same vertical component $(1/6)c$ but different horizontal components

$$\{(\tfrac{1}{3})^2 + \tfrac{1}{3}\cdot 2u + (2u)^2\}^{\frac{1}{2}}a \qquad\qquad (11.11a)$$

and

$$\{(\tfrac{1}{3} - u)^2 - (\tfrac{1}{3} - u)u + u^2\}^{\frac{1}{2}}a \qquad\qquad (11.11b)$$

Neglecting second-order small quantities, their lengths are

$$\frac{a}{\sqrt{3}}\,(1 + u + \tfrac{2}{3}\eta) = 0.95\,\frac{a}{\sqrt{3}} = 2.61\ \text{Å} \qquad\qquad (11.12a)$$

and

$$\frac{a}{\sqrt{3}}\,(1 - \tfrac{1}{2}u + \tfrac{2}{3}\eta) = 0.99\,\frac{a}{\sqrt{3}} = 2.72\ \text{Å} \qquad\qquad (11.12b)$$

Each oxygen atom has four Al neighbours, two below and two above—cf. Figure 11-5(b). The sum of the bond strengths is thus $4 \times (3/6) = 2$, satisfying Pauling's rule.

Oxides of other trivalent metals have the corundum structure, notably Cr_2O_3 and Fe_2O_3. Since some of these oxides possess alternative structures, the corundum structure is given the distinctive prefix α.

The next two structures to be considered, ilmenite and lithium niobate, are sometimes classed as members of the corundum family. This is legitimate for ilmenite, but lithium niobate is fundamentally different, and though comparisons between the three are useful, the structures are best treated separately.

11.5 THE ILMENITE STRUCTURE

Essentially, the ilmenite structure is a relation of corundum in which there are two different cations playing rather similar chemical roles; i.e., they are not too unequal in bond lengths and bond strengths. The departures from ideal close-packing are similar to those in corundum.

> General formula: ABO_3; example, $FeTiO_3$ (the mineral *ilmenite*)
> Lattice: Rhombohedrally-centred hexagonal, with $a = 5.082$ Å, $c = 14.026$ Å
> Number of formula units per unit cell: 6
>
> Space group: $R\bar{3}$ (No. 148)
> Atomic coordinates:
>   Fe in $6(c)$: $0, 0, \frac{1}{3} + w_1$, with $w_1 = 0.025$
>   Ti in $6(c)$: $0, 0, \frac{1}{6} + w_2$, with $w_2 = -0.025$
>   O in $18(f)$: x, y, z, with $x = 0.305$, $y = 0.015$, $z = 0.250$

As for corundum, it is convenient to express the parameters in terms of small quantities defining the differences from close-packing. Writing

$$x = \tfrac{1}{3} + u, \qquad y = v, \qquad z = \tfrac{1}{4} + w_0, \qquad \frac{c}{\sqrt{8}\,a} = 1 + \eta$$

we have

$$u = -0.028, \qquad v = 0.015, \qquad w_0 = 0, \qquad \eta = -0.024$$

The close relationship to the corundum structure is obvious. The differences are that the two Al atoms related by the c-glide plane have been replaced by unrelated atoms Fe and Ti (whose w parameters have, however, accidentally the same magnitude); and that the O atoms have moved away from the plane $y = 0$. Either of these differences would by itself have destroyed the c-glide plane; but the centre of symmetry is retained.

To construct the array of oxygens, we first operate with $R3$, obtaining the set

$$(0, 0, 0; \tfrac{2}{3}, \tfrac{1}{3}, \tfrac{1}{3}; \tfrac{1}{3}, \tfrac{2}{3}, \tfrac{2}{3}) + (x, y, z; \bar{y}, x - y, z; y - x, \bar{x}, z)$$

This resembles the set shown in Figure 11-5(a), but with each group of three atoms at the heights 3/12, 7/12, 11/12, rotated slightly clockwise about its triad axis. Operation of the centre of symmetry gives the complete set shown in Figure 11-7(a). Fe and Ti lie on the triad axes, at heights shown in Figure 11-8(a).

The same octahedral interstices are occupied as in Al_2O_3, but now one of the pair which share a face contains Ti and the other contains Fe. One sheet of octahedra parallel to (0001) contains only Ti or only Fe, as can be seen from Figure 11-8(a). The octahedra are rather more distorted than in Al_2O_3, partly because of the larger value of $|u|$, and partly because of the additional twist due to the v parameter. The effect of the latter is to increase the size of the octahedra occupied by Fe relative to those occupied by Ti. This can be verified by joining oxygen atoms at appropriate levels in Figure 11-7(a) so as to outline the octahedra, as was done in Figure 11-5(c) and (d).* The fact that the w_0 parameter happens to be zero means that the layers of O atoms parallel to (0001) are plane, and not puckered, as would be permissible in this space group.

Each O has two Fe and two Ti neighbours; the sum of the bond strengths is $2 \times (2/6) + 2 \times (4/6) = 2$, as expected.

Isomorphous materials, which again can have the more general formula ABX_3, require cations that are the right size for 6-coordination and that have valencies satisfying the relation $q_A + q_B = -3q_X$. Since there is large choice of cations of this size range, we might expect many isomorphous compounds—more, in fact, than are found. The reason for their non-occurrence is probably (at least in part) that changes in the parameters u and v have a large effect on the regularity of the octahedra and a relatively small effect on their volumes, and that therefore the permissible cation sizes are more narrowly restricted than would appear at first sight.

Among the reported isomorphous materials, we can be sure only of those whose structures have been analysed in some detail, namely $NiTiO_3$, $MnTiO_3$, $NiMnO_3$, and $CoMnO_3$. Table 11-1 gives their parameters, and the Goldschmidt radii of the cations (counting A as 2-valent and B as 4-valent). It is noteworthy that for the $AMnO_3$ compounds (as for Al_2O_3) v is zero, and hence that the two kinds of octahedra are identical—a result which would be slightly surprising if we did not realise that tabulated ionic radii are not reliable guides when A and B are similar transition metals, each capable of showing different valencies. For the $ATiO_3$ compounds, the non-zero v allows a rather smaller octahedron for Ti than for the A cation.

11.6 LITHIUM NIOBATE STRUCTURE

In contrast to ilmenite, the two cations here are very different, and consequently distort the hexagonal close-packing in very different ways in their immediate environment. The resulting structure is quite different from that of ilmenite and also from that of corundum.

> General formula: ABO_3; example, $LiNbO_3$
> Lattice: Rhombohedrally-centred hexagonal, with $a = 5.148$ Å, $c = 13.863$ Å
> Number of formula units per unit cell: 6
> Space group: $R3c$ (No. 161)
> Atomic coordinates:
> Nb in $6(a)$: $0, 0, w_1$, with $w_1 = 0.0186$
> Li in $6(a)$: $0, 0, \frac{1}{3} + w_2$, with $w_2 = -0.0318$
> O in $18(b)$: x, y, z, with $x = 0.0492$, $y = 0.3446$, $z = 0.0833$

---

* In all figures, the values of u and v have been exaggerated to display their effects better; comparison of Figure 11-5 and Figure 11-7 shows only the part of the effect due to v, since u has been kept the same in both.

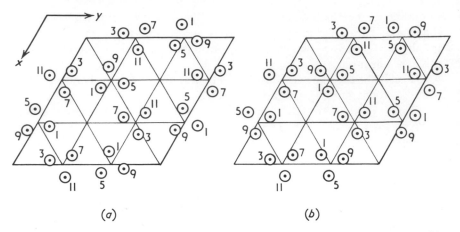

Figure 11-7 (a) FeTiO$_3$ structure: projection on (0001) of array of oxygen atoms with symmetry $R\overline{3}$ (drawn with $u = -0.067$, $v = 0.033$). (b) Projection on (0001) of array of oxygen atoms with symmetry $R3c$ (the symmetry of LiNbO$_3$) and parameters with the same signs and magnitudes as in (a). Heights are in units of $c/12$.

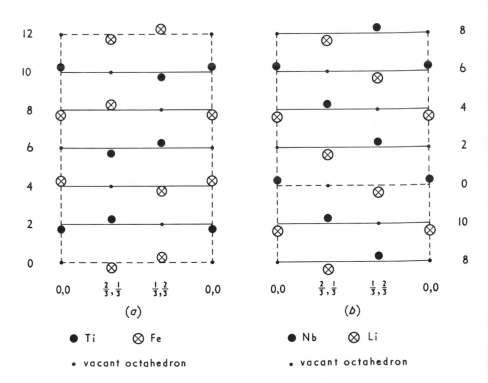

Figure 11-8 Sections parallel to (11$\overline{2}$0), showing cation arrangement: (a) FeTiO$_3$; (b) LiNbO$_3$. The choice of origin in (b), at a different level from that in (a), matches the choice in the text, which is that in the literature.

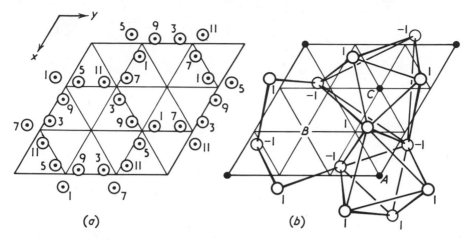

Figure 11-9 LiNbO$_3$: projection of parts of structure on (0001) (drawn with $u = 0.100$, $v = 0.033$). Heights are in units of $c/12$. (a) Complete array of oxygens; (b) octahedra surrounding cations at heights near zero.

The structure has obvious resemblances to that of ilmenite, and for many years was wrongly thought to be identical with it. The differences are fundamental, however, and not of a kind that allow the ilmenite structure to be taken as a good approximation. The array of oxygens is approximately the same and so are the sets of interstices occupied by cations, but there are important differences both in the detailed distortion of the oxygen array and in the distribution of cations among the sites.

The atomic parameters of LiNbO$_3$ appear to be different from those of FeTiO$_3$, but part of this is only due to a different choice of origin, and we must look more carefully to find the real distinction.

Suppose we had operated with $R3c$ on the prototype O atom of ilmenite, we should have obtained the result shown in Figure 11-7(b). Comparing it with Figure 11-7(a), we see that alternate sets of three O atoms, those at heights 1/12, 5/12, 9/12, are rotated counterclockwise in (b), whereas all sets of three are rotated clockwise in (a). The effect on the octahedral interstices is that, while the two smaller ones are individually of the same size as in ilmenite, they now occur in the same (0001) sheet. There are now only three kinds of interstices, and each sheet in fact contains all three kinds, the two small ones and an unsymmetrical large one; whereas in ilmenite each sheet contains two small equal interstices and one symmetrical large one, but both the small and the large differ from sheet to sheet. This statement can be verified by joining up oxygen atoms at appropriate heights in Figure 11-7(b), as was done in Figure 11-5—say those at heights 1/12 and 3/12.

We now compare parameters in the two structures. Operating on the prototype O atom of ilmenite, that at $1/3 + u$, v, $1/4$, with the rhombohedral lattice vector 2/3, 1/3, 1/3, and changing the origin by $c/2$, we obtain coordinates u, $1/3 + v$, $1/12$. These have the same form as the coordinates of

Table 11-1 Parameters of Some ABO_3 Compounds Based on Hexagonal Close-Packing

STRUCTURE TYPE	FORMULA	IONIC RADII*		LATTICE PARAMETERS			ATOMIC POSITION PARAMETERS				
		r_A Å	r_B Å	c Å	a Å	η	u	v	w_A	w_B	w_0
Corundum	Al_2O_3	0.57	0.57	13.00	4.76	−0.035	−0.027	—	0.014	—	—
Ilmenite	$NiTiO_3$	0.78	0.64	13.82	5.04	−0.031	−0.026	0.013	0.020	−0.020	−0.002
	$FeTiO_3$	0.83	0.64	14.06	5.08	−0.025	−0.028	0.015	0.025	−0.025	0
	$MnTiO_3$	0.91	0.64	14.28	5.14	−0.017	−0.016	0.023	0.024	−0.024	0
	$NiMnO_3$	0.78	0.52	13.59	4.90	−0.020	−0.023	0	0	0	0
	$CoMnO_3$	0.82	0.52	13.91	4.93	−0.003	−0.013	0	0	0	0
Lithium niobate	$LiNbO_3$	0.78	(0.64)	13.86	5.15	−0.048	+0.049	0.011	−0.032	0.019	—
	$LiTaO_3$	0.78	(0.64)	13.78	5.15	−0.054	+0.057	0.011	−0.040	0.014	—
Calcite	$CaCO_3$	1.06	(0.15)	17.02	4.99	+0.21	−0.076	—	—	—	—
	$MgCO_3$	0.78	(0.15)	15.01	4.63	+0.15	?	—	—	—	—
	$FeCO_3$	0.83	(0.15)	15.29	4.63	+0.17	−0.06	—	—	—	—
	$MnCO_3$	0.91	(0.15)	15.66	4.73	+0.16	−0.06	—	—	—	—
	$NaNO_3$	0.98	(∼0)	16.82	5.07	+0.18	−0.08	—	—	—	—

The lattice parameter $\eta(= c/(\sqrt{8}a) - 1)$ measures the departure of the axial ratio from that for ideal packing. The atomic position parameter $u(= x - 1/3)$ measures the angle of rotation of the octahedra from their ideal orientation. The lattice parameter η depends on the position parameter u and the distortion of the octahedra; if the octahedra are regular, η can be calculated geometrically from u.

* Ionic radii are those of Goldschmidt (Table 2-2) except those in round brackets, for which there are no Goldschmidt values.

the chosen prototype atom in $LiNbO_3$, with $x = u$, $y = 1/3 + v$. It can now be recognised that, whereas u is negative for ilmenite, it is positive for $LiNbO_3$. This difference of sign is important. Its effect can be seen in Figure 11-9(a), the actual $LiNbO_3$ structure. Each projected triangle of O atoms has turned inside out. (The reversal would have been exact if v were zero.) The effect of this on the octahedral interstices is very important, and is shown in Figure 11-9(b). They are still of three kinds, which we may refer to as small (at A), medium (at C), and large (at B), but now the small ones remain nearly regular (if v were zero, they would be perfectly regular), while the medium ones are both distorted and considerably enlarged. In the column of sites projecting at $(0, 0)$, small interstices occur at heights 0 and $6/12$, medium interstices at $4/12$ and $10/12$, and large interstices at $2/12$ and $8/12$. Obviously, the small interstices are suitable sites for Nb, the medium ones for Li, and the large ones remain empty.

The pattern of cation arrangement in the vertical plane is shown in Figure 11-8(b). Because of the different symmetry, each layer now contains both Nb and Li cations; indeed, were it not for the oxygen atoms, the cell height would be half that for $FeTiO_3$. There is no centre of symmetry, and therefore the origin may be chosen at any height along the triad axis; but since the oxygen layers are truly plane and equidistant, it is convenient to choose the origin midway between them, at the centre of an octahedron. The octahedron chosen is that containing Nb, as shown in the figure.

The cations, like the oxygen atoms, are displaced from their ideal positions midway between layers. They are displaced away from each other and towards the empty octahedron of the same column, just as they were in $FeTiO_3$; but in $LiNbO_3$ this implies that all Nb atoms are displaced in the same direction as each other, and all Li atoms in the opposite direction. This important feature of the structure is a consequence of its polar symmetry. Another way of considering it is to compare the cation-cation vectors in pairs of octahedra sharing faces; whereas in $FeTiO_3$ half the Fe $\rightarrow$ Ti vectors point along $+z$ and half along $-z$, in $LiNbO_3$ *all* the Li $\rightarrow$ Nb vectors point along $+z$.

Electrostatically, Pauling's rule is satisfied; each O atom has two Nb neighbours with bond strength $5/6$ and two Li neighbours with bond strength $1/6$. In Figure 11-9(b), for example, the oxygen atom at height $1/12$, which is joined to two cations at height zero, Nb in column A and Li in column C, has also two neighbours at height $2/12$, Nb in column C and Li in column B.

The great discrepancy in bond strength between Nb—O and Li—O suggests that the former will play a much larger part in determining the geometry of the structure, and that it would be worth while to consider the Nb—O octahedra alone, completely ignoring Li. In support of this is the fact that the face of the Nb octahedron shared with the Li octahedron—the lower face of the octahedron at A in Figure 11-9(b)—is only slightly smaller than the unshared face parallel to it.

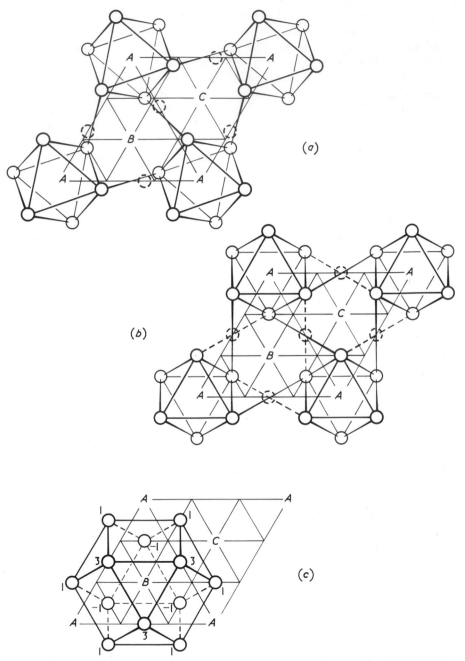

Figure 11-10 Relation between LiNbO$_3$ and perovskite framework. (a) LiNbO$_3$: same part of structure as in Figure 11-9(b), but with $v = 0$. The O atoms at $(1/12)c$ are shown as heavy circles, those at $-(1/12)c$ as light circles, and those at $(3/12)c$ as broken circles. (b) Part of perovskite framework. The O atoms at $(1/6)c'$ are shown as heavy circles, those at $-(1/6)c'$ as light circles, and those at $(3/6)c'$ as broken circles, where $c' = (1/2)c$. (c) Part of (b), with O atoms joined to outline the cavity at height 1. (Heights are in units of $c'/6$ or $c/12$.)

Each corner of the Nb octahedron is shared with another octahedron, the whole set forming a corner-linked network. This at once suggests that there may be a relationship to the perovskite octahedral framework, and so it turns out. Imagine a model of the perovskite framework with rigid octahedra but free hinges at the corners. Set it with a triad axis vertical, so that each octahedron has a pair of faces in the horizontal plane. Now compress it gently in the horizontal plane without allowing it to expand vertically. The octahedra rotate about their triad axes, alternate octahedra in opposite directions, keeping their upper and lower faces horizontal, until all the O atoms are close-packed. This is the idealised $LiNbO_3$ structure.

The process can be followed in the reverse direction in Figure 11-10. We begin with the $LiNbO_3$ structure, putting $v = 0$, as in Figure 11-10(a). This shows the same part of the structure as in Figure 11-9(b), with additional atoms at height 3/12 inserted as broken circles. The octahedra of this sheet at B and C are centrosymmetrically related. As the parameter increases, the octahedra at A rotate but remain regular, while those at B and C become greatly distorted. Round B, for example, the three nearest oxygen atoms at height 1/12 move out, and the three next-nearest close in, till all are equidistant. This situation is shown in 11-10(b), and represents the ideal perovskite framework. The new positions of the oxygen atoms at height 3/12 are also shown. It can be seen that the centre of the cavity projecting at B is no longer at height zero but at height 1/12. This cavity is shown in Figure 11-10(c); it is 12-coordinated. What has happened is that the two similar cavities above B at heights zero and 2/12 have merged to one large cavity at height 1/12, while the octahedron at height 4/12 (like that at A at height zero) remains regular but rotated. It can be seen by constructing further layers that the oxygen layer at height 7/12 is an exact repeat of that at height 1/12, and the c-repeat distance is therefore half its value in the original $LiNbO_3$ structure.

In practice, of course, we cannot expect that any structure will undergo such large volume changes (1/4 of the perovskite volume, or 1/3 of the $LiNbO_3$ volume), without breaking up. The relationship is interesting, however, and will be discussed further in §12.9.

In any case, the picture of $LiNbO_3$ as a framework of corner-linked Nb octahedra with Li filling suitable interstices is a more realistic one for many purposes than the picture involving face-sharing pairs of octahedra used successfully for Al_2O_3 and $FeTiO_3$. The much greater regularity of the Nb octahedra confirms this.

We may write down expressions for some of the interatomic distances as follows:

$$Nb\text{—}O \quad \text{(i)} \quad a\{(\tfrac{1}{3} + v)^2 - (\tfrac{1}{3} + v)u + u^2 + 8(1 + \eta)^2(\tfrac{1}{12} - w_1)^2\}^{\frac{1}{2}}$$

$$= 0.900 \frac{a}{\sqrt{6}} = 1.89 \text{ Å} \quad \textbf{(11.13a)}$$

(ii) $a\{(\frac{1}{3} - v)^2 - (\frac{1}{3} - v)u + u^2 + 8(1 + \eta)^2(\frac{1}{12} + w_1)^2\}^{\frac{1}{2}}$

$$= 1.005 \frac{a}{\sqrt{6}} = 2.11 \text{ Å} \quad (11.13b)$$

Li—O   (i) $a\{(\frac{1}{3} - v)^2 - (\frac{1}{3} - v)u + u^2 + 8(1 + \eta)^2(\frac{1}{12} - w_2)^2\}^{\frac{1}{2}}$

$$= 1.067 \frac{a}{\sqrt{6}} = 2.24 \text{ Å} \quad (11.13c)$$

(ii) $a\{(\frac{1}{3} + u)^2 - (\frac{1}{3} + u)v + v^2 + 8(1 + \eta)^2(\frac{1}{12} + w_2)^2\}^{\frac{1}{2}}$

$$= 0.986 \frac{a}{\sqrt{6}} = 2.07 \text{ Å} \quad (11.13d)$$

(In each case, the longer bond is to the oxygen atoms shared between the two octahedra.)

O—O   (i) in (0001) plane, common to Nb and Li octahedra

$$\frac{a}{\sqrt{3}} \{1 - 3(u + v) + 9(u^2 - uv + v^2)\}^{\frac{1}{2}}$$

$$= 0.915 \frac{a}{\sqrt{3}} = 2.72 \text{ Å} \quad (11.13e)$$

(ii) in (0001) plane, Nb octahedron, not shared

$$\frac{a}{\sqrt{3}} \{1 - 3(u - 2v) + 9(u^2 - uv + v^2)\}^{\frac{1}{2}}$$

$$= 0.968 \frac{a}{\sqrt{3}} = 2.88 \text{ Å} \quad (11.13f)$$

(iii) in (0001) plane, Li octahedron, not shared

$$\frac{a}{\sqrt{3}} \{1 + 3(2u - v) + 9(u^2 - uv + v^2)\}^{\frac{1}{2}}$$

$$= 1.131 \frac{a}{\sqrt{3}} = 3.36 \text{ Å} \quad (11.13g)$$

(iv) sloping edge

$$\frac{a}{\sqrt{3}} \{1 - (u + v) + 3(u + v)^2 + \frac{4}{3}\eta + \frac{2}{3}\eta^2\}^{\frac{1}{2}}$$

$$= 0.942 \frac{a}{\sqrt{3}} = 2.80 \text{ Å} \quad (11.13h)$$

(v) sloping edge

$$\frac{a}{\sqrt{3}} \{1 - (u - 2v) + 3(u - 2v)^2 + \tfrac{4}{3}\eta + \tfrac{2}{3}\eta^2\}^{\frac{1}{2}}$$

$$= 0.955 \frac{a}{\sqrt{3}} = 2.84 \text{ Å} \quad \text{(11.13i)}$$

The relative differences between pairs of O—O edges (i) and (ii) or (iv) and (v) depend essentially on v and are small, whereas the difference between (i) and (iii) depends on u and is conspicuously large, implying a very large distortion of the Li octahedron from regularity.

The only isomorphous structure at present known is $LiTaO_3$.

Strictly speaking, in giving a description here of the room-temperature $LiNbO_3$ structure we have broken the rule of choosing, for this chapter, the simplest and most symmetrical member of a structure family. The simplest member of the family is the form found above 1200° C, which differs in having a higher space group $R\bar{3}c$, oxygen parameter $v = 0$, Nb parameter $w_1 = 0$, and Li parameter $w_2 = 0.25$. The Nb is now central in its octahedron, while Li has moved to the centre of the triangle between the two other octahedra, which has become larger to hold it.* This will be discussed further in §12.9.

11.7 THE CALCITE STRUCTURE

Calcite, $CaCO_3$, is one of the best-known minerals. Large well-formed crystals are reasonably common, and cleavage fragments with nearly perfect faces are familiar from their use in teaching to illustrate the double refraction of light by anisotropic materials.

The formal description of the structure is as follows:

Lattice: Rhombohedrally-centred hexagonal, with $a = 4.990$ Å, $c = 17.002$ Å
Number of formula units per unit cell: 6
Space group: $R\bar{3}c$ (No. 167)
Atomic coordinates:
 Ca in 6(b): 0, 0, 0
 C in 6(a): 0, 0, $\frac{1}{4}$
 O in 18(e): x, 0, $\frac{1}{4}$, with $x = 0.257 = \frac{1}{3} - 0.076$

As with all structures having a rhombohedral lattice, rhombohedral axes of reference giving a primitive unit cell could have been chosen. They are less convenient to use, in spite of the smaller number of formula units per cell, but, because they have been much used in the past for calcite, the alternative

---

* The centre of the triangle is the *average* position of Li. More experimental evidence is needed before we can say for certain whether the atom is ordered in this position or disordered between a pair of positions on either side of the triangle.

description is given. For a discussion of the geometrical relationships between the two, see §10.3(iii).

> Lattice: rhombohedral, with $a = 6.36 \text{Å}$, $\alpha = 46°5'$
> Number of formula units per unit cell: 2
> Atomic coordinates:
> Ca in 2(b): 0, 0, 0
> C in 2(a): $\frac{1}{4}$, $\frac{1}{4}$, $\frac{1}{4}$
> O in 6(e): x', $\frac{1}{2} - x'$, $\frac{1}{4}$, with $x' = \frac{1}{4} + x = 0.507$

We use hexagonal axes of reference in the discussion that follows.

If we compare the formal description of calcite with that of corundum (§11.14), we see that the oxygen positions are exactly similar except for the numerical value of the x-parameter. Figure 11-5(b) illustrates this oxygen arrangement. If x were exactly 1/3, the oxygen would be in ideal hexagonal packing; the greater value of $1/3 - x$ for calcite—0.076 as compared with 0.027 for corundum—indicates a greater deviation from the ideal.

The vital difference between the two structures is in the placing of the cations. In both structures, they lie on the triad axes. In §11.14 we saw that there are six octahedral interstices on each triad axis within the height of the unit cell. Two of these, at heights 0 and 6/12, coincide with centres of symmetry, whereas the others, at $\pm 2/12$, $\pm 4/12$, do not. In corundum, the Al atoms were placed at interstices of the latter kind; in calcite, the Ca atoms are at the centres of symmetry, and the other octahedra are empty. To construct a projection of calcite comparable to Figure 11-6, we should have to remove cations from all the occupied octahedra and put half as many into the empty octahedra; and then we should have to find sites for the C atoms. There are suitable special positions at the centre of the shared face between what are now empty octahedra. The three oxygen-oxygen edges of this face were already somewhat shortened in corundum; they are further shortened in calcite as the strong C—O bonds pull the oxygen atoms together. This explains the greater departure from ideal close packing in calcite.

The arrangement of the CO_3 groups is shown in Figure 11-11(a) and (b). All groups at the same height—for example, at $(3/12)c$—lie in the same plane and have the same orientation. Groups at the next height level—a difference of $(2/12)c$—have the opposite orientation. It can be seen at once from the symmetry that this must be so, since layers at height differences of $(4/12)c$ are related by the rhombohedral lattice translation and must be identical in orientation, while layers at height differences of $(6/12)c$ are related by the c-glide plane and thus reversed in orientation.

The arrangement of the Ca octahedra is shown in Figure 11-11(c) and (d). We see the same difference between the top and bottom halves of the unit cell as for the CO_3 groups: octahedra centred on the same triad axis are not parallel in orientation, but are related by the c-glide plane.

The Ca atom is thus 6-coordinated, and the C atom is 3-coordinated, giving electrostatic valences for Ca—O and C—O of 2/6 and 4/3 respectively. Each O atom has one C neighbour and two Ca neighbours.

Pauling's rule is thus satisfied, since $(2/6) \times 2 + (4/3) \times 1 = 2$. Of course, it is unrealistic to treat the C—O bond as if it were electrostatic—it is much more nearly homopolar. However, this approach leads to the same result as more sophisticated theories of bonding, namely, that the C—O bonds are much the strongest in the structure, and the $(CO_3)^{2-}$ groups thus behave as "molecule-ions." Both chemically and physically these $(CO_3)^{2-}$ groups have vital effects on the properties.

We may calculate bond lengths in the same way as we did for corundum, using Figure 11-5(b). We have

$$C—O = \frac{a}{3}(1 - 3u) = 0.77\frac{a}{3} = 1.28 \text{ Å} \qquad (11.14a)$$

$$Ca—O = \frac{a}{3}[1 + 3u + 9u^2 + \tfrac{1}{2}(1 - \eta)^2]^{\frac{1}{2}} = 1.16\frac{a}{\sqrt{6}} = 2.35 \text{ Å} \qquad (11.14b)$$

The expression for Ca—O is exactly the same as that for Al—O$_{\text{unshared}}$ in corundum—equation (11.8b)—with $w \equiv 0$.

The four different O—O distances can be calculated from the general expressions given in §11.4, but now the first (there called l_{shared}) is the edge of the CO$_3$ group, the second and fourth are the edges of the Ca octahedron in and inclined to the (0001) plane respectively, and the third (here written last) is a van der Waals contact between O atoms not bonded to the same cation. The numerical values are:

O—O (CO$_3$ group) $= 0.77a/\sqrt{3} = 2.22$ Å
O—O (octahedron edge perpendicular to triad axis) $= 1.13a/\sqrt{3} = 3.25$ Å
O—O (van der Waals) $= 1.10a/\sqrt{3} = 3.17$ Å

We notice how much longer the octahedron edges are than in corundum—a fact attributable to the much larger size of the cation. They are even a little longer than the van der Waals contact.

There is a different way of considering the structure, due to W. L. Bragg, who first determined the structure in 1914. He pointed out that if we treat the $(CO_3)^{2-}$ ion as a rigid unit, and disregard its shape, the structure of calcite formally resembles that of rock salt, NaCl (§4.7). Though in practice this is less useful than an approach which gives separate consideration to the oxygen atoms, like that used in the foregoing, it is worth tracing the connection between the two.

We begin with a simple cube projected down its triad axis, as shown in Figure 11-12(a). We shall want to use hexagonal axes of reference, and these are shown in their conventional relation to the rhombohedral (or cubic) axes. However, the rock-salt structure has a face-centred lattice, and this means we can choose a primitive unit cell which is a rhombohedron of the same height as the original cube but one quarter the projected area—cf. §10.10(ii) and

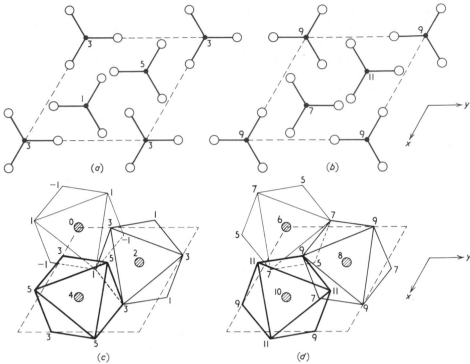

Figure 11-11 Calcite. (a) and (b) Projections showing arrangement of CO_3 groups in the two halves of the unit cell from heights 0 to $(1/2)c$ and $(1/2)c$ to c respectively. (c) and (d) Projections showing arrangement of CaO_6 octahedra from heights $-(1/12)c$ to $(5/12)c$ and $(5/12)c$ to $(11/12)c$ respectively. The base of the unit cell is shown by a broken outline. Heights of the atoms are marked in units of $c/12$.

Figure 10-9(a). This new cell is shown in Figure 11-12(b) on the same scale as in (a). (The change of direction of the hexagonal axes is necessary to keep the rhombohedron an obverse one, according to convention.) If we now put Ca atoms at the lattice corners, and spherical $(CO_3)^{2-}$ groups at height $\frac{1}{2}c$ above each, we have a structure identical with NaCl and representing a first approximation to calcite. The $(CO_3)^{2-}$ groups are, of course, not spherical but planar, and their plane is normal to the triad axis; hence we must flatten the structure, compressing it down the triad axis and extending it in the plane at right angles, to allow for this. But this is not all, because the $(CO_3)^{2-}$ groups are not in fact circular discs, but triangles. The sites allotted to them in the distorted NaCl structure are centres of symmetry, and they themselves are not centrosymmetric. To allow for this, the height of the unit cell must be doubled, successive groups now being related by a c-glide plane instead of a simple translation. Correspondingly, the z coordinates of atomic positions must be halved; we replace z in Figure 11-12(b) by $\frac{1}{2}z$ and $\frac{1}{2} + \frac{1}{2}z$ in Figure 11.12(c). This necessitates a further change of direction of hexagonal axes, back to that of Figure 11-12(a). (In each of these figures it

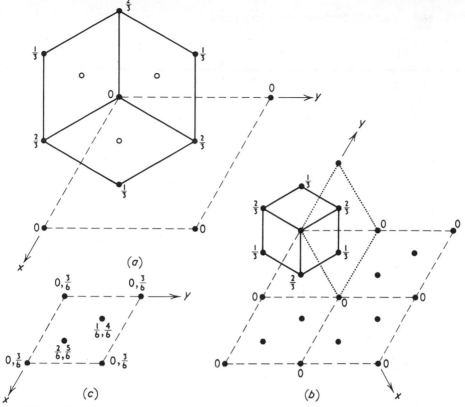

Figure 11-12 Relation of calcite and rock salt structures. (a) Projection of simple cube down its triad axis. Black circles represent atoms at points of a simple cubic lattice, solid lines the outline of the projected cube and the edges meeting in its upper corner; small open circles mark points at the centres of the upper faces, at height $(2/3)c$. Broken lines show the base of the hexagonal unit cell. (b) Projection of primitive unit cell of a face-centred cube, in the same orientation and to the same scale. Dotted lines show the base of the hexagonal unit cell. [Note: The open circles of (a) are not lattice points; the hexagonal lattice translation a is therefore halved in (b); the hexagonal axial directions are rotated through 60°.] (c) Effect of doubling the height c of the hexagonal (or rhombohedral) unit cell. Broken lines show the base of the hexagonal unit cell, which is the same as that used in Figure 11-11(a) to (d).

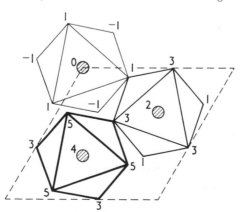

Figure 11-13 Octahedral arrangement in LiNbO$_3$, drawn for comparison with that in calcite, Figure 11-11(d). To see the resemblance most clearly, the origin in LiNbO$_3$ must be moved by $c/2$; i.e., all heights marked in this figure must be increased by 6.

should be noted that the projection of a rhombohedral cell edge joining the origin to a lattice point at height $\frac{1}{3}c$ lies 30° anticlockwise from the X or Y axis, in accordance with convention.) If we now orient the CO_3 groups at height $(3/12)c$ so that C—O bonds point along $+OX$ and $+OY$, and those at $(5/12)c$ so that C—O bonds point along $-OX$ and $-OY$, application of the lattice translations of the true rhombohedral cell gives us the complete structure derived earlier, and shown in Figure 11-11(a) to (d).

For discussion of the oxygen arrangement, we have been making much use of comparison with corundum. The structure as a whole, however, has a much more important relationship to $LiNbO_3$, and especially the high-temperature form of the latter which is mentioned at the end of §11.6. There is a one-to-one correspondence for all atoms. The only geometrical difference, in fact, is in the values of the arbitrary parameters. If we had chosen the same origin and the same prototype atom for $LiNbO_3$ as for calcite, we should have had $x = 1/3 + 0.049$ where calcite has $x = 1/3 - 0.074$. The consequence of this difference of sign is that, whereas in calcite the octahedra overlap in projection, as shown in Figure 11-11(d), in $LiNbO_3$ they have gaps between them, as shown in Figure 11-13.

Geometrically speaking, we might say that calcite is isomorphous with high-temperature $LiNbO_3$, since both structures have the same formal composition ABO_3 and the same symmetry, not only macroscopic symmetry but also symmetry of detailed atomic positions, differing only in the values of their arbitrary parameters, c, a, and x. However, from a physical standpoint the big difference in the *magnitude* of the x parameter matters, because it is associated with a big difference in the relative strengths of the bonds. In $LiNbO_3$, the bonds in the octahedron are the strongest, and the Li—O bonds are weak; if there are changes in the structure, as for example may be caused by thermal expansion, we expect the octahedra to remain nearly regular while the Li environment changes. In calcite, the reverse is the case; the CO_3 groups are likely to remain rigid, while the octahedra become distorted.

It is worth noticing the shape and size of the CaO_6 octahedron. We have already seen that the O—O edges are 3.25 and 3.40 Å respectively; this means that the octahedron is slightly elongated along the triad axis. The angle its edge makes with the axis is in fact 53.1°, as compared with 54.7° for an ideal octahedron. There are two ways of looking at this fact. On the one hand, we may say it is not surprising, for we know that Ca is rather large for 6-coordination (cf. Table 2-6) and the O—O distances are greater than van der Waals contact distances, so there is nothing to prevent distortion. On the other hand, we may ask why it should occur at all, since it is perfectly possible, geometrically, to construct a structure without any distortion. What we have overlooked, however, in our treatment so far, is the role of forces which are not nearest-neighbour forces. In this case, we notice that there can be electrostatic repulsion between Ca and C along the triad axis (even though C is not an ion in a strict sense)—they "see" each other through a face of the octahedron—and this, though small, is big enough to produce distortion when the octahedron is, as here, "soft."

There are many materials with isomorphous structures. Among the carbonates we may note $MgCO_3$, $FeCO_3$, and $MnCO_3$, for which data are given in Table 11-1. Another isomorphous structure listed in the table is $NaNO_3$. This resembles $CaCO_3$ particularly closely because of the similarity of ionic size between Na and Ca, NO_3 and CO_3; the forces within the octahedra are of course much weaker.

It is interesting to notice that dolomite, $CaMg(CO_3)_2$, has a structure which, while not isomorphous with calcite, is closely related to it, in much the same way that ilmenite is related to corundum. As in ilmenite, the presence of two different octahedral cations destroys the c-glide plane but not the centre of symmetry, and the space group becomes $R\bar{3}$.

The elements with the closest chemical resemblance to Ca, namely Sr and Ba, do *not* give carbonates with the calcite structure. We can easily understand this from a knowledge of their sizes. The Ca atom is already rather on the large side for 6-coordination by oxygen, and Sr and Ba are so large that they cannot tolerate it. We shall come back to this point in the next section.

11.8 THE ARAGONITE STRUCTURE

There are other structures besides that of calcite possible for $CaCO_3$, and the most important of them is found in the mineral aragonite. Its description is as follows:*

> Lattice: Orthorhombic, primitive, with $a = 4.94$ Å, $b = 7.94$ Å, $c = 5.72$ Å
> Space group: *Pmcn* (No. 62)
> Number of formula units per unit cell: 4
> Atomic coordinates:
> Ca   in 4(c): $\frac{1}{4}$, 0.08, 0.25
> C   in 4(c): $\frac{1}{4}$, 0.75, 0.08
> O(1) in 4(c): $\frac{1}{4}$, 0.59, 0.08
> O(2) in 8(d):   0.02, 0.82, 0.08

Here again the origin is taken at the centre of symmetry, and hence the mirror plane, the c-glide plane, and the n-glide plane are at $x = 1/4$, $y = 1/4$, $z = 1/4$ respectively.

The structure is as shown in Figure 11-14(a) and (b). In projection on (001) it looks very nearly hexagonal, and this is physically an important

---

* The choice of axes here (though not of origin) is that made in the original structure determination by W. L. Bragg, which is in fact the most convenient for showing structural relationships. Later workers have sometimes used different axes, and have not always been consistent in changing the space-group symbol and atomic parameters to match their new choice of axes.

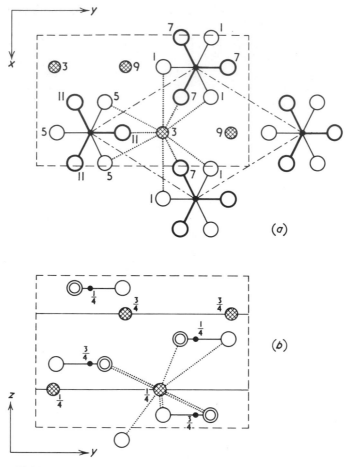

Figure 11-14 Aragonite structure. (a) Projection on (001), with heights marked in units of $c/12$. (b) Projection on (100), with heights marked as fractions of a: hatched circles Ca; small black circles C; large open circles O; double circles two O atoms superposed in projection. Dotted lines are Ca—O bonds; broken lines outline the unit cell; dot-dash lines in (a) show the pseudohexagonal unit cell; thin solid lines in (b) show planes containing cations.

characteristic. The base of the pseudohexagonal cell is outlined. If we consider the array of Ca atoms alone, they are approximately in a hexagonal close-packed formation; ideally, this would require an axial ratio of $b/a = 1.73$, whereas the observed value is 1.69. (Of course the Ca atoms are not actually in contact, being well separated by the O atoms; however, "hexagonal close packing" is a phrase often used, as here, to describe the pattern of centres rather than the physical packing.) If we replaced the CO_3 groups by spheres, we should have the NiAs structure (cf. §4.15), just as the same process in calcite would give us the NaCl structure.

The CO_3 groups all lie parallel to (001), as in calcite; but whereas, in calcite, midway between each pair of planes of Ca atoms there was a single plane of CO_3 groups all pointing in the same direction, in aragonite there are two planes, staggered, as in Figure 11-14(b), and with CO_3 groups pointing in opposite directions. The arrangement is such that each Ca atom has now 9 oxygen neighbours, at distances between 2.3 and 2.7 Å. The up-and-down staggering of the layers of CO_3 groups, and adjustment of height of the Ca atoms, allows the different Ca—O distances to be made roughly equal.

Though Ca is rather large for 6-coordination, it is rather small, at room temperature, for 9-coordination; hence it is not surprising that aragonite is less stable than calcite. On the other hand, carbonates and nitrates with larger cations—$SrCO_3$, $BaCO_3$, $PbCO_3$, and KNO_3—have the aragonite structure as their stable room-temperature structure.

11.9 THE OLIVINE* STRUCTURE

Like the preceding structures, the olivine structure is based on a hexagonal close packing of anions, but now the cations occupy different sets of interstices, including tetrahedral as well as octahedral sites, and chosen in such a way as to make the symmetry orthorhombic. There are many more variable atomic position parameters, allowing considerable freedom of adjustment.

General formula: A_2BX_4; example, Mg_2SiO_4 (forsterite)
Lattice:† Orthorhombic, primitive, with $a = 4.75$ Å, $b = 10.21$ Å, $c = 5.98$ Å
Number of formula units per unit cell: 4
Space group: *Pbnm* (No. 62)
Atomic coordinates:
Mg(1) in $4(a)$: 0, 0, 0
Mg(2) in $4(c)$: $x, y, \frac{1}{4}$, with $x = 0.990$, $y = 0.277$
Si in $4(c)$: $x, y, \frac{1}{4}$, with $x = 0.427$, $y = 0.094$
O(1) in $4(c)$: $x, y, \frac{1}{4}$, with $x = 0.766$, $y = 0.092$
O(2) in $4(c)$: $x, y, \frac{1}{4}$, with $x = 0.220$, $y = 0.448$
O(3) in $8(d)$: x, y, z, with $x = 0.278$, $y = 0.163$, $z = 0.034$

If the origin is taken at a centre of symmetry, the *b*-glide plane is at $x = 1/4$, the *n*-glide plane at $y = 1/4$, and the mirror plane at $z = 1/4$; thus, for the general position we obtain the set of points

---

* *Olivine* is the general name for a range of minerals of approximate composition (Mg, Fe)$_2$SiO$_4$, with the structure here described. The names *forsterite* and *fayalite* apply to the magnesium-rich end and the iron-rich end respectively. The data given here are for a forsterite of composition about 90 percent Mg and 10 percent Fe, which is believed to differ very little from pure Mg_2SiO_4.

† Some authors use different axes of reference; e.g., Wyckoff takes a_w, b_w, c_w equal to our b, c, a respectively, getting the space group in the form *Pnma*. It is preferable to stick to the axes which were already well-established in the literature, as we have done.

$(x, y, z; 1/2 - x, 1/2 + y, 1/2 - z; x, y, 1/2 - z; 1/2 - x, 1/2 - y, z)$.
Most of the atoms are in special positions on the mirror planes.

The lattice parameters give a hint that the oxygen atoms may approximate to a hexagonally close-packed array. The argument is as follows. From such an array we could pick out a large orthogonal repeat unit with cell sides $\sqrt{8/3}l$, $2\sqrt{3}l$, and $2l$, where l is the nearest-neighbour distance. For O atoms in contact, l is about 2.8 Å, giving cell sides of 4.6, 9.7, and 5.6 Å respectively. With a uniform increase of 5 percent to allow for the insertion of the cation, these become 4.8, 10.2, and 5.9 Å, which are very near the observed values. Moreover, the symmetry of such an array includes all the elements of the observed space group *Pbnm*.

To compare the actual structure with the ideal, we may (as in previous examples) write the atomic parameters in terms of small deviations from the ideal, as follows:

Mg(1) in 0, 0, 0

Mg(2) in $u_{\mathrm{Mg}}, \frac{1}{4} + v_{\mathrm{Mg}}, \frac{1}{4}$, with $u_{\mathrm{Mg}} = -0.010$, $v_{\mathrm{Mg}} = 0.027$

Si   in $\frac{5}{12} + u_{\mathrm{Si}}, \frac{1}{12} + v_{\mathrm{Si}}, \frac{1}{4}$, with $u_{\mathrm{Si}} = 0.009$, $v_{\mathrm{Si}} = 0.011$

O(1) in $\frac{3}{4} + u_1, \frac{1}{12} + v_1, \frac{1}{4}$, with $u_1 = 0.016$, $v_1 = 0.009$

O(2) in $\frac{1}{4} + u_2, \frac{5}{12} + v_2, \frac{1}{4}$, with $u_2 = -0.030$, $v_2 = 0.033$

O(3) in $\frac{1}{4} + u_3, \frac{2}{12} + v_3, w_3$, with $u_3 = 0.029$, $v_3 = -0.004$, $w_3 = 0.034$

Then if all the u, v, w values had been zero, we should have had a regular hexagonal array with Mg in octahedral interstices and Si in tetrahedral interstices.

Figure 11-15(a), (b),.. (c) shows projections of parts of the structure on (100). In (a), the oxygen atoms alone are shown, with one layer joined up to indicate how they approximate to close packing. In (b), the Mg atoms are shown within their octahedra. In (c), the Si atoms are added, giving the complete structure; the tetrahedra round the Si atoms are indicated by solid lines.

There are two kinds of octahedra, as indicated by giving the Mg atoms occupying them the symbols Mg(1) and Mg(2). The Mg(1) octahedra share edges with one another, to form ribbons parallel to [001], alternate ribbons lying at heights $x = 0$ and $x = 1/2$. The ribbons are linked to one another by Mg(2) octahedra, sharing edges with them, as shown by the heavy sloping lines in Figure 11-15(b). Together the Mg(1) and Mg(2) octahedra build up a three-dimensional framework, as shown in projection on (001) in Figure 11-15(d), where the ribbons are perpendicular to the paper. The Si atoms fit into suitable interstices in the framework, where their tetrahedra help to link the ribbons laterally in the (010) plane, as shown in Figure 11-15(e).

The deviations from ideal hexagonal packing of the oxygen atoms have two important consequences: they shorten the shared edges of octahedra, and at the same time shorten all the edges of the tetrahedra relative to those

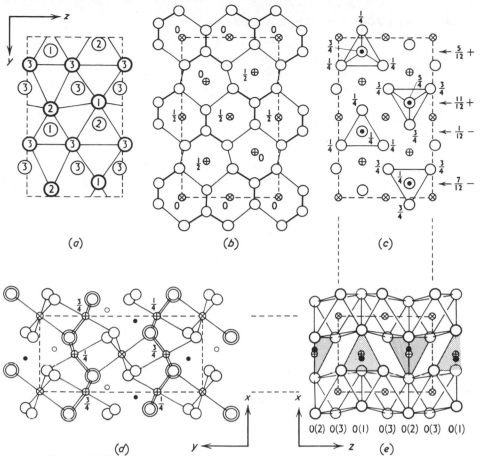

Figure 11-15 Structure of forsterite. The Mg atoms are shown as crossed circles, with a diagonal cross for Mg(1) and an erect cross for Mg(2); Si atoms are very small black or open circles; O atoms are large or medium sized open circles. The unit cell is shown by a broken line.

(a), (b), and (c) Projections on (100). In (a), only oxygen atoms are shown, numbered (as in the text) to indicate the three symmetry-independent sets. Atoms joined by lines are near $x = 1/4$; others are near $x = -1/4$. In (b), Mg atoms are included, with approximate heights marked as fractions of a, and their surrounding octahedra are outlined. Heavy lines show edges shared between octahedra. In (c), all atoms are shown, and SiO_4 tetrahedra are picked out. For downward-pointing tetrahedra, only the triangular base is shown; for upward-pointing ones, the apical edges are also shown. Heights marked within the unit cell refer to corners of tetrahedra; figures with arrows at the side refer to the height of Si atoms— all as fractions of a.

(d) Projection on (001) of the whole cell. Heights of Mg atoms are marked as fractions of c. Double lines are pairs of bonds in the same octahedron related by the mirror planes at $z = 1/4$ or $3/4$. Double circles are oxygen atoms at $z = 0$ and $z = 1/2$, coinciding in projection; other oxygen atoms are at the same heights as the Mg(2) atoms to which they are joined. The Si atoms are at $z = 1/4$ (black circles) and $z = 3/4$ (small open circles); their tetrahedra have not been outlined.

(e) Projection on (010) of the part of the structure with y between $1/4$ and $-1/4$. The O atoms drawn as heavy circles lie in a double layer with y values from 0 to 0.17; the others are below, with corresponding negative values of y. The Mg(2) atoms are above and below the ribbons of Mg(1) octahedra and the Si tetrahedra linking them; their octahedra have not been shown, to avoid confusing the diagram. The SiO_4 tetrahedra (shaded) point up and down alternately.

The other part of the unit cell ($y = 1/4$ to $3/4$) is similar but displaced by $a/2$, so that ribbons of Mg(1) octahedra lie above the rows of tetrahedra, and vice versa.

251

of the octahedra. Arguing physically, we should of course have inverted this statement—it is the need for shortening the shared edges of octahedra and all the edges of Si tetrahedra that *causes* the deviations. It is the possibility of meeting both requirements simultaneously that helps to secure the stability of the structure. A byproduct of the deviations is an increase in size of the Mg(2) octahedron relative to Mg(1), but the discrepancy is not great.

The sharing of edges can be seen in the diagram in Figure 11-15(b). Those shared between Mg(1) and Mg(2) are shown with heavy lines. They coincide in projection with three edges of the tetrahedron—Figure 11-15(c)—which share only corners with octahedra. The other three edges of the tetrahedron (those outlining the triangles) are each shared with one octahedron.

It follows that each O atom has one Si and three Mg neighbours, which may comprise either two Mg(1) and one Mg(2) atom, or one Mg(1) and two Mg(2) atoms. Pauling's electrostatic valence rule is satisfied, with $1 \times (4/4) + 3 \times (2/6) = 2$.

Because of the lower symmetry, there are more independent bond lengths to be calculated. We may illustrate this for Si—O. The neighbours of the Si atoms at $0.427, 0.094, \frac{1}{4}$ are as follows:

$$
\begin{aligned}
&O(1) \text{ at } 0.766, 0.092, \tfrac{1}{4} \\
&O(2) \text{ at } 0.280, \bar{1}.948, \tfrac{1}{4} \\
&O(3) \text{ at } \begin{cases} 0.278, 0.163, 0.034 \\ 0.278, 0.163, 0.466 \end{cases}
\end{aligned}
$$

The components of the bonds are therefore:

$$
\begin{aligned}
&\text{Si—O(1): } 0.339a, 0.002b, 0 \\
&\text{Si—O(2): } 0.147a, 0.146b, 0 \\
&\text{Si—O(3): } 0.149a, 0.069b, 0.216c \text{ (two equal bonds):}
\end{aligned}
$$

Squaring and adding, we find resultant bond lengths:

$$
\begin{aligned}
&\text{Si—O(1)} = 1.61 \text{ Å} \\
&\text{Si—O(2)} = 1.65 \text{ Å} \\
&\text{Si—O(3)} = 1.63 \text{ Å}
\end{aligned}
$$

In the same way, we can calculate the O—O edges of the tetrahedron, obtaining values 2.74 Å, 2.76 Å [2] for the unshared edges, and 2.59 Å, 2.56 Å [2] for the edges shared with octahedra. Here [2] means that there are two symmetry-equivalent bonds of the length specified. We can also calculate the three independent Mg(1)—O bonds, the four independent Mg(2)—O bonds, the six independent O—O edges of the Mg(1) octahedron, and the seven independent O—O edges of the Mg(2) octahedron.

Values worked out from the parameters are given in Table 11-2. There are interesting points of detail to be seen. Octahedron (1) is more regular than octahedron (2), if judged by the difference between extreme values of Mg—O, but less regular if judged by the difference between extreme values of

Table 11-2 Forsterite: Bond Lengths in Å

OCTAHEDRON (1)			OCTAHEDRON (2)		
Mg—O(1)	[2]	2.09	Mg—O(1)	[1]	2.18
Mg—O(2)	[2]	2.07	Mg—O(2)	[1]	2.06
Mg—O(3)	[2]	2.14	Mg—O(3)	[2]	2.07
			Mg—O(3)	[2]	2.22
Mean [6]		2.10	Mean [6]		2.14
O(1)—O(2)	[2]	3.03	O(1)—O(3)	[2]	3.03
O(1)—O(3)	[2]	3.12	O(2)—O(3)	[2]	2.94
O(2)—O(3)	[2]	3.36	O(2)—O(3)	[2]	3.19
O(1)—O(2)	[2]	2.86*	O(3)—O(3)	[2]	3.00
O(1)—O(3)	[2]	2.86†	O(3)—O(3)	[1]	3.41
O(2)—O(3)	[2]	2.56‡	O(1)—O(3)	[2]	2.86†
			O(3)—O(3)	[1]	2.59‡
Mean [12]		2.96	Mean [12]		3.00
TETRAHEDRON			TETRAHEDRON		
Si—O(1)	[1]	1.614	O(1)—O(2)	[1]	2.74
Si—O(2)	[1]	1.654	O(1)—O(3)	[2]	2.76
Si—O(3)	[2]	1.635	O(2)—O(3)	[2]	2.56‡
			O(3)—O(3)	[1]	2.59‡
Mean [4]		1.634	Mean [6]		2.66

* Shared between two octahedra of type (1).
† Shared between two octahedra of different types.
‡ Shared between octahedron and tetrahedron.

O—O edges. Some empirical reasons for differences of individual Mg—O and Si—O bonds from their mean values can be recognised. Thus, the long Si—O bonds of 1.63 and 1.65 Å, and the long Mg(2)—O bonds of 2.18 and 2.22 Å, are to O atoms of an edge shared between octahedron and tetrahedron, or between two octahedra. Again, where Mg(2)—O bonds are to O atoms not involved in shared edges, the bond is short if the Si—O bond to the same atom is long, as to O(2) and O(3); and where Si—O is short, as to O(1), Mg(2)—O is long.

Examples of isomorphous compounds are Fe_2SiO_4 (fayalite), γ-Ca_2SiO_4, Mg_2GeO_4, Al_2BeO_4 (chrysoberyl), Na_2BeF_4, $CaMgSiO_4$ (monticellite), and $LiMnPO_4$. Various interesting points may be noticed.

(i) Na_2BeF_4 is a "model structure," the atoms resembling those of Mg_2SiO_4 in size, but carrying half the charge.

(ii) The valencies of the two cations in Al_2BeO_4 and in Mg_2SiO_4 are different, but isomorphism is nevertheless possible if Pauling's rule is obeyed. The electrostatic valences are different, but their sum is the same: $3 \times (3/6) + (2/4) = 2$. The strong morphological resemblance between chrysoberyl and olivine was well-known to mineralogists long before structure determination became a possibility, and it was puzzling because of the apparently quite different chemistry of the two substances.

(iii) Because the two octahedral sites are symmetry-independent, they can be occupied in an ordered way by two different cations without destroying the isomorphism. The sizes of the two octahedra are not very different, even allowing for possible adjustments of position parameters. In some compounds, such as $LiMnPO_4$, the two kinds of A cation occupy different kinds of octahedra; but in others, for example some specimens of olivine itself, $(Mg, Fe)_2SiO_4$, they occupy both at random, even for the stoichiometric $1:1$ ratio, which would have allowed perfect ordering. The compound $CaMgSiO_4$ (monticellite) is however ordered, with Mg at the centres of symmetry.

(iv) Very interesting detailed comparisons are possible between structures for which exact parameters have been determined. These are given in Tables 11-3 and 11-4.*

Table 11-3 shows the lattice parameters and atomic position parameters, and their deviations from the ideal close-packed values. In almost every case they are of the same sign and roughly the same magnitude for all the compounds listed—as we should expect since they have the same physical cause. In consequence, individual interatomic distances deviate from their mean in much the same way for all the compounds—the individual long and short bonds correspond to those we noted in forsterite.

A more general comparison of interatomic distances is given in Table 11-4. Here we see that in all the compounds, as in forsterite, octahedron (1) is rather smaller, and has more nearly equal A—O bonds, than octahedron (2); and that for both octahedra the extreme difference of their unshared edges increases with the mean edge length, and therefore with the size of the A cation. The mean edge length of the *tetrahedron* also increases with the size of A, even when B is the same, as in the three silicates listed. (The difference is close to the limit of error of measurement of individual bond lengths, but it is probably significant for their means.) This serves as a warning against the too-glib assumption that interatomic distances—or even their mean values—can be predicted accurately by addition of radii taken from tables; the character of the neighbours may, as here, have a modifying influence.

---

* The columns headed "Fe_2SiO_4" and "Mg_2SiO_4" really refer to naturally occurring crystals with about 90 percent Fe plus 10 percent Mg, and 90 percent Mg plus 10 percent Fe, respectively. The differences between the parameters and interatomic distances quoted and the values for the pure materials are very small.

Table II-3 Parameters of Compounds with the Olivine Structure, A_2BO_4*

		γ-Ca$_2$SiO$_4$	Fe$_2$SiO$_4$	Mg$_2$SiO$_4$	Al$_2$BeO$_4$	LiMnPO$_4$
a (Å)		5.09	4.82	4.76	4.24	4.74
b (Å)		11.37	10.47	10.22	9.39	10.46
c (Å)		6.78	6.10	5.99	5.47	6.10
$c/2$ (Å)		3.39	3.05	3.00	2.73	3.05
$(\sqrt{3}/\sqrt{2})(a/c) - 1$		-0.08	-0.03_5	-0.02_5	-0.05	-0.05
$(\sqrt{3}/3)(b/c) - 1$		-0.03_5	-0.010	-0.01_5	-0.01_5	-0.01_5
ATOM	x_0, y_0, z_0	$u = x - x_0,$	$v = y - y_0,$	$w = z - z_0$		
$A(2)$	0	-0.012	-0.014	-0.010	-0.006	-0.028
	$\frac{1}{4} = 0.250$	0.030	0.030	0.027	0.023	0.032
	$\frac{1}{4}$	—	—	—	—	—
B	$\frac{5}{12} = 0.417$	0.010	0.014	0.010	0.016	-0.009
	$\frac{1}{12} = 0.083$	0.016	0.014	0.011	0.010	0.009
	$\frac{1}{6}$	—	—	—	—	—
O(1)	$\frac{3}{4} = 0.750$	-0.012	0.017	0.016	0.040	-0.016
	$\frac{1}{12} = 0.083$	0.004	0.014	0.009	0.008	0.013
	$\frac{1}{4}$	—	—	—	—	—
O(2)	$\frac{1}{4} = 0.250$	-0.052	-0.040	-0.030	-0.009	-0.043
	$\frac{5}{12} = 0.417$	0.041	0.036	0.031	0.016	0.039
	$\frac{1}{4}$	—	—	—	—	—
O(3)	$\frac{1}{4} = 0.250$	0.042	0.038	0.028	0.008	0.028
	$\frac{2}{12} = 0.167$	-0.004	-0.002	-0.004	-0.004	-0.006
	0	0.060	0.036	0.034	0.017	0.049

* The first part of this table gives the lattice parameters and some related quantities; the second part gives the ideal atomic position parameters x_0, y_0, z_0, and the displacement parameters u, v, w, derived from the actual atomic position parameters x, y, z.

The shortening of shared edges is apparent for all compounds. The two kinds of shared edges, between two octahedra and between octahedron and tetrahedron, are nearly equal in Al$_2$BeO$_4$, where the valency of the octahedral cation is higher than that of the tetrahedral; elsewhere the edges between two octahedra are longer, and the difference becomes progressively greater, for constant valency, as the size of the octahedron increases.

Table 11-4 Average Bond Lengths (in Å) in Compounds with the Olivine Structure, A_2BO_4

	γ-Ca_2SiO_4	Fe_2SiO_4	Mg_2SiO_4	Al_2BeO_4	$LiMnPO_4$
Mean distances					
A(1)—O	2.38	2.16	2.10	1.89	2.17
A(2)—O	2.41	2.18	2.14	1.94	2.20
B—O	1.64_4	1.63_8	1.63_4	1.62_2	1.54_1
O—O					
Octahedron (1)	3.34	3.04	2.96	2.67	2.99
Octahedron (2)	3.39	3.07	3.01	2.72	3.09
Tetrahedron	2.68	2.66	2.64	2.64	2.51
Variations of distance					
A(1)—O, greatest − least	0.08	0.10	0.05	0.06	0.11
A(2)—O, greatest − least	0.25	0.22	0.16	0.16	0.15
B—O, greatest − least	0.14	0.04	0.04	0.11	0.02
O—O					
Mean shared octahedron/ tetrahedron	2.62	2.59	2.57	2.53	2.46
Mean shared octahedron/ octahedron	3.22	2.92	2.86	2.54	2.98
Mean unshared tetrahedron	2.71	2.75	2.75	2.77	2.57
Mean unshared octahedron	3.57	3.19	3.13	2.80	3.18
Unshared, greatest − least:					
Octahedron (1)	0.50	0.37	0.32	0.11	0.14
Octahedron (2)	0.95	0.54	0.47	0.18	0.65

In general, the scatter of cation-oxygen distances within either kind of octahedron (indicated in Table 11-4 by the difference between their greatest and least values) also increases with cation size. We note, however, that in Al_2BeO_4 the Al octahedra, which are very regular when judged by the near uniformity of their unshared edges, are less so when judged by the scatter of their Al—O distances; this may be associated with the rather large scatter of the Be—O distances. A simpler argument could explain the scatter of Si—O distances in γ-Ca_2SiO_4, where the O atoms of the tetrahedron are held rather far apart by the large Ca. If Si stayed central, the tension in the Si—O bonds would be correspondingly high; the Si atom therefore moves off centre, as described in §2.8, and relieves the tension by making the bonds unequal.

This somewhat detailed discussion of the olivine structure has been included to illustrate how a study of the differences within an isomorphous series, when the atomic parameters are sufficiently well known, can throw light on the nature of the structural forces, and help to sort out the different effects of the independent factors involved.

We may notice that γ-Ca_2SiO_4, with its rather loose, irregular octahedra and too-large tetrahedra, easily undergoes transformation to different polymorphic forms (see §15.2).

11.10 THE RUTILE STRUCTURE

This is our first example of a structure which cannot naturally be described in terms of anion packing. From the outset, we must think of it in terms of linkage of octahedra.

General formula: AX_2; example, TiO_2 (rutile)*
Lattice: Tetragonal, primitive, with $a = 4.594$ Å, $c = 2.958$ Å
Number of formula units per unit cell: 2
Space group: $P4_2/mnm$ (often written $P4/mnm$) (No. 136)
Atomic coordinates:
  Ti in $2(a)$: $0, 0, 0$
  O in $4(b)$: $x, x, 0$, with $x = 0.305$

With the origin at the centre of symmetry, the tetrad axes are at $1/2, 0, z$, and $0, 1/2, z$; operating with them on the prototype atoms, we obtain the structure shown in Figure 11-16(a). The set of coordinates so found is

Ti at $0, 0, 0$; $\frac{1}{2}, \frac{1}{2}, \frac{1}{2}$
O at $\pm(x, x, 0; \frac{1}{2} + x, \frac{1}{2} - x, 0)$

Because the atoms were inserted in special positions, the rest of the symmetry has been generated, and the structure is complete.

It can be seen from Figure 11-16(a) that there are TiO_6 octahedra, and that they share edges parallel to (001) (the plane of projection) to form ribbons extending perpendicular to this plane. The ribbons are connected laterally by sharing corners, as shown in Figure 11-16(b). The pattern is thus a very simple one: it is one unit cell high in the [001] direction, and contains only two identical ribbons oriented with the shared edges of the one at right angles to those of the other.

Each O atom has three Ti neighbours, and thus receives electrostatic bonds of total strength $3 \times (4/6) = 2$. The three bonds are coplanar, an arrangement suggestive of homopolar rather than ionic character.

The octahedra are nearly regular. There are two different Ti—O bonds, of lengths $\sqrt{2}xa = 1.98$ Å and $\{[\sqrt{2}(1/2 - x)a]^2 + [\frac{1}{2}c]^2\}^{\frac{1}{2}} = 1.95$ Å. There are three different O—O lengths, $c = 2.96$ Å, $\sqrt{2}(1 - 2x)a = 2.55$ Å, and

---

* TiO_2 is polymorphic, its other forms occurring as the minerals *anatase* and *brookite*. Anatase is also tetragonal, with space group $I4_1/amd$ and atomic coordinates different from rutile; brookite is orthorhombic.

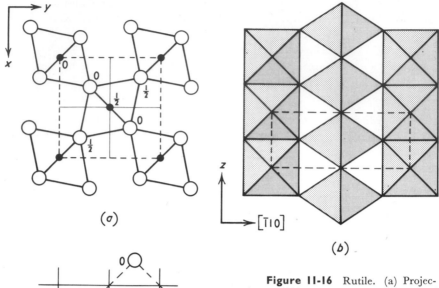

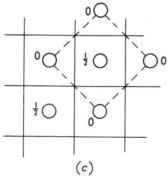

Figure 11-16 Rutile. (a) Projection on (001). (b) Projection on (110) of the sheet of octahedra shown in (a) running from bottom left to top right. The faces of the octahedra are shaded as if it were not a projection but a three-dimensional model viewed directly normal to (110) from the bottom right-hand corner of (a). (c) Projection corresponding to (a) with oxygen atoms given a parameter 0.25, showing the relation to body-centred cubic packing.

$\{[(1/2 - 2x)a]^2 + [\frac{1}{2}a]^2 + [\frac{1}{2}c]^2\}^{\frac{1}{2}} = 2.78$ Å. The shortening of the shared edge is noticeable.

The structures described earlier in this chapter could all be discussed in terms of packing of oxygen atoms; whether or not their stability was really determined primarily by packing, it was easy to think of them in that way. Rutile is different; it is an example of a *linkage structure*. If we try to find the kind of packing of oxygens, the best we can get is a rather coarse approximation to the body-centred cubic array. Thus, if x had been 0.25 instead of 0.30, we should have got the arrangement shown in Figure 11-16(c). Clearly, in this structure, the maintenance of nearly regular octahedra in a simple linkage pattern is a more important consideration than the achievement of close packing.

Other substances strictly isomorphous with TiO_2 are SnO_2 (cassiterite), MgF_2, and ZnF_2. All have similar values of x, namely 0.305 ± 0.003.

Another very interesting example is one of the high-pressure forms of SiO_2, a material first synthesised in the laboratory, and later found to occur as a rare mineral, given the name *stishovite*. It is one of the very few known structures in which Si occurs in 6-coordination with oxygen.

11.11 FORMS OF SILICA (SiO₂)

Three polymorphic crystalline forms of silica, *quartz, tridymite,* and *cristobalite,* have been known to mineralogists for many years. As detailed studies increased, it became clear that each name was being used for at least two different forms, stable in different temperature ranges. The high and low forms have truly different structures, but the differences are so small that careful work is needed to detect them.

There are other, rarer, crystalline polymorphs of silica. *Coesite, stishovite,* and *keatite* were first made in the laboratory, in high-pressure studies, and coesite and stishovite were afterwards discovered occurring naturally in conditions where geological evidence suggested that the impact of meteors had caused high pressures during their formation. At the other extreme, *melanophlogite* is a low-temperature low-pressure form growing in specialized conditions, and always containing inclusions of organic matter, which, however, do not affect the structure.

Apart from stishovite, which has the rutile structure (§11.10) with Si in 6-coordination, all the forms are built from SiO_4 tetrahedra sharing each of their corners with one other tetrahedron in a continuous three-dimensional network. We cannot assume that all possible polymorphs have been discovered. It is a characteristic of linkage or framework structures that the same building units may be used in an immense variety of structural patterns; some of the range of possibilities will be considered later in describing the zeolites (§12.18–20).

In this chapter we shall describe the three common polymorphic forms of SiO_2, and in doing so illustrate some of the important general features of linkage or framework structures.

11.12 CRISTOBALITE

The earliest determination of the cristobalite structure led to the following description.

Lattice: Cubic face-centred, with $a = 7.2$ Å
Number of formula units per unit cell: 8
Space group: $Fd3m$ (No. 227)
Atomic coordinates:
  Si in $8(a)$: $0, 0, 0$
  O in $16(c)$: $\frac{1}{8}, \frac{1}{8}, \frac{1}{8}$

More recent work has shown this to be an approximation only, with over-simplified coordinates, but it is nevertheless a useful starting point for thinking about it. We may in fact call it the *idealised high-cristobalite structure*.

To draw the structure diagram, we begin by using the d-glide plane (as we did for the spinel structure, §11.3). We obtain sets of atoms as follows:

$$\text{Si} \quad \text{at} \quad 0, 0, 0; \; \tfrac{1}{4}, \tfrac{1}{4}, \tfrac{1}{4}$$
$$\text{O} \quad \text{at} \quad \tfrac{1}{8}, \tfrac{1}{8}, \tfrac{1}{8}; \; \tfrac{1}{8}, \tfrac{3}{8}, \tfrac{3}{8}; \; \tfrac{3}{8}, \tfrac{1}{8}, \tfrac{3}{8}; \; \tfrac{3}{8}, \tfrac{3}{8}, \tfrac{1}{8}$$

Adding the lattice face-centring, we obtain the complete structure, shown in Figure 11-17(a). We notice that the Si and O positions are the same as the Mg and Al positions in spinel. Each Si tetrahedron shares each of its corners with another tetrahedron, and the centres of its four neighbouring tetrahedra are also arranged tetrahedrally round it. If we ignore the O atoms, the arrangement of Si atoms is *exactly* like that of the carbon atoms in diamond, or of the Si atoms in silicon.

It is often very convenient, in drawing diagrams of SiO_4 linkage structures, to leave out the O atoms and consider the pattern of linkage as displayed by the Si—Si vectors. This is done for cristobalite in Figure 11-17(b). The diagram can, as we have seen, be regarded as merely a different way of drawing the diamond structure.

In idealised high cristobalite, the O atoms are midway between two Si; the angle Si—O—Si is 180°. This is not true for the structure that has actually been found for high cristobalite. The feature that remains the same for all cristobalite structures, whatever details may change, is the pattern of linkages.

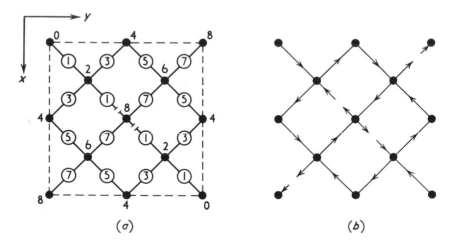

(a)   *(b)*

Figure 11-17 High cristobalite. (a) Idealised structure, projected on (001). Heights are marked in units of $c/8$. (b) Linkage scheme of tetrahedra, projected on (001). Lines show the Si—Si joins; arrows indicate the upward direction. Breaks in the joins occur where a tetrahedron in the repeat unit shown joins another tetrahedron above or below that shown.

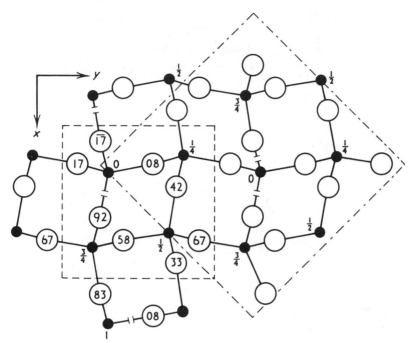

Figure II-18 Low cristobalite, projected on (001). The primitive unit cell is shown by broken lines, and the base-centred unit cell (comparable to the unit cell of high cristobalite) by dot-dash lines.

The revised structure of high cristobalite has the same unit cell and cell content, but reduced symmetry. The space group becomes $P2_13$ (No. 198), and the atomic coordinates are as follows:

Si(1) in $4(a)$: u_1, u_1, u_1,   with $u_1 = -0.008$
Si(2) in $4(a)$: $\frac{1}{4} + u_2, \frac{1}{4} + u_2, \frac{1}{4} + u_2$, with $u_2 = 0.005$
O(1) in $4(a)$: $\frac{1}{8} + u_3, \frac{1}{8} + u_3, \frac{1}{8} + u_3$, with $u_3 = 0$
O(2) in $12(b)$: $(\frac{3}{8} + u_0, \frac{3}{8} + v_0, \frac{1}{8} + w_0)$ with $u_0 = 0.065$, $v_0 = -0.035$,
$w_0 = 0.035$

(To find from these the complete set of atomic coordinates, note that the triad axis passes through $(0, 0, 0)$, and screw diads are at $x, 1/4, 0$; $0, y, 1/4$; $1/4, 0, z$.)

The Si—Si vectors are no longer all exactly equal. The O(1)-atoms lie almost midway between two Si atoms, but the O(2) atoms are considerably displaced—about 0.6 Å, in a direction nearly at right angles to the Si—Si vector—and in consequence the Si—O—Si angle is reduced to about 140°.

Low cristobalite, which occurs at temperatures below about 200° C, retains the same pattern of linkages, but has a still lower symmetry. Its description is as follows:

> Lattice: Tetragonal, primitive, with $a = 4.97$ Å, $c = 6.93$ Å
> Number of formula units per unit cell: 4
> Space group: $P4_12_12$ (No. 92)
> Atomic coordinates:
> Si in $4(a)$: $\frac{1}{4} + u_1, \frac{1}{4} + u_1, 0$, with $u_1 = 0.05$
> O in $8(b)$: $\frac{1}{4} + u_0, \frac{1}{2} + v_0, \frac{1}{8} + w_0$, with $u_0 = 0.01$, $v_0 = 0.10$, $w_0 = 0.045$

(The tetrad axes are at $1/2, 0, z$, and $0, 1/2, z$; the diads are at $x, 1/4, 3/8$, and $1/4, y, 1/8$.)

The structure is shown in Figure 11-18. Its relation to that of high cristobalite is much more obvious if we choose the larger unit cell with axes at 45° to the original, for which $a' = \sqrt{2}\, a = 7.02$ Å. The same linkage pattern is seen, but now all the O atoms are displaced from the Si—Si lines to give Si—O—Si angles of roughly 140°. All Si—O bonds remain very nearly equal, with a length of 1.59 Å.

11.13 TRIDYMITE

The structure of tridymite is only approximately known. Its linkage pattern bears the same relation to that of cristobalite as does hexagonal close-packing to cubic close-packing, or as does the wurtzite structure of ZnS (§4.11) to the zinc-blende structure (§4.10). If, in zinc blende, we make both atoms identical, we have the diamond structure; if we make them all Si atoms, and insert O atoms midway between pairs of them, we have (idealised) cristobalite. In the same way, if in the wurtzite structure we make both kinds of atoms into Si atoms, and insert O atoms midway between pairs of them, we have (idealised) tridymite. This gives the structure shown in Figure 11-19. Its

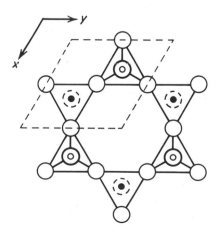

Figure 11-19 Part of the idealised tridymite structure projected on (0001). The unit cell is shown by broken lines. Large light circles are O atoms at height 1/2; large heavy circles O atoms at height 3/4, above Si atoms at height 5/16 (small open circles); and large dashed circles are O atoms at height 1/4, below Si atoms at height 3/16 (small black circles). Other atoms in the unit cell project in coincidence with those shown (because of mirror planes at $x = 1/4, 3/4$).

formal description (believed to be that of high tridymite) is as follows:

> Lattice: Hexagonal, primitive, with $a = 5.03$ Å, $c = 8.22$ Å
> Number of formula units per unit cell: 4
> Space group: $P6_3/mmc$ (sometimes written $P6/mmc$) (No. 194)
> Atomic coordinates:
> Si in $4(f)$: $\frac{1}{3}, \frac{2}{3}, z$, with $z = 0.44$
> O(1) in $2(c)$: $\frac{1}{3}, \frac{2}{3}, \frac{1}{4}$
> O(2) in $6(g)$: $\frac{1}{2}, 0, \frac{1}{2}$

As in cristobalite, triad axes of the tetrahedron are parallel to triad axes of the structure. Pairs of tetrahedra share corners lying on the (0001) mirror plane, and their triangular bases are linked in a puckered hexagonal ring about the hexad axis, with alternate vertices pointing up and down. There are thus rather large unoccupied channels extending through the structure parallel to the hexad axis.

11.14 QUARTZ

This is the commonest and the most important of the crystalline polymorphs of silica, SiO_2. Like tridymite and cristobalite, it possesses high and low forms.* Both forms of quartz have enantiomorphous forms, and therefore right-handed and left-handed crystals can occur, which are mirror images of one another. Here we shall describe right-handed crystals.†

High quartz, stable above 600° C, is described as follows:

> Lattice: Hexagonal, primitive, with $a = 5.01$ Å, $c = 5.47$ Å
> Number of formula units per unit cell: 3
> Space group: $P6_222$ (No. 180)
> Atomic coordinates:
> Si in $3(c)$: $\frac{1}{2}, 0, 0$
> O in $6(j)$: $x, \bar{x}, \frac{5}{6}$, with $x = 0.20$

Operation with the 6_2 axis at 0, 0, z gives the complete set of atoms as follows:

> Si: $\frac{1}{2}, 0, 0$; $\frac{1}{2}, \frac{1}{2}, \frac{1}{3}$; $0, \frac{1}{2}, \frac{2}{3}$
> O: $x, \bar{x}, \frac{5}{6}$; $2x, x, \frac{1}{6}$; $x, 2x, \frac{3}{6}$; $\bar{x}, x, \frac{5}{6}$; $2\bar{x}, \bar{x}, \frac{1}{6}$; $\bar{x}, 2\bar{x}, \frac{3}{6}$

All atoms are in special positions on diad axes perpendicular to z.

Figure 11-20(a) shows the set of general positions, and most of the symmetry elements of the space group. (Diad screw axes lying in the (0001)

---

* Low quartz and high quartz were called α-quartz and β-quartz respectively in the original structure work by W. H. Bragg and his coworkers, and this usage has been followed by some later workers. Others, however, have reversed it, taking α for high quartz and β for low quartz; consequently, it is safest to abandon the use of α and β.

† There has been a good deal of confusion in the literature about the choice of axes of reference for high and for low quartz, and serious mistakes have resulted. (See footnote on pp. 267–268.)

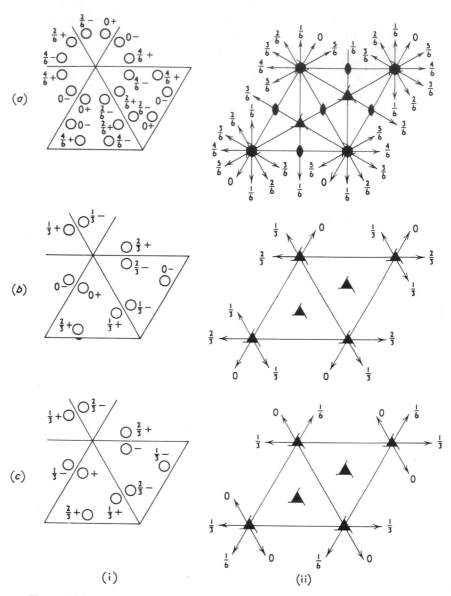

Figure 11-20 (i) General equivalent positions, (ii) important symmetry elements, in space groups $P6_222$ and $P3_221$. (Diad screw axes midway between parallel diad rotation axes in (0001) have been omitted from all diagrams to avoid confusing them.) (a) $P6_222$, as in *International Tables for X-ray Crystallography*. (b) $P3_221$, with origin chosen to show resemblance to (a). (c) $P3_221$, with origin chosen as in *International Tables for X-ray Crystallography*, at (0, 0, 1/3) relative to that of (b).

plane, parallel to diad rotation axes and midway between them, have been left out to avoid confusing the figure). Figure 11-21 gives a projection of the structure.

To describe left-handed crystals, without changing our axes of reference, we should *either* reverse the sign of all z coordinates *or* interchange the x and y coordinates and move the origin by $\frac{1}{3}c$. We then find that we have also changed the relative positions of the diad axes in the z direction, and that in fact the symmetry is that of the enantiomorphous space group $P6_422$ (No. 181).

The linkage pattern of quartz is quite different from that of tridymite or cristobalite. Triad axes of tetrahedra do not lie along the hexad or triad axis of the structure; they are equally inclined to it. Figure 11-22(a), which indicates only centres of tetrahedra, shows how they are joined in tight threefold spirals round lines perpendicular to the paper through the points marked B and C in Figure 11-21, and also in a double spiral of larger cross-section, with two nonintersecting threads, round the hexad axis at the point marked A in the same Figure. These general features are retained in low quartz. It has in fact been possible to show that crystals which are optically and morphologically right-handed have *left-handed* threefold spirals; this was predicted from the sense of their rotary polarisation of light long before a direct proof from diffraction analysis was possible.

We notice once again how the Si—O—Si angles are not 180°, but appreciably less, as in other forms of silica.

Low quartz is closely related to high quartz; the transition from one to the other occurs at about 580° C, and can take place reversibly without damaging the crystal (cf. §15.9).

The description of low quartz (right-handed enantiomorph) is as follows:

Lattice: Hexagonal, primitive, with $a = 4.91$ Å, $c = 5.40$ Å
Number of formula units per unit cell: 3
Space group: $P3_221$ (No. 154)
Atomic coordinates:
  Si: $\frac{1}{2} - u_1, 0, 0$, with $u_1 = 0.03$
  O: x, y, z, with $x = x_0 - u_0$, $y = -x_0 - u_0$, $z = \frac{5}{6} + w_0$,
      and $x_0 = 0.21$, $u_0 = 0.06$, $w_0 = 0.05$

The symmetry elements are as shown in Figure 11-20(b). Our choice of origin is not that of the *International Tables for X-ray Crystallography*, which would be at height $\frac{1}{3}c$ on our axes, placing the diad axes at the heights shown in Figure 11-20(c). Our complete set of atomic positions is as follows:

Si: $\frac{1}{2} - u_1, 0, 0$; $\frac{1}{2} + u_1, \frac{1}{2} + u_1, \frac{1}{3}$; $0, \frac{1}{2} - u_1, \frac{2}{3}$
O: x, y, z; $\bar{y}, x - y, \frac{2}{3} + z$; $y - x, \bar{x}, \frac{1}{3} + z$;
    $x - y, \bar{y}, \bar{z}$; $\bar{x}, y - x, \frac{1}{3} - z$; $y, x, \frac{2}{3} - z$

The structure is as shown in Figure 11-23.

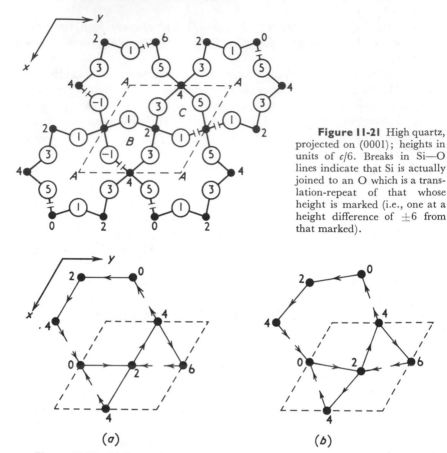

Figure 11-21 High quartz, projected on (0001); heights in units of $c/6$. Breaks in Si—O lines indicate that Si is actually joined to an O which is a translation-repeat of that whose height is marked (i.e., one at a height difference of ± 6 from that marked).

(a)

(b)

Figure 11-22 Linkage scheme of tetrahedra in (a) high quartz, (b) low quartz, using same conventions as in Figure 11-16(b). The Si positions are as in Figures 11-21 and 11-23 respectively. Notice the distortion of the framework in (b) as compared with (a). Notice also the *left-handed* spirals round (2/3, 1/3) and (1/3, 2/3) in both, corresponding to optically *right-handed* crystals.

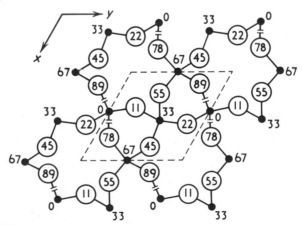

Figure 11-23 Low quartz, projected on (0001). Heights are in units of $c/100$; otherwise the conventions used are the same as in Figure 11-21.

Comparison of Figure 11-20(a) and (b) shows that all the symmetry elements of low quartz were already present in the space group of high quartz; some symmetry elements have been lost in the transition, and none have been gained. It is less easily seen with the choice of origin for $P3_221$ used in Figure 11-20(c); hence our choice of that in (b). Because of this relationship, the choice of the 6_2 enantiomorph for high quartz determines the choice of the 3_2 enantiomorph for low quartz, whether or not we know the relation between optical handedness and structural handedness, and whatever our choice of axes. In fact, the 6_2 and 3_2 enantiomorphs (space groups 180 and 154) correspond to macroscopically right-handed crystals, and the 6_4 and 3_1 enantiomorphs (space groups 181 and 152) to macroscopically left-handed crystals. Handedness is not lost at the transition.

The relation between the atomic positions in high and low quartz is shown up by the way we have chosen to define position parameters in terms of the displacement parameters u_1, u_0, and w_0. The remaining parameter x_0 in low quartz is close to its value in high quartz, and if u_1, u_0, w_0 were all zero the two structures would be the same. In fact the quantities are small, and the linkage pattern is unchanged, as we have already noted—cf. Figure 11-22(a) and (b). What has taken place is a slight crumpling of the structure, not changing the shapes of the tetrahedra significantly (they remain nearly regular throughout) nor even requiring large changes in the hinge angles Si—O—Si (though these may nevertheless be significant—see §15.9). The main change is in the placing and orientation of the tetrahedra. In high quartz they had one diad axis parallel to the hexad axis, as well as one coinciding with a diad rotation axis, and their centres, marked by Si atoms, were midway between hexad axes. In low quartz, alternate Si atoms have moved towards the hexad axis, the others away from it, keeping triad screw symmetry; moreover, each tetrahedron has rotated through an appreciable angle about its diad axis in (0001), and no longer has a diad axis of the tetrahedron parallel to the triad axis of the structure.

Here we may note the importance of choosing the *sense* of the x and y axes in the conventional way in low quartz, if confusion is to be avoided.* The

---

* From the symmetry it can be seen that there is a physical difference in low quartz (though not in high quartz) between directions at 60° to one another in the (0001) projection. The unconventional choice of sense of x and y axes therefore interchanges the indices of physically different planes; this can create great confusion. Unfortunately, it has been adopted by a number of authors, and copied by others not realising its implications. The trouble is that unless one is careful when describing crystals of different handedness, one changes the sense of the x and y axes, thus confusing two physically distinct features, the handedness of the screw triad axis and the polarity of the diad axes. The history of the situation has been summarised by A. Lang (*Acta Cryst.*, **19**: 290, 1965), whose recommendations are followed here.

The confusion arose partly from the fact that the first formal description by Wyckoff (*The Structure of Crystals*, 2nd ed., 1931), based on the determination by Bragg and Gibbs (1925), was published before the agreement on standard settings adopted in the *International Tables for X-ray Crystallography* (1952), which differed in some respects from Wyckoff's conventions; partly from the rather inconvenient choice of origin in the *International Tables for X-ray Crystallography* (referred to in the text); and partly from the fact that the relation

correct choice consistent with the older mineralogical conventions, and following the original structure determination by W. H. Bragg and R. E. Gibbs, is to take as positive the ends pointing, in the projection on (0001), from the intersections of triad and diad axes towards the three nearest of the six surrounding Si atoms, i.e., towards the more obtuse of the two different O—Si—O angles facing the intersection of triad and diad axes.

It is obvious from the diagrams that high quartz has a very open structure (though not quite so open as tridymite or cristobalite) and that low quartz, though more compact, is still by no means close-packed. The crumpling to the low form can occur without encountering any steric difficulty from close oxygen-oxygen contacts.

When the stability of a structure is not determined by the packing of its atoms, why should one structure be stable rather than another? There can be no general answer yet to this question, but we can see some factors that must be taken into account.

We have already recognised the importance of a regular (or nearly regular) tetrahedron as a structure-building unit. In quartz, we are brought face to face with the problem of how the tetrahedra are joined up. We notice that the bond angle at oxygen is about 145°, and remember that large departures from 180° are characteristic of semipolar bonding—an approach towards the smaller covalent bond angle. The corner linkage between tetrahedra cannot be considered as a freely movable hinge, adapting itself as required for good packing. Instead, we should picture the pair of Si—O bonds as constituting a sort of angle bracket, with an equilibrium angle of about 145°, but capable of being strained somewhat from its ideal value to fit in with other structural requirements. This picture helps us to understand not only quartz but the great variety of other silicate frameworks, some of which are described later in this and the following chapter. The ideal or unstrained value may differ from one structure to another—actual values often range from 120° to 160°—but differences within a structure can often be

---

between macroscopic and structural handedness was not established till much later (A. de Vries, *Nature*, **181:** 1193, 1958). This last fact meant that until then no one could say that a particular way of drawing the structure corresponded to a crystal which was physically left-handed or right-handed, and it was easy to overlook the implications of different choices of axes of reference. Wyckoff's later description (*Crystal Structures*, Volume I, 1968), while intended to conform to the standard setting, only increased the confusion; the coordinates for a structure said to belong to $P3_121$ in fact belong to $P3_221$, the diagram does not match the coordinates but reverses the senses of x and y, and the relationship between the high and the low symmetry is assumed to be $P3_121 \rightleftarrows P6_222$, rather than $P3_121 \rightleftarrows P6_422$, as we have seen it must be.

The only safe rules are (i) to use right-handed axes of reference, as is now standard practice, whether the crystals are right-handed or left-handed, (ii) to fix the sense of x and y for low quartz by the convention given in the text, and (iii) to express the difference of hand by reversing the sense of the z coordinates and changing to the enantiomorphous space group. With these rules, we now know that a macroscopically right-handed crystal, which possesses left-handed single-strand (threefold) spirals, will belong to $P3_221$ or $P6_422$ as set out in the *International Tables for X-ray Crystallography* (except for a possible change of origin which we may make to suit our convenience).

recognised as due to stresses exerted by other parts of the structure. Though 180° angles do occur—as in high cristobalite—they are rare.

We can see the effect of hinge-angle rigidity in the high-quartz structure. Suppose the tetrahedra were perfectly regular: then, since they are fixed in orientation, and joined into the quartz linkage pattern, the hinge angle is geometrically fixed. If this imposes too great a strain on the preferred Si—O—Si angle, the only way to alleviate it without loss of symmetry is to allow the tetrahedron to change shape, by compression or extension in the c-direction. This is what actually happens. A lower overall potential energy is achieved if the strains—departures from the ideal shapes preferred by the tetrahedron and the Si—O—Si bond angle separately—are distributed between the two of them.

Stability of a structure depends on the possibility of reaching a reasonable compromise between the conflicting requirements of different features of the structure. If the strains involved in building a particular structure are too great, the required linkage pattern can sometimes be kept if the symmetry demands are lessened. This is what has happened in low quartz. (Why strains that are acceptable above 600° C cannot be tolerated at room temperature is a point we must leave for discussion in Chapters 14 and 15.) By allowing the tetrahedra to tilt, low quartz has obtained an additional adjustable parameter, and can alter tetrahedron shape and Si—O—Si angles independently. (For a detailed discussion see §15.9.)

The relationship between high and low quartz is representative of what is often found in linkage structures. It seems as if, on the whole, those patterns are favoured that have a rather small repeat unit and high symmetry, at least in their idealised forms. The simpler the ideal pattern, the greater the variety of ways in which it can lose symmetry and increase its adjustable parameters. There will thus be *families of structures*, not very different in energy, not generally predictable from the ideal structure, but often capable of explanation when they occur. This subject will be discussed further in Chapter 12.

11.15 THE FELSPAR STRUCTURE

Tetrahedral framework structures can be built using other cations instead of some of the Si atoms provided that something is done to maintain electrostatic neutrality. Consider, for example, the replacement of some Si atoms by Al. Geometrically this makes little difference. True, the Al tetrahedron is slightly larger, but its increased size can generally be accommodated without much trouble by slight changes of tilt. Electrostatically, the balance can be restored either by replacing other Si atoms by 5-valent cations or by introducing large low-valent cations into interstices of the structure. It is this latter effect which occurs in the felspars. If we consider the structure in terms of electrostatic valence, each O atom still has two framework neighbours linked to it by strong bonds—strength 1 for Si, 3/4 for Al—and the deficit is partly or wholly made up by weak bonds to the larger low-valent cations.

The atoms Si and Al are so alike in size and structural role that disorder between them can often occur, with very little effect on the structure as a whole. Even when they are fully ordered, rather careful structure analysis is required to distinguish which sites are occupied by which species of atom. It is therefore often a useful approximation to treat them all as identical "tetrahedral cations," with the symbol "T," and we shall do so here.

The characteristic feature of a felspar is its linkage pattern. As with the various forms of SiO_2, the framework can undergo different kinds of crumpling without losing its identity. In this chapter we consider the simplest and most symmetrical variant, leaving others to §12.17.

The general formula of a felspar is MT_4O_8. We consider the particular example *sanidine*, $KAlSi_3O_8$, in which Al plus 3Si are distributed at random among the T sites, so that each T atom is in effect an average atom of composition $Al_{1/4}Si_{3/4}$. The structure is described as follows:

Lattice: Monoclinic C-face-centred, with $a = 8.56$ Å, $b = 13.03$ Å, $c = 7.17$ Å, $\beta = 116.0°$

Number of formula units per unit cell: 2

Space group: $C2/m$ (No. 12)

Atomic coordinates:

		x	y	z
K	in 4(i):	0.2857	0	0.1379
T_1	in 8(j):	0.0097	0.1850	0.2233
T_2	in 8(j):	0.7089	0.1178	0.3444
$O_A(1)$	in 4(g):	0	0.1472	0
$O_A(2)$	in 4(i):	0.6347	0	0.2858
O_B	in 8(j):	0.8278	0.1469	0.2244
O_C	in 8(j):	0.0341	0.3100	0.2575
O_D	in 8(j):	0.1792	0.1269	0.4025

(The distinguishing subscripts for the atoms were given to them in the original structure analysis, and have been used in all subsequent work.) We note that K and $O_A(2)$ are in special positions on mirror planes, $O_A(1)$ in a special position on a rotation diad, and all other atoms in general positions.

For a complicated structure like this, a stylised diagram is generally most helpful, and we can in the first instance consider the T atoms only, each representing the centre of a tetrahedron. Figure 11.24(a) shows the projection on (010) of the T atoms in the lower half of the unit cell, i.e., the bit of the structure between heights $y = 0$ and $y = 1/2$. Four tetrahedra form a ring at a height of about $y = 0.15$; actually the ring is not plane but slightly puckered. Operation of the C-face centring followed by the mirror plane at $y = 1/2$ repeats the ring at about $y = 0.35$. The two rings are connected up so as to form other four-rings lying nearly perpendicular to the plane of the diagram. This two-level band, which has been likened to a crankshaft,

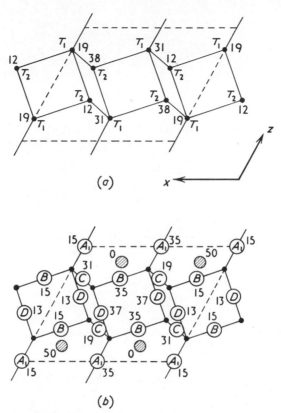

Figure 11-24 Felspar: projection of part of structure on (010). (a) Schematic diagram to show linkage of tetrahedra in the "crankshaft." Heights are in units of $b/100$. (b) Schematic, like (a) but with addition of O atoms (large circles) and K atoms (shaded circles), whose heights are marked. All atoms at height y are repeated by mirror plane at height $1 - y$ (except those already on the mirror plane, at heights 0 and 1/2).

extends parallel to the x axis. There is one such band per c-repeat, and successive bands are linked through their adjacent T_1 vertices in the [001] direction. The sheet so formed constitutes the lower half of the unit cell contents. The upper half is obtained by reflecting the lower in the mirror plane at $y = 1/2$; the two halves are linked by connecting the four-ring at 0.35 to the four-ring at 0.65 through their T_2 vertices.

Though each T atom is joined to four other T atoms, the environments of T_1 and T_2 differ in detail. Neighbours of the T_1 atoms are three T_2 and one T_1 atom; neighbours of the T_2 atoms are three T_1 and one T_2 atom. The T_1—T_1 join is bisected by a rotation diad, the T_2–T_2 join by a mirror plane.

A perspective view of the linkage scheme is given in Figure 11.25. The T_1 atoms are those shown with one unfinished link; the T_2 atoms are those with one dotted link. The piece of structure shown represents one repeat unit in the c direction, and is joined to an identical unit at the points marked by.

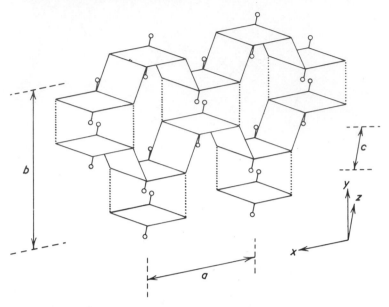

Figure 11-25 Felspar: perspective diagram (schematic) of framework, showing "crankshafts" reflected by the mirror plane. Solid lines join T atoms, as in Figure 11-24(a). The O_A (2) atoms lie on the dotted lines joining the "crankshaft" to its mirror image; $O_A(1)$ atoms are represented by circles, and join the slab shown to its translation-repeat in the z direction.

small circles—as can be more easily seen in Figure 11-24, where the joins lie in the plane of projection.

The joining lines in Figures 11-24(a) and 11-25 in fact represent T—O—T links, and we may now insert one O in each of them, as shown in Figure 11-24(b). We notice that O_B, O_C, O_D, in general positions, form joins within the four-rings of the "crankshaft"; $O_A(2)$ atoms, in special positions on the mirror plane, come at the centres of dotted lines in Figure 11-25, joining "crankshafts" into a slab; and $O_A(1)$ atoms, in special positions on the diad axis, coincide with the small circles in the same figure, where they hold slabs together.

(Actually the O atoms are not exactly on the T—T lines. If we had used actual parameters for T and O instead of drawing stylised diagrams, we should have found T—O—T angles with an average value of about 140°.)

It remains to put in the K atoms. They lie in the mirror planes, occupying the largest cavities in the framework. Figure 11-26 shows those in the plane $y = 0$, with the neighbouring parts of the T—O framework. Each K has 9 neighbours, one $O_A(2)$ in the mirror plane, and a pair each of $O_A(1)$, O_B, O_C, O_D above and below it, at heights marked in the figure. The lengths are rather unequal: 2.70 Å for K—$O_A(2)$, 2.93, 3.03, 3.13, 2.95 Å for the others, in the order named. There are two other $O_A(2)$ atoms in the mirror plane which, though considerably more distant than the first, might in some circumstances be classed as neighbours.

In detail, the T—O distances are not all exactly equal either; the tetrahedra, though nearly regular, are not perfectly so. If, however, one takes the average of the four T—O bonds in a tetrahedron, the value is very nearly equal for T_1 and T_2 tetrahedra. This is an example of a very common effect in strongly-bonded polyhedra: whatever the individual variations of bond length, the average for the polyhedron is very nearly constant for a given kind of atom (or average-atom)—more nearly so than would have been expected statistically, assuming the variations to be random.

Moreover, one can account for some of the individual differences in O—T—O bond angles by considering the kinds of stress inherent in the linkage pattern. The argument is as follows.

Suppose all bond lengths and all bond angles were rigidly determined. Then the repeat period a would be fixed by the length of the "crankshaft" repeat, and the repeat period c by the width of the "crankshaft" plus the length of the T_1—T_1 link. This part of the structure lies at height $y = 0.25 \pm 0.10$, and has a mirror image at height $y = 0.75 \pm 0.10$, the two being rigidly fixed to one another by $O_A(2)$ atoms at $y = 0$ and $y = 1/2$. Now the M cations have also to be fitted into the planes at $y = 0$ and $1/2$, and since they are in fact rather close together, their electrostatic repulsions are not negligible.

This can be seen in Figure 11-26, and more clearly in Figure 11-27, where most of the K—O bonds have been omitted, and the upper part of the T—O framework has been added. Consider the line of atoms $RUVS$; R and S are $O_A(2)$ atoms rigidly attached to the framework, and U, V are potassium atoms held tightly between them, and repelling each other. The framework is compressing the link RS and its separate parts RU, UV, VS; RS in turn is exerting a force on the framework tending to stretch it. Suppose these were the only structural stresses: then they must be equal, according to Newton's third law.

We started by assuming that the framework was rigid; now let it be elastic. The stresses produce strains in bond lengths and bond angles. We therefore expect to find compression in the K—$O_A(2)$ bonds, and changes in the bond lengths and bond angles of the framework corresponding to an elongation in the general direction of RS. In fact, as we saw, the K—$O_A(2)$ bond is particularly short, which supports the assumption that it is being compressed by the other parts of the structure; and the assumption of tension in the "crankshaft" allows us to predict correctly the largest deviations of its O—T—O angles from the ideal tetrahedral angle. This set of stresses turns out to be an effect characteristic of the felspar linkage pattern, not seriously altered by the differences between sanidine and other felspars, which we shall consider in §12.17. There will be other stress systems, less fully studied as yet, accounting for some of the other strains, but their effects can be considered separately, and this one we have discussed seems to be the largest. The stresses and strains *are built into the structure*; they are not imperfections, but repeat perfectly with the periodicity of the structure.

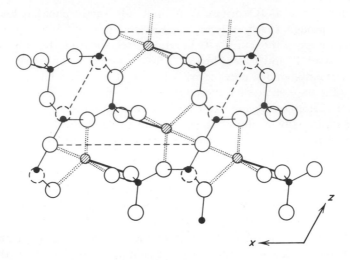

Figure 11-26 Felspar: projection of part of structure on (010). This is the sa:
projection as in Figure 11-24, but with the atoms given their actual coordinates; only those
heights between $y = 0$ and $y = 0.25$ are shown (except for a few at $y = 0.31$, shown
broken circles). Full thin lines are T—O bonds; dotted double lines are K—O bonds to t.
oxygen atom shown and its mirror image; heavy lines are K—$O_A(2)$ bonds.

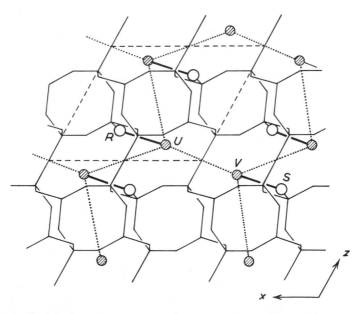

Figure 11-27 Felspar: same projection as in Figure 11-26, but showing the complete
T—O framework and the set of K atoms at $y = 0$. Dotted lines show the direction of K—K
repulsions. Circles show the position of $O_A(2)$ atoms; K—$O_A(2)$ bonds are marked by
heavy lines. Note the almost collinear K—K repulsions and K—$O_A(2)$ bonds along $RUVS$
and in symmetry-equivalent places.

The role of the M cation is thus partly to secure electrical neutrality, but partly also to act as a spacer. The K cation is just the right size to fit neatly into the available cavity and keep the structure fully stretched. For M cations which are not such a good fit, modifications of the structure occur, which will be considered in §12.17.

11.16 SUMMARY

The oxides considered in this chapter have illustrated a number of important general principles.

(a) Many oxides can be considered as derived from a close-packed array of oxygens, with cations occupying octahedral and/or tetrahedral interstices.

(b) In such structures, polyhedra commonly share edges, and shortening of shared edges relative to unshared edges is to be expected.

(c) Where polyhedra of the same kind, with approximately equal bond strengths, are linked together to give a three-dimensional framework, any other cations in the structure must be of the correct size to fit into the available interstices or cavities. Small amounts of misfit can be accommodated, either by adjustment of parameters or by lowering of symmetry (consideration of the latter being generally left to Chapter 12).

(d) High electrostatic valence generally implies bonds with considerable covalent character; consequently, the bond angle between two strong bonds to an oxygen atom becomes a more important feature of the structure than the regular packing of other atoms around the oxygen.

(e) Tetrahedral framework structures characterised by nearly regular tetrahedra sharing each corner with one other tetrahedron tend to have an Si—O—Si angle in the range 130° to 150°, but they can have a great variety of linkage patterns; they tend to be open structures.

(f) Linkage patterns may remain unchanged even though the structure changes; they retain their identity in spite of any loss of symmetry.

(g) The framework of a linkage structure need not be electrically neutral; large low-valent cations in the cavities may compensate for a net negative charge on the framework.

(h) Structures with semipolar bonds can generally not satisfy all bond-length and bond-angle requirements simultaneously. There are therefore built-in stresses and strains, having the periodicity of the structure, which represent the energetically best compromise.

PROBLEMS AND EXERCISES

1. Compounds of composition KBF_3 have the ideal perovskite structure, where B is Mn, Fe, Co, Ni, or Zn. Predict the lattice parameters and the tolerance factor t, assuming any consistent set of radii from Table 2-2. The observed lattice parameters are 4.182, 4.120, 4.071, 4.012, and 4.055 Å respectively. Using only the radii for K^+ and F^-, find the effective radii of the octahedral cations, and the corresponding values of t.

2. Predict the lattice parameters and values of u you would expect for compounds Mg_2TiO_4, Na_2MoO_4, and Li_2NiF_4, with the normal spinel structure, assuming ionic radii from Table 2-2.

3. (a) For ilmenite, $FeTiO_3$, calculate the lengths of the edges shared between two octahedron faces, and the edges of the opposite faces of the octahedra.

(b) Using the data of Table 11-1, repeat the procedure for the iso-morphous structures $NiTiO_3$, $MnTiO_3$, and $NiMnO_3$. Are there any significant differences between the effects in these structures and in $FeTiO_3$, or in Al_2O_3?

4. Lanthanum sesquioxide, La_2O_3, is trigonal, with $a = 3.94$ Å, $c = 6.13$ Å, and space group $P\bar{3}m1$ (No. 164). There is one formula unit per unit cell, with La at $1/3, 2/3, 0.245$, and oxygen atoms at $0, 0, 0$ and $1/3, 2/3, 0.645$. Draw diagrams of the structure. Describe the environment of each independent atom, and calculate the La—O bond lengths and the O—O edges of the polyhedron.

5. (a) Using the data of Table 11-1, calculate the angle between the A—O bond and the triad axis for each of the compounds isomorphous with calcite.

(b) How far can the departures from the value for a regular octahedron be correlated with the differences of axial ratio between the compounds? What other factors might affect the correlation?

(c) Discuss the possible ways of arranging the four carbonates in a logical sequence according to one or other of their geometrical param-eters. Look up the other isomorphous carbonates in *Crystal Structures* and see whether they fit into your sequence.

6. The structure of dolomite, $CaMg(CO_3)_2$, is like that of calcite except that A cations on the same triad axis are alternately Ca and Mg. Using ionic radii from Table 2-2, predict the lattice parameters and the atomic positions.

7. (a) Calculate the individual Ca—O bond lengths in aragonite, and the K—O bond lengths in isomorphous KNO_3—see Question 4(b), Chapter 10. Is there any significant difference in the range of lengths in the two materials?

(b) Comment on the variation of lattice parameters and atomic position parameters with cation size.

(c) Predict the lattice parameters and atomic position parameters in the isomorphous $BaCO_3$ and $PbCO_3$, using radii from Table 2-2.

8. (a) Predict, from the A-cation radius, the lattice parameters of the compounds with the olivine structure given in Table 11-3, and compare them with those observed.

(b) Find the volume per oxygen atom in each of these structures, and compare it with the volume in a close-packed array of atoms whose diameter is (i) 2.8 Å, (ii) the mean octahedron edge length in the structure (see Table 11-4).

9. The high-pressure (δ) form of Mn_2GeO_4 is orthorhombic, with $a = 5.26$ Å, $b = 9.27$ Å, $c = 2.95$ Å; the space group is $Pbam$ (No. 55) and there are two formula units per unit cell. Atomic positions are as follows: Ge, 0, 0, 0; Mn, 0.07, 0.32, 1/2; O(1), 0.24, 0.04, 1/2; O(2), 0.36, 0.32, 0. Draw projections of the structure on (001) and (100). Describe and discuss the environments of the atoms, listing the numbers of neighbours of each cation and their distances from it. Are the distances those that would be expected from the radii in Table 2-2? Compare the structure with those of spinel and olivine, which have similar chemical formulae. In fact α-Mn_2GeO_4 has the olivine structure. Can you suggest why the δ-structure described above should be favoured by high pressure?

10. One form of MnO_2H is orthorhombic, with $a = 4.58$ Å, $b = 10.76$ Å, $c = 2.89$ Å; the space group is $Pbnm$ (No. 62), and there are four formula units per unit cell. One oxygen atom, O(2), has position parameters $x = 0.71$, $y = 0.45$. Assuming it is legitimate, at a first approximation, to ignore the hydrogen atom, use this information to predict the structure from the general principles of packing and a knowledge of ionic radii.

11. Some compounds of composition A_2BO_4 with the olivine structure (α-structure) also have high-pressure forms with the spinel structure (γ-structure); others have a somewhat different structure (β-structure). It has been suggested that the spinel structure tends to be unstable unless the edges shared between octahedra are shortened. Using the bond lengths given below (which have been either directly observed in β-structures or γ-structures or derived by comparison from other structures), predict the lattice parameter and the position parameter u of a possible normal spinel structure for the following compounds: Mg_2SiO_4, Mn_2SiO_4, Fe_2SiO_4, Co_2SiO_4, Ni_2SiO_4, Mg_2GeO_4, Mn_2GeO_4. Compare the lengths of shared and unshared octahedron edges. Comment on the hypothesis in relation to the observation that at high pressures the β-structure is found for Mg/Si, Co/Si, and Mn/Ge compounds, the γ-structure for Fe/Si, Co/Si, Ni/Si, and Mg/Ge compounds. (Mg—O $= 2.09$ Å, Mn—O $= 2.19$ Å, Fe—O $= 2.15$ Å, Co—O $= 2.11$ Å, Ni—O $= 2.08$ Å, Si—O $= 1.63$ Å, Ge—O $= 1.76$ Å.)

12. Anatase (one form of TiO_2) is tetragonal, with $a = 3.776$ Å, $c = 9.486$ Å, space group $I4_1/amd$ (No. 141), and four formula units per unit cell. The Ti atoms are at 0, 0, 0, and oxygen atoms at 0, 0, 0.21. Describe the structure, noting in particular the Ti—O bond lengths, the linkage

pattern of the octahedra, and their detailed shape. [Note: the origin in this space group is chosen at a centre of symmetry on the line of intersection of the mirror planes; the 4_1 axis lies at $x = 1/4$, $y = 1/4$, and the a-glide plane at $z = 3/8$.] Compare the structure with that of rutile.

13. $KAlF_4$ is tetragonal, with $a = 3.55$ Å, $c = 6.139$ Å; its space group is $P4/mmm$ (No. 123), and there is one formula unit per unit cell. The K atoms are at $0, 0, 0$; Al at $1/2, 1/2, 1/2$; F(1) at $1/2, 0, 1/2$, and F(2) at $1/2, 1/2, u$, with $u = 0.21$. Describe the structure. Calculate the different K—F and Al—F distances, and describe the environment of each independent atom listed. Comment on the linkage of polyhedra.

14. Zircon, $ZrSiO_4$, is tetragonal, with $a = 6.60$ Å, $c = 5.88$ Å, and four formula units per unit cell. The space group is $I4_1/amd$ (cf. Question 12). The Zr atoms are at $0, 0, 0$; Si atoms at $0, 0, 1/2$; and O atoms at $0, 0.20, 0.34$.

    Draw projections of the structure on (001) and (100), marking the heights of the atoms. Calculate all distances from Zr and Si to neighbouring anions, and the angle between the Zr—O bonds and the tetrad axis; describe and comment on the environments of these "cations" and the linkage scheme of their polyhedra. Are the polyhedra those to be expected from the radii in Table 2-2? Is Pauling's electrostatic valence rule obeyed?

15. Draw a diagram, from the data in Chapter 10, Question 14, showing the part of the structure of BPO_4 between $z = -1/4$ and $z = 1/4$. Calculate the B—O and P—O distances. How do they compare with predictions from the radius sums? Comment on the linkage pattern of the structure.

16. Using the data of Chapter 10, Question 15, draw diagrams projected on the symmetry plane showing the linkage pattern of the tetrahedra in coesite, (a) in the part of the unit cell between heights 0 and 1/2, (b) in the complete structure. Compare it with the linkage pattern in sanidine.

17. Keatite is tetragonal, with $a = 7.456$ Å, $c = 8.604$ Å, and 12 formula units of SiO_2 in the unit cell. The space group is $P4_32_12$ (No. 96). Atomic positions are as follows:

	x	y	z
Si(1) in 8(b):	0.326	0.120	0.248
Si(2) in 4(a):	0.410	0.410	0
O(1) in 8(b):	0.445	0.132	0.400
O(2) in 8(b):	0.117	0.123	0.296
O(3) in 8(b):	0.344	0.297	0.143

Construct the projection on (001). Show that there are corner-sharing SiO_4 tetrahedra arranged in spirals around the lines $x = 0$, $y = 1/2$, and $x = 1/2$, $y = 0$, and that these are linked by Si atoms lying on the diagonal diads, each bonded to O atoms of four different spirals.

Calculate all eight Si—O distances to three-figure accuracy, and show that they lie in the range 1.57 Å to 1.61 Å.

Show similarly that the mean tetrahedron edge is 2.60 Å, with an extreme range of 0.16 Å between the greatest and the least.

18. Beryl, $Be_3Al_2(SiO_3)_6$, is hexagonal, with $a = 9.21$ Å, $c = 9.20$ Å. Its space group is $P6/mcc$ (No. 192), and it has two formula units per unit cell. Assuming it to be ideal in composition and perfectly ordered, predict the structure from a knowledge of the above facts, the approximate packing diameter of the oxygen atom, and the usual coordinations of the cations. Develop the argument in steps as follows:

(a) From consideration of the space-group symmetry and special positions (cf. Chapter 8, Question 5), the numbers of Al and Be atoms, and the ideal symmetry of their usual coordinations, show that Al must be octahedral rather than tetrahedral. Hence show that Al is at 1/3, 2/3, 1/4, and Be at 1/2, 0, 1/4.

(b) Draw a scale diagram of the unit cell (most conveniently on hexagonal graph paper). Insert enough of the Al and Be atoms in the lower half of the unit cell to allow the rest to be easily visualised with the help of the hexad axis. Mark their heights.

(c) Assuming that the Al polyhedron is a regular octahedron with an O—O edge of 2.8 Å, find the positions of its corners, and show that these are symmetry-related oxygen atoms in a general position, with one of them at approximately 0.48, 0.15, 0.15. Mark on your diagram all such oxygen atoms at heights between 0 and 1/2.

(d) Show that the remaining atoms lie on mirror planes, and make deductions about the placing of the SiO_4 tetrahedra. Which, if any, of the oxygen atoms previously inserted form parts of these tetrahedra—corners or edges?

(e) What restrictions are imposed on the placing of the mirror-plane oxygen atoms by their closeness of approach to (i) others of their own symmetry-related set, (ii) others in general positions in planes above and below? (This is best answered with the help of your diagram; begin by assuming a packing diameter of 2.8 Å, but remember that tetrahedron edges may be shorter than this by as much as 10 or 15 percent when a compromise with the requirements of other polyhedra is necessary.)

Show that the use of these restrictions defines tetrahedra which can be made most nearly regular by placing one oxygen atom at approximately 0.30, 0.24, 0.

(f) Hence show that Si is at approximately 0.39, 0.12, 0.

(g) Complete the structure diagram, outlining the polyhedra round Al, Be, and Si.

19. Using the data and results of Question 18, discuss the structure of beryl.

    (a) Comment on the general linkage pattern. Is it what you would expect? Is Pauling's rule obeyed?

    (b) Calculate the different cation-anion distances, and compare them with those expected from Table 2-2.

    (c) How regular are the Be tetrahedra, the Si tetrahedra, and the Al octahedra? For each, try to define the nature and calculate the magnitude of its greatest departure from the ideal shape.

    (d) Can you suggest, qualitatively, any changes of parameters which would improve the regularity of one kind of polyhedron without making another worse? If not, comment on the mutual effect of the distortions in achieving a compromise structure.

20. One kind of garnet, grossularite, has the composition $Ca_3Al_2Si_3O_{12}$. It is cubic, with $a = 11.85$ Å, space group $Ia3d$ (No. 230), and eight formula units per unit cell.

    (a) Using the above information together with the facts about the space group taken from Chapter 8, Question 6 and its answer, show that possible positions for the atoms are: Al in $16(a)$: $0, 0, 0$; Ca in $24(c)$: $1/8, 0, 1/4$; Si in $24(d)$: $3/8, 0, 1/4$; and O in $96(h)$: u, v, w. Are there any possible alternatives, or are all these uniquely determined?

    (b) Show that if u, v, w are fairly small, there is a group of six oxygen atoms round Al.

    (c) From the lattice parameter and an approximate knowledge of interatomic distances, show that the only oxygen atoms which are possible neighbours to Ca at $1/8, 0, 1/4$, are those which are also neighbours to an Al at $0, 0, 0$; $0, 1/2, 0$; $1/4, 1/4, 1/4$; or $1/4, 3/4, 1/4$.

    (d) Given that $u = -0.04$, $v = 0.05$, $w = 0.15$, show that Ca has eight neighbours; find the Ca—O bond lengths and the shape of the polyhedron.

    (e) Find the Al—O bond lengths.

    (f) Write down the coordinates of the oxygen atoms adjacent to Si at $3/8, 0, 1/4$, and calculate the Si—O bond lengths.

    (g) Write down the coordinates of all the cation neighbours of any chosen oxygen atom. Draw a sketch stereogram of their arrangement. Is Pauling's rule obeyed?

    (h) What edges are shared in common between SiO_4, AlO_6, and CaO_8 groups? Calculate the lengths of all O—O edges, and comment on them.

    (i) How far can any departure from the ideal shape of SiO_4 groups or AlO_6 groups be explained?

    (j) Compounds known to have the garnet structure include those of composition $Mg_3Al_2Si_3O_{12}$, $Y_3Fe_5O_{12}$, $Na_3Al_2Li_3O_{12}$, and $Y_3Al_4GaO_{12}$.

Were all these to be expected, or are they surprising in any way? Comment in particular on the question of ionic radii, and on the application of Pauling's rule.

21. A high-pressure form of $KAlSi_3O_8$ is tetragonal, with $a = 9.38$ Å, $c = 2.74$ Å, space group $I4/m$, and two formula units per unit cell. It is believed to have the hollandite structure, and to be quasiperfect, with Al and Si distributed at random over one set of symmetry-equivalent sites. The atomic positions are then as follows: K in $2(b)$: $0, 0, 1/2$; Al/Si in $8(h)$: $0.167, 0.384, 0$; O(1) in $8(h)$: $0.208, 0.152, 0$; O(2) in $8(h)$: $0.152, 0.542, 0$. Describe the structure, and comment on the environments of the atoms.

    Compounds $KAlGe_3O_8$, $KAlTi_3O_8$, and $RbAlSi_3O_8$ can all occur with this structure. Do you consider it more or less likely to be the stable low-pressure structure for these than for $KAlSi_3O_8$? Can you predict the lattice parameters?

Chapter 12   FAMILIES OF
STRUCTURES

12.1   DEFINITION OF A FAMILY

In the last chapter, an account was given of some representative structures, and materials strictly isomorphous with the type structures were included. Strict isomorphism between structures requires that they have the same symmetry, the same sets of occupied sites, and approximately the same parameters. In consequence, corresponding atoms will have closely similar environments and linkages, and this is responsible for a close similarity in chemical and physical properties. On the other hand, closely similar environments and linkages may still be found even when the symmetry is *not* the same, as we saw in comparing high quartz and low quartz (§11.12). This qualitative resemblance is in some ways more important than strict isomorphism, and it is useful to define it more carefully.

Structures may be said to belong to the *same family* if there is a one-to-one correspondence between all their atoms, and between all their interatomic bonds. This does not mean that all bonds remain of the same relative strengths from one member of the family to another, but that they must be formally recognisable, however weak. The symmetries of the structures may differ, and so may the relation of the periodic repeat lengths to the structure-building units, as well as bond lengths and bond angles. Thus, high and low quartz belong to one family, high and low cristobalite to a different family; zinc blende (ZnS) belongs to the diamond family, but its polymorph wurtzite does not, because its linkage pattern is topologically different.

Differences among members of a family may be due to a distortion or crumpling of the structure, like that in quartz; or it may be due to different kinds of occupation of the same atomic sites, like that of zinc blende as compared with diamond. In this connection, we may widen our definition so that the family includes structures in which a regular selection of the atomic sites are unoccupied, provided that they are not an essential part of the structural framework; they might, for example, be interstitial or cavity atoms—cf. §11.16(c). We can do this formally by considering a vacancy as a species of atom with which to fill these sites. Examples will follow in §12.11 and 12.15.

The simplest and most symmetrical member of any family is called the *aristotype* (from the Greek, $\alpha\rho\iota\sigma\tau\sigma\varsigma$, "best," referring to the symmetry). A member of lower symmetry is called a *hettotype* ($\eta\tau\tau\sigma\varsigma$, "inferior").

12.2 GEOMETRICAL RELATIONS

The loss of symmetry in changing from the aristotype to the hettotype may be of various kinds, with different effects on the unit cell. Sometimes it may leave the unit cell unaltered (except for small changes in lattice parameters), as happens in the change from high to low quartz. More commonly, the shape of the unit cell is distorted, again without much change of size; examples of this are found among the perovskites (§12.5–12.11) and rutiles (§12.14). Yet again, the translation-repeat length may be doubled, or multiplied by some other small integer, with or without accompanying changes of interaxial angles.

Let us consider this formation of a *multipartite* cell in more detail. Suppose that in the aristotype the repeat unit is a_0, and in the hettotype $a_1 \simeq 2a_0$. Then the part of the hettotype structure bounded by planes $x = 0$ and $x = a_0$ is nearly but not quite equal to that bounded by planes $x = a_0$ and $x = 2a_0$. The true repeat unit consists of two very similar halves, and we may refer to it as *bipartite* in the x direction. In more complicated cases, the unit cell may be bipartite, or multipartite, in more than one direction. It can be subdivided into parallelepipeds that are very similar to the unit cell of the aristotype, and are identical with each other in size and shape, but have contents differing slightly, either in the nature of the atoms or in their exact positions, or in both. They are referred to as *subcells*.

The doubling of the translation-repeat distance is itself a lowering of symmetry, even if the space group remains unaltered. Suppose, for example, that the unit cell of the aristotype is $ABCD$, as in Figure 12-1(a), with mirror planes at AB and DC; from §8.6, we have seen that this implies another mirror plane EF. Now if the unit cell of the hettotype is $ABGH$, as in Figure 12.1(b), with mirror planes at AB and HG, the plane at DC is retained but that at EF is lost. If the aristotype has atoms at P and Q, as in Figure 12.1(c), the hettotype has corresponding atoms P' and Q', but they are no longer symmetry-related. Thus the hettotype may or may not have less symmetry per unit cell than the aristotype, but it certainly has less symmetry per formula unit.

In this example, P and Q in the aristotype must be the same species of atom, because of the symmetry. On the other hand, P' and R' in the hetto-type may be different species (though having similar structural roles), and

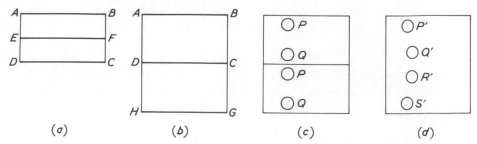

Figure 12-1 Relations between (hypothetical) aristotype and hettotype. (a) Unit cell of aristotype, with mirror planes at AB and EF; (b) unit cell of hettotype, with mirror planes twice the distance apart, at AB and DC; (c) two unit cells of aristotype, with symmetry-related atoms P and Q, repeated by the lattice translation; (d) one unit cell of hettotype, with atoms related in pairs $P'S'$, $Q'R'$, by the mirror planes.

even if they are the same species their coordinates are not crystallographically related.

It is helpful sometimes to be able to use the same set of axes of reference for all members of the same structure family, even if they are not the axes which are conventional, and would otherwise be most convenient, for some of them. For this reason, it is always desirable to give the matrix for transforming the axes of reference chosen for any individual member of the family to the axes which are standard for the family as a whole, i.e., the axes of the aristotype.

The number of formula units per unit cell (conventionally denoted by Z) depends of course on the choice of axes of reference.

12.3 SUBSTITUTION EFFECTS

As we saw in Chapter 11, replacement of one kind of atom by another, in a complete equipoint, often results in a structure isomorphous with the first (e.g., $FeTiO_3$, $MgTiO_3$; Mg_2SiO_4, $CaMgSiO_4$). Sometimes, however, it results in a geometrically modified structure belonging to the same family (e.g., $SrTiO_3$, $CaTiO_3$). We shall have many examples of such modified structures in this chapter. Sometimes, of course, a change of cation leads to a wholly different structure (e.g., $SrTiO_3$ and $MgTiO_3$).

We must also consider what happens when only part of the atoms in the equipoint are replaced by a different species. Suppose exactly half (or some other simple fraction) are replaced, in a regularly repeating pattern of sites. Then the new structure is a hettotype of the old. In this way, for example, the zinc-blende structure is a hettotype of that of diamond, and the ilmenite structure a hettotype of that of corundum.

On the other hand, replacement of one kind of atom A_1 by another kind A_2 may be at random, so that the chance of finding A_2 at any of the original A_1 sites is the same, with no kind of periodic pattern. Then all sites of the equipoint are still equivalent, and we may think of them all as occupied by an "average atom," intermediate between A_1 and A_2. The structure can be dealt with as if it were a perfect structure built from such average atoms. (Structures with unoccupied sites may again be included by treating these sites as occupied by zero atoms.) For random replacement, the exact proportions of A_1 and A_2 are unimportant provided they are such as allow the structure to be formed, and the result is a *solid solution* (cf. §1.9).

In this chapter, for the most part, we shall consider fully ordered structures. Families of structures which contain many members of different chemical composition generally include also an immense variety of solid solutions, and we shall mention only a few which are of particular interest. Partly ordered structures sometimes occur, but they are generally best understood by considering the fully ordered and disordered extremes between which they lie.

12.4 THE PEROVSKITE FAMILY

The perovskite structure, with general formula ABX_3, consists essentially (as we saw in §11.2) of a framework of BX_6 octahedra linked by their corners, with a large A cation occupying a cavity of the same size as an X anion. The aristotype (the structure of $SrTiO_3$) is very simple but very demanding, because it has only one variable parameter, the unit-cell side a. If the atoms A, B, and X are not of the ideal relative sizes, there is no adjustable atomic position parameter by which the misfit can be remedied, as there was in the spinel structure. Instead, the aristotype is replaced by a hettotype.

Most of the important perovskite hettotypes involve displacement of atomic positions rather than ordered substitution processes. There are a great variety of such hettotypes, and they can be most easily understood by considering separately three different kinds of distortion.

(i) The X_6 group round a B cation may be distorted in shape, becoming, for example, a tetragonal bipyramid rather than an octahedron.

(ii) The B cation may be displaced from the centre of its surrounding X_6 group, which remains a regular octahedron (or very nearly so).

(iii) The BX_6 octahedra, while remaining regular, may tilt relative to one another, reducing the size of the cavity occupied by A.

The magnitude of each kind of distortion may separately range from zero up to quite large values, and the different kinds may occur independently or in combination. If tilts occur, the resultant unit cell must be at least bipartite; if the octahedra all undergo the same distortion, and remain in parallel orientation, a unipartite structure results.

The kinds of distortion depend on the sizes and bonding character of the cations.

Large distortions of the octahedra are generally due to nonoctahedral covalent bonding by B cations, or to pronounced covalent character in the bonds to A (cf. §2.10). For example, Cu^{2+} in 6-coordination tends to form a tetragonal bipyramid, elongated along its tetrad axis, rather than an octahedron; again, when Pb^{2+} is the A cation, its tendency to covalent bond formation distorts the octahedron from the outside, in ways which are not easily predictable.

Off-centre displacements of B tend to occur when B is rather too small for octahedral coordination, for example Nb^{5+} surrounded by O atoms. The mechanism is that discussed in §2.8, enhanced by the high polarisation of the anion in the field of a high-valent cation. Systematic distortions of the octahedra accompany the displacements, but they are much smaller than those due to pronounced covalent character, and for many purposes may be ignored completely.

Tilting of octahedra occurs when the A cation is too small for the volume available in the aristotype, for example, in the oxide perovskites when A is Ca or Na rather than Sr or K (cf. §11.2). If the octahedra can be assumed regular, it is possible to deduce the tilt system and the magnitudes of the tilt angles from the lattice parameters and symmetry. Tilting can occur in various ways, which are considered in §12.8 to 12.10. The A cation is displaced from the geometrical centre of the cavity to a position where it has less than twelve reasonably close neighbours.

Many materials belonging to the perovskite family are polymorphic, possessing different hettotype structures in different temperature ranges. In this chapter we shall not be concerned with the reasons for this, nor with the character of the transitions between them; temperature ranges are merely mentioned for purposes of identification and reference. The physical principles underlying the geometrical relationships will be discussed in Chapter 15.

12.5 DISTORTED OCTAHEDRA

An example of a structure with covalently distorted octahedra is $KCuF_3$, described as follows:

> Lattice: Tetragonal body-centred, with $a = 5.85$ Å, $c = 7.85$ Å.
> (Relation to aristotype: $a = \sqrt{2}a_p, c \simeq 2a_p$.)
> Number of formula units per unit cell: 4
> Space group: $I4/mcm$ (No. 140)
> Atomic coordinates:
> K   in $4(a)$: $0, 0, \frac{1}{4}$
> Cu   in $4(d)$: $0, \frac{1}{2}, 0$
> F(1) in $4(b)$: $0, \frac{1}{2}, \frac{1}{4}$
> F(2) in $8(h)$: $\frac{1}{4} - u, \frac{3}{4} - u, 0$,   with $u = 0.022$

The K atoms are on intersections of glide planes, and the Cu atoms on intersections of mirror planes.

A projection of the structure is shown in Figure 12-2(a). There are two sheets of octahedra, all octahedra being of equal height $c/2$ and point-group

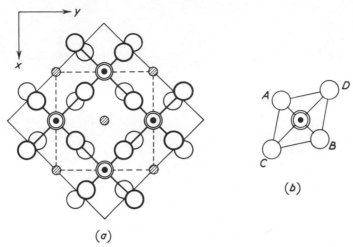

(a)

(b)

Figure 12-2 Structure of $KCuF_3$. (a) Projection on (001): small black circles, Cu at heights 0 and 1/2; hatched circles, K at heights 1/4 and 3/4; large double circles, F at heights 1/4 and 3/4; large heavy circles, F at height 1/2; large light circles, F at height 0. Heavy solid lines are Cu—F bonds at height 1/2; broken line is outline of unit cell; fine solid line is outline of all-face-centred unit cell corresponding to pseudocubic axes of reference. (b) Projection on (001) of single octahedron (or tetragonal bipyramid); long Cu—F bonds lie along CD, short Cu—F bonds along AB and perpendicular to the paper.

symmetry *mmm*. All are alike in shape, elongated along one corner-to-corner diameter in accordance with the covalent-bond requirements of the Cu^{2+} atom. They are not, however, all in the same orientation. All lie with one short diameter parallel to [001], but within a sheet parallel to (001) each has its long diameter at right angles to that of its neighbours. A single octahedron is shown in Figure 12-2(b). The short Cu—F bond along [001] is $c/4 =$ 1.96 Å, the short bond in (001) is $\sqrt{2}a(1/4 - u) = 1.88$ Å, and the long bond in (001) is $\sqrt{2}a(1/4 + u) = 2.25$ Å; the elongation is thus about 20 percent.

On pseudocubic axes the unit cell is all-face-centred, and bipartite in each axial direction, with an axial ratio $c_p/a_p = 0.95$.

Both the cations, K and Cu, remain central in their polyhedra.

At low temperatures the crystal becomes antiferromagnetic (Néel temperature $-30°$ C), but no change of crystal structure has been detected.

12.6 B-CATION DISPLACEMENT IN UNIPARTITE PEROVSKITE STRUCTURES

Displacement of B cations from their octahedron centres can only result in unipartite structures if all displacements are identical. Unless all symmetry is to be lost, there are only then three possibilities: displacement along a tetrad axis of the octahedron, along a diad axis, or along a triad axis. The symmetry axis of the structure which lies in the direction of the displacement is retained,

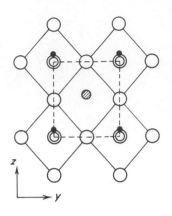

Figure 12-3 Tetragonal BaTiO$_3$, projected on (100). Small black circles are Ti at height 0; single open circles are oxygen at height 0; double open circles are oxygen at height 1/2; hatched circles are Ba at height 1/2. Unit cell is shown by broken line. (All difference parameters are exaggerated.)

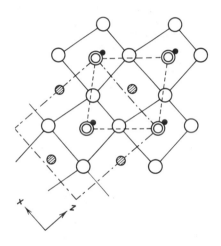

Figure 12-4 Orthorhombic BaTiO$_3$, projected on (010). Symbols for atoms are as in Figure 12-3; primitive unit cell is shown by broken line, B-face-centred unit cell by dot-dash line. (All difference parameters are exaggerated.)

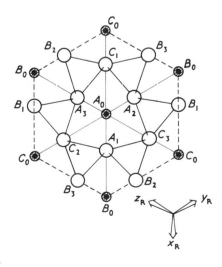

Figure 12-5 Rhombohedral CsGeCl$_3$, projected on (111). Small black circles are Ge just above corners of rhombohedron A_0, B_0, C_0. Hatched circles are Cs just above body centre of rhombohedron. Open circles are Cl at height midway between heights of rhombohedron corners: A_1, A_2, A_3 between A_0 and B_0; B_1, B_2, B_3 between B_0 and C_0; C_1, C_2, C_3 between C_0 and A_0. Expressed as fractions of the rhombohedron height, the heights above the plane of the paper are: Ge, 0.075; Cs, 0.558; Cl (A_1, A_2, A_3,), 1/6; Cl (B_1, B_2, B_3), 1/2; Cl (C_1, C_2, C_3), −1/6. Outline of projection of rhombohedron is shown by broken line. (Difference parameters are exaggerated.)

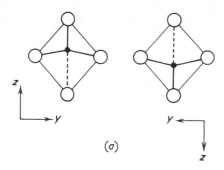

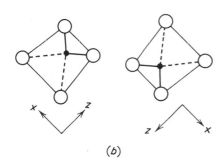

Figure 12-6 Reversal of octahedral dipoles. (a) Tetragonal $BaTiO_3$: section of octahedron parallel to (100). In the left-hand diagram, the dipole is oriented as in Figure 12-3; in the right-hand diagram, the dipole is reversed. (The polar axis is [001].) (b) Orthorhombic $BaTiO_3$: section of octahedron parallel to (010). In the left-hand diagram, the dipole is oriented as in Figure 12-4; in the right-hand diagram, the dipole is reversed. (The polar axis is [001].) (c) Rhombohedral $CsGeCl_3$: projection of octahedron on basal plane. Heavy lines mark the upper face. In the left-hand diagram, the dipole is oriented as in Figure 12-5; in the right-hand diagram, the dipole is reversed. (The polar axis is normal to the plane of projection.) Hexagonal axes of reference are shown. [Note: in (a) and (b) the reversal of sense of y or x is a purely formal accompaniment of the reversal of sense of z, to keep the axes right-handed; in (c), the x_H and y_H axes, as well as z_H, are reversed physically, and formality requires that the names of the x_H and y_H axes should be interchanged.]

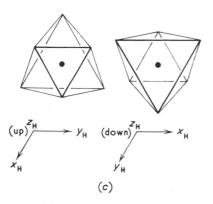

and so are the planes parallel to it; all other symmetry is lost. The unit cell becomes a tetragonal prism in the first case, a rhombic prism in the second, a rhombohedron in the third (cf. §6.11). The distortion of the cube, and of the octahedron inscribed in it, is typically very small, of the order of one percent or less. Since there is no centre of symmetry, the choice of origin along the symmetry axis is arbitrary; it is generally most convenient to take it at the geometrical centre of the octahedron. The off-centre displacement of B is then directly proportional to the atomic parameter of B. The A cations are also displaced parallel to the same direction. Examples of the three possible hettotypes follow.

Tetragonal BaTiO$_3$ (about 0° to 120° C) (Figure 12-3).

Lattice: Tetragonal, primitive, with $a = 3.99$ Å, $c = 4.03$ Å, $c/a = 1.010$
Number of formula units per unit cell: 1
Space group: $P4mm$ (No. 99)
Atomic coordinates:
  Ti   in $1(a)$: $0, 0, z_{Ti}$, with $z_{Ti} = 0.032$
  Ba   in $1(b)$: $\frac{1}{2}, \frac{1}{2}, \frac{1}{2} + z_{Ba}$, with $z_{Ba} = 0.018$
  O(1) in $1(a)$: $0, 0, \frac{1}{2} + z_1$, with $z_1 = -0.007$
  O(2) in $2(c)$: $\frac{1}{2}, 0, z_2$, with $z_2 = 0.003$

Other structures strictly isomorphous with this are KNbO$_3$ (225° to 450° C) with $c/a = 1.017$, and PbTiO$_3$ (room temperature) with $c/a = 1.063$. The greater distortion of PbTiO$_3$ is to be attributed to the covalent bonding of Pb.*

Orthorhombic BaTiO$_3$ (−90° to 0° C) (Figure 12-4).

Lattice: Orthorhombic, B-face-centred, with $a = 5.669$ Å, $b = 3.990$ Å,
     $c = 5.682$ Å
Number of formula units per unit cell: 2
Space group: $Bmm2$ (No. 38)
Atomic coordinates:
  Ti   in $2(a)$: $0, 0, z_{Ti}$, with $z_{Ti} = -0.02$
  Ba   in $2(b)$: $0, \frac{1}{2}, \frac{1}{2} + z_{Ba}$, with $z_{Ba} = -0.01$
  O(1) in $2(b)$: $0, \frac{1}{2}, z_1$, with $z_1 = 0$
  O(2) in $4(d)$: $\frac{1}{4} + x_2, 0, \frac{1}{4} + z_2$, with $x_2 = 0.03$, $z_2 = 0$

(The primitive unit cell has $a_p = c_p = 4.012$ Å, $b_p = 3.990$ Å, $\beta = 90°8'$, $b_p/a_p = 0.994$, $Z = 1$.) Strictly isomorphous with this is KNbO$_3$ (−50° to

---

* One might ask why, if the off-centring of the B cation depends entirely on its size relative to O, does Ti behave differently in SrTiO$_3$ on the one hand, BaTiO$_3$ and PbTiO$_3$ on the other ? The answer seems to be that Ti is just large enough to remain central if there are no disturbing factors. In PbTiO$_3$ the disturbing factor is the polarising power of Pb, and in BaTiO$_3$ the large size of the Ba atom, which pushes the O atoms apart and makes the octahedron too large for a central Ti.

220° C), with $b_p/a_p = 0.985$, $\beta = 90°14'$, and nearly all its deviations from the aristotype very slightly greater than those of $BaTiO_3$.

Rhombohedral CsGeCl₃ (Figure 12-5).

Lattice: Rhombohedral, primitive, with $a = 5.44$ Å, $\alpha = 89°40'$
Number of formula units per unit cell: 1
Space group: $R3m$ (No. 160)
Atomic coordinates:
Ge in $1(a)$: x_{Ge}, x_{Ge}, x_{Ge}, with $x_{Ge} = 0.03$
Cs in $1(a)$: $\frac{1}{2} + x_{Cs}$, $\frac{1}{2} + x_{Cs}$, $\frac{1}{2} + x_{Cs}$, with $x_{Cs} = 0.01$
Cl in $3(b)$: x_{Cl}, x_{Cl}, $\frac{1}{2} - 2x_{Cl}$, with $x_{Cl} = -0.02$

It is probable that $BaTiO_3$ below $-90°$ C and $KNbO_3$ below $-50°$ C are strictly isomorphous with this structure.

In all three structures, displacement of the B cation results in inequality in the six B—O bond lengths. In tetragonal $BaTiO_3$ we have one short and one long bond in the direction of the displacement, and four of intermediate length nearly at right angles to it—1.87 Å, 2.17 Å, and 2.00 Å respectively. Since the difference between short and long bonds is about 15 percent of their mean, it is clear that the effect of B-cation displacement on the bond lengths is an order of magnitude greater than its effect on the regularity of the octahedron, which was about one percent. In orthorhombic $BaTiO_3$ we have two short and two long bonds in a plane containing the displacement vector, and two intermediate ones nearly at right angles to that plane; the lengths, 1.90 Å, 2.11 Å, and 2.00 Å, are not unlike those in the tetragonal structure for the same material. Unfortunately the atomic parameters of rhombohedral $BaTiO_3$ are not known, for comparison with the other two polymorphic forms.

In rhombohedral $CsGeCl_3$, there are three short and three long Ge—Cl bonds, of lengths 2.31 Å and 3.13 Å. The long Ge—Cl bonds are of course much weaker than the short bonds. If we supposed them to be weaker than the Cs—Cl bonds, it would be permissible to ignore them and think of the short bonds as forming $GeCl_3$ molecule-ions. Though this picture is almost certainly too extreme, it is useful to keep it in mind.

One very important feature of these structures concerns the two B—X bonds to the same anion. They are either one long and one short bond, or two bonds of intermediate length. This situation is understandable in terms of polarisation of the anion. A cation approaching an anion to form a short bond polarises the anion, and in this way, so to speak, ties down a substantial part of the electron cloud into the short bond; there is disproportionately less of the cloud available to form a bond on the other side of the anion, and the bond is therefore weaker than normal and the cation linked by it is more remote. We find the same feature recurring in other more complicated hettotypes (§12.10).

2.7 DISPLACEMENTS AND DIPOLES

The fact that all displacements of the B cation, in any of these structures, are in the same direction has very important physical consequences. Though the centres of the anions form nearly regular octahedra, the polarisation of their electron clouds by the nearer approach of the B cation to some of them has the consequence that cation and surrounding anions together constitute an electric dipole. All the dipoles point in the same direction, and their effects are additive; the structure therefore possesses a *spontaneous polarisation*.

Spontaneous polarisation, the possession of an externally observable dipole moment, is formally possible in crystals of any point group which allows a nonzero resultant vector. These are sometimes called the *polar point groups*. For those with a symmetry axis—namely, $6mm$, 6, $4mm$, 4, $3m$, 3, $2mm$, and 2—the resultant polarisation must be parallel to the axis though the individual dipoles need not be. (In the other two polar groups, m and 1, the polar direction is not fixed, except that in m it must be in the symmetry plane.) However, if, as in the hettotypes described in the last section, there is only one formula unit per primitive unit cell, the individual dipoles do lie parallel to the symmetry axis.

In an applied electric field, dipoles tend to orient themselves in the direction of the field. In the structures we are considering, the dipoles cannot do so by rotating, but they *can* turn themselves inside out (if the displacements involved are not too large). No breaking up of polyhedra is involved. Each cation simply moves within its own polyhedron, the polarisation of all the anions changes to correspond, and the octahedral distortions (which, as we saw, were very small) change in sign. The resulting structure is the same as the original, but in a new orientation (Figure 12-6).

We have assumed that all changes of atomic positions occur together. The effect must be cooperative, because any octahedron which has reversed creates a strained region in which it is easier for its neighbours to reverse. This property of reversal of direction of a spontaneously polarised structure in an electric field is known as *ferroelectricity*.

In general, for chemically similar structures, the smaller the deviations from the aristotype the easier it is for a structure to reverse.

12.8 PEROVSKITES WITH TILTED OCTAHEDRA

Many tilt patterns are possible; Figure 12-7 illustrates three of the simplest.

In Figure 12-7(a), each octahedron is rotated in the plane of the diagram about its tetrad axis normal to the paper, alternate octahedra twisting in opposite directions. The primitive unit cell is tetragonal, and contains two formula units; its base is the square $EFGH$, and it is one octahedron high. The original cube has become the pseudocubic subcell, of side a_p, and the

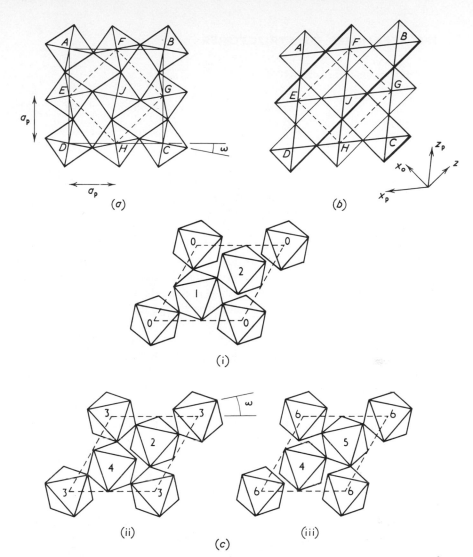

Figure 12-7 (a) Tilting of octahedra about an axis perpendicular to the plane of projection, parallel to the tetrad axis of the octahedron. The unit cell of the aristotype corresponds to a small square such as $AFJE$; the primitive unit cell of the hettotype is $EFGH$.

(b) Tilting about an axis parallel to BD, parallel to the diad axis of the octahedron. The unit cell of the aristotype corresponds to a subcell such as $CHJG$; the primitive unit cell of the hettotype is $EFGH$. Edges parallel to BD shown by heavy lines are parallel to the plane of projection, at a higher level than the parallel edges shown by light lines; edges projecting parallel to AC are inclined to the plane of projection.

(c) Tilting about the triad axis of the octahedron, perpendicular to the plane of projection. Heights of centres of octahedra above a plane perpendicular to the triad axis through the corner of the unit cell are marked. Only part of the unit cell is shown in (i), (ii), and (iii); (ii) fits on top of (i), with octahedra at height 2 in common, and (iii) on top of (ii), with octahedra at height 4 in common. Note that if the tilt ω were zero, the octahedron at height 3 would project to coincide with that at height 0, giving the ideal perovskite structure; as it is, the first true repeat is at height 6. The pseudocubic subcell is defined by vectors from the origin to the centre of octahedron 1, the primitive rhombohedral true unit cell by vectors from the origin to the centre of octahedron 2. The latter is therefore in the *reverse* setting (cf. §7.6); we ought to have chosen hexagonal axes of reference rotated 60° anticlockwise from those shown, to obtain the conventional obverse setting.

293

new lattice has dimensions

$$a = b \simeq \sqrt{2}a_p, \qquad c = a_p \qquad (12.1)$$

The octahedra remain very nearly regular, and to this approximation we may express the lattice dimensions more precisely in terms of the octahedron edge length l and the tilt angle ω:

$$a = b = 2l \cos \omega, \qquad c = \sqrt{2}l \qquad (12.2)$$

In Figure 12-7(b) the tilt is about a diad axis of the octahedron, which is parallel to a pair of edges, and to the face diagonal GH of the original unit cell. Alternate rows of octahedra are tilted in opposite directions; edges shown by heavy lines are at a slightly higher level than other edges parallel to them. Edges parallel to AC are inclined to the plane of the projection, and therefore shorter in projection than those parallel to BD. As a consequence, the original square $CHJG$ has turned into a rhombus, and the symmetry directions which are retained must lie parallel and perpendicular to its long diagonals. Choosing the plane of the diagram as (010), and the long diagonal of the rhombus as the z axis, we have

$$a \simeq c \simeq \sqrt{2}a_p, \qquad b_p \simeq 2a_p \qquad (12.3)$$

or, more exactly, for a tilt angle φ,

$$a = 2l \cos \varphi, \qquad b = 2\sqrt{2}l \cos \varphi, \qquad c = 2l \qquad (12.4)$$

These two kinds of tilts, about tetrad and diad axes respectively, were originally described as hypothetical possibilities, but can now be illustrated from structures to be considered in §15.12 and 12.10.

Both kinds of tilt can occur together without loss of orthorhombic symmetry; the combination is, in fact, one of the commonest among known structures. An interesting way of looking at the combination is as follows. First, we consider the φ tilt as the resultant of two tilts of equal magnitude $\varphi/\sqrt{2}$ about tetrad axes (of the octahedron) in the plane of the layer. Now we add a tilt of similar magnitude about the third tetrad axis, perpendicular to the layer. The resultant is a tilt about a triad axis inclined to the layer. This can be seen in Figure 12-8(a), in which the tilt axis comes up from the plane of the diagram as it extends from the lower left to the top right corner. To build an orthorhombic structure, we use a mirror plane or a diad axis parallel to the layer to generate the next layer, which therefore has tilt axes all parallel to a different triad axis. For orthorhombic symmetry, the first two tilt components must be equal in magnitude, but the third is independent. It is interesting to find in practice that it is often nearly equal to the others, and much more rarely zero. Examples are given in §12.10 and discussed further in §15.14 and 15.15.

Now suppose that we had made the same layer as before, with three tilt components of equal magnitude, but instead of using a mirror plane to generate the next layer we repeated the first layer, making the repeat distance the same as the shortest repeat distances within the layer. In this case all octahedra keep the same direction for their tilt axis, and it is a triad axis of the structure as a whole.

For many purposes it is easier to look at such a structure down the triad axis, and this is shown in Figure 12-7(c). Starting with the ideal cubic structure, successive octahedra are rotated in opposite directions about their triad axes. The height of the primitive unit cell is now twice the body diagonal of the cubic subcell. This kind of crumpling of the framework is particularly interesting, because if the A cation were absent and the hinges at the shared anions were completely flexible, the distortion could be carried out continuously until, with a tilt angle of 30°—instead of the 10° shown in Figure 12-7(c)—the anions were in hexagonal close packing and the 12-coordinated cavities had been reduced to a pair of octahedral interstices. The BX_6 framework would then be identical with a slightly idealised form of the NbO_6 framework in $LiNbO_3$ (cf. p. 239).

Comparisons between the distorted frameworks of the different hettotypes and the aristotype are most easily made if we use pseudocubic axes of reference and a unit cell with $a \simeq b \simeq c \simeq 2a_p$, $\alpha \simeq \beta \simeq \gamma \simeq 90°$. This cell contains 8 formula units, and the pseudocubic subcell is one of its octants. We denote its edge lengths by a_p, b_p, c_p.

The orthorhombic structure has then a B-face-centred unit cell; its b axis is a symmetry direction though its a and c axes are not, and the formulae appropriate to monoclinic symmetry can therefore be used, with the additional condition that a is identically equal to c. The geometry is like that of orthorhombic $BaTiO_3$, but the physical reason for the departure from a 90° interaxial angle is wholly different.

The rhombohedral structure has an all-face-centred unit cell. The primitive unit cell is related to this larger cell as the primitive unit cell of a face-centred cubic lattice is to the face-centred cube—see §7.6, Figure 7-8, and also Figure 3-6(d).

Other more complicated combinations of tilts can occur, though the systems described above are the commonest and most important. For a systematic treatment, it is easiest to think of a general tilt as made up of component tilts about the three tetrad axes of the aristotype, and note whether successive octahedra along each axis have tilts in the same sense or opposite senses about that axis.

Structures with tilted octahedra must have multipartite unit cells. The B cations may remain central in their octahedra, or they may be displaced, as in the unipartite hettotypes. The off-centring of B and the octahedral tilts are independent features of the hettotype.

If, however, off-centre B displacements occur in a multipartite structure with tilted octahedra, they need not all be parallel—they may lie in general directions. It depends on the space-group symmetry whether or not they add up vectorially to zero for the unit cell.

When the B cations are central, the octahedra do not constitute dipoles, and the structures are often called *paraelectric*. When the B cations are non-central, but their displacements (and therefore the dipoles) add up vectorially to zero within the unit cell, the materials are referred to as *antiferroelectric*. When the B cations are noncentral and their displacements do not add up to zero (which is possible only in a polar space group), the structures may be ferroelectric if the displacements are sufficiently small to be reversed by a field; otherwise they are *pyroelectric*. Cases where the vector sum of the displacements is zero are much commoner than those where it is not zero. (It is not true, however, as sometimes assumed in the literature, that evidence for the multipartite character of a structure proves that it cannot be ferro-electric.)

12.9 EXAMPLES OF THE RHOMBOHEDRAL TILT SYSTEM

An example of the rhombohedral tilt system occurs in $LaAlO_3$. Its primitive rhombohedral unit cell has

$$a = 5.36 \text{ Å}, \quad \alpha = 60.1°, \quad Z = 2$$

The larger face-centred rhombohedral unit cell has

$$a = 7.58 \text{ Å}, \quad \alpha = 90.1°, \quad Z = 8$$

Using hexagonal axes, $a_H = 5.365$ Å, $c_H = 13.11$ Å, $Z = 6$. The space group is $R\bar{3}c$ (No. 167), and the cations are at centres of symmetry, but whereas in ideal perovskite, referred to the hexagonal axes, one oxygen atom would lie at 1/6, 1/3, 1/12, in $LaAlO_3$ it has moved to $1/6 + 0.03$, 1/3, 1/12, with the others moving to correspond. The tilt angle is

$$0.03 a_H \Big/ \left(\frac{1}{3} \cdot \frac{\sqrt{3}}{2} \, a_H \right) \simeq 0.10 \text{ radians} \simeq 6° \tag{12.5}$$

If we had been able to assume perfectly regular octahedra, we should have been able to calculate the tilt angle from the axial ratio, since c is determined by l, the cation-oxygen distance in the octahedron, and a by l and ω, the tilt angle. We should then have

$$c_H = \frac{12l}{\sqrt{3}}, \qquad \frac{a_H}{\sqrt{3}} = \frac{2\sqrt{2}\, l}{\sqrt{3}} \cos \omega \tag{12.6}$$

hence

$$\frac{c_H}{a_H} = \frac{12}{2\sqrt{6} \cos \omega} = \frac{\sqrt{6}}{\cos \omega} \tag{12.7}$$

Thus we should expect c_H/a_H to be greater than for ideal perovskite, which means that the rhombohedron must be steeper, i.e., the angle α less, than its ideal value of 60° or 90°. We find that the contrary is true. The octahedra must be slightly flattened down the triad axis.

Both $PrAlO_3$ and $NdAlO_3$ have structures isomorphous with $LaAlO_3$, but rather more distorted from the ideal, with α values (for the face-centred rhombohedron) of 90.3° and 90.4°, respectively. The tilt angle in $PrAlO_3$ is about 8°.

We may digress slightly at this point to consider more carefully the relation of these rhombohedral structures to $LiNbO_3$. As we saw in §11.6, the room-temperature structure of $LiNbO_3$ has the polar space group $R3c$, with Nb and Li atoms off-centre in their octahedra; the material is therefore pyroelectric. At high temperatures, above 1000° C, it becomes ferroelectric. The necessary reversal of the structure is effected by correlated movements of Nb and Li atoms in the c direction, and of O atoms in the plane perpendicular to this. The movement of Nb is within its octahedron, but that of Li is through the face of its octahedron into the empty octahedron.

We may ignore the Li movement for the moment and consider only that of the NbO_3 framework. Geometrically the change is a reversal of the signs of the displacement parameters w_{Nb} and v_O. We might therefore expect to find a simpler member of the family for which these two parameters are zero. Such a structure actually exists above 1250° C. It is paraelectric, with space group $R\bar{3}c$, and it still possesses one variable atomic position parameter u. Formally, its description is exactly the same as that of $LaAlO_3$, but the value of u (assuming it to be nearly the same as in the ferroelectric form) corresponds to a much larger rotation about the triad axis than is likely to occur in $LaAlO_3$, $PrAlO_3$, or $NdAlO_3$. A rotation of 30° from the ideal perovskite configuration would lead to hexagonal close packing of anions. In high-temperature $LiNbO_3$ the rotation is probably about 24°, making the anions approximate to close packing, whereas in the aluminates the rotations are probably less than about 10°, retaining the resemblance of the structure to the perovskite aristotype. Figure 12-7(c) can thus be taken as a representation of the $LaAlO_3$ framework, except that the rotation (shown as 10°) is in this case too large.

Rotation of the octahedra decreases the volume available to the A cation. In $LaAlO_3$, the twelve A—O bonds are divided into three short, six intermediate, and three long bonds, and a possible explanation of the increase of interaxial angle from 90° might be its effect of making the three short and six intermediate bonds more nearly equal, at the expense of a slight deformation of the octahedron. In high-temperature $LiNbO_3$ there are the same three groups, but the short bonds are so much shorter than the rest that it is reasonable to think of Li as 3-coordinated.*

The rhombohedral tilt system is combined with triad-axis displacement of the B cation in low-temperature $NaNbO_3$. Here the rotation is 12°,

---

* See footnote on p. 241.

almost half way between the zero value for ideal perovskite and the 30° value for hexagonal close packing. This results in a very irregular environment for the Na atom, which, like the Nb atom, is displaced along the triad axis. The six nearest oxygen neighbours of Na, at a distance of about 2.5 Å, form a figure resembling a trigonal prism with one of its triangular faces considerably enlarged and slightly rotated with respect to the other.

12.10 EXAMPLES OF THE ORTHORHOMBIC TILT SYSTEM

Calcium Titanate. An example of a crystal with an orthorhombic tilt system and central B cations is $CaTiO_3$ (the mineral perovskite). Its description is as follows:

Lattice: Orthorhombic, primitive, with $a = 5.37$ Å, $b = 7.64$ Å, $c = 5.44$ Å
(approximately $\sqrt{2}a_p$, $2a_p$, $\sqrt{2}a_p$)

Number of formula units per unit cell: 4

Space group: $Pcmn$ (No. 62)

Atomic coordinates:

Ti in $4(a)$: $\frac{1}{2}, 0, 0$

Ca in $4(c)$: $x_{Ca}, \frac{1}{4}, z_{Ca}$, with $x_{Ca} = 0$, $z_{Ca} = 0.03$

O(1) in $4(c)$: $\frac{1}{2} - x_1, \frac{1}{4}, z_1$, with $x_1 = 0.037$, $z_1 = -0.018$

O(2) in $8(d)$: $\frac{1}{4} + x_2, y_2, \frac{1}{4} + z_2$, with $x_2 = -0.018$, $y_2 = -0.026$, $z_2 = -0.018$

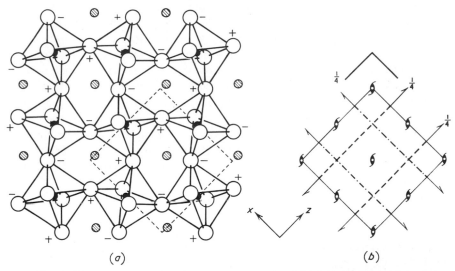

(a)　　　　　　　　　(b)

Figure 12-8 Perovskite, $CaTiO_3$ (isomorphous with $GdFeO_3$). (a) Layer of octahedra in lower half of unit cell, projected on (010). Small black circles are Ti at height 0, hatched circles Ca at height 1/4, open circles oxygens at heights 1/4, −1/4, and $\pm v$ (marked as + and −) where $v \simeq 0.03$. All atomic displacements are approximately doubled, to show their effect more clearly. The unit cell has a broken outline. Notice the Ti atoms central in the octahedra; the tilts of octahedra about [010]; the tilts about [001], giving alternate rows of atoms in this direction at heights $+v$ and $-v$; and the displacement of the Ca atom within its cavity to give it fewer but nearer neighbours. (b) Symmetry elements of the unit cell.

The parameters of the pseudocubic subcell are $a_p \equiv c_p = 3.83$ Å, $b_p = 3.82$ Å, $\beta = 90°48'$. The relation between the orthorhombic unit cell and the pseudocubic subcell is given by:

	a	b	c
a_p	$\frac{1}{2}$	0	$\frac{1}{2}$
b_p	0	$\frac{1}{2}$	0
c_p	$\frac{1}{2}$	0	$-\frac{1}{2}$

The structure is shown in Figure 12-8(a) and its symmetry elements in Figure 12-8(b).

It can be seen that the Ti atoms here lie centrally in their octahedra (as in $SrTiO_3$, but in contrast to $BaTiO_3$). The octahedra are tilted about both the y and the z axes, and are only slightly distorted in shape. The Ca atoms lie in the mirror planes, but not at symmetry centres; they have moved within their cavities so that they now have, instead of 12 equidistant neighbours, 8 neighbours at distances from about 2.4 Å to 2.7 Å, the next nearest being at 2.9 Å.

Isomorphs of Calcium Titanate. There are many compounds strictly isomorphous with $CaTiO_3$. One is $NaMgF_3$ (room-temperature form),

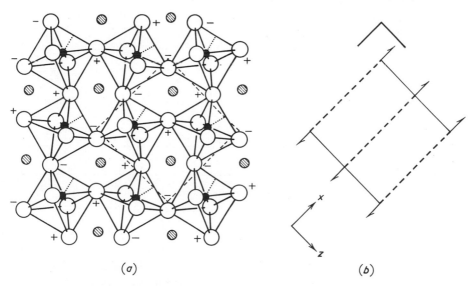

(a) (b)

Figure 12-9 $NaNbO_3$, Phase Q. (a) Layer of octahedra in upper half of unit cell, projected on (010). The part of the structure shown, and its orientation on the page, are chosen to allow comparison with Figure 12-8, in spite of the different choice of axes of reference and origin. Small black circles are Nb at height 3/4; hatched circles, Na at height 1; open circles, oxygen at heights 1/2, 1, and $3/4 \pm v$ (marked as $+$ and $-$). Atomic displacements are only approximate and slightly exaggerated. The unit cell has a broken outline. Notice the Nb atoms off-centre in the octahedra, with direction of displacement shown by the dotted line; also notice the tilts of the octahedra, and displacements of Na atoms within the cavities, very much as in Figure 12-8. (b) Symmetry elements of the unit cell.

found as the rare mineral *neighbourite*, and interesting as an example of a *model structure*—a structure where the valency of each ion is the same multiple or submultiple of the corresponding ion in an isomorphous structure, in this case $CaTiO_3$. Dimensionally, the resemblance between them is very close.

Another important series isomorphous with $CaTiO_3$ has trivalent cations for both A and B; A is Y, La, or one of the rare earths, and B is Al, Ga, or one of the transition elements from Sc to Co. Sometimes $GdFeO_3$ is regarded as the type member. This series shows very interesting size regularities. If we take the angle between the pseudocubic axes as a measure of the distortion from the aristotype, we find that it increases with increasing atomic number of the A cation from La to Gd, when B is kept constant (see Table 12-1); and if A is kept constant and the B cations are arranged in

Table 12-1 Distortion of Interaxial Angle in Some Compounds ABO_3 Isomorphous with $CaTiO_3$: Values of $\beta - 90°$ (in Degrees)

B / A	Sc	Fe	V	Ga	Cr	Al
La	1.1	0.2		0.3	0.4	*
Pr	1.6	0.8	0.7	0.3	0.4	*
Nd	2.0	1.4	1.6	1.0	0.9	*
Sm		2.0			1.4	0.1
Eu		2.5				0.2
Gd	2.7	2.8	2.9		2.1	0.6

* Different structure found at room temperature.

order of increasing distortion, the order is nearly the same for all A cations. The order is approximately that of the ionic radii, the smallest distortion being found for large A and small B cations. This series of compounds has helped to establish values for the ionic radii of the atoms concerned.

Detailed structural studies have been made of the rare-earth ortho-ferrites. The octahedra are not perfectly regular, but nearly enough so to allow tilt angles to be calculated. Both the diad-axis tilt and the tetrad-axis tilt vary smoothly with the atomic number of the A cation, the two tilt angles being nearly in the ratio $\sqrt{2}:1$ throughout. The distortion of the octahedra does not change greatly. If we had attempted to predict the tilt angles from the lattice parameters, ignoring distortion of the octahedra, we should have obtained a reasonable, though rather rough, approximation to the correct values.

Low-Temperature Praseodymium Aluminate. A low-temperature form of $PrAlO_3$, occurring between 151° and 205° K, provides an example with tilts about the diad axis only and none about the tetrad axis. Its lattice is orthogonal with lattice parameters $a = 5.34$ Å, $b = 7.48$ Å, $c = 5.32$ Å— very like those of $CaTiO_3$—and 4 formula units per unit cell. It is body-centred. Atomic positions are as follows, together with others related by the body-centring operator $(1/2, 1/2, 1/2) +$:

Al(1): $0, 0, 0$

Al(2): $0, \frac{1}{2}, 0$

O(1): $u_1, \frac{1}{4}, 0;\ u_1, \frac{3}{4}, 0;\ $ with $\ u_1 = 0.04$

O(2): $\frac{1}{4}, v_2, \frac{1}{4};\ \frac{1}{4}, v_2, \frac{3}{4};\ \frac{3}{4}, \bar{v}_2, \frac{1}{4};\ \frac{3}{4}, \bar{v}_2, \frac{3}{4};\ $ with $\ v_2 = -0.02$

Pr: $\quad x_p, y_p, \frac{1}{2};\ \bar{x}_p, \bar{y}_p, \frac{1}{2};\ $ with $\ x_p = -0.002, y_p = \frac{1}{4} + 0.002$

The Pr atoms are displaced from ideal positions, but in a very low-symmetry way; their only symmetry is a diad axis parallel to [001] and a mirror plane normal to it, with the accompanying centre. This, of course, represents the true symmetry of the structure as a whole. Nevertheless, in thinking about the AlO_3 framework, it is helpful to consider its symmetry in isolation, as if the Pr atoms were not there at all or were undisplaced.

If we compare the position parameters with those of $CaTiO_3$, we see that they could be made to match as follows. Put the z_1, x_2, and z_2 parameters of $CaTiO_3$ equal to zero, which is equivalent to abolishing the tilt about the tetrad axis of the octahedron; move the origin to $0, 0, 1/2$; compare the remaining parameters, x_1 and y_2 for $CaTiO_3$, with u_1 and v_2 for $PrAlO_3$—they are alike in sign and very similar in magnitude. Obviously, the two frameworks are closely similar, except for the tilt about the tetrad axis in $CaTiO_3$. The loss of this tilt in $PrAlO_3$ has created new mirror planes perpendicular to [001], interleaving the existing glide planes, and this results in body-centring. The new space group, *Icmm*, contains all the symmetry of the $CaTiO_3$ space group, *Pcmn*, as shown in Figure 12-8(b), with additions.

In both structures, the B cations remain centrosymmetric; in both, the A cations are displaced, but the character of the displacements is very different. In $CaTiO_3$ the displacements are large and symmetrically arranged with respect to the framework; in $PrAlO_3$, they are about one tenth as great and of much lower symmetry than the framework.

Sodium Niobate. Examples of structures with a combination of tilted octahedra and off-centre cations are found among the various polymorphs of $NaNbO_3$. The Nb atom, unlike Ti, is normally off-centre in its octahedron, and Na is too small to allow the octahedra to remain parallel, as in $KNbO_3$. The interplay of these two factors allows the existence of several structures which are not very different in energy and are stable under different conditions. We have already described the low-temperature form, Phase N, in §12.9.

The normal room-temperature form of $NaNbO_3$ (Phase P) has tilts very much like those in $CaTiO_3$ for a double sheet of octahedra parallel to (010) possessing a mirror plane, but this double sheet is then operated on by a diad screw axis to give a unit cell four octahedra high. The Nb displacements lie very nearly in the (010) plane (though not constrained to do so by symmetry, which allows their direction to be a general one). Their largest components are along [100], all those in one double sheet being in the same direction, and in the next double sheet in the opposite or antiparallel direction. This form is therefore antiferroelectric.

A third form of $NaNbO_3$ (Phase Q) occurs stably in solid solutions with about 2.5 percent of the Na replaced by K, and can also be prepared (with difficulty) from pure $NaNbO_3$ by the action of an electric field on the first form. The octahedra in the double layer have almost the same tilts as before, but this layer (still keeping its mirror plane) now constitutes the complete repeat unit, which is shown in Figure 12-9(a) and (b). The Nb displacement within its octahedron is also nearly the same (though detailed examination shows it to be more nearly parallel to an octahedron edge) but all the large components are now in the same direction, and the material is ferroelectric.

A fourth form (Phase R), found at about 400° C, has a tilt system of more complicated symmetry, whose general effect is nevertheless very like that in the room-temperature form; once again the Nb displacements are parallel within a double sheet of octahedra, but antiparallel in the structure as a whole.

In all tilted forms the same general rule is obeyed as in the unipartite hettotypes (cf. §12.6)—that the two Nb—O bonds to any O atom comprise either one short and one long bond or two bonds of intermediate lengths.

Above 480° C, there are other phases in which there is no longer a displacement of Nb, but different tilt systems. The relationship between the phases will be considered in §15.13.

12.11 OTHER PEROVSKITE STRUCTURES

We may mention some ordered structures with more complex formula units which are sometimes classed with the perovskite family. An example is Ba_2CaWO_6, which is cubic face-centred, with $a \simeq 8$ Å, and Ca and W occupying centres of alternate octahedra. It is strictly isomorphous with $(NH_4)_3FeF_6$ (if we can assume NH_4^+ to be spherical—cf. §15.24). Geometrical distortions of the kind we have been considering for the ABX_3 compounds can occur in these other compounds, in addition to ordering effects—Ca_3WO_6 and Na_3AlF_6 (cryolite), for example, are noncubic. In all of them, however, there are very big differences in electrostatic bond strength between the cations in adjacent octahedra, and the lower valencies are associated with larger size. The structure no longer possesses a continuous strongly-bonded octahedral framework; the weak octahedral links are not much stronger than those of the polyhedra of the cavity cation, and they are easily distorted, not nearly rigid as in a true perovskite. These differences are dynamically so important that the compounds should not be regarded as perovskites.

On the other hand, we must include with the perovskites those BX_3 compounds which have the true octahedral corner-linked framework but lack a large cation to fill out its cavities. The simplest example is ReO_3; it is cubic, with one formula unit per unit cell, exactly like the aristotype except for the absence of the 12-coordinated cation. Indeed, we may include it with the aristotype if we allow A to represent a zero atom. Another example is VF_3, which has a primitive rhombohedral unit cell, with $a = 5.37$ Å, $\alpha = 57°31'$, $Z = 2$; its space group is $R\bar{3}c$ (No. 167). It may be more convenient to refer it to pseudocubic axes and use the face-centred cell for which $Z = 8$. Each octant of this is like the ReO_3 cell, slightly elongated along its triad axis, but alternate octahedra along this direction are rotated in opposite directions relative to one another. The relation of VF_3 to ReO_3 is in fact exactly the same as that of $LaAlO_3$ to cubic $SrTiO_3$.

In this connection we may also consider RhF_3, though its octahedra are so far rotated (through 30°) that the F atoms have moved into positions of hexagonal close packing (slightly flattened down the triad axis). It is thus the analogue of idealised high-temperature $LiNbO_3$. Of the other trivalent fluorides, NbF_3, TaF_3, and MoF_3 are cubic and strictly isomorphous with ReO_3; PdF_3 and IrF_3 are isomorphous with RhF_3 and have 30° tilt angles; RuF_3 has a slightly smaller value; and VF_3, FeF_3, CoF_3, while technically isomorphous with RhF_3 (as $LaAlO_3$ is with high $LiNbO_3$), are in fact quite sharply distinguished by the much lower values of their tilt angles. The related MnF_3 has tilt angles like FeF_3, but the Mn is off-centre, and the symmetry is lowered; the monoclinic unit cell contains 12 formula units.

The structures of WO_3, though not fully known, are probably perovskite hettotypes. The structure reported at about 900° C is like that of $BaTiO_3$ except that the W displacements in adjacent octahedra are in opposite directions; thus the unit cell dimensions are $a = b \simeq \sqrt{2}\, a_p$, $c \simeq a_p$. Another complicated hettotype is found for $PbZrO_3$, but its details are not reliably known; others probably occur among solid solutions of compounds of Pb. The difficulty with these materials is that of locating oxygen atoms by X-ray diffraction in the presence of the heavy Pb or W (though with careful work it would not be impossible); for single-crystal neutron diffraction, larger crystals would be needed than are readily available.

Though we are not generally concerned with disordered structures, one such group must be mentioned—the "perovskite bronzes," sometimes called "tungsten bronzes."* Examples are Na_xWO_3, with x lying between 0.3 and 0.9, and K_xWO_3, with $x > 0.6$. They have an octahedral WO_3 framework like that of perovskite, with the cavities occupied at random by the alkali metal. One explanation of the nonstoichiometric composition attributes it to the variable valency of tungsten, and assumes the octahedral sites are occupied at random by $xW^{5+} + (1 - x)W^{6+}$; an alternative suggestion is that all the octahedral atoms are W^{6+}, and the Na or K metallic.

12.12 THE SPINEL FAMILY (COMPOSITION B_2AO_4)

The spinel structure, as we saw in §11.3, is derived from a cubic close packing of anions. The unit cell contains 32 anions, and there are 16 cations occupying octahedral interstices and 8 occupying tetrahedral interstices. For ideal close packing, the tetrahedral interstice would be smaller than the octahedral one, but the anions have an adjustable position parameter by which the sizes can be made more nearly equal, at the same time as the shared edges of the octahedra are shortened. If we allow for the correction of ionic radius for coordination number summarised in Table 2-3, the value of the parameter to allow occupation by cations of identical size is 0.385, as compared with 0.375 for ideal close packing; all recorded values of parameters for different spinels (with one doubtful exception) are between 0.375 and 0.392. There is thus plenty of latitude for accommodating a variety of different cations while retaining strict isomorphism with Al_2MgO_4, where Al is octahedral and Mg tetrahedral. This is the *normal spinel structure*, and there are plenty of examples, including such varied oxides as Al_2CoO_4, Co_2GeO_4, Ti_2MgO_4, and V_2MnO_4, as well as sulphides such as Cr_2CdS_4, Cr_2HgS_4, and Ti_2CuS_4, and also selenides and tellurides, Cr_2CdSe_4, and Cr_2CuTe_4.

It was also noted in §11.3 that the second cation does not always occupy the tetrahedral site, but that *inverse spinels* occur in which half of the more numerous kind of cation occupy tetrahedral or A sites, the other half being distributed, in a random mixture with the second kind, over the octahedral or B sites. The original example was Fe_2MgO_4; there are other oxides such as Mn_2NiO_4, Zn_2SnO_4, Mg_2TiO_4, and Zn_2TiO_4, sulphides such as In_2FeS_4, and at least one fluoride, Li_2NiF_4. Another important example is Fe_3O_4 (the mineral *magnetite*), which may be written $Fe_2^{3+}Fe^{2+}O_4$, and which illustrates how atoms in different valency states are considered, for our purposes, as different atomic species.

---

* The name "tungsten bronzes" is used with a confusing variety of meanings. It was given, because of their appearance, to the original tungsten-based materials M_xWO_3, where M is an alkali metal and $x < 1$; later it was extended to crystallographically similar compounds with other B cations instead of W. It is now known, however, that the series includes several distinct structure families, and unfortunately the name "tungsten bronze" has been used indiscriminately for all. For discussion of the nonperovskite bronzes, distinguished essentially by their linkage pattern rather than their symmetry (which is always noncubic), see §12.21.

The occurrence of some of the inverse spinels can be explained simply on the basis of the relative sizes of the cations. In Fe_3O_4, for example, it is not surprising to find Fe^{3+} ions (radius 0.67 Å) in tetrahedral sites in preference to Fe^{2+} (radius 0.83 Å). Others are less obvious, and others are quite surprising if only packing considerations are used—notably Zn_2TiO_4, in which the radius of Zn^{2+} is 0.83 Å and that of Ti^{4+} is 0.64 Å. Moreover, by the same arguments some of the normal spinels would have been expected to be inverse, for example Al_2ZnO_4, in which Zn^{2+} is much larger than Al^{3+}.

Clearly the size effect is not by itself the criterion. It is necessary also to consider covalent bond character. Indeed, we might have been led to this conclusion otherwise by noticing the resemblance of the oxides to the sulphides and selenides, which cannot reasonably be considered as mainly ionic. In fact the theory has been worked out, showing how the relative energies of octahedral bonding and tetrahedral bonding vary from one kind of ion to another, and the predictions of theory agree reasonably well with the facts. When there is a large energy difference in favour of occupation of an octahedral site by the more numerous cation, a normal spinel is formed. When the energy difference favours occupation of an octahedral site by the less numerous cation, an inverse spinel is formed. If both cations are equally suited for the octahedral site, a disordered structure may occur, in which each kind of site contains, on the average, two thirds of the first plus one third of the second. Intermediate cases can occur, in which both kinds of cation occupy both kinds of site, but with an unequal preference.

If we consider spinels in terms of cation valency, they fall into two groups. In both, of course, the sum of the cation valencies (for oxides) is 8, but the individual values are 3, 2 in one group (represented by Al_2MgO_4) and 2, 4 in the other (represented by Zn_2TiO_4).

Spinel structures may be formed when the ratio of the two kinds of cation is not $2:1$, though the total cation:anion ratio remains $3:4$, and electrostatic balance is preserved. For example, $LiFe_5O_8$ is a spinel, and can occur in a disordered form, with Li^+ and Fe^+ occupying tetrahedral and octahedral sites at random. It can occur also in an ordered form, keeping the same unit cell but with lower symmetry corresponding to the new space group $P4_332$ (No. 212). Here the 16 octahedral sites are divided into two independent sets, a set of 4 occupied by Li^+ and a set of 12 occupied by Fe^{3+}.

There are also spinels in which the cation:anion ratio is less than $3:4$. An important example is γ-Fe_2O_3. If, in Fe_3O_4, we remove one third of the Fe^{2+} atoms, we can restore electrical neutrality by replacing the remaining Fe^{2+} atoms by Fe^{3+}. The composition is now Fe_2O_3, but the structure remains that of spinel with one ninth of its cation sites vacant, and the formula could be written $(Fe^{3+}_{8/9})_3O_4$. It is possible but not certain that the vacancies are ordered: if so, the hettotype has a unit cell larger than that of the aristotype. Intermediate structures occur, with compositions between Fe_2O_3 and Fe_3O_4. Isomorphous with γ-Fe_2O_3 is γ-Al_2O_3.

There are, of course, extensive ranges of solid solutions among the spinels, and the various effects we have considered separately may occur in combination. In each case, one must consider the distribution of each kind of cation between the octahedral and the tetrahedral sites. Even if the pure end-member, the host material, has fully ordered cations in a normal arrangement, it does not follow that the guest atoms will distinguish perfectly between the octahedral and tetrahedral sites; if the guest cations are statistically distributed, it will tend to encourage disorder among the host cations. Many of the most representative spinels contain transition metals capable of different valencies, and since vacant sites (behaving like zero atoms) can occur in many others as in Fe_3O_4–Fe_2O_3, the variety of possibilities is immense.

12.13 GEOMETRICALLY DISTORTED SPINELS

Though cation substitution is the most important characteristic of the spinel family, geometrical distortions do occur. One important example is Mn_3O_4 (the mineral *hausmannite*):

> Lattice: Tetragonal body-centred, with $a = 5.75$ Å, $c = 9.42$ Å
> Number of formula units per unit cell: 4
> Space group: $I4_1/amd$ (No. 141)
> Atomic coordinates:
>   Mn(1) in 4(a): 0, 0, 0
>   Mn(2) in 8(d): $0, \frac{1}{4}, \frac{5}{8}$
>   O   in 16(h): $0, u, v$, with $u = 0.25, v = 0.375$.

Its relation to the ideal spinel structure can most easily be seen by choosing new axes of reference with OX and OY rotated by 45° relative to the old axes (cf. §10.9). This gives a pseudocubic all-face-centred cell with $a' = 8.13$ Å and axial ratio $c/a' = 1.15$. The space-group symbol referred to the new axes is $F\bar{4}/ddm$—a subgroup of $Fd3m$ (No. 227) obtained by omitting the triad axis.

To the accuracy within which the structure has been determined, the atomic position parameters are the same as those of ideal spinel, and the MnO_6 octahedron has therefore undergone the same distortion as the unit cell, namely a 15 percent extension along its tetrad axis. There are four short Mn—O bonds in the (001) plane, and two long bonds perpendicular to it. This bond system is frequently found for Mn, as it is for Cu; the shape of the MnO_6 octahedron may be compared with that of the CuF_6 octahedron in $KCuF_3$ (§12.5). Obviously the geometry of the bond system determines the distortion of the structure.

The same effect is seen in Mn_2CdO_4, which is strictly isomorphous with Mn_3O_4, but even more elongated ($a = 5.81$ Å, $c = 9.87$ Å, pseudocubic axial ratio 1.20) and with nonideal atomic position parameters ($u = 0.20$, $v = 0.40$). However, the ratio of bond lengths is sensitive to the relative departures of u and v from their ideal values, and one must know the accuracy with which these parameters have been determined before basing any arguments on them. This caution is particularly necessary in considering the other isomorphous compounds with pseudocubic axial ratios nearer to unity, namely, Cr_2NiO_4 (a normal spinel) and Fe_2CuO_4 (an inverse spinel), with values of 1.04 and 1.03 respectively.

Several other modifications of the spinel structure have been reported. Cr_2CuO_4 has a different tetragonal structure, with approximately the same unit cell as Cr_2NiO_4 but space group $I\bar{4}2d$ (No. 122). Low-temperature Fe_3O_4 (below 120° K) has been reported as orthorhombic, with lattice parameters 5.91 Å, 5.94 Å, 8.89 Å, and space group *Imma* (No. 74); the octahedral sites are now divided into two independent sets, of which one holds Fe^{2+} and the other Fe^{3+}, giving a fully ordered structure even though it remains an inverse spinel. A different low-temperature form has also been reported, which is rhombohedral, and corresponds to a very slight extension (about 0.03 percent) of the structure along its cube body diagonal.

12.14   THE RUTILE FAMILY

As we saw in §11.5, there are a number of compounds strictly isomorphous with rutile, TiO_2. They include not only oxides such as SrO_2 (the mineral *cassiterite*) but also fluorides such as MgF_2 and ZnF_2. The structure has an adjustable axial ratio c/a as well as one adjustable position parameter, but in fact neither changes greatly, all recorded values of the latter lying in the range 0.303 ± 0.004. The reason is that the parameters are dependent on the shape of the octahedron, which is regular except for a compression parallel to the shared edge, a feature common to all members of the family.

In the aristotype, there is only one kind of cation, and its valency must be double that of the anion; for the oxides, it must be 4-valent. It is possible, however, to build oxides with the same structure using equal numbers of 3-valent and 5-valent cations. Examples of such materials are $AlSbO_4$ and $CrNbO_4$. If the two kinds of cation are statistically distributed (as they are reported to be in these examples), the structure is still isomorphous with the aristotype, and the structural formula may be written $(Al_{1/2}Sb_{1/2})O_2$. If ordering occurred, it would result in a hettotype of lower symmetry.

Geometrical effects are important in this family. There are different hettotypes illustrating the separate effects of the same three factors we saw operating in the perovskite family, namely, distortion of the octahedron while the cation remains central, off-centre displacement of the cation, and tilting of the octahedra without distortion. These factors are less easy to recognise in this family than in the perovskite family, because the symmetry is not cubic to start with. The three structures described below illustrate the three separate effects.

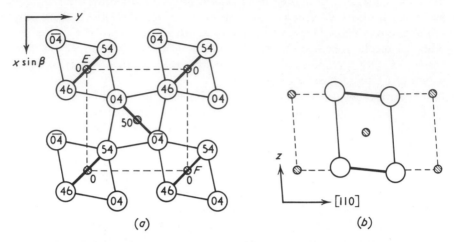

Figure 12-10 CuF_2. (a) Projection down [001]. Small hatched circles are Cu; large open circles are F; heights are given in units of $c/100$. Shared edges of octahedra are shown by heavy lines. (b) Section in the plane whose trace in (a) is *EF*. Short and long Cu—F bonds lie along the short and long diagonal respectively of the parallelogram. Shared edges are shown by heavy lines. (Both diagrams are drawn to scale.)

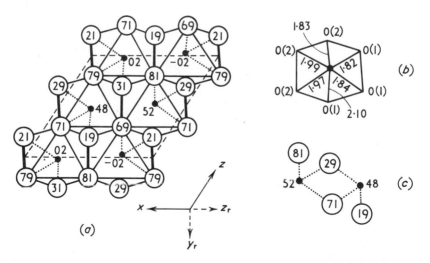

Figure 12-11 VO_2. Projection on (010), corresponding to (100) in rutile, for which the axial directions are indicated by y_r, z_r. Small black circles are V; large open circles are O. (a) Complete structure. Heights are in units of $b/100$. Octahedra are outlined (except for omission of the edges of the lower faces); shared edges are shown by heavy lines; short V—O bonds are shown by dotted lines. (b) Part of (a), octahedron with V at height 52/100, showing V—O bond lengths, and identifying O(1) and O(2). (c) Part of (a) continued into adjoining unit cell, showing strongly bonded V_2O_4 group. (All diagrams are drawn to scale.)

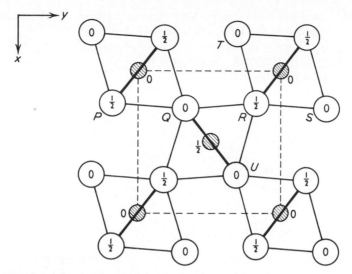

Figure 12-12 CaCl₂. Projection on (001). Hatched circles are Ca; large open circles are Cl. Fractional heights are marked. Shared edges are shown by heavy lines. The diagram is drawn to the same scale as Figures 12-10 and 12-11; note the larger size due to the greater radius of Cl.

Cupric Fluoride. An example of a structure with distorted octahedra is CuF_2. Knowing the usual bond system of Cu^{2+} (cf. §12.5), we expect the octahedron to be drawn out into a tetragonal bipyramid. At first sight it would seem easy to incorporate this in the rutile aristotype, shown in Figure 11-16(a), with the long diameter lying in the (001) plane and the square section perpendicular to it. This arrangement would mean, however, that the tightly bonded CuF_4 squares must share edges, which seems unlikely, and in fact it is not found. Instead, the structure becomes monoclinic, as follows:

Lattice: Monoclinic, primitive, with $a = 4.59$ Å, $b = 4.54$ Å, $c = 3.32$ Å, $\beta = 96°40'$

Number of formula units per unit cell: 2

Space group: $P2_1/n$ (No. 14)

Atomic coordinates:

   Cu in $2(a)$: 0, 0, 0

   F in $4(e)$: $x, y, z,$ with $x = y = 0.30$, $z = 0.044$

Figure 12-10(a) shows this in projection down the [001] direction. The projection is almost the same as that of the aristotype, but the height differences from 0 and 1/2 represent a distortion of the octahedron combined with a tilt. Figure 12-10(b) shows the section parallel to ($1\bar{1}0$), the plane of the shared edges. The length of the short Cu—F bonds in this section is 1.93 Å, nearly equal to that of the bonds in the (001) plane, while the long bonds are 2.27 Å. The octahedra are tilted about their [110] axis as well as their $[1\bar{1}0]$ axis, but these tilts are a consequence of the distortion and not an independent effect as in $CaCl_2$ (to be considered below).

The compound CrF_2 is strictly isomorphous with CuF_2, with Cr—F bond lengths 1.98 Å, 2.01 Å, and 2.43 Å.

Vanadium Dioxide An example of a structure with off-centre cations is room-temperature VO_2. This is again monoclinic, but the geometry is completely different from that of CuF_2:

> Lattice parameters: $a = 5.74$ Å, $b = 4.52$ Å, $c = 5.37$ Å, $\beta = 122°36'$
> Number of formula units per unit cell: 4
> Space group: $P2_1/c$ (No. 14)
> Atomic coordinates (all atoms in general positions):
> V in $4(e)$: $\frac{1}{4} - 0.008$, -0.025, 0.025
> O(1) in $4(e)$: $0.10, 0.21, 0.20$
> O(2) in $4(e)$: $0.39, 0.69, 0.29$

The unit cell can be derived from a subcell like the unit cell of the aristotype, with dimensions $a_s \simeq b_s \simeq 4.54$ Å, $c_s = 2.88$ Å, $\alpha \simeq \beta \simeq \gamma \simeq 90°$, by the following matrix:

	a_s	b_s	c_s
a	0	0	2
b	1	0	0
c	0	-1	1

Figure 12-11(a) shows the projection of the unit cell on (010), corresponding to (100) of the aristotype. The shared edges of the octahedra (heavy lines) lie nearly perpendicular to [100], which corresponds to [001] of the aristotype. In Figure 12-11(b) one octahedron is drawn, with V—O distances marked; three are long, and three are short. The V atom is displaced from its octahedron centre by about 0.06 Å in the direction $[\bar{1}01]$. Each O atom is involved in at least one short bond, and the O atoms of the shared edge (the O(1) atoms) in two (an exception to the general rule of off-centre displacement, that the same O is not involved in two short bonds). We could if we wished think of the structure as built from strongly-bonded V_2O_4 units, as shown in Figure 12-11(c), joined to each other by weaker V—O bonds. Such a picture would exaggerate the difference between the roles of the long and short V—O bonds, but it is useful to illustrate the sort of intermediate case that can occur between a completely isodesmic structure and a molecular structure. (Compare the very similar way of looking at $CsGeCl_3$ so as to pick out discrete $GeCl_3$ groups—§12.6).

At temperatures above 340° C, VO_2 has the structure of the aristotype—the ordinary rutile structure. It has a high electrical conductivity, and is to be classed as a metal. Room-temperature VO_2 is a semiconductor.

Other examples isomorphous with room-temperature VO_2 are MoO_2 and probably WO_2; the off-centre displacement of Mo is rather greater than that of V.

Calcium Chloride. An example of a structure with tilted octahedra is provided by $CaCl_2$.

Lattice: Orthorhombic primitive, with $a = 6.24$ Å, $b = 6.43$ Å, $c = 4.20$ Å
Number of formula units per unit cell: 2
Space group: *Pnnm* (No. 58)
Atomic coordinates:
  Ca in $2(a)$: 0, 0, 0
  Cl in $4(g)$: $x, y, 0$, with $x = 0.300 + u$, $y = 0.300 - u$, $u = 0.025$

The *n*-glide planes are at $x = 1/4$ and $y = 1/4$. The projection on (001) is shown in Figure 12-12. The unit cell is nearly the same shape as in rutile, but larger, because of the large Cl ion. If a and b were equal, and u were zero, we should have a structure isomorphous with rutile. The effect of the parameter u is to move all the Cl atoms off the lines joining Ca atoms in such a way that the octahedron remains almost undistorted. By tilting like this, a much better approximation to close packing is achieved, at the expense of more irregularity in the environment of the Cl atom. In TiO_2 the three nearest cation neighbours of the anion are coplanar with the anion; in $CaCl_2$ they all lie slightly to one side of it. This kind of difference is what we should expect from the greater tendency of O to form covalent bonds. With the more ionic Cl, the bond angles at the anions are of less importance and the packing is of greater importance. In Figure 12-12 the projected row *PQRS* is more nearly a straight line than the corresponding line in Figure 12-10 (or Figure 11-16); if it were exactly a straight line, the anion packing would be a cubic body-centred array flattened by about 6 percent down the z axis.

$CaBr_2$ has the same structure as $CaCl_2$.

12.15 THE QUARTZ, CRISTOBALITE, AND TRIDYMITE FAMILIES

We have seen in §11.11 to 11.14 that the structures of these forms of SiO_2 are built up from corner-linked tetrahedra, the pattern of linkages being different in each of the three. Though they are different structures, belonging to different families, the same sort of generalisations may be made about all three. We may take quartz as representative.

Quartz has a very open structure. Its bonding is certainly semipolar, as shown by the fact that the angle at oxygen is roughly 145° rather than 180°. It is a linkage structure, its configuration being determined by bond-angle requirements rather than by packing considerations. In such structures it is easy to see that there may be slightly different configurations of the framework with nearly the same energy, even though we cannot generally predict them or explain them in detail. We saw, in fact, in §11.14, that there are two distinct quartz structures, stable above and below 550° C; high quartz is the aristotype, and low quartz the hettotype.

Other members of the quartz family can be derived by substitution or addition of atoms.

One such structure is that of $AlAsO_4$. Electrical neutrality is preserved by the substitution of one Al^{3+} and one As^{5+} for two Si^{4+} atoms. The Al and

As atoms are regularly ordered in alternate tetrahedra, and hence each O atom receives one electrostatic valence of 3/4 and one of 5/4. Since, however, there are only three tetrahedra per unit cell of quartz, the repeat unit cannot remain as in quartz, but must double in height. In Figure 11-21, for example, if the atoms at height 0 are Âl, those at height 6 must be As. The symmetry is that of low quartz, with space group $P3_22$ (No. 154), or $P3_12$ (No. 152). Distortion of the framework is not large, because the Al—O and As—O distances are nearly alike, 1.70 Å and 1.62 Å, respectively, and not much greater than Si—O, which is 1.60 Å.

Several other phosphates and arsenates of trivalent metals are isomorphous with $AlAsO_4$.

There are comparable hettotypes in the cristobalite family, some related to high cristobalite, some to low cristobalite. Some of the compounds concerned are polymorphic. For example, $GaPO_4$ has one form isomorphous with $AlAsO_4$ belonging to the quartz family, and two forms belonging to the cristobalite family, related to high and low cristobalite respectively; BPO_4 has one $AlAsO_4$ form, and one form related to high cristobalite. The high-cristobalite form of BPO_4 is isomorphous with $BeSO_4$, in which the valencies of the two cations differ even more, but their distances to oxygen atoms are nearly equal (Be—O = 1.56 Å, S—O = 1.50 Å).

The tridymite family gives examples of yet another kind of variation. As we saw in §11.13, there are large open channels in the framework, and these provide suitable sites for large cations if the charge on the framework is decreased by replacing some Si^{4+} atoms by Al^{3+} atoms. This kind of structure has been called a "stuffed tridymite structure." There are so many possible variations, however, that it is more convenient to consider them as constituting a separate family, called after one important member, the mineral *nepheline*. It would in fact be legitimate to classify tridymite as a member of the nepheline family in which the framework cavities commonly occupied by large cations are empty, just as we classified ReO_3 as a member of the perovskite family with its 12-coordinated cavities empty.

The structure of low tridymite is not reliably known, and there may be more than one hettotype. And just as we can have nonstoichiometric perovskites, the perovskite bronzes, of composition $M_x(W_x^{5+}W_{1-x}^{6+})O_3$, so we may possibly have nonstoichiometric tridymites, $M_x(Al_xSi_{1-x})O_2$.

12.16 THE NEPHELINE FAMILY

Structures of the nepheline family consist of a tetrahedral framework with a tridymite linkage, slightly distorted in different ways by tilting or crumpling to reduce the volume of the cavities, which are occupied by large cations. The effect is analogous to the crumpling of the perovskite structure round the A cations, achieved by tilting its BO_6 octahedra, except that in the nephelines the cavity is larger.

Figure 12-13 BaAl₂O₄: projection on (0001). The outline of the unit cell is shown by broken lines. Small black circles are Al, hatched circles Ba, open circles O. Thin solid lines join O atoms at height zero, heavy solid lines O atoms at height 1/2, and dotted lines O atoms at height −1/2. Within the heavy triangle is the projection of O at height 1/4 and Al at heights 0.05 and 0.45, and within the dotted triangle that of O at height −1/4 and Al at −0.05 and −0.45. The Ba atoms are at ±1/4.

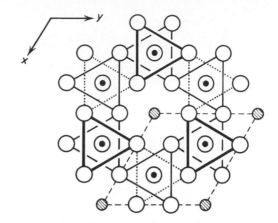

We begin by considering BaAl₂O₄ (though it might be argued that this cannot strictly be called a nepheline). All the framework cations are here Al. The unit cell is similar in size to that of tridymite, with $a = 5.21$ Å, $c = 8.76$ Å, but the symmetry is $P6_322$ (No. 182), with 2 formula units per unit cell, instead of $P6_3/mmc$ (No. 194). The atomic coordinates are:

Ba in 2(b) : 0, 0, ¼
Al in 4(f): ⅓, ⅔, u, with $u = 0.05$
O(1) in 2(c) : ⅓, ⅔, ¼
O(2) in 6(g) : x, 0, 0, with $x = 0.67$

The structure is shown in Figure 12-13.

The framework has changed from that of ideal tridymite (Figure 11-19) by rotation of the tetrahedra through 30° about their triad axes, as indicated in Figure 12-14. This kind of rotation of the tetrahedra is closely analogous to the rotation of octahedra about their triad axes in the perovskite family (§12.9); BaAl₂O₄ corresponds to ideal perovskite, and tridymite to idealised LiNbO₃. Here, however, the tridymite end of the series is the one of greater interest. If we use it as the starting point, the changes involved in arriving at BaAl₂O₄ leave the tetrahedral cations in their special positions on triad axes, but all the mirror planes are destroyed. The Ba atoms are on the hexad

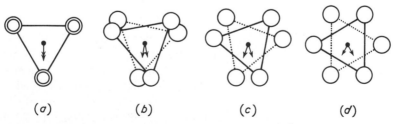

| (a) | (b) | (c) | (d) |

Figure 12-14 Relation between tridymite and BaAl₂O₄. (a) Pair of tetrahedra in tridymite (cf. Figure 11-19). (b) and (c) Corresponding pair of tetrahedra in hypothetical structures intermediate between tridymite and BaAl₂O₄. (d) Corresponding pair of tetrahedra in BaAl₂O₄. Arrows give orientation of tetrahedron, as marked by T—O bond.

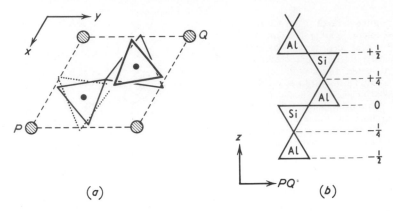

Figure 12-15 Kalsilite, KAlSiO₄. (a) Projection on (0001). Small black circles are Al and Si atoms superposed; hatched circles are K atoms; O atoms are at corners of triangles and superposed on the projection of Al and Si atoms. Heights of atoms are the same as in Figure 12-13. (Atomic displacements are not drawn to scale.) (b) Schematic diagram of the sequence of tetrahedra near the plane whose trace in (a) is *PQ*. Coordinates of oxygen atoms at the tetrahedron corners are marked as fractions of c.

axes, in 12-coordinated sites, their cuboctahedra sharing triangular faces with similar cuboctahedra below and above them.

A closely related structure, much nearer the tridymite end of the range, is *kalsilite*, $KAlSiO_4$ (Figure 12-15). The unit cell has again about the same size, with $a = 5.15$ Å, $c = 8.67$ Å, but the space group is now $P6_3$ (No. 173). The T_2O_4 framework is not very different from that of $BaAl_2O_4$, with Si and Al still on special positions on triad axes, but they are ordered, occupying alternate tetrahedra. This destroys the diad axes of symmetry, and allows the two kinds of tetrahedra to be different in size. Every O atom has one Al neighbour and one Si neighbour. This is an example of the *aluminium avoidance rule* (§12.17). The electrostatic strengths of the two framework bonds to the oxygen atom are 3/4 and 1/4 respectively, leaving a balance of 1/4 for the total contribution of K—O bonds. Another consequence of the particular Si/Al ordering is that, of each pair of tetrahedra on the same triad axis, sharing a corner in what used to be the (0001) mirror plane, the upper is always Si and the lower Al, as shown in Figure 12-15(b). The structure is therefore polar. The K atoms are on the hexad axis, though the point symmetry of their environment is not hexagonal.

A structure isomorphous with this is $KLiSO_4$, where the difference in size of the two kinds of tetrahedra is more obvious.

Nepheline itself ($KNa_3Al_4Si_4O_{16}$) has a more complicated distortion of the framework. Its space group is again $P6_3$ (No. 173), but its *a* parameter is twice that of tridymite (whence $Z = 2$). The tetrahedra have nearly the same tilt about the triad axes as in tridymite, but they are also tilted about a horizontal axis so that they no longer point exactly up or down; moreover, the T atoms, Si and Al, have moved off the triad axes. Figure 12-16 shows the arrangement of tetrahedral centres resulting from this. Round every

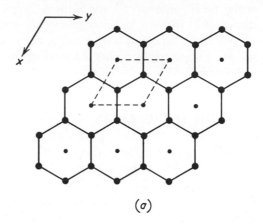

(a)

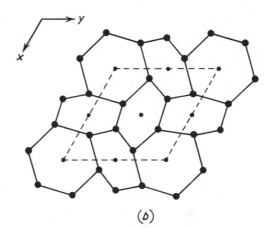

Figure 12-16 Comparison of tetrahedral framework in (a) tridymite, (b) nepheline (KNa$_3$Al$_4$Si$_4$O$_{16}$). The base of the unit cell is shown by the broken line; small black circles mark the projection of T cations. Oxygen atoms are omitted; ideal projections of large cations are shown by dots. The displacements from the ideal in (b) are schematic, not drawn to scale.

(b)

alternate hexad axis the ring of six T atoms has rotated slightly in the same direction; round the intervening axes which were originally hexads, the ring has become elongated. There are thus three narrowed channels for every one that retains nearly its original width. The K atoms are located on hexad axes in the wide channels, and the Na atoms in less special positions in the narrow channels.

Because of the lower symmetry, there are four independent sets of TO$_4$ tetrahedra, two sets of six shared by a wide channel and two narrow ones, and two sets of two located on triad axes and shared by three narrow channels. The evidence suggests that atoms in the two latter sets are ordered, one set being Si and one Al, so that Si and Al alternate in the structure; but that the two former sets are occupied at random by Si or Al.

It is likely that some of the materials to which the broad name of nepheline is given are different hettotypes belonging to the same family. "Nephelines" include solid solutions with varying Na:K ratio, and with Ca replacing K; moreover, some of them have vacant cation sites, and some are disordered. It might be wiser to give the ordered structure with a 3:1 ratio for Na:K a name of its own, keeping the name nepheline for the family as a whole. This would be analogous to using the name felspar for the family next to be described.

12.17 THE FELSPAR FAMILY

The felspars are stuffed framework structures, like those we have just been considering. Their linkage pattern is different from that of tridymite and is the characteristic feature of a felspar; it was described in §11.15. The framework has never been found without its large cation, but otherwise many of the same generalisations can be made about the felspars as about the nephelines.

In the felspars, a K or Ba atom occupying the cavity is sufficiently large to keep the framework extended in its most symmetrical configuration, displayed in sanidine. A smaller atom, Na or Ca, allows the framework to collapse, reducing the size of the cavity and destroying the monoclinic symmetry. With still smaller atoms, no stable configuration apparently exists.

Because of the low symmetry, it is not easy to define or describe the kinds of rotations of tetrahedra which occur in going from one hettotype to another, as we were able to do in the nepheline family, but physically it is the same sort of effect. It seems that there is probably more than one distinct low-symmetry configuration of the framework, and, though each can be changed continuously to a certain extent by changes of composition or of Si/Al ordering (to be discussed further below), or by thermal expansion, they do not merge into one another but are separated by a clean break—as are the different tilted configurations of octahedra in the perovskites. They are hard to distinguish because they often do not involve changes of symmetry, and also because they are partly masked by differences of Si/Al ordering, which is a separate effect but one on which much more attention has been focussed in the literature.

In many felspars, Si and Al are fully ordered, and this has the effect of lowering the symmetry even when a large cation is present. We proceed in the next few paragraphs to show why this must be so. The kind of ordering, and therefore the kind of symmetry change, depends on the Si:Al ratio. We may compare the structures of *microcline* (ordered $KAlSi_3O_8$) and *celsian* ($BaAl_2Si_2O_8$), with the aristotype, sanidine (disordered $KAlSi_3O_8$, or $K(Al_{1/4}Si_{3/4})_4O_8$).

We saw in §11.15 that sanidine has a unit cell with a 7 Å c-axis, containing two formula units, and its space group is $C2/m$ (No. 12). There are two independent sites for T atoms, called T_1 and T_2, and each equipoint, corresponding to a general position, is a set of eight atoms. With the Si:Al

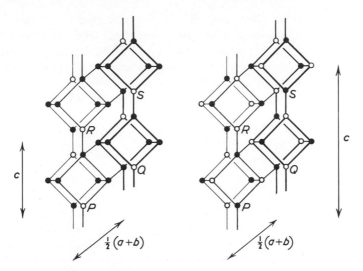

Figure 12-17 Schematic diagram of microcline or low albite, showing the pattern of Si/Al order. Black circles are Si; open circles are Al.

Figure 12-18 Schematic diagram of celsian or anorthite, showing the pattern of Si/Al order. Black circles are Si; open circles are Al.

ratio of 3:1, it is obviously impossible to distribute atoms between these sites in an ordered way. What happens in the ordered structure of microcline is that the monoclinic symmetry is lost, but the lattice translations remain nearly the same, and the centre of symmetry is kept. Using the same axes of reference, the new space group is $C\bar{1}$ (No. 2). (We *could* have chosen new axes to give it the conventional form $P\bar{1}$, but this would be inconvenient, because these axes fail to display the resemblance to the aristotype.) There are now only four atoms in each equipoint instead of eight, and thus there are four independent sets of T atoms instead of two—two with the same sort of linkage as the original T_1, and two with the same sort as the original T_2. They have been called $T_1(0)$, $T_1(m)$, $T_2(0)$, and $T_2(m)$: the names indicating that, in the aristotype, pairs of sites which are now different were related by a mirror plane. One of the four can be filled by Al—in fact, the $T_1(0)$ site is chosen—and the other three are filled by Si, in a fully ordered way. This is shown schematically in Figure 12-17. There is some crumpling of the structure, but not much, because the K atom acting as a spacer keeps the cavity large.

For celsian, with the Si:Al ratio of 1:1, the same difficulty need not arise. It would be possible to put all the Al into T_1, for example, and all the Si into T_2. This does not happen. The ordering takes place in such a way that Al and Si always occupy alternate tetrahedra (Figure 12-18). An example of the same sort of ordering was seen in kalsilite (§12.16), and it is in fact a very common feature of framework silicates, so common that it probably represents the most energetically favourable arrangement whenever the Si:Al ratio allows it. The empirical observations are summed up in the aluminium avoidance rule: *In an ordered tetrahedral framework structure containing Si and Al as tetrahedral atoms, if the Si : Al ratio is greater than 1, it is unusual for two Al tetrahedra to share a corner.* One can see a general reason for this on an electrostatic basis. If O were shared between two Al tetrahedra, a total electrostatic valence of 1/2 would be needed from the cavity cations to satisfy Pauling's rule, and this would imply a closer packing of such cations round that O than can generally be achieved, and a corresponding moving away of the cations from O atoms shared between two Si atoms. Obviously a more balanced arrangement can be obtained by making as many Al—O—Si links as possible.

The alternation of Si and Al results in a doubling of the c-dimension of the unit cell, as can be seen from comparison of Figures 12-17 and 12-18. Because there are three tetrahedra in any part of a chain with a 7 Å repeat in the c-direction, and an even number is required for alternation, the unit cell length is doubled and the repeat distance along c becomes 14 Å. The cell is now bipartite. The tetrahedra in the second half are very similar to those in the first half, except for the interchange of Si and Al and the small consequent size changes. If we operated on any Si atom in the first half with the original translation vector $c_{sanidine}$, we should find an Al atom very near the correct place in the second half. If we operated with the original C-face centring, the same thing would happen. But operating with both, one after another, which is equivalent to a body-centring operation, we find Si at the *exact* right place. The new space group is in fact $I2/c$ (No. 15). The symmetry remains monoclinic, though the mirror plane has been replaced by a glide plane.

In *anorthite*, $CaAl_2Si_2O_8$, the same Si/Al alternation is found, and the same doubling of the unit cell to give a 14 Å repeat. But the smaller Ca atom allows a crumpling of the structure, and this destroys not only the monoclinic symmetry but also the body centring. We are left with a unit cell of twice the volume of the aristotype, and space group $P\bar{1}$ (No. 2) (with axes of reference corresponding in directions to those of the aristotype). There are thus eight formula units in the unit cell, i.e., 104 atoms, and since the only symmetry is a centre, 52 are independent by symmetry—though perfectly ordered and repeating their position exactly from one unit cell to the next.

The crumpling effect produces larger changes of atomic positions than does the substitution effect, as can be seen by comparing the shapes of 4-membered rings related by the pseudosymmetry (Figure 12-19). Rings P and S have the same Si/Al distributions but markedly different shapes; rings P and Q have opposite Si/Al distributions but surprisingly similar shapes, and the same is true for R and S. On the other hand, changes of framework strains resulting from temperature changes may remove the difference between P and S, whereas P and Q must always differ in their Si/Al content.

The sodium felspar, *albite* ($NaAlSi_3O_8$), may be expected, in its ordered form, to have lost its monoclinic symmetry for a double reason: because of its 3 : 1 ratio of Si : Al, and because of crumpling round the moderately small

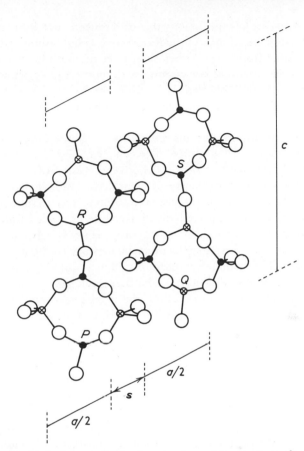

Figure 12-19 Projection on (010) of parts of anorthite structure, drawn to scale. Black circles are Si, small crossed circles Al, open circles O. The rings are lettered to correspond with Figure 12-18. Rings P and R are at a height near $y = 0.35$, rings Q and S near $y = 0.85$. For clarity, Q and S have been displaced in the x direction, relative to P and R, by the distance marked s. In the structure, ring P is not joined to ring Q but to its near-mirror-image at height $y \sim 0.65$; similarly for R and S.

Na atom. The structure found, *low albite*, is in fact isomorphous with microcline, but with slightly different interaxial angles.

Another structure isomorphous with microcline and low albite is that of *reedmergnerite*, $NaBSi_3O_8$. In this B replaces Al in the $T_1(0)$ sites. Not only is the pattern of ordering the same as in low albite, but the detailed distortions of the tetrahedral framework are surprisingly alike.

There is, however, a second albite structure, known as *high albite*, of the same composition and the same symmetry as low albite, but with markedly different interaxial angles and different optical properties. It is found in albites synthesised at high temperatures rather than in naturally occurring albites, and it is difficult to convert one into the other—indeed, no reasonable-sized crystals of low albite have ever been made artificially. In high albite, there appears to be complete Al/Si disorder, as contrasted with complete order in low albite, but this difference is not in itself sufficient to explain the different shapes of the framework. High albite and low albite in fact differ more from one another than does high albite from anorthite. It seems likely that there are two essentially different framework configurations, which can be altered only slightly by different atomic occupations, but that their relative stabilities may depend on the character of the occupation.

At high temperatures there appears to be a complete range of solid solution between all felspars, but this is not true for those formed, or annealed, at lower temperature. The range between $NaAlSi_3O_8$ and $CaAl_2Si_2O_8$ is more nearly complete than that between $NaAlSi_3O_8$ and $KAlSi_3O_8$; obviously the chemical similarity of K and Na, and the similar pattern of ordering in the $3:1$ compounds, are of less importance than the size of the M cation and the consequent configuration of the crumpled framework. It is probable that at room temperature the range is not nearly complete even for the Na—Ca series. There may be other framework configurations which are more stable at intermediate compositions. The possibility of Si/Al disorder is always present, and is influenced by the thermal history of the material. It is important to realise, however, that the framework configuration is not merely a consequence of the disorder, but a factor to be considered independently of it.

Though, in this chapter, we are not concerned with the physical processes by which one structure can be converted into another, it is easier to understand relationships in the felspar family if one or two principles are accepted. In particular, we must distinguish between a purely geometrical process in which the framework changes shape by a tilting of the polyhedra and displacement of the cations within them, and a substitution process by which cations are interchanged between tetrahedra. The latter cannot be done without breaking chemical bonds; it becomes easier at high temperatures, but is always slower than the geometrical change. (We shall return to this subject in §15.1 to 15.3 and 15.7 to 15.8.).

In felspars, Si/Al interchange needs long annealing at high temperatures. Since disorder is also favoured by high temperatures, it may prove difficult to find a set of conditions of time and temperature which allows a felspar to anneal to the ordered state. This in fact is the case for the $3:1$ felspars. Microcline has never been synthesised in the laboratory, and low albite with very great difficulty if at all (the products of long annealing of high albite being at best so poorly crystallised that no direct confirmation of their Si/Al distribution has been possible.) Fortunately, however, geological conditions of growth or annealing have been more effective, and it is among natural materials that we find the perfectly ordered structures described above for microcline* and low albite. For the $1:1$ felspars, ordering is much easier, and can readily be achieved in the laboratory. For Si/Al ratios between these extremes, the

---

* Not all natural microclines are perfectly ordered. What we have described should strictly have been called "maximum microcline," to distinguish it from the partly disordered "intermediate microclines" which also occur.

situation is much more complicated, and it would be out of place to discuss it here.

Replacement of the cavity cation is a much easier process than interchange of framework cations. It is possible to convert microcline into low albite, and low albite into microcline, by heating at about 700° to 900° C, with sodium or potassium halides respectively. If replacement of the alkali-metal cations had required a break-up of the framework, even locally and temporarily, we should expect the product to be sanidine or high albite, because of the difficulty of restoring Si/Al order once it is lost. Hence we see that the K or Na ions must be able to move through the framework without breaking it. (This is true only when the experiments are done without any water vapour present: with water, sanidine and high albite result. Obviously, water vapour must help in the breaking and rejoining of framework bonds.)

Felspars have been synthesised with Ga replacing Al, and Ge replacing Si, and also with Fe replacing Al.

12.18 FAMILIES OF ZEOLITES

A *zeolite* is a material with a tetrahedral alumino-silicate framework having a rather elaborate linkage pattern and large cavities between tetrahedra, holding ions and small molecules which can move in and out easily. It is this openness of the framework (illustrated in Table 12-2 in terms of the volume per oxygen atom of the framework), together with the mobility of the cavity contents, which characterises a zeolite. The frameworks have many of the features of those we have been discussing in the last sections, but an immensely greater variety of linkage patterns.

The shape and contents of the cavities determine the uses of zeolites. These are of three kinds:

(a) *Action as ion exchangers.* If the cavity ion is Na, for example, it may be readily replaceable by Ca when the zeolite is bathed in a solution of a Ca salt. This is the principle of water softening by zeolites. Bathing the zeolite in a solution of NaCl restores it to its original state.

(b) *Action as catalysts.* A zeolite with a large cavity may attach ions or small molecules loosely to the cavity walls in specific ways, and two kinds of ions or molecules can encounter each other spread out over a large effective surface.

(c) *Action as "molecular sieves."* Zeolites can be made with cavities and tunnels of particular sizes which will allow the passage of small molecules but exclude larger ones, and this can be used to separate the different products of a reaction.

The ion-exchange effect is similar in principle to that which we saw in §12.17 could occur in the felspars, but whereas for microcline it needed

Table 12-2 Volume Per Oxygen Atom (in $\mathring{A}^3$) in Various Structures

CLOSE-PACKED SPHERES		MISCELLANEOUS COMPOUNDS		FORMS OF SILICA		FRAMEWORK SILICATES	
2.6 Å radius	12.5	Corundum	14.2	Coesite	17.2	Albite (Na felspar)	20.7
2.8 Å radius	16.0	Rutile	15.5				
		Spinel	16.6			Sanidine (K felspar)	22.6
		Lithium niobate	17.8	Low quartz	18.8	Nepheline	23.4
		Lanthanum aluminate	18.1	High quartz	19.7		
		Forsterite	18.1			Natrolite*	28.1
		Calcium titanate	18.6	Low cristobalite	21.4	Chabazite*	34.2
		Aragonite	18.7				
		Strontium titanate	19.8	High cristobalite	23.4		
		Calcite	20.3	Melanophlogite	25.1		
		Sodium nitrate	20.6				

* Counting framework oxygen atoms only

molten halides at several hundred degrees C, for the zeolites it happens in aqueous solution at room temperature.

The variety of zeolite frameworks is so great that classification is not easy. An early classification divided them simply into fibrous and non-fibrous zeolites; more modern work recognises five broad groups, and includes in them materials such as *ultramarine* and the mineral *sodalite* which were not originally counted as zeolites. Classification, by drawing attention to the building blocks which constantly recur in the different frameworks, suggests possibilities for synthesising new zeolites with properties to meet particular needs. It is, however, too specialised a subject to discuss here, and we shall limit ourselves to considering two representative structures and their near relations.

A particular zeolite family, like any other family, is characterised by its linkage pattern. But just as the framework can collapse round the cavity in different ways in felspars or perovskites, so it can in a zeolite. Its detailed configuration and its symmetry depend on the contents of the cavity. Again, just as different Si:Al ratios can be compensated by different cavity contents in the felspars, so they can in the zeolites. The question of Si/Al ordering also arises for the zeolites. Early work could not distinguish between Si and Al, and assumed disorder; but it now seems certain that there is a very strong tendency to order, the aluminium avoidance rule (§12.17) applying to zeolites as to other frameworks. These are points to be noticed in dealing with each particular example.

We shall generally show the linkage patterns schematically, leaving out the oxygen atoms. It has to be remembered, however, that (as in other frameworks) O atoms do not lie on the T—T join but to one side of it, giving a T—O—T angle of about 130° to 150°. This tends to make all channels and windows smaller than one pictures from such a diagram. They are large nevertheless, relative to openings and channels in most other structures; diameters range from about 2.5 Å up to 7 Å. The examples illustrate this.

12.19 THE NATROLITE FAMILY

Natrolite, a "fibrous zeolite," was one of the first to be analysed. Its ideal composition is $Na_2T_5O_{10} \cdot 2H_2O$, with Si:Al = 3:2. There are five tetrahedra in the unit of pattern, or building block, which is repeated by symmetry eight times in the orthorhombic unit cell (characterised by $a = 18.3$ Å, $b = 18.6$ Å, $c = 6.6$ Å); space group $Fdd2$ (No. 43).

The building block is shown schematically in Figure 12-20. Four tetrahedra, centred at P_1, Q_1, P_2, Q_2, form a puckered ring, and a fifth, R, connects P_1 and P_2. There is Si/Al order, of such a kind that two opposite corners of the ring (for example, Q_1 and Q_2) are Al, and the others, together with the unique atom R, are Si. The whole unit, of length 6.6 Å, is repeated by translation, to bring R to R', which is joined to Q_1 and Q_2. Alternatively,

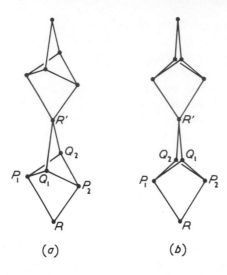

Figure 12-20 Chain of tetrahedra in natrolite (schematic): (a) perspective view; (b) projection almost normal to symmetry plane. Small black circles represent centres of tetrahedra; oxygen atoms are not shown. The building block comprises the five tetrahedra $P_1P_2Q_1Q_2R$.

and perhaps more easily, we can think of the block as consisting of the ring $P_1Q_1P_2Q_2$ bridged above and below by half tetrahedra at R and R'. In its idealised form, ignoring differences between Si and Al, the block of five tetrahedra has point-group symmetry $\bar{4}m2$.

Chains of these building blocks parallel to [001] are cross-linked by their P, Q corners to form the structure. In a single sheet of chains, there are two possibilities, shown in Figure 12-21(a) and (b). In (a), all chains are in parallel orientation, and all corresponding atoms R (or R') at the same height. In (b), each alternate chain is rotated by 90° about its own axis and displaced by $\pm c/4$ relative to its neighbour. Natrolite embodies the second scheme, and in so doing satisfies the aluminium avoidance rule. It is shown schematicall·

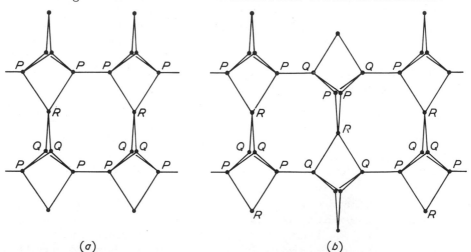

Figure 12-21 Possibilities for lateral linkages of chains. In (a), all chains are in the same orientation; in (b), alternate chains are rotated through 90° about their axis and displaced parallel to it by $c/4$.

in Figure 12-22(a). However, the chains do not remain exactly at 90°, but rotate relative to one another as in Figure 12-22(b), without destroying the linkage scheme. This figure is still oversimplified, because actually the building-block is not exactly tetragonal, and the symmetry of the structure is reduced to orthorhombic, with symmetry directions at 45° to the planes of chains.

There are quite wide channels perpendicular to [001]. A cross-section of one is shown in Figure 12-21(b). There are in fact two intersecting sets of these channels, parallel to [110] and [1$\bar{1}$0] respectively. A narrower collapsed set runs parallel to [001], shown in cross-section in Figure 12-22(b). The Na atoms are located in the channels at specific sites such that each Na has six neighbours at about 2.5 Å distance—four oxygen atoms of the framework, and two water molecules.

A zeolite of the same family is *scolecite*, $CaAl_2Si_3O_{10}\cdot3H_2O$. The framework has an additional slight distortion that lowers the symmetry to monoclinic (but keeps the linkage scheme unchanged). Each Ca has seven neighbours—four framework oxygen atoms and three water molecules. *Mesolite*, $Na_2Ca_2(Al_2Si_3O_{10})_3\cdot8H_2O$, has a unit cell of three times the volume, with ordered arrangement of Na and Ca (and their associated water molecules) in different channels.

There are related zeolites which do not belong to the same family, though they have the same kind of chains. We may perhaps call them members of the same *tribe*. They make use of the different kind of lateral linkage shown in Figure 12-21(a). *Edingtonite*, $BaAl_2Si_3O_{10}\cdot4H_2O$, embodies this kind of linkage only, and would have only one chain per unit cell were it not for the tilting of the chains about their own axis. *Thomsonite*, $NaCa_2Al_5Si_5O_{20}\cdot6H_2O$, makes use of the fact that the kind of linkage in one plane of chains is independent of that in a plane at right angles; it is linked as in Figure 12-21(a) in one plane, and as in Figure 12-21(b) in a plane at right angles. This would give two chains per cell if they were all parallel, or four when tilt is introduced. Actually there are also distortions that double the length in the chain direction. The relationships are shown in Figure 12-23.

In each case, the cavities are of the right shape and size to give satisfactory accommodation to the cations and their associated water molecules. The free openings in the channels are wide enough ($\sim$3 Å diameter) to allow easy communication between the cavities and the surrounding medium.

12.20 CHABAZITE

The tribe of zeolites to which chabazite belongs is built up from rings of six tetrahedra, the rings all lying parallel to one another, and linked to each other through four-membered rings. The linkage pattern is best described by comparison with close-packed arrays of single atoms. In zeolites of this tribe, each atom of a close-packed sheet is replaced by a six-membered ring, and

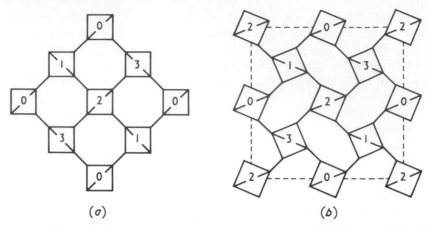

(a) (b)

Figure 12-22 Lateral linkage of chains in natrolite, projected down the length of the chain. In (a) the orientations are idealised so that each chain is at 90° to its neighbours; in (b) they are rotated from the ideal as in the actual structure. The diagonal drawn for each square is the direction of the edge of tetrahedron R lying above the PQ ring, and hence joining the Q corners. Figures are heights of R above the origin, in units of $c/4$. In (b), the unit cell is shown by a broken line; in (a), it is the square whose corners are marked 0.

successive sheets follow the cubic sequence . . . ABC . . . or the hexagonal sequence . . . AB . . . (cf. §4.5) or some more complicated close-packed sequence; or two adjacent sheets (not more) may be directly superposed, giving for example the sequences . . . $AABAAC$ . . . or . . . $AABCCABBC$ . . . . When such superposition occurs, the six-membered rings combine to form hexagonal prisms. In chabazite, all sheets are paired in this way, the pairs following the cubic sequence to give . . . $AABBCC$ . . . . Each prism is linked to its three neighbours below and three above by sloping four membered rings, as shown schematically in Figure 12-24.

Chabazite has the ideal composition $Ca_2Al_4Si_8O_{24}\cdot13H_2O$, and its aristotype has a rhombohedral unit cell with $a = 9.44$ Å, $\alpha = 94\frac{1}{2}°$, space group $R\bar{3}m$ (No. 166), $Z = 1$. The unit cell is thus nearly a cube, in spite of the fact that its corners are not occupied by spheres but by hexagonal prisms (which are considerably distorted, however, and have become ditrigonal rather than hexagonal). Figure 12-25 is a projection down one rhombohedral cell edge, showing centres of tetrahedra and their linkage scheme. It can be seen that most of the framework atoms (including the oxygen atoms, which lie close to, though not on, the lines shown joining tetrahedral atoms) are concentrated near the corners and the edges. The body centre of the cell makes a large empty cavity, and the centres of the faces are also open spaces free from atoms. In Figure 12-26(a), the circumferences of the oxygen atoms forming the cavity walls have been drawn to show the size of the opening. Its diameter is about 4 Å.

Into each cavity—one per unit cell—we must fit two Ca atoms and thirteen H_2O molecules. The number of H_2O molecules seems surprisingly large, but a rough calculation can show that there is plenty of room. We

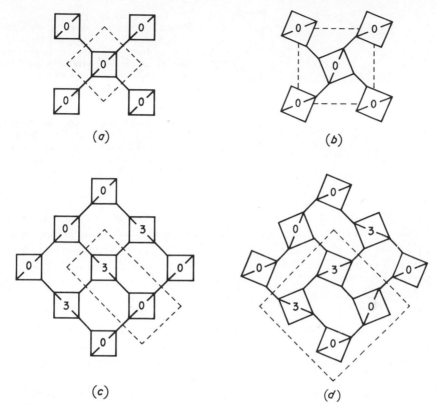

Figure 12-23 Projection of related zeolites for comparison with Figure 12-22: (a) and (b) edingtonite; (c) and (d) thomsonite. The left-hand diagrams (a) and (c) are idealised, but have the same linkage patterns as the right-hand diagrams (b) and (d) respectively. Note that the patterns in (a) and (c) differ essentially from each other and from that of Figure 12-22(a), because of the different relative heights of the chains as well as their 90° rotations, but that the deviations from the ideal are alike in all three. The primitive unit cells, shown by broken lines, are smaller for the idealised structure than for the simple rotated structure in edingtonite and thomsonite, but in natrolite those for the idealised and simple rotated structures are equal (the unit cell outlined in Figure 12-22(b) is C-face-centred.)

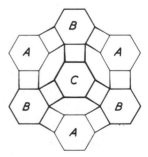

Figure 12-24 Stylised view of chabazite down the triad axis, showing the linkage pattern of the tetrahedra. The lines join the centres of tetrahedra; the oxygen atoms are not shown. Hexagonal prisms appearing as hexagons A, B, C are at heights 0, 1/3, 2/3, respectively.

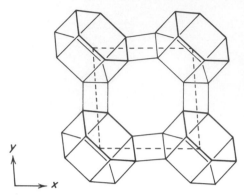

Figure 12-25 Projection of the chabazite structure down a rhombohedron edge direction (slightly simplified from an exact scale diagram) showing the linkage pattern of tetrahedra by lines joining their centres; the oxygen atoms are not shown. The hexagonal prisms, or double six-membered rings, can be seen at each corner; the eight-membered ring at the centre of this diagram occurs also in each other face of the rhombohedron. The unit cell is shown by a broken line. The projection of the triad axis runs from the lower left to the upper right-hand corner.

argue as follows. First, we make an approximation to the framework by building a skeleton cube of side 9.4 Å, as in Figure 12-27, from beams with a square cross-section of side 2.7 Å. (This size of beam is chosen so as to leave square openings of side 4 Å in each face.) The empty volume is $(9.4^3 - 8 \times 2.7^3 - 2 \times 2.7^2 \times 4)$ Å$^3 \simeq 320$ Å^3. The volume per water molecule is thus about 25 Å^3. Now the volume per close-packed oxygen is about 16 Å^3; the volume per water molecule in ice (a relatively open structure—cf. §13.18) is about 32 Å^3. There is thus reasonable room to fit in all the water molecules. In fact they are found to form a loose but specific structure filling the cavity, the linkage to each other and to the framework being provided by hydrogen bonds (§13.23). The two Ca atoms fill interstices of suitable size.

Electrostatically, we may think of the arrangement of water molecules as a system of charge distributors between Ca^{2+} and the O^{2-} ions of the cavity walls. The framework oxygens bonded to water molecules will be those with electrostatic valence to spare, i.e., those whose two tetrahedral neighbours are not both Si^{4+}. (If, as we expect, the aluminium avoidance rule is observed—cf. §12.17—they cannot both be Al^{3+}; one must be Si^{4+} and one Al^{3+}, and the electrostatic valence sum is therefore 7/4). An alternative way of looking at the arrangement is to consider the Ca^{2+} surrounded by its

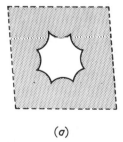

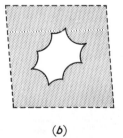

(a)   (b)

Figure 12-26 Same projection of the unit cell as in Figure 12-25, but with arcs showing the inner surfaces of the oxygen atoms, which define the limits of the framework and frame the eight-sided window. (The radius of oxygen is taken as 1.35 Å.) The shaded region covers the projection of all framework atoms. (a) Hydrated chabazite; (b) dehydrated chabazite. Note the different shapes of the window in (a) and (b).

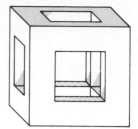

Figure 12-27 Perspective view of a skeleton cube, with solid regions approximating to those occupied by framework atoms in the actual structure of chabazite, showing the central cavity and windows.

sheath of water molecules as a large hydrated ion, linked through water molecules to the framework.

The ordered arrangement of Ca and water molecules cannot be achieved without destroying the triad symmetry of the aristotype framework; probably the Al/Si ordering pattern would independently require a lowering of the symmetry. In the actual structure, this results in minor adjustments of the shape of the framework to adapt it to the cavity contents.

Chabazite can be dehydrated without destroying it, and the original structure restored on rehydration. The structure of the dehydrated form has been studied. With the loss of the water molecules, the calcium atoms have moved to new sites where they are directly attached to oxygen atoms of the framework; some occupy the centres of the hexagonal prisms, and the others line the cavity walls. Not surprisingly, the shape of the framework has changed with the change of cavity content, as can be seen in Figure 12-26(b).

Other molecules besides water can be made to move in and out of the cavity quite easily, for example, O_2, N_2, CO_2. A chabazite containing molecular Cl_2 has been examined in detail, and its framework shape is slightly different from both the forms already described. In general, not only the size of the window but its shape may be important. On the other hand, the oxygen atoms are not rigid spheres, and the window openings may allow through larger molecules than Figure 12-26(a) or (b) would apparently suggest. Moreover, the smaller windows not shown in Figures 12-26 and 12-27—those formed by six-membered and four-membered rings of tetrahedra—may have an important influence on the movement of atoms or molecules into or out of the structure. Clearly, however, the changes in framework shape will affect the behaviour of the zeolite, even when the linkage scheme remains unaltered. The possibility of wedging the framework open with one kind of small molecule so that the windows are large enough to allow the passage of a second kind of molecule which would otherwise be excluded suggests interesting ideas for the development of fine controls.

12.21 OTHER OCTAHEDRAL FRAMEWORKS

Just as frameworks with a variety of linkage patterns can be built from corner-linked tetrahedra, so they can be built from corner-linked octahedra, though in this case there are fewer kinds. We have already considered fully the simplest and most important pattern, that of the perovskite family. Another occurs in the family of what are sometimes called "tetragonal tungsten bronzes"—a confusing name, since the octahedral cation need not be W, and the symmetry may be lower than tetragonal; moreover, bronzes of the perovskite family (cf. §12.11) might possibly be tetragonal. They are better called *tunnel bronzes* or *5-4-3 bronzes*, as we shall see.

In the tunnel-bronze aristotype, all octahedra have their tetrad axes parallel to z, in which direction the repeat unit is the height of one octahedron. Sheets of octahedra in the (001) plane are formed from a regular repetition of rings of five, four, and three octahedra. We may construct it as follows (using cations only, in the first instance, and adding oxygens later, as we have done for silicates).

Join four cations into a square. Use each corner of this as the apex of an equilateral triangle. Add four more cations so that each completes a five-ring with one side of the square and one side of each of two triangles already in position. Up to this point we have been able to keep $4mm$ symmetry. At the next step, the mirror planes must be abandoned, but the tetrad axis retained. Subject to this, use *one* side of each of the five-rings as the side of a new square. The centres of each of these new squares are the corners of the unit cell.

There is one problem with this construction. The structure is over-determined: we cannot keep the high local symmetry we started with if we want to make all the four-rings the same size. What in fact happens is a compromise. The tetragonal symmetry is kept; the four-rings at the centre and the corners are made the same size as each other; but the triangles and five-rings become slightly distorted in angles and edge-lengths. The oxygen atoms shared between two cations need not lie on the cation-cation join, any more than they do in silicates; but even with the consequent possibility of tilt, the octahedra are not perfectly regular. All the adjustments needed are, however, small.

There are ten octahedra in the unit cell, four each belonging to the four-rings centred at $(0, 0)$ and $(1/2, 1/2)$, and two with their edges belonging only to three-rings and five-rings. The former set are slightly larger and more irregular than the latter.

Possible sites for the A cations are above and below the centres of the rings, at heights midway between the planes of B cations. To examine them, we must insert the oxygen atoms roughly midway along the B-B joins. The five-rings provide the largest cavities; they form *tunnels* with periodic constrictions, running continuously through the structure. The four-rings form 12-coordinated *cages*, exactly like the A sites in perovskite. The three-rings form small trigonal prisms, capped on their three vertical sides. There are four tunnel sites (A_2), two cage sites (A_1), and four trigonal-prism sites in the unit cell. Different members of the family differ greatly in the use made of these sites, as well as in the detailed distortion of the framework and possible off-centring of B cations. If all sites were filled, the composition would be

$X_2YZ_2B_5O_{15}$, but some sites are often vacant, especially the small trigonal-prism sites.

One structurally simple example is $Ba_3TiNb_4O_{15}$, with lattice parameters $a = 12.54$ Å, $c = 4.01$ Å, space group $P4bm$ (No. 110), $Z = 2$. Here the Ba atoms occupy the tunnel and cage sites; the trigonal-prism sites are vacant. The Ti and Nb atoms occupy both kinds of octahedral site at random. They have off-centre displacements in the z direction which are different for the two kinds: 0.08 Å for the two smaller octahedra, and 0.17 Å for the eight larger octahedra.

If the cations had not been off-centre, the structure would have been the aristotype, with space group $P4/mbm$ (No. 127). This is the structure of $PbNb_2O_6$ above 560° C, except that there are only five Pb atoms occupying six A sites, which they do at random. Below this temperature, $PbNb_2O_6$ forms a different hettotype. It is orthorhombic, with $a = 17.65$ Å, $b = 17.91$ Å, $c = 7.73$ Å. There is a polar axis—and therefore presumably off-centring of Nb cations—in the direction parallel to y, *in* the (001) plane (not perpendicular to it, as in $Ba_3TiNb_4O_{15}$); the near-doubling of the c-axis represents a crumpling of the framework.

The different roles of the A and B off-centrings are illustrated in a non-stoichiometric compound of approximate composition $(Ba_{1.4}Sr_{3.6})Nb_{10}O_{30}$. Its room-temperature form is very similar to that of $Ba_3TiNb_4O_{15}$, though with Ba restricted to the tunnel sites, and vacancies distributed about equally between cage sites and tunnel sites. Above about 67° C there is a different form, with space group $P\bar{4}b2$, which still allows off-centre displacements to eight Nb atoms, though the other Nb atoms and the A cations are now in special positions. The aristotype may be formed above 135° C.

Though few hettotypes in the family have been studied in detail, many are believed to occur. Those described above serve as an indication how our analysis of the various factors occurring in the perovskite family can be used to sort out and clarify effects in a structure with a very different linkage pattern. The same principles apply, though the detailed applications are different, and the more complex linkage pattern allows new and more difficult combinations of them.

Besides the "tetragonal" or 5-4-3 bronzes, there are "hexagonal" or 6-3 bronzes, whose linkage pattern consists of six-rings sharing each side with a three-ring, and three-rings sharing each side with a six-ring. This of course constitutes a different family. An example is K_xWO_3 with x less than about 0.3.

Some of the same general principles apply to another large group of compounds, the 'perovskite-raft' group. Their building units are rectangular rafts of perovskite structure, one octahedron thick, cut parallel to (001), and displaced so as to share edges with identical rafts in a regular (though a very long-period) repetition. At these joins they create suitable interstices for small cations. There is an immense variety of such structures; one particularly elegant example is PNb_9O_{25}, with a 3×3 raft, and off-centre displacements of Nb in the (001) plane like those in perovskites. These compounds have well-defined chemical composition, though the formulae look so complicated that they might be assumed nonstoichiometric.

12.22 STRUCTURES WITHOUT THREE-DIMENSIONAL FRAMEWORKS

The structures considered in this chapter and Chapter 11 are all homodesmic, in the sense that forces of the same character and approximately the same strength operate in a continuously joined system in three dimensions. There are of course other very important structure families for which this is not true.

In the *micas*, for example, sheets of edge-sharing octahedra of cations such as Al, Mg, Fe are linked by their corners to Si and Al tetrahedra in a complex two-dimensional slab, and the slabs are held together by weaker bonds through OH, Na, and K ions. In the *pyroxenes*, the strongly-bonded units are chains. There are comparable effects in octahedral structures. One interesting group may be considered as slabs of perovskite structure cut parallel to (100) or (110), with the outer corners of their octahedra linked only to the inter-slab A cations and through them to the next slab. Examples are K_2NiF_4 with a (100) cut, $BaMnF_4$, $NaNbO_2F$, and $Ca_2Nb_2O_7$ with a (110) cut. Sometimes the inter-slab A cations are replaced by a tightly bonded cation-anion sandwich, with cations outside, as in Bi_2NbO_5F with a (100) cut. Many of these materials are of considerable technological importance. Again, there are compounds with large and complex but finite groups, generally roughly spherical in shape, held together internally by strong covalent or semipolar bonds, and linked to each other through weaker bonds to single ions of opposite sign.

It is outside the scope of this book to discuss any of these structures in detail. Many of the general principles noted for three-dimensional framework structures apply to them, except that their "frameworks" are continuous in only two dimensions, one dimension, or no dimension. For example, in $K_2Zn_2V_{10}O_{28}\cdot16H_2O$, where there are isolated tightly bonded ions of composition $V_{10}O_{28}$, these are, to a first approximation, like pieces of rock-salt structure, with VO_6 octahedra sharing edges; to a second approximation, we note the off-centring of V atoms within their octahedra, resembling the off-centring of Ti in the perovskites ($\S12.6$) but very much larger. Here, however (as in many of these structures), the off-centring is also favoured because of the provision it makes for inequalities between bond lengths to bridging and nonbridging oxygens (cf. $\S2.10$). The types of environment of the K and Zn atoms and H_2O molecules which link the $V_{10}O_{28}$ groups to one another are what we should expect for them in simpler compounds. In a way, these compounds with huge finite molecule-ions are like zeolites turned inside out: the zeolites have an infinite strongly bonded framework with large cavities holding finite groups of weakly bonded cations, while these materials have an infinite system of weak ionic bonds holding finite large strongly bonded groups.

12.23 SURVEY OF PRINCIPLES

(a) Families of structures are essentially structures with the same building units—octahedra or tetrahedra—and the same linkage pattern.

(b) Framework cations are to be distinguished from interstitial or cavity cations. Framework cations may be all alike or they may be of different kinds having nearly the same size; but they are only counted as framework cations if their cation-anion bond strengths are of the same order of magnitude, so that it is sensible to think of the framework as a continuous entity. Their polyhedra share corners, edges, or (more rarely) faces, to form a three-dimensional network.

(c) Frameworks built from octahedra sharing edges commonly form close-packed, fairly rigid, structures with small tetrahedral interstices only (e.g., spinel). Frameworks built from corner-linked octahedra or tetrahedra are more open and flexible, with large cavities of varying size and shape (e.g., perovskite, tridymite, chabazite); they tend to have a greater variety of hettotypes. Frameworks with octahedra sharing some corners and some edges are both common and important (e.g., olivine, rutile). Table 12-2 shows the volume per oxygen atom in some representative structures; the larger this volume, the larger the cavities available for nonframework atoms.

(d) When framework atoms are of more than one kind, they may occupy different framework sites in an ordered way or at random, but an ordered arrangement is more typical. As compared with random arrangement, ordering generally involves a lowering of the symmetry, whether by changing the shape of the unit cell to something less symmetrical or by doubling one or more of its edge lengths.

(e) Interstitial or cavity atoms must be of the right size for the sites provided. This can be a fairly restricting condition when the sites are small and the bonds strong, (e.g., for the tetrahedral atom in olivines and spinels). Some structures (e.g., rutile) have no cavity sites. In other families (e.g., perovskites, nepheline) cavity sites which are usually occupied may sometimes be left empty (e.g., ReO_3, tridymite), but this is not always possible (e.g., felspar).

(f) When cavities are so large that they are occupied by groups of atoms or molecules rather than single atoms (for example, in the zeolites), the occupation is nevertheless at quite specific ordered sites, satisfying the usual structure-building rules. Minor changes of framework shape, without change of linkage pattern, can provide for different sets of occupying atoms or molecules (e.g., chabazite). Because of the large volume and weak non-framework bonds, the variety of possible compositions and arrangements is large, and disorder between them can occur.

(g) If the linkage pattern of a structure is overdetermined, and cannot at the same time satisfy ideal bond-length and ideal bond-angle requirements, distortions of polyhedra may represent a compromise solution, and may be characteristic of the family as a whole, i.e., they may be similar in character for all hettotypes (e.g., felspars, 5-4-3 bronzes).

(h) Subject to this, variations of shape of the framework can be grouped under four headings:

(1) alternation of polyhedra of different sizes, due to ordered occupation by cations of two or more kinds;

(2) distortions of shape of polyhedra due to covalent bond systems of symmetry less than that of the regular polyhedron;

(3) small distortions of shape of polyhedra due to the off-centring of the cation resulting from its small size and high charge;

(4) crumpling of the framework by tilting or rotation of polyhedra, without distortion of their regularity.

(i) Of these effects, crumpling, when present, generally produces the largest distortions both of lattice parameters and atomic position parameters. It can, however, effectively occur only in corner-linked structures. Its detailed character depends on the size and environmental requirements of the cavity-filling cations; or if these are absent (as in SiO_2) on effects such as the preferred value of the hinge angle, and the repulsive forces between atoms of different polyhedra forming the cavity walls. It tends to form bipartite structures in which polyhedra in successive subcells have tilts of opposite sign.

(j) Of the three kinds of distortion of the polyhedra themselves, the effect on the structure as a whole of the size of different substituents is generally fairly small. It tends to affect all lattice parameters uniformly, and in the presence of crumpling it is smoothed out and rendered inconspicuous. Covalent bonding effects are commonly an order of magnitude greater than off-centre effects, but of much less frequent occurrence, being characteristic of fewer species of ions—Cu^{2+} and Mn^{2+} are the most obvious examples. Neither effect is very noticeable except when the symmetry of the aristotype is high; in such a case, even the small distortions caused by off-centring become conspicuous and important, as in the various forms of $BaTiO_3$.

(k) Off-centring of framework cations is associated with polarisation of the anions, which keeps the direction of the off-centre displacement the same throughout a complete chain or sheet of the framework, irrespective of whether there are also crumpling effects present.

(l) Off-centring of cavity cations is only a useful concept when the cavity itself remains fairly regular (e.g., in $BaTiO_3$ or Al_2BeO_4); even so, it is not easy to make generalisations.

(m) The foregoing discussion refers to homodesmic structures, where bonds of comparable strength provide a continuous linkage in three dimensions. Many of the principles also apply to heterodesmic structures (considered only briefly in §12.22), where the continuous linkage is in 2, 1, or 0 dimensions.

PROBLEMS AND EXERCISES

1. Lead titanate, $PbTiO_3$, is isomorphous with $BaTiO_3$, with $a = 3.99$ Å, $c = 4.03$ Å; if the origin is taken on Pb, then the z coordinates of Ti, $O(1)$, and $O(2)$ are 0.541, 0.112, and 0.612 respectively. Comment on the regularity of the octahedron, the displacement of Ti from its centre, and the range of A—O and B—O distances, comparing them with those in $BaTiO_3$.

2. Use the data of Table 12-1 to estimate the tilts of the octahedra (assuming them to be regular) about [001] for the series of compounds $MFeO_3$.

3. Using the information about $NaNbO_3$, Phase Q, given in Chapter 10, Question 5, (a) calculate the Nb—O bond lengths and O—O edge lengths of the octahedron, so far as is possible from the incomplete data, (b) predict the positions of the other atoms, (c) describe the environment of the Na atoms, with the help of sketch stereograms and calculations of Na—O bond lengths.

4. The structure of elpasolite, K_2NaAlF_6, can be thought of as derived from a perovskite by making Na and Al occupy alternate octahedra. The lattice parameter is 8.11 Å, and the space group $Pa3$. What can you deduce about the cation positions and the F positions?

Show that such a structure is compatible with the Goldschmidt radii of the framework atoms, assuming undistorted octahedra. Calculate the tilt, and hence suggest possible position parameters for F.

Are the K—F distances compatible with the Goldschmidt radii for K and F, or can any adjustment of atomic positions allowed by the space group make them so?

Would any different conclusions be reached if the Shannon and Prewitt radii were used?

Apply Pauling's rule to this structure. Do you consider that it ought to be classified as a perovskite structure, or are there reasons against it?

[Note: in space group $Pa3$, the origin is taken at a centre of symmetry on a triad axis parallel to [111], and the a-glide plane is then at $y = 1/4$.]

5. The series of trifluorides MF_3 with space group $R\bar{3}c$, referred to in §12.11, have been described in terms of a primitive rhombohedral unit cell containing two formula units, with M at 0, 0, 0, and F at x, $1/2 - x$, $1/4$. The values of the lattice parameters and x are given in the table shown on page 336.

(a) Calculate the angle of tilt of the octahedra about the triad axis, measured from a hypothetical state in which they are all parallel.

(b) Is the octahedron itself regular in each case, or in what way and by how much is it distorted?

(c) Calculate the *M*—F bond lengths, and compare them with those expected from Table 2-2.

(d) How large a cation could be inserted into the cavity of the framework of each, assuming all atoms to be rigid and spherical? Find the coordinates giving its position, and describe the arrangement of anions round it.

CATION *M*	*a* (Å)	α (°)	*x*
Ti	5.52	58.9	−0.183
V	5.37	57.5	−0.145
Cr	5.26	56.6	−0.136
Fe	5.36	58.0	−0.164
Co	5.28	57.0	−0.15
Ga	5.20	57.5	−0.136
Mo	5.67	54.7	−0.25
Ru	5.41	54.7	−0.100
Rh	5.31	54.4	−0.083

6. The structure of MnF_3 is a distorted version of that of VF_3. It is monoclinic, with $a = 8.90$ Å, $b = 5.04$ Å, $c = 13.45$ Å, $\beta = 92.7°$, space group $C2/c$. The Mn atoms are in 4(a), 0, 0, 0, and 8(f), x, y, z, with $x = 0.167$, $y = 0.500$, $z = 0.333$. Show the relationship of the unit cell and the cation positions to those of VF_3, and describe the nature and magnitude of the distortion so far as these are concerned.

If the only difference between the two structures were the small homogeneous distortion indicated by the difference in lattice parameters, show that the F atoms would be in positions 4(e), 0, 0.586, 1/4, and 8(f), 0.293, 0.707, 0.250, together with other general positions related to these by the non-space-group translations (1/6, 1/2, 1/3).

In fact the positions are 0, 0.617, 1/4, and 0.310, 0.714, 0.24, though the non-space-group translations still hold good. Show that this changes the shape of the octahedra but not their linkage pattern. Calculate the Mn—F distances and comment on the Mn environment.

Consider the relation of this structure to that of $KCuF_3$.

7. (a) Using the information about $LaAlO_3$ given in §12.9, calculate the fractional increase in *c* which would be needed to make the AlO_6 octahedron regular, assuming no change in *a* or the position parameter.

(b) Assuming the distortion of the octahedron to remain constant throughout the series, calculate the tilt angles about the triad axis of the isomorphous compounds $PrAlO_3$, $NdAlO_3$, high-temperature $SmAlO_3$ ($a = 5.341$ Å, $c = 12.99$ Å) and high-temperature $LaGaO_3$ ($a = 5.579$ Å, $c = 13.54$ Å), and predict position parameters for their oxygen atoms.

(c) Repeat the procedure using the assumption that the distortion of the octahedron increases proportionately to the tilt angle, the separate contributions of the two factors to the observed departure from ideal hexagonal close-packing being in the same ratio as in $LaAlO_3$.

(d) For each of the aluminates, calculate the Al—O distances. Are there any significant differences?

(e) For each of the compounds, calculate all twelve distances from the large cation to oxygen atoms that would be neighbours in the aristotype. Consider which of them should actually be counted as neighbours, and compare the mean value for this set, and the difference between extreme values, with the corresponding figures for the set of twelve distances. Describe the environment of the large cation.

8. Gadolinium ferrite, $GdFeO_3$, described with the same choice of axes as $CaTiO_3$, has $a = 5.35$ Å, $b = 7.67$ Å, $c = 5.61$ Å; Gd is at 0.984, 1/4, 0.063; O(1) is at 0.400, 1/4, 0.967; O(2) is at 0.196, 0.949, 0.198.

(a) Find the distortions of the octahedra along directions [010] and [001].

(b) Find the tilts of the octahedra about these axes.

(c) How far would these tilts have been predictable from the evidence in Table 12-1? (Cf. Question 2.)

(d) Find the Gd—O distances and discuss the environment of Gd.

(e) Repeat these calculations for the isomorphous $PrFeO_3$ and $LuFeO_3$. Comment on any differences, noting in particular whether the environment of the large cation is consistent with a smooth change with atomic number.

	LATTICE PARAMETERS (Å)			POSITION PARAMETERS								
				M			O(1)			O(2)		
	a	b	c	x	y	z	x	y	z	x	y	z
$PrFeO_3$	5.48	7.79	5.58	0.991	1/4	0.044	0.418	1/4	0.979	0.208	0.952	0.208
$LuFeO_3$	5.21	7.56	5.55	0.980	1/4	0.071	0.380	1/4	0.954	0.189	0.938	0.193

9. Sodium niobate, $NaNbO_3$, Phase P, has space group $Pbma$ (No. 57), with $a = 5.57$ Å, $b = 15.52$ Å, $c = 5.51$ Å. The atomic positions are as follows: Nb: $1/4 + 0.022$, $1/8$, $1/4 + 0.007$; Na(1): $3/4$, 0, $1/4 - 0.007$; Na(2): $3/4 + 0.032$, $1/4$, $1/4 - 0.011$; O(1): $1/4$, 0, $1/4 + 0.054$; O(2): $1/4 - 0.017$, $1/4$, $1/4 - 0.059$; O(3): 0.032, $1/8 + 0.015$, $1/2 + 0.036$; O(4): $1/2 - 0.033$, $1/8 - 0.015$, $1 - 0.034$.

(a) Show that the O_6 octahedron is nearly regular.

(b) Find the displacement of Nb from the centre of the octahedron, noting its magnitude and its direction relative to the local symmetry of the octahedron.

(c) Find the tilt of the octahedron about [010] and [100]. How does the latter compare with what would have been predicted from the axial ratio?

(d) Describe and note the differences between the environments of the two sodium atoms.

(e) Compare the part of the structure between $y = 0$ and $y = 1/2$ with the structure of Phase Q for which the data are given in Chapter 10, Question 5. (Cf. also Question 3 of this chapter.)

(f) Compare the same part of the structure with the structure of $GdFeO_3$ and its isomorphous compounds (Question 8), considering in particular the shape of the octahedra, the set of tilts, and the environments of the B cations and the A cations.

10. Predict a simple structure for the low-temperature form of sodium niobate (Phase N), using evidence from the known structures of Phases P and Q, the three forms of $KNbO_3$, $LiNbO_3$, and any other structures you consider relevant.

11. Scheelite, $CaWO_4$, is tetragonal, with $a = 5.24$ Å, $c = 11.38$ Å. The space group is $I4_1/a$ (No. 88), and there are four formula units per unit cell. The Ca atoms are at 0, 0, 1/2, the W atoms at 0, 0, 0, and the oxygen atoms in general positions x, y, z, with $x = 0.25$, $y = 0.15$, $z = 0.075$. Draw diagrams of the structure, and calculate the interatomic distances. Find the angle between the x axis and the projection on (001) of the Ca—O bond. Comment on the relationship to zircon (Chapter 11, Question 14) and to potassium dihydrogen phosphate (§13.12).

12. The compound Sb_2ZnO_6 is tetragonal, with $a = 4.66$ Å, $c = 9.24$ Å, space group $P4_2/mnm$, (No. 136), and two formula units per unit cell. Atomic positions are as follows: Zn at 0, 0, 0; Sb at 0, 0, 1/3; O(1) at 0.30, 0.30, 0; O(2) at 0.30, 0.30, 0.33. Describe and discuss the environments of the two cations, and the linkage scheme of their polyhedra. Is there a relationship to any simpler structure described in Chapter 11?

[Note: the origin in this space group is chosen at the intersection of the mirror planes, and the 4_2 axes lie midway between edges of the unit cell.]

13. Tapiolite, Ta_2FeO_6, is isomorphous with Sb_2ZnO_6, with $a = 4.67$ Å, $c = 9.14$ Å, and is reported to have the same atomic position parameters. Is this to be expected, or is it surprising?

14. The high-temperature phase of $LiAlSiO_4$ has an ordered hexagonal structure with $a = 5.27$ Å, $c = 11.25$ Å, related to that of quartz; its space group is that of high quartz. Describe qualitatively the main features of the structure. Show that all cations must lie in special positions without variable parameters; give the coordinates of these positions. How far, and along what lines, is it possible to estimate the parameters of the oxygen atoms?

15. The high-temperature phases of $LiAlSi_2O_6$, $LiGaSi_2O_6$, and $LiAlGe_2O_6$, all have a "stuffed keatite" structure, which is tetragonal, with space group $P4_32_12$; $LiAlSi_2O_6$ has $a = 7.53$ Å, $c = 9.14$ Å. Using the data and results of Chapter 11, Question 17, suggest how the atoms are placed. How would you expect the lattice parameters of $LiGaSi_2O_6$ and $LiAlGe_2O_6$ to differ from those of $LiAlSi_2O_6$?

16. With which other cations would you expect to be able to make synthetic compounds having a felspar structure? For each composition, consider which of the known structures of $KAlSi_3O_8$, $NaAlSi_3O_8$, $CaAl_2Si_2O_8$, or $BaAl_2Si_2O_8$ it is likely to resemble most closely.

17. Calculate the radius of the largest cation which it would be possible to insert into the cavity in (a) beryl (Chapter 11, Question 18), (b) tridymite (§11.13), (c) high quartz (§11.14), (d) sanidine (§11.15), assuming that the ions are rigid and spherical and that there is no change in any structural parameter.

    For beryl, calculate also the radius of the largest cation which could pass through the O_6 ring constituting the window of the cavity. How do these cavity sizes compare with cavity sizes in zeolites?

18. Scapolite is a tetragonal zeolite with lattice parameters $a = 12.06$ Å, $c = 7.57$ Å, space group $I4/m$. The unit cell content is approximately $M_8T_{24}O_{48}Cl_2$; for the purposes of the present question, this may be taken as exact and the structure as quasiperfect. (In fact M is roughly $Na_{\frac{3}{4}}Ca_{\frac{1}{4}}$ and T is roughly $Al_{\frac{1}{3}}Si_{\frac{2}{3}}$.)

    (a) Which of the atoms are in special positions and of what kinds?

    (b) Tetrahedra form four-rings of two kinds, both lying parallel to the (001) plane. Tetrahedra in rings of type (1) all have one edge perpendicular to (001); those in rings of type (2) all have one face nearly parallel to (001), but the remaining corner points upwards for two and downwards for two. Show that, in projection on (001), type (1) rings must have their centres at 0, 0, and type (2) rings at 0, 1/2, and that each ring is linked by each of its free corners to a different ring of the other kind. What is the point-group symmetry of each type of ring?

    (c) Predict the value of the lattice parameter c from the known dimensions of the SiO_4 tetrahedron, and compare it with the observed value.

    (d) Using the known value of the lattice parameter a, and the known dimensions of the tetrahedron (assumed regular), deduce the orientation of the four-rings. (This is perhaps most easily done by cutting out pieces of paper on which the projections of the two types of ring are drawn to scale, and tilting them in ways consistent with the symmetry to obtain the correct value of a.)

    (e) Draw a diagram of the TO_2 framework, marking approximate heights of the atoms.

(f) Show that there is room to insert the M cations and the Cl anions at sites consistent with the symmetry. Discuss their environments.

(g) Using the results of (f), suggest an explanation for the observed fact that Al atoms are found preferentially in the type (2) rings, type (1) being entirely Si.

19. Compounds with the following compositions are believed to have the 5-4-3 bronze structure:

$$BaTa_2O_6, \quad KW_2O_6, \quad Ba_2NaNb_5O_{15}, \quad Sr_2KNb_5O_{15},$$

$$Sr_4KLiNb_{10}O_{30}, \quad K_3LiNb_5O_{15}, \quad K_3Li_2(Ta_xNb_{1-x})_5O_{15}$$

How far, in each case, can you predict the distribution of cations between octahedral (framework) sites, tunnel sites, cage sites, and trigonal-prism sites? Is any kind of site only partly occupied?

20. Extend Table 12-2 to include the volume per oxygen atom in structures described in Chapter 11, Questions 9, 12, 14, 15, 17, 18, 20; and in the current chapter, Questions 8, 9, 11, 12, 13, 14, 15, 18. Comment in particular on the values for δ-Mn_2GeO_4 (Chapter 11, Question 9), keatite (Chapter 11, Question 17), beryl (Chapter 11, Question 18), garnet (Chapter 11, Question 20), and scapolite (Chapter 12, Question 18).

Chapter 13 STRUCTURES OF COMPOUNDS CONTAINING HYDROGEN

13.1 INTRODUCTION

The special characteristic of the hydrogen atom is that it has only one valency electron. We may therefore expect it to play three different roles. (i) By adding one electron to form the first complete rare-gas shell, it will form a negative ion H^- roughly comparable in size to Li^+. (ii) By losing its electron it will form a positive ion H^+ consisting of a bare proton, negligible in size. (iii) It can use its orbitals to form one covalent bond with one neighbouring atom; in nonbond directions, it will be held to other neighbours by van der Waals forces only, and its effective "radius" in such directions (1.2 Å) will be much larger than in the bond direction (0.3 Å).

This classification is of course oversimplified. With hydrogen, as with other atoms, there are intermediate states between the purely ionic and the purely covalent. Nevertheless, it serves as a useful starting-point.

Most of the examples with which we shall be concerned in this chapter can best be considered in terms of the H^+ ion, though the bonds are far from pure ionic, and of a wide range of strengths. They are discussed in §13.2 onwards.

Examples involving the H^- ion are much less common. We shall describe only one of them, LiH.

Compounds where the hydrogen forms one nearly pure covalent bond are very common. Since the hydrogen has only one such bond to a neighbour it must terminate the molecule; hence hydrogen-rich compounds tend to give molecules sheathed in hydrogen that fit together by simple packing. Ethane, C_2H_6, and naphthalene, $C_{10}H_8$, are examples.

13.2 THE HYDROGEN BOND

Suppose, in a compound of H^+ and anions X^-, H^+ behaved as a very small but otherwise normal cation, obeying the Goldschmidt rules. Only two X^- ions would be in contact with it, and it would lie symmetrically between them. It is not possible to build up a three-dimensional structure using only 2-coordinated atoms, but if there are other larger cations A, which of course form A—X bonds, we may expect X—H—X bonds to be available in addition to A—X bonds to hold the structure together. Examples of this simple kind are not common, but one occurs in KHF_2 (§13.7). The X—H—X link is known as *the hydrogen bond*.

Experimentally, it is much commoner to find that, when H^+ lies between two anions, it is noncentral—just as other cations take up noncentral positions when they are rather too small for the coordination polyhedron (cf. §2.5). The bond to the nearer anion is strengthened, and becomes more covalent; the bond to the more distant anion is weakened, and becomes more like a van der Waals bond.

The name *hydrogen bond* is generally given to any group *where a hydrogen atom lying between two more electronegative atoms exerts attractive forces of more than van der Waals strength on both*. It covers a great range of cases. The atoms joined by it may be F, O, N, Cl, or S; they may be alike or different; they may be covalently bonded to the other neighbours or be nearly ionic, or they may be the oxygen atoms of water molecules held in position by hydrogen bonds only. Examples where the hydrogen atom lies midway between two neighbours are believed to occur (though not all cases where this has been claimed in the past have been substantiated by more recent work). Much more commonly, however, the hydrogen atom is asymmetrically placed between its neighbours, and it is useful to consider the nearer neighbour as the *donor* of hydrogen to the bond, the more distant neighbour as *acceptor* of the bond. The bond is then written X—$H \cdots X'$, where X is donor and X' acceptor.

In further general discussion, we shall restrict ourselves to the very important case when both anions are oxygen, and the bond is O—$H \cdots O$. An example involving N—$H \cdots O$ bonds will, however, be described in §13.14, and other kinds of hydrogen bonds will be mentioned in Chapter 15.

If H^+ behaved like an ordinary cation, except for having a negligibly small radius, we should expect the O—O distance in the hydrogen bond to be equal to the diameter of O^{2-}—about 2.6 to 2.8 Å—and the O—H distance to be half the O—O distance. The experimental values of O—O agree with this, but not those of O—H. As previously noted, the H atom is nearly always considerably off-centre, and the O—H distance is in fact about 0.9 to 1.1 Å. In consequence, *the proton is actually embedded inside the packing sphere of an oxygen ion.* Obviously the proton must have a considerable effect on the polarisation of the oxygen; it is therefore surprising that it has so little effect on its radius. The O—O distance in a hydrogen bond, in fact, is not very different from that in the edge of an octahedron or tetrahedron round a small cation. This means that, for packing purposes, OH⁻ groups (and also water molecules) may be treated as if they were O^{2-} ions. (Packing, of course, is not the only consideration; the bond directions are also important, and will be considered in §13.3.)

We must remember here that X-ray diffraction methods "see" the lump of electron density associated with an atom, and not, like neutron diffraction, its nucleus. For hydrogen covalently bonded to its donor, the difference in position may be detectable. The O—H distances measured by X-rays are often 0.1 to 0.2 Å shorter than those measured by neutrons. We shall use neutron measurements wherever possible for bond lengths and bond angles, and remember that X-ray measurements are subject to corrections of this order of magnitude when so used (though of great potential interest in showing the redistribution of electron density). For atoms other than H, the discrepancy between the electron-density maximum and the position of the nucleus is generally too small to observe, though it has been detected in at least one case (See Chapter 13, reference 39).

Variations in the strength of the hydrogen bond, and correspondingly in its length, are to be expected from one compound to another. We may make predictions as follows. When the bonds of the donor O to its other neighbours are weak, O—H will be strong and short. In consequence, H · · · O will be weak—and it is the strength of this link which determines the overall strength of the bond. Because of the nonlinear variation of bond length with bond strength (cf. §2.8), the overall O—O distance will be long. By contrast, when the bonds of the donor O to its other neighbours are strong, O—H will be relatively weak and long, and H · · · O relatively strong and short, resulting in a smaller difference between O—H and H · · · O; the OH · · · O bond as a whole is strong and short. A centrally placed H atom represents the extreme case, and is only possible in structures where the O—O distance is exceptionally short.

Experimental values of the O—H and O—O distances in representative compounds are plotted in Figure 13-1. The general trend is obvious. The scatter need not be surprising, because hydrogen bonds, like any other bonds, are subject to stresses from other parts of the structure, making them longer or shorter than we should otherwise expect. Empirically, the separation between the two curves in Figure 13-1 (in so far as it is not an allowance for experimental error) is a measure of the bond-length strain due to such structural stresses or "steric effects" as can be tolerated without rendering the structure unstable. (Effects on bond angles are likely to be greater, and will be discussed later in this section.)

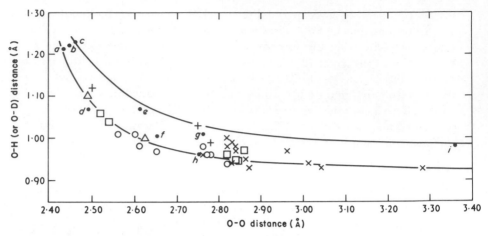

Figure 13–1 Relation between O—H (or O—D) and O—O in hydrogen bond. Black dots: *a*, sodium hydrogen acetate; *b*, potassium bisphenyl acetate; *c*, potassium hydrogen malonate; *d*, KH₂PO₄; *e*, NaHCO₃; *f*, AlOOH; *g*, hexagonal ice; *h*, cubic ice; *i*, Ca(OH)₂. Open circles: alums. Squares: oxalic acid dihydrate and its deuterium analogue. Triangles: HNO₃·3H₂O (short bonds: X-ray data). Upright crosses: sodium sesqui-carbonate. Diagonal crosses: hydrates listed by Baur (cf. Chapter 13, reference 5).

Curves show empirically the general trend and the scatter. Experimental error in O—O is generally less than 0.01 Å; in O—H or O—D, it varies from about 0.01 Å to 0.05 Å for different investigations. For *a*, *b*, and *c* a truly central H atom has been assumed.

Though there is no sharp break at any point in the series, it is convenient to think of hydrogen bonds in categories roughly dependent on their O—O length.*

> *Very short bonds:* about 2.4 to 2.45 Å (H atom believed to be central)
> *Medium-short bonds:* about 2.5 to 2.65 Å
> *Medium-long bonds:* about 2.7 to 2.9 Å (at one time called hydroxyl bonds)
> *Very long bonds:* from 3.0 Å upward.

For lengths above 3 Å, it is not always easy to be sure which contacts should be counted as bonds, since there is no sharp distinction between very long, very weak, bonds and van der Waals contacts.

The correlation of decreasing O—O length with increasing polarising power of the cation neighbour of the donor can be seen in Figure 13-1, and more clearly in Table 13-1, which includes some of the same examples together with others for which the hydrogen parameters have not yet been determined. Electrostatic valence is a measure of bond strength and hence (in a rough general way) of polarising power; as it increases from 1/6 for KOH to 5/4 for KH₂PO₄, the O—O length decreases. (At higher values, it is not really surprising that a comparison based like this on electrostatic

---

* The terms "long hydrogen bonds" and "short hydrogen bonds" are used differently by different authors. Some call all bonds up to, say, 2.9 Å short, while others call all exceeding, say, 2.7 Å long; this can lead to misunderstanding.

Table 13-1 Dependence of Hydrogen Bond Length on Electrostatic Valence of Bond to Donor Oxygen Atom

COMPOUND	CATION NEIGHBOUR	ELECTROSTATIC VALENCE	O—O LENGTH (Å) Bond	Nonbonding	COMMENT	REFERENCE
KOH	K	1/6	3.4	>4.0		Ibers, Kumamoto, and Snyder, 1960
MgSO$_4$·4H$_2$O	Mg	2/6	{ 2.83 2.84 2.86	3.04 3.28	Long distances may be bifurcated bonds	Baur, 1965
Diaspore, AlOOH	Al	3/6	2.65			Busing and Levy, 1958
Afwillite, Ca$_3$(SiO$_3$OH)$_2$·2H$_2$O	Si	4/4	2.53		Mean of two; experimental values less accurate	Megaw, 1952
KH$_2$PO$_4$	P	5/4	2.49			Bacon and Pease, 1953
H$_2$SO$_4$	S	6/4	2.63		See discussion, §13.15	Pascard-Billy, 1965
NaHCO$_3$	C	4/3	2.61		See discussion, §13.16	Sharma, 1965

345

analogies may prove unhelpful.) The same effect is shown in more detail by the series of alums listed in Table 13-2, which will be discussed in §13.24.

Table 13-2 O—H $\cdots$ O Bond Lengths in Alums (Å)

CATION M	BONDS FROM $O(W1)$	BONDS FROM $O(W2)$
Na	2.75⎫ 2.78 2.82⎭	2.65⎫ 2.63 2.62⎭
Rb	2.86⎫ 2.80 2.73⎭	2.60⎫ 2.58 2.55⎭
Cs	2.82⎫ 2.79 2.76⎭	2.61⎫ 2.63 2.65⎭
Electrostatic valence of bond from cation to O(W)	1/6	3/6

At one time, a lot of discussion centred on the question whether O—H $\cdots$ O bonds are necessarily linear. As the number of exact structure analyses reporting hydrogen positions increased, it became quite clear that in general the two links O—H and H $\cdots$ O are *not* in the same straight line. The angle between them commonly deviates from 180° by 10° to 20°, and sometimes by as much as 40° or 50°. In an extreme case we may get the *bifurcated* hydrogen bond, in which H attached to one donor O is appreciably attracted to two acceptor O atoms. This can easily be understood if it is recognised that the H $\cdots$ O link is mainly electrostatic, and that its angular requirements are therefore flexible. If, sterically, two O atoms which are possible acceptors can get equally close to H on the side remote from its donor O, it will be equally attracted to both; but each attraction will be weaker than when there is only one acceptor involved. Bifurcated bonds therefore tend to be longer than others with the same O—H length. Carrying the argument further, we see that if H has only rather distant neighbours (apart from its donor O), it will be weakly attracted to all, and if we choose to call it nonbonding we merely mean that the bonds are too weak, and too random in their O—H—O angles, to be taken into consideration for our immediate purposes. An example occurs in $Ca(OH)_2$ (§13.8).

The effect of substituting deuterium for hydrogen (the isotope effect) makes little difference to most structures. The O—D $\cdots$ O bonds are expected to be slightly weaker, and therefore slightly longer, than corresponding O—H $\cdots$ O bonds, but the difference is only detectable in very accurate analyses. (Bond angles are more sensitive, as we shall see in §13.26.)

Bond lengths OD and O(D)O have been included in Figure 13-1, and do not show any different behaviour from those for OH and O(H)O.

As a structure-building force, the hydrogen bond plays the part of a relatively strong bond in molecular structures, where the comparison is with materials held together by van der Waals bonds, but of a relatively weak bond in structures where the other bonds are ionic or semipolar. Indeed, in these latter structures, a region where hydrogen bonds preponderate is very often marked by a cleavage plane.

13.3 ENVIRONMENT OF OH⁻ AND H₂O

We saw in Chapter 11 that in silicates, where the O—Si bonds have a considerable amount of covalent character, the two bonds from an oxygen atom are inclined at an angle of about 130° to 150°; for fully covalent oxygen, a rather smaller angle might be expected. In the same way, we may expect the angle made by an O—H bond with the oxygen—cation bond to be markedly different from 180°.

Without going into detailed theory, we can get a good picture of the structural behaviour of OH⁻ and H₂O groups from a simple model, put forward originally by J. D. Bernal about 1933. Let the effective point charge at the centre of an O^{2-} ion be replaced by four charges of $-1/2$ electron units, placed at the corners of a tetrahedron, at about 1 Å from the centre (the packing radius of ∼1.4 Å remaining unchanged). Let protons be added at one or two corners, making the net charge at each $+1/2$. Thus OH⁻ has one corner $+1/2$ and three corners $-1/2$; H₂O has two $+1/2$ and two $-1/2$. Corners of either sign are attracted electrostatically by oppositely charged corners of neighbouring groups, or by ions of appropriate sign. The typical environment of OH⁻ or H₂O thus consists of four neighbours, arranged in a tetrahedron. This is the same coordination commonly found for O^{2-}, for example in spinel (§11.3), olivine (§11.9), and zinc oxide (§4.11).

Sometimes, however, OH⁻ or H₂O groups have three neighbours lying in the same plane. The same simple model will account for this if we assume that two negative corners are equally attracted towards the same cation or the same positive corner of another group. This planar configuration is also sometimes found for O^{2-}, notably in rutile (§11.10).

Less symmetrical configurations around OH⁻ or H₂O can generally be seen as more-or-less distorted versions of tetrahedra with one corner missing, implying that the corresponding corner of the tetrahedral model is nonbonding. This may occur because the rest of the structure allows no neighbour to be fitted in at a suitable distance from the group in question.

In all these configurations, it follows from the model that the bond angle at the oxygen atom of the group is never 180° but ideally in the range of about 110° to 125°; allowing for the easy distortion of bond angles by structural stresses, any angle from 90° to 130° is reasonable. It is this fact which gives rise to the characteristic features of hydroxide and hydrate structures.

More sophisticated models can, of course, be devised. For example, in water vapour the H—O—H angle is 104.5°, and though the "unstrained" bond angle in the crystal may not be the same, since it depends on the polarisation of O by its environment (just as "unstrained" bond lengths depend on the polarisation, cf. §2.9), it is probably a better approximation to use 104.5° rather than 109.5° (the tetrahedral angle) in our model. A rough measure of the bond angle strain in a particular water molecule O_W can be obtained, without knowledge of the exact hydrogen positions, from the difference between 105° and the observed angle O_1—O_W—O_2 made by the bonds to the two acceptors.

Again, the description of the negative end of the group by two point charges of $-1/2$ does not adequately explain the observed behaviour of incoming bonds in all cases. The direction of these bonds, to which the oxygen atom is acceptor, has more effect on the energy, and therefore more effect in determining the structure, than the simple model would suggest. An example is given in §13.26.

Nevertheless, for qualitative understanding of hydroxides and hydrates, the simple model takes us a very long way. It represents an excellent approximation to the truth.

13.4 WATER MOLECULES OR HYDROXYL GROUPS?

The tetrahedral model gives an explanation of the fact that water molecules and OH⁻ groups play very similar roles in a structure, and are equally able to form long or short hydrogen bonds. In consequence, it is often hard to distinguish, in the absence of really careful structure determination, whether a compound whose formula can be written as if it were a hydrate is in fact a hydroxide. For example, the mineral diaspore, AlOOH, (§13.11), was at one time assumed to be aluminium oxide monohydrate, $Al_2O_3 \cdot H_2O$. Again, a compound whose formula used to be written as $NaBO_2 \cdot 4H_2O$ is now known from structure analysis to be $NaB(OH)_4 \cdot 2H_2O$. The so-called hydrogarnets, in which $(OH^-)_4$ replaces SiO_4 in the parent formula $Al_2Ca_3(SiO_4)_3$, are structurally $Ca_3[Al(OH)_6]_2$, the positions of cations and oxygen atoms remaining unchanged when Si^{4+} is removed and $4H^+$ added. Even with silicates, one may find OH⁻ groups as part of the tetrahedron round Si: an example is the mineral afwillite, whose empirical formula may be written as $3CaO \cdot 2SiO_2 \cdot 3H_2O$, but which proves structurally to be $Ca_3(SiO_3OH)_2 \cdot 2H_2O$.

Spectroscopic methods—infrared, Raman, nuclear magnetic resonance—can sometimes clarify the role of H in a given structure, but the arguments are indirect, and sometimes depend on assumptions about the character of the structure that are only justifiable when the structure is already fairly well-known; in these latter cases they may be able to provide very useful refinements. An alternative approach, from diffraction analysis, is possible even when the hydrogen atoms cannot thus be located directly. For a trial model, it is possible to ignore the hydrogen atoms, since they do not affect the packing radius of their donor oxygen atoms; in subsequent refinement, if their contribution to the observed intensities of diffraction maxima is too small to indicate their positions, it is also too small to introduce serious errors into the positions of other atoms, and in particular into those of the O atoms. Once we know the rest of the structure, there are several ways in which we can proceed.

First, inspecting the bond lengths, we look for short O—O distances between O atoms not coordinated to the same cation, and consider them as possible hydrogen bonds. In favourable cases, when there are as many such short distances as H atoms, and nothing else nearly as short, we can confidently locate one hydrogen on each.

The second argument is from electrostatic valence. A group can be tentatively identified as O^{2-}, OH^-, or H_2O according to whether the electrostatic valence sum of its bonds from cations is 2, 1, or 0. In practice, intermediate values may occur and make identification difficult. This is partly because (as we saw in §2.6) Pauling's rule is only rigorously true when the cations are in positions of high point symmetry; and partly because, in a hydrogen bond, the strength of the O—H link is less than 1 (but more than 1/2) while that of the H $\cdots$ O link is more than zero (but less than 1/2).*

A third approach is to consider the other neighbours of the O atoms. Both lengths and directions of bonds are relevant. In general, bond lengths increase in the order cation—O^{2-}, cation—OH^-, cation—H_2O, but the differences are often small and sometimes masked by other causes. The directions, however, can be very informative. For example, if all the cation neighbours of an oxygen lie to one side of it, there is a strong suggestion that it must be bonded to hydrogen on its other side.

These arguments can only be used safely when the positions of all atoms other than hydrogen are reliably and fairly precisely known. In such circumstances, if all the independent lines of evidence lead to the same conclusion, we can be as sure of the whereabouts of the hydrogen atoms as if we had found them directly, like the other atoms, from their diffraction effects. (Deductions from ill-refined structures can, however, be misleading.) For exact positions, and values of O—H bond lengths and H—O—H bond angles, we must, of course, await further experimental studies, preferably by neutron diffraction.

---

* It has been suggested, and found empirically useful (W. H. Baur and A. A. Khan: *Acta Cryst.*, **B26:** 1584, 1970), that an atom involved in hydrogen bonds should be taken as receiving 1/6 valency unit (v.u.) from a bond to which it is acceptor, and 5/6 valency unit from a bond to which it is donor.

13.5 H_3O^+ AND $H_5O_2^+$

In strong acids, the proton tends to be dissociated from the anion, and may attach itself to a water molecule to form an oxonium group. The tetrahedral model still holds good, but now the group can act as donor to three hydrogen bonds, and acceptor to one.

Instead of attaching itself to one H_2O group, the proton can sometimes place itself midway between two H_2O groups forming a very short hydrogen bond. The new unit is $H_5O_2^+$. Each end can still act as donor to two hydrogen bonds, and acceptor to one. It is possible to envisage the building up of still more complex groups using the same principles.

Examples of some of these groups will be given in §13.27 to 13.29. Oxonium groups have sometimes been reported as occurring in compounds other than strong acids, but the evidence is generally rather slender or capable of other interpretation.

13.6 LITHIUM HYDRIDE, LiH

Lithium hydride has the NaCl structure, with $a = 4.085$ Å. The hydrogen serves as the anion. If we take the radius of Li^+ as 0.68 Å, that of H^- is $2.04 - 0.68 = 1.36$ Å.

13.7 POTASSIUM HYDROGEN FLUORIDE, KHF_2

Lattice: Tetragonal, with $a = 5.67$ Å, $c = 5.81$ Å
Number of formula units per unit cell: 4
Space group: $I4/mcm$ (No. 140)
Atomic coordinates:
  K in $4(a)$: $0, 0, \frac{1}{4}$
  H in $4(d)$: $0, \frac{1}{2}, 0$
  F in $8(h)$: $u, \frac{1}{2} + u, 0$, with $u = 0.14$

There are eight F atoms round each K, not in a cube, but in a square antiprism (Figure 13-2). Each F atom is shared between four K atoms, two lying in polyhedra below it and two in polyhedra above it, but all within one hemisphere. For example, the F atom labelled P is bonded to the two K atoms which project at Q and the two which project at R. On its other side is the hydrogen bond joining it to S. Within the accuracy of measurement, (about 0.1 Å) H lies centrally in the bond. The bond length, SP, is 2.26 Å, as compared with the F diameter of 2.66 Å; the attractive forces in the bond are thus strong enough to produce a very considerable shortening.

Of all known hydrogen bonds, this is probably the nearest approximation to the pure ionic bond.

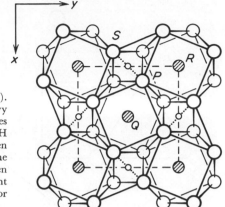

Figure 13-2 Projection of KHF_2 on (001). Hatched circles are K at heights $\pm 1/4$; heavy open circles are F at height 0; light open circles are F at height $-1/2$; small black circles are H at heights $0, -1/2$. Unit cell is shown by broken lines; solid lines (heavy and light) outline the polyhedra round K; dotted lines are hydrogen bonds at height 0. (Hydrogen bonds at height $-1/2$, and all K—F bonds, are omitted for clarity.)

13.8 POTASSIUM HYDROXIDE, KOH

Lattice: Monoclinic, primitive, with $a = 3.95$ Å, $b = 4.00$ Å, $c = 5.73$ Å, $\beta = 103°36'$

Number of formula units per unit cell: 2

Space group: $P2_1$ (No. 4)

Atomic positions:

    K in $2(a)$: x_K, y_K, z_K, with $x_K = 0.175$, $y_K = 0.250$, $z_K = 0.288$

    O in $2(a)$: x_O, y_O, z_O, with $x_O = 0.318$, $y_O = 0.250$, $z_O = 0.770$

The positions of K and O atoms satisfy the requirements of a higher space group, $P2_1/m$. They are shown in Figure 13-3. Each K is surrounded by six O atoms in a distorted octahedron, and all the octahedra share all their edges. The structure is in fact a distorted version of the rock-salt structure. The K—O distances range from about 2.6 Å to 3.2 Å, and the O—O edges from 3.4 Å to 4.6 Å.

We may try to predict the H positions, which have not been directly determined, by the methods of §13.4. The edges projecting at PQ in Figure 13-3 have a length of 3.4 Å, none of the others being less than about 4 Å; these seem likely to be hydrogen bonds (though very long ones). Application of Pauling's rule gives us an electrostatic valence sum of $6 \times (1/6)$ for each O, implying that each must carry one H—which was known already from the chemistry, and independently from the symmetry. If, however, we examine the environment of an oxygen atom, such as P, we find three close cation neighbours on one side—two superposed cations labelled R at 2.8 Å and one cation labelled S at 2.6 Å—and three more distant neighbours on the other—two superposed cations labelled T at 3.0 Å and one cation labelled U at 3.2 Å. These last three define a larger solid angle than do the others; it is therefore here that we should expect to find the hydrogen atom. In fact, the

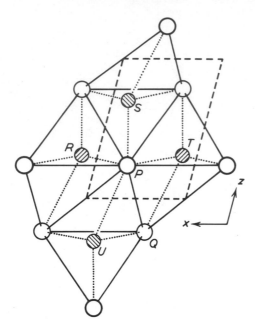

Figure 13-3 Projection of KOH on (010). Hatched circles are K atoms (R and T are at height 1/4, S and U at height 3/4); heavy open circles are O atoms at height 3/4; light open circles are O atoms at height 1/4. The unit cell is shown by broken lines; solid lines are O—O octahedron edges; dotted lines are K—O bonds.

central line of the solid angle coincides roughly in projection with the projection of the short O—O edges such as PQ, which we had already picked out as the hydrogen bonds.

There is no reason to think that H lies midway between two O atoms; that would be most unlikely in a long hydrogen bond. Instead, each O is donor to one hydrogen bond and acceptor for another, and the bonds thus form zig-zag chains parallel to the y axis. Without the hydrogen, the bond is centrosymmetric, and there is no reason to choose one O rather than the the other to represent the donor; when this choice has been made, and H inserted accordingly, the centrosymmetry is destroyed, and the space group is reduced to $P2_1$, in agreement with experiment.

Assuming an O—H distance of 0.8 Å, and a linear O—H $\cdots$ O bond, the predicted parameters of H are approximately 0.395, 0.375, 0.885. Though there are no direct determinations by which these values can be checked, there is no reason to doubt their approximate correctness provided the other atomic positions are correct.

The high-temperature form of KOH, above 248° C, is reported to have the NaCl structure, with a lattice parameter of 5.78 Å. Here, the OH⁻ is at a position of high point symmetry, equidistant from 12 neighbouring OH groups. Either the groups are rotating, and thus effectively spherical, or (perhaps more likely) each OH group directs its H towards one of the 12 neighbours at random, subject to the condition that no two H atoms lie on the same O—O line (cf. §15.18).

13.9 CALCIUM HYDROXIDE, Ca(OH)$_2$

Lattice: Hexagonal, primitive, with $a = 3.59$ Å, $c = 4.91$ Å,
Number of formula units per unit cell: 1

Space group: $P\bar{3}m1$ (No. 164)

Atomic coordinates:

Ca in 1(a): 0, 0, 0.
O in 2(d): $\frac{2}{3}, \frac{1}{3}, u$, with $u = 0.234$
H in 2(d): $\frac{2}{3}, \frac{1}{3}, u'$, with $u' = 0.425$

The structure, shown in projection on (0001) in Figure 13-4(a), is that of CdI$_2$ (described in §4.16) provided we treat OH$^-$ groups as single atoms, i.e., neglect the H atoms. The O atoms lie in close-packed layers on either side of a layer of Ca atoms, which are octahedrally coordinated between them, forming a sandwich sheet. The next sandwich sheet lies vertically above the first, so that all the O atoms are in positions of hexagonal close packing (expanded in the c direction). The H atoms lie vertically above and below the O atoms, on the outside of the sandwich. Thus each O has three equi-distant Ca atoms on one side, and its own H atom on the other, giving it an approximately tetrahedral environment. (The Ca—O—Ca angles are about 100° and the Ca—O—H angles about 120°.)

To see what hydrogen bonds are possible, we look for the shortest O—O contacts not in octahedron edges. We find three, of equal length 3.33 Å, between an oxygen such as that marked in P, Figure 13.4(a), and its neigh-bours Q, R, and S in the next sandwich. If these are bonds (and comparison

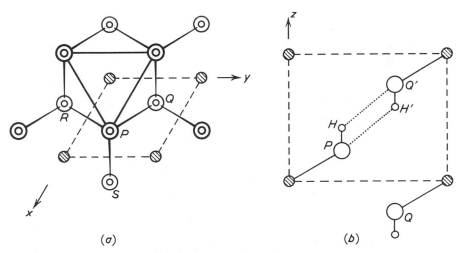

(a) (b)

Figure 13-4 (a) Projection of Ca(OH)$_2$ on (0001). Hatched circles are Ca at height zero; large and small heavy open circles are O and H above Ca; large and small light open circles are O and H below Ca. The unit cell is shown by broken lines. (b) Section of Ca(OH)$_2$ in the plane whose trace is along PQ in (a). Hatched circles are Ca; large and small circles are O and H. Dotted lines are the H-acceptor part of the hydrogen bond. [Note: two other such links from each H lie outside this plane of section.]

with KOH would suggest that they are), they must be trifurcate, but not much stronger than van der Waals bonds. Figure 13-4(b) shows a section parallel to $(11\bar{2}0)$. Between two oxygen atoms P and Q' there are two fractional bonds, $P—H \cdots Q'$ and $Q'—H' \cdots P$, each representing one-third of a bond because its hydrogen atom is equally related to three oxygen atoms. It is more usual, however, to regard the OH groups as nonbonding.

We may ask: What happens to the electron belonging to H when an OH^- ion is formed? Is it given up altogether by its own nucleus, the proton, and allowed to merge indistinguishably in the electron cloud of O? Experimental evidence does not support this. X-ray diffraction analysis has shown that there is a small lump of electron density lying inside the radius of the oxygen atom and near the position of the proton, as determined by neutron diffraction analysis. The lump is not exactly centred on the proton; its position parameter is 0.395 instead of 0.425. (There is no such discrepancy for O, whose position parameter is found to be the same by X-ray and neutron analyses, in agreement with our usual assumption that the electron cloud of an atom is centred round its nucleus.) The distances from O to the peak of electron density and to the proton are 0.79 Å and 0.98 Å respectively. The difference is greater than would be expected from experimental error, but it is qualitatively what we should expect for O—H covalent bond formation.

A number of other hydroxides are isomorphous with $Ca(OH)_2$, notably $Mg(OH)_2$, $Ni(OH)_2$, $Co(OH)_2$, $Fe(OH)_2$, $Mn(OH)_2$, and $Cd(OH)_2$.

13.10 GIBBSITE* Al(OH)₃

This is a structure with individual sandwich sheets resembling those in $Ca(OH)_2$ except that one octahedral site is systematically vacant. In the filled octahedra, the electrostatic valence is $3/6$, in contrast to $2/6$ in $Ca(OH)_2$.

> Lattice: Monoclinic primitive, with $a = 8.62$ Å, $b = 5.06$ Å, $c = 9.70$ Å, $\beta = 94°34'$; $a/b = 1.70 = 0.98\sqrt{3}$
>
> Number of formula units per unit cell: 8
>
> Space group: $P2_1/n$ (No. 14)
>
> Atomic coordinates (all atoms are in general positions):

	x	y	z
Al(1):	$\frac{1}{6} + 0.010$	$\frac{3}{6} + 0.020$	-0.005
Al(2):	$\frac{2}{6} + 0$	0.020	-0.005
O(1):	$\frac{1}{6} + 0.015$	$\frac{1}{6} + 0.035$	-0.110
O(2):	$\frac{4}{6} + 0.015$	$\frac{4}{6} + 0.005$	-0.110
O(3):	$\frac{3}{6} + 0.015$	$\frac{1}{6} - 0.035$	-0.110
O(4):	-0.015	$\frac{4}{6} - 0.035$	-0.110
O(5):	$\frac{2}{6} - 0.035$	$\frac{4}{6} + 0.035$	-0.100
O(6):	$\frac{5}{6} + 0.005$	$\frac{1}{6} + 0.005$	-0.100

---

* Formerly known by the alternative name of hydrargillite.

In spite of the large unit cell and low symmetry, the resemblance to the much simpler $Ca(OH)_2$ structure can be seen. Consider an idealisation of the structure in which $a/b = \sqrt{3}$, $\beta = 90°$, all x and y parameters are multiples of $1/6$, and the z parameters are equal for all O atoms and zero for all Al atoms. This would give the sandwich sheet shown in Figure 13-5(a), which has six octahedra within the area of the unit cell, four filled with Al and two empty. There are two such sheets within the repeat distance c; and operation of the n-glide plane (at $y = 1/4$) means that the second sheet must lie with its lower O atoms exactly above the upper O atoms of the first sheet. This unusual head-to-tail arrangement remains nearly true when we take into account the actual departures from the ideal; it cannot be explained by packing effects, and must be due to hydrogen bonds.

As in previous examples, we look for short O—O contacts which are not edges of Al octahedra. There are nine of these in the asymmetric unit—three head-to-tail between adjacent sandwich sheets, and three each within the top or bottom layers of a sheet, outlining the face of an empty octahedron (shown, for the top layer, by the dotted lines in Figure 13-5[a]); but there are only six H atoms to allocate among them. Moreover, each O must, from its electrostatic valence, have one H attached to it as donor. The placing of the H atoms cannot be done to satisfy these conditions and retain the idealised symmetry; this explains the lower symmetry of the actual structure.

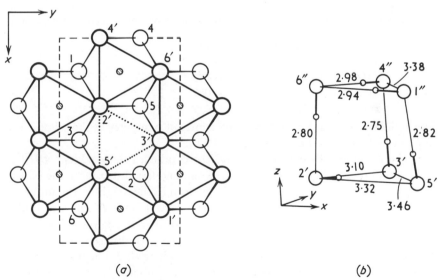

(a)   (b)

Figure 13-5 (a) Idealised sandwich sheet of gibbsite, $Al(OH)_3$. Small hatched circles are Al atoms; large heavy and light open circles are O atoms above and below the layer of Al atoms (at heights about ± 0.10). The numbering matches the text. Broken lines show the unit cell, dotted lines show short O—O contacts which are not octahedron edges. (The H atoms are not shown.)

(b) Perspective view of a group of O atoms joined by hydrogen bonds. Large circles are O; small circles are approximate positions of H. Atoms 2′, 3′, and 5′ are those shown in (a); atoms 6″, 4″, and 1″ belong to the sandwich sheet with Al atoms at height $z \simeq 1/2$. The O—O lengths are marked (in angstroms).

In the actual structure, the nine O—O contacts are of different lengths, ranging from 2.75 Å to 3.46 Å,* as shown in Figure 13-5(b). Taking the five shortest lengths (all less than 3 Å) as hydrogen bonds, and allowing the sixth bond to be bifurcated—from O(2') to O(3') and O(5') in Figure 13-5(b)—we obtain the scheme shown. The requirement of a tetrahedral configuration for O—all Al—O—H angles about 110°—is very roughly satisfied and could be better fitted by allowing the H atoms to move off the O—O lines. Two of the interlayer bonds point in the opposite direction from the third; two intralayer bonds point towards the same acceptor, and the one bond in the other layer is bifurcated between two acceptors. It seems likely that the detailed departures from ideal parameters can be explained in terms of bond-angle requirements and perhaps H—H repulsions between different bonds.

13.11 DIASPORE AlOOH, AND RELATED COMPOUNDS

Diaspore is another compound in which we can see the role of hydrogen bonds in determining the structure.

Lattice: Orthorhombic, primitive, with $a = 4.40$ Å, $b = 9.43$ Å, $c = 2.84$ Å
Number of formula units per unit cell: 4
Space group: $Pbnm$ (No. 62)
Atomic coordinates (all atoms are in special positions 4(c) on mirror planes):

	x	y	z
Al	-0.045	0.145	$\frac{1}{4}$
O(1)	$\frac{1}{4} + 0.038$	$\frac{3}{4} + 0.051$	$\frac{1}{4}$
O(2)	$\frac{3}{4} + 0.053$	-0.053	$\frac{1}{4}$
H	$\frac{1}{2} + 0.090$	-0.088	$\frac{1}{4}$

The b-glide plane is at $x = 1/4$, and the centre of symmetry at the origin; the complete structure is shown in Figure 13-6.

The Al atom is 6-coordinated, and the octahedra share edges. $ABEF$ is the projection of a ribbon of octahedra, extending perpendicular to the paper, with shared edges AE. There is one octahedron in the height of the unit cell, its Al at height 3/4 and its shared edges at 1/4 (and 5/4). $BCDE$ is a similar ribbon, but with Al atoms at height 1/4 and shared edges at 3/4. Each octahedron of the ribbon $ABEF$ shares two edges with different octahedra of the ribbon $BCDE$; they coincide in projection, along BE. The double ribbon is linked by octahedron corners to other double ribbons. The lengths of the shared edges are $AE = BD = 2.54$ Å, $BE = 2.45$ Å, significantly shorter than those of the unshared edges, $AB = ED = 2.80$ Å, $BC = EF = 2.75$ Å, $CD = AF = 2.79$ Å.

---

* This structure has not been refined by modern methods, but the error in bond lengths is not likely to be much more than 0.1 Å.

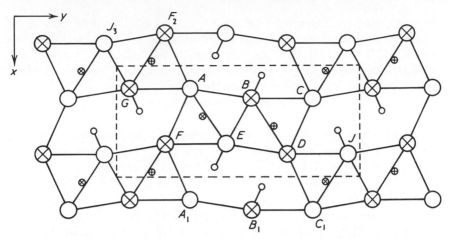

Figure 13-6 Projection of diaspore, AlOOH, on (001). Small circles with upright cross or sloping cross are Al at height 1/4 or 3/4; large open circles and large circles with sloping cross are O at height 1/4 or 3/4; small open circles are H, at the same height as the neighbouring O. Broken lines show the unit cell.

It is easy to predict the position of H. There is one H per formula unit, and two different O—O lengths which are not part of an octahedron edge, namely EA_1, 2.65 Å, and EB_1, 3.09 Å. Obviously the former is the hydrogen bond. To decide whether oxygen atom A or E is the donor, we consider their environments. Each has three cation neighbours, and an electrostatic valence sum of 3/2; we cannot distinguish between them on these grounds.

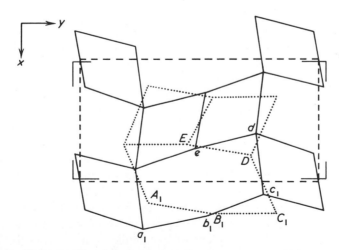

Figure 13-7 Outline of the octahedra of paramontroseite, VO_2, projected on (001) (solid lines), with octahedra of montroseite, VOOH (dotted lines), superposed. Broken lines show the unit cell of paramontroseite; the corners of the unit cell of montroseite are marked. The labelling of atoms A_1, B_1, C_1, D, E of montroseite matches that in Figure 13-6; a_1, b_1, c_1, d, e are the corresponding atoms of paramontroseite.

But the spatial arrangement of the three cations is quite different: E, at height 1/4, has one Al neighbour in $BCDE$ at height 1/4 and two in $ABEF$ at heights −1/4 and 3/4, all lying very much to one side of it; while A, also at height 1/4, has one Al neighbour in AGJ_3F_2 at height 1/4, and two in $ABEF$ at height 3/4, much more nearly coplanar with it. Hence we expect E— which is $O(2)$, like its prototype J—to be the donor, and A—which is $O(1)$, like its prototype C—the acceptor. We therefore predict that the hydrogen atom will lie nearly on EA_1, nearer to E than to A_1. (By symmetry, of course, there will also be one on B_1D, nearer B_1.) A prediction of this kind was in fact made before any direct experimental test was possible. Experiment has confirmed it, and shown that the $O(2)$—H distance is 1.005 Å, inclined at 19° to the line $O(2)$—$O(1)$.

A number of other structures are strictly isomorphous with diaspore: for example, goethite (α-FeOOH), groutite (α-MnOOH), and montroseite (VOOH).

In addition, there are structures of the same family lacking the hydrogen atoms, for example ramsdellite (γ-MnO$_2$) and paramontroseite (a form of VO$_2$ which is structurally quite different from that described in §12.14). Comparison of montroseite and paramontroseite is interesting in showing the difference made to the structure by the hydrogen bond. Both have the same space group and rather similar atomic positions, but the shape of the octahedra constituting the double ribbons has changed, and with it the hinge angle between the ribbons and also the shape of the space not occupied by cations.

Figure 13-7 shows the oxygen array of paramontroseite with part of the montroseite structure (dotted lines) superposed on it. The only big displacement affects the $O(1)$ atoms; for example, those marked A_1 and C_1 for montroseite, a_1 and c_1 for paramontroseite (cf. the parameters in Table 13-3). The $O(2)$—$O(1)$ distances across the cavity are much more nearly equal for paramontroseite—ea_1 and ec_1 are about 3.8 Å and 3.3 Å respectively, where EA_1 and EC_1 were 2.6 Å and 4.4 Å. It is as if the cation-oxygen framework in montroseite were held together by elastic ties such as EA_1 and B_1D, and the removal of these allowed it to take up a new shape in which the oxygens lining the cavity have normal van der Waals contact distances.

Paramontroseite is in fact formed from montroseite without break-up of the structure. But if we picture the change in this way as due to breaking of the hydrogen bond, we have to remember that the energetically favourable situation in the cavity has been achieved at the expense of considerable distortion of the octahedra. If, on the other hand, we had begun by considering the octahedra, we might have predicted the kind of distortion observed, because a change from V^{3+} to V^{4+} implies a decrease of the "ionic" radius and a tendency towards 5-coordination, i.e., a shortening of five of the V—O bonds at the expense of the sixth, which would elongate the octahedron. The change in cavity shape would follow as a consequence. However, we are not really entitled to think of one effect as cause and the

Table 13-3 Comparison of Structures Related to Diaspore

	DIASPORE (AlOOH)	MONTROSEITE (VOOH)	PARAMONTROSEITE (VO$_2$)
Lattice parameters (Å)			
a	4.40	4.54	4.89
b	9.43	9.97	9.39
c	2.84	3.03	2.93
Position parameters			
Cation x	−0.045	−0.052	0.088
y	0.145	0.145	0.143
O(1) x	$\frac{1}{4}$ + 0.038	$\frac{1}{4}$ + 0.051	$\frac{1}{4}$ − 0.144
y	$\frac{3}{4}$ + 0.051	$\frac{3}{4}$ + 0.053	$\frac{3}{4}$ + 0.015
O(2) x	$\frac{3}{4}$ + 0.053	$\frac{3}{4}$ + 0.052	$\frac{3}{4}$ + 0.023
y	− 0.053	− 0.054	− 0.013
H x	$\frac{1}{2}$ + 0.090	?	—
y	− 0.088	?	—
Interatomic distances (Å)			
V—O(1)		1.94	1.88
		1.96 [2]	1.91 [2]
V—O(2)		2.10 [2]	2.00 [2]
		2.10	2.13
O(1)—O(2) (cavity)		2.63	3.87
		∼4.4	∼3.3

other as consequence; both happen together. It is because they cooperate—because the energetically favourable shape of the octahedra and the energetically favourable shape of the cavity require more or less the same set of atomic positions—that the structure can exist. Most probably neither requirement can be perfectly fulfilled; each must adjust slightly to accommodate the other. The greater the adjustment needed, the less stable will be the structure as a whole. There is evidence that paramontroseite is not very stable and it may perhaps only exist when it is formed from the more stable montroseite.

13.12 POTASSIUM DIHYDROGEN PHOSPHATE, KH$_2$PO$_4$ (ROOM-TEMPERATURE FORM)

In this material, the electrostatic valence is higher than in the previous examples—5/4 as compared with 3/6 in Al(OH)$_3$ and AlOOH—and we therefore expect stronger hydrogen bonds with weaker O—H links.

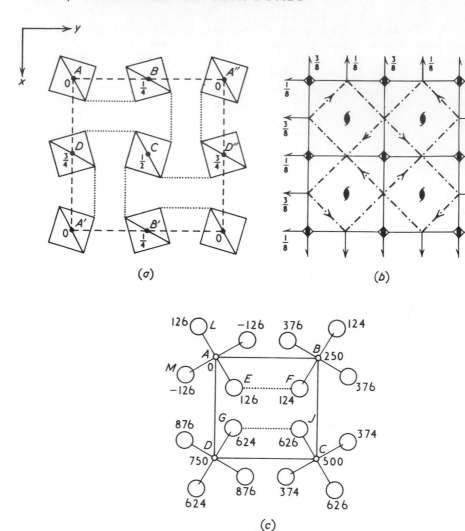

Figure 13-8 (a) Projection on (001) of PO_4 tetrahedra of KH_2PO_4 (room-temperature structure). The upper edge of each tetrahedron is shown; P atoms lie at the centres, at heights marked, and K atoms at height $c/2$ above them. Dotted lines show hydrogen bonds, broken lines the outline of the unit cell. (b) Symmetry elements of (a). (c) Part of projection drawn on a larger scale, showing heights of O atoms in units of $c/1000$. The letters *ABCD* refer to the same points as in (a).

Lattice: Tetragonal, body-centred, with $a = 7.45$ Å, $c = 6.98$ Å
Number of formula units per unit cell: 4
Space group $I\bar{4}2d$ (No. 122)
Atomic coordinates:
 P in 4(a): 0, 0, 0
 K in 4(b): 0, 0, $\frac{1}{2}$
 O in 16(e): x, y, z, with $x = 0.149$, $y = 0.083$, $z = 0.126$
 H in 8(d): $x_H, \frac{1}{4}, \frac{1}{8}$, with $x_H = 0.148$

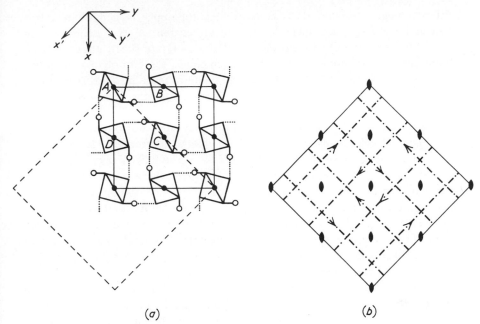

(a)

(b)

Figure 13-9 (a) Projection on (001) of low-temperature structure of KH_2PO_4. Tetrahedra are shown as in Figure 13-8(a), with letters $ABCD$ to correspond. Small circles mark the positions of H. The thin solid line is the outline of the translation-repeat unit; broken lines show the conventional unit cell. (b) Symmetry elements of (a).

A projection of the structure is shown in Figure 13-8(a), and its symmetry elements in Figure 13-8(b). Part of the same projection is shown in more detail in Figure 13-8(c). The structure is built from PO_4 tetrahedra. If we think of them as first all arranged parallel, with their upper edges at 45° to the a and b edges of the unit cell, we must then rotate alternate tetrahedra in opposite directions about the c axis by approximately 15°. They are at

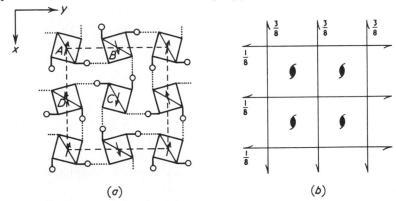

(a)

(b)

Figure 13-10 (a) Projection on (001) of low-temperature structure of $NH_4H_2PO_4$. Small black arrows show the approximate direction of the dipoles; otherwise the figure is drawn in the same way as Figure 13-9(a). (b) Symmetry elements of (a).

different heights, as marked. In consequence, the upper oxygens of one tetrahedron lie at the same level as the lower oxygens of two neighbouring tetrahedra, and the O—O distances (*EF* and *GJ*) are short, about 2.49 Å. These are obvious positions for the hydrogen bond. The complete set of bonds is as shown in Figure 13-8(a). They link the tetrahedra into a three-dimensional framework. The K atoms lie vertically above the P atoms, midway between them. They are 8-coordinated: each has four neighbours belonging to the tetrahedra immediately below and above it, and four belonging to tetrahedra of neighbouring columns at a smaller difference of height. The K—O bond lengths for the two sets are 2.89 Å and 2.82 Å respectively.

Suppose we tried to build this structure, given the size and shape of the PO_4 tetrahedron and the length of the hydrogen bond. We do not lay down any requirements for the lattice parameters a and c beforehand, but wait and see what we get. Without damaging the general tie-up, we could vary the angle of tilt φ of the tetrahedra about their tetrad axes, or the inclination ψ of the hydrogen bonds to the plane of the diagram. By doing so we would alter the shape and size of the site to be occupied by K, as well as the P—O—H bond angles. By choosing suitable values for φ and ψ we would hope to be able to provide K with eight nearly equidistant neighbours, while keeping the P—O—H angles in the range 100° to 120°. This is what is achieved in the actual structure. The values of a and c can be deduced from the dimensions of the PO_4 tetrahedron, the length of the hydrogen bond, and the values of φ and ψ needed to give the correct K—O lengths. If the hydrogen did not form bonds, the K ions could hold the tetrahedra together, but the oxygen atoms are by no means close packed and the structure would be too loose to be very stable.

We have still to think where to place H on the bonds—to decide, for example, whether E is donor or acceptor. This presents a difficulty, because E and F are related by a diad axis at height 1/8 perpendicular to EF, as shown in Figure 13-8(b), and to allot H to E and not to F would destroy the symmetry. If the symmetry is correct, H must lie at the midpoint of each bond. But we believe that centrosymmetric hydrogen bonds are rare (cf. §13.2), so this conclusion needs careful experimental testing before it can be accepted. In fact, experiment shows that there is some abnormality about the H; either the symmetry requirement is only satisfied statistically, with H off-centre towards E or F at random, or H is vibrating with rather a large amplitude in the EF direction about a symmetrical mean position. The former possibility is regarded as the more likely. We shall consider it further in §15.18. For our present purposes, either result means that, placing the hydrogen at its average position at the midpoint of the bond, we must allocate half a hydrogen atom to each oxygen, and write the complete tetrahedral group as $P(OH_{1/2})_4$.

Some of the important interatomic distances are given in Table 13-4. The tetrahedron is very nearly regular, but slightly compressed in the c direction. The angle P—O—H is about 110°.

Table 13-4 Interatomic Distances (Å) in KH$_2$PO$_4$ and H$_2$SO$_4$

	KH$_2$PO$_4^*$			H$_2$SO$_4$
	~293° K	132° K	77° K	
P—O	1.538	1.538	{1.508 P—O 1.583 P—O(H)	1.426 S—O 1.535 S—O(H)
O—O ⊥ to axis	2.528	2.528	{2.549 O—O 2.521 O(H)—O(H)	2.450 O—O 2.419 O(H)—O(H)
Sloping	2.503	2.503	{2.519 O(H)—O 2.512 O(H)—O	2.433 O(H)—O† 2.363 O(H)—O‡
O—H···O	2.487	2.475	2.486	
O—H	1.07	1.07	1.05	

* For KH$_2$PO$_4$, the standard deviation is of order of magnitude 0.001 Å, except for O—H $\cdots$ O, where it is about 0.005 Å.

† Nearly perpendicular to a.

‡ Nearly parallel to a.

Potassium dihydrogen phosphate is a member of a series of strictly isomorphous compounds in which K can be replaced by Rb, Cs, or NH$_4$; P by As; and H by its isotope, D. Geometrically there is very little difference, if any, between members of the series containing H and those containing D.

13.13 POTASSIUM DIHYDROGEN PHOSPHATE, KH$_2$PO$_4$ (LOW-TEMPERATURE FORM)

Suppose that, in Figure 13-8(a), we try to insert our H atoms in an ordered way. We have to abandon our previous demands about symmetry, but we still assume that each short O—O contact, such as *EF*, contains one H atom, and that each PO$_4$ group has two H atoms; the donor and acceptor O atoms must now be distinguished, and the group written PO$_2$(OH)$_2$. There are only two ways of allocating H atoms to satisfy these assumptions: we may attach the hydrogens to the two upper O atoms of each tetrahedron—*E* and *L* in Figure 13-8(c)—or to one upper and one lower O (for example, *E* and *M*). Either initial choice enables us to locate all the rest of the atoms; any other initial choice would only give us the same structure as one of these two, though perhaps in a different orientation.

The first choice turns out to be that adopted by KH_2PO_4 below $-123°$ C. The completed structure is shown schematically in Figure 13-9(a). Its symmetry is shown in Figure 13-9(b). Comparing this with Figure 13-8(b), we see that it retains the original d-glide planes, but its $\bar{4}$ axes are reduced to rotation diads, and all the rotation and screw diads perpendicular to them have disappeared. Though the portion corresponding to Figure 13-8(a) is still a translation-repeat unit, its edges are no longer parallel to the remaining symmetry directions. We have to choose the conventional unit cell as marked by the dashed lines, which shows up the true orthorhombic symmetry.

We can now describe the structure formally.

Lattice: Orthorhombic, all-face-centred, with $a = 10.54$ Å, $b = 10.46$ Å, $c = 6.92$ Å

Number of formula units per unit cell: 8

Space group: $Fdd2$ (No. 43)

Atomic coordinates:*

K in $8(a)$: $0, 0, \frac{1}{2} + w_K$, with $w_K = 0.006$

P in $8(a)$: $0, 0, w_P$, with $w_P = -0.011$

O(1) in $16(b)$: x_1, y_1, z_1, with $x_1 = 0.116$, $y_1 = -0.034$, $z_1 = -0.128$

O(2) in $16(b)$: x_2, y_2, z_2, with $x_2 = 0.034$, $y_2 = 0.116$, $z_2 = 0.128$

H in $16(b)$: x_H, y_H, z_H, with $x_H = -0.037, y_H = 0.187, z_H = 0.131$

We note that the space group is polar, allowing free choice of origin along the z axis; here it is chosen at the geometrical centre of the tetrahedron, which is physically more meaningful as a reference point than is the P atom.

Interatomic distances are given in Table 13-3. The O(2) is the donor atom, forming the OH group. As compared with the high-temperature form, the most striking feature is that the O_4 group is almost unchanged not only in size and shape but even in its angle of tilt about the c axis. There appears to be a very slight increase in the O—H $\cdots$ O bond length and a decrease in the O—H length in the low-temperature structure, but though these changes would agree with the general rule illustrated in Figure 13-1 they are too close to the limits of experimental error to be reliable.

The biggest differences are in the P—O distances: P—O(2), which we might alternatively write as P—(OH), is markedly longer than P—O(1). This is what we should have expected; indeed, we could have used the fact to predict the position of H. The $PO_2(OH)_2$ groups (Figure 13-11) are obviously dipoles, and KH_2PO_4 is thus an example of a polar structure where the dipoles are all in parallel alignment.

It has been shown experimentally that the positive end of the dipole is *that towards which the P atom is displaced* (i.e., downwards in Figure 13-11). This is the direction *away* from the H atoms. If one had argued naïvely from the addition of positive H atoms to a rigid negative PO_4 group, one would have predicted the opposite result. On the other hand, if we think of the

---

* These are the values at $77°$ K, which differ slightly from those at $113°$ K.

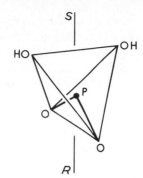

Figure 13-11 Perspective view of PO$_2$(OH)$_2$ tetra-hedron, with displacement of P exaggerated. Direction of dipole is along symmetry axis *SR*.

electron clouds of the oxygen atoms being distorted by the close approach of P, and the charges being redistributed accordingly, the experimental result is more readily predictable. A similar result was found in LiNbO$_3$, where the dipoles are NbO$_6$ octahedra; the positive end is that towards which Nb was displaced. It is the group of oxygen atoms that makes the dipole; the role of the hydrogen is in its effect on the polarisation of the oxygens, in which it cooperates with the central cation P.

A polar structure is of interest as a ferroelectric if its dipoles are reversible. We therefore ask: is it structurally possible for these dipoles to reverse their orientation? They do not need to swing round bodily. It is obviously easy for P to move within the tetrahedron, but movement of H from an upper to a lower corner of the same tetrahedron is much harder. On the other hand, movement of H within a hydrogen bond is not difficult, and this will achieve the same result. Figure 13-12 shows it schematically. The dipole direction is

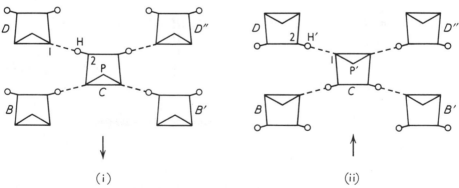

(i)   (ii)

Figure 13-12 Schematic diagram of ferroelectric reversal. The tetrahedra are shown in outline, with their short P—O bonds inserted. The H atoms are shown by small circles, the hydrogen bonds by broken lines. The direction of the dipole is shown by the arrow; it is the negative sense of the z axis. During reversal, from (i) to (ii), P moves to P′, H to H′; and oxygens (1) and (2), while moving very little, interchange their environments and thereby become (2) and (1) respectively.

The letters *C, B, B′, D, D″* refer to the same tetrahedra as in Figure 13-8(a) if the structure is viewed along *A′A″* in that figure.

indicated by the arrow. In (i) it points downward. When a field pointing upward is applied, P moves upward, and in so doing it repels the H atoms at the upper corners, which move out along the hydrogen bond and become attached to the lower corners of adjacent tetrahedra, made available by the upward movement of P atoms in these tetrahedra. Meanwhile, H atoms from other neighbours move in to the lower corners of the original tetrahedron. There are small lateral movements of O atoms, resulting from their interchange of roles as donor and acceptor. Without inquiring in detail into the mechanism of the movement, we can see that the result, propagated from one tetrahedron to the next, is a reversal of direction of the whole structure, as shown in Figure 13-12(ii). Not only does the z axis change its direction, but the structural details associated with the original x and y axis are interchanged, and therefore the names of the axes must be interchanged. The matrix describing the change of orientation in the ferroelectric reversal is thus

$$\begin{pmatrix} 0 & 1 & 0 \\ 1 & 0 & 0 \\ 0 & 0 & -1 \end{pmatrix}$$

Many of the compounds which were isomorphous with KH_2PO_4 at room temperature have a low-temperature form isomorphous with that just described, though the transition temperatures from one form to the other vary over a wide range. On the other hand, $NH_4H_2PO_4$, which was isomorphous with KH_2PO_4 at room temperature, has a different low-temperature form—a different hettotype of the same family—which we shall consider next.

13.14 AMMONIUM DIHYDROGEN PHOSPHATE, $NH_4H_2PO_4$

The room-temperature isomorphism of $NH_4H_2PO_4$ with KH_2PO_4 is an example of how NH_4^+ can play the same structural role as K^+. To a first approximation, it behaves as a spherical atom, with a packing radius slightly larger than that of K^+. Assuming the PO_4 group to remain unaltered, we therefore expect larger unit-cell dimensions to accommodate the larger cation, and a larger tilt of the tetrahedron about [001] in order to keep the hydrogen bond length constant. Both these effects are found. The lattice parameters are $a = 7.51$ Å, $c = 7.56$ Å, and the position parameters of O are $x = 0.146$, $y = 0.085$, $z = 0.115$. The tilt angle is $\tan^{-1} (0.085/0.146)$ instead of $\tan^{-1} (0.083/0.149)$, an increase of about 1°.

In more detail, however, we expect the NH_4^+ group to be tetrahedral. It is capable of adopting a random orientation in some structures (cf. §15.24), but here it is ordered. Four of the neighbouring O atoms form a flattened tetrahedron around the cation, and with these the NH_4 group can form N—H $\cdots$ O hydrogen bonds, which are slightly bent, with N—H = 1.0 Å,

H $\cdots$ O $= 1.95$ Å, and N—O $= 2.91$ Å. The other four O atoms are at a rather greater distance, 3.17 Å. In KH_2PO_4, the difference between the corresponding K—O bond lengths is only 0.07 Å as compared with 0.26 Å here. The concentration of N—O bond strength into the set of bonds less steeply inclined to (001) shortens a relative to c, giving an axial ratio of about 1.01 as compared with 0.93 for KH_2PO_4.

The low-temperature form of $NH_4H_2PO_4$, which is found below $150°$ K, differs from that of KH_2PO_4 in two important ways, which may be seen by comparison of Figure 13.10 and Figure 13.9.

(i) Though its symmetry is again orthorhombic, this is achieved without changing the size or orientation of its unit cell. Whereas KH_2PO_4 loses all its diads in the (001) plane but keeps its d-glide planes, $NH_4H_2PO_4$ loses its d-glide planes but keeps its screw diads and thus has space group $P2_12_12_1$. The symmetry elements of the three different structures of the family may be compared in Figures 13-8(b), 13-9(b), and 13-10(b).

(ii) The ordering of the H atoms follows a different pattern; each tetrahedron in the low-temperature $NH_4H_2PO_4$ structure has one H attached to an upper corner and one to a lower corner. (This is the second possible arrangement referred to in §13.13.) In consequence, the dipole lies approximately in the (001) plane. Figure 13-10(a) shows the complete unit cell. The four dipoles are not arranged in antiparallel pairs in three dimensions, though they appear to be so in projection. Nevertheless, their vector sum is zero (as it must be in this space group). The crystal is therefore *antiferroelectric*, since there are recognisable dipoles adding up to zero by symmetry.*

The reason why this ordering scheme is adopted rather than the simpler scheme of KH_2PO_4 is not immediately obvious. Almost certainly it is due to the part played by N—H $\cdots$ O bonds, but this cannot be properly understood without a detailed determination of the low-temperature structure.

13.15 SULPHURIC ACID, H₂SO₄ (LOW-TEMPERATURE FORM)

This structure is interesting in its resemblances to, and differences from, low-temperature KH_2PO_4.

> Lattice: Monoclinic, C-face-centred, with $a = 8.14$ Å, $b = 4.70$ Å, $c = 8.54$ Å, $\beta = 111.5°$
> Number of formula units per unit cell: 4
> Space group: $C2/c$ (No. 15)
> Atomic coordinates:
> S   in 4(e): $0, y_s, \frac{1}{4}$, with $y_s = 0.075$
> O(1) in 8(f): x_1, y_1, z_1, with $x_1 = 0.990$, $y_1 = 0.276$, $z_1 = 0.338$
> O(2) in 8(f): x_2, y_2, z_2, with $x_2 = 0.160$, $y_2 = 0.920$, $z_2 = 0.322$

---

* It is sometimes assumed that in an antiferroelectric the dipoles must be antiparallel. This example shows that it is not necessarily so.

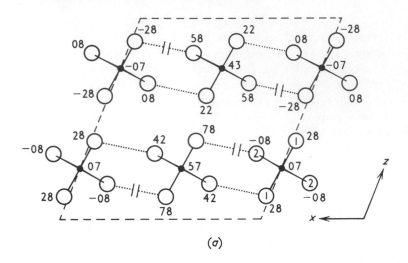

(a)

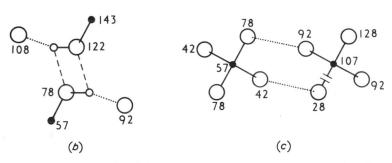

(b)   (c)

Figure 13-13 Sulphuric acid, H_2SO_4 (low-temperature form), projected on (010). Large open circles are O; small black circles are S; heights are in units of $b/100$. The dotted lines show hydrogen bonds; bars across a bond indicate that it ends on an atom at height b above or below that marked. (a) Complete unit cell. (b) Detail of region surrounding the point 1/2, 1, 1/2. Small open circles are H; the O—H link of the hydrogen bond is shown by a solid line, H · · · O by a dotted line, other short O—H contacts by broken lines. (c) Detail of region near $z = 1/4$, showing two columns of tetrahedra joined by pairs of hydrogen bonds into a spiral parallel to [010].

The structure is shown in projection in Figure 13-13(a). There is a diad axis at $x = 0$, $z = 1/4$, and the S atom lies on it. The O atoms form a tetrahedron round it, but S lies below their geometrical centre, giving short S—O(2) bonds of 1.42 Å and longer S—O(1) bonds of 1.54 Å. The shortest O—O distances between tetrahedra, shown by the dotted lines, are 2.63 Å. They link the tetrahedra two-dimensionally into sheets parallel to (001). They are obviously the hydrogen bonds. Though the H atoms have not been found directly, it is clear from the S—O distances that the O(1) atoms are the donors. The next shortest distance, 2.81 Å, connects two sheets through their O(1) atoms. It seems possible that some O—H attraction is effective here, like a weak bifurcation of the bond. Figure 13-13(b) shows this schematically, with possible (approximate) positions for H, displaced a little from the line of the 2.63 Å bond, towards the O of the next layer. Such a bifurcation would account for the fact that 2.63 Å is longer than we should expect for an electrostatic valence of 6/4 (cf. Table 13-1).

The tetrahedra themselves closely resemble those in KH_2PO_4, though they are slightly smaller and their distortion is greater. Distances are given in Table 13-4.

The hydrogen bonds (if we neglect the weaker prongs of the forks and consider only the 2.63 Å distances) resemble those in KH_2PO_4 in joining upper corners of each tetrahedron to lower corners of two tetrahedra at a higher level, but the network as a whole is markedly different. In H_2SO_4, the bonds connect two columns of tetrahedra, shown in Figure 13-13(c), in a two-membered spiral round the diad axis, each tetrahedron taking part in two such spirals; in KH_2PO_4, each spiral involves four columns of tetrahedra, and each tetrahedron takes part in four spirals. By this means, the short bonds in KH_2PO_4 build up a three-dimensional network, while those in H_2SO_4 build up only a two-dimensional sheet.

13.16 SODIUM HYDROGEN CARBONATE, NaHCO₃

This was one of the first crystals in which a short hydrogen bond was observed.

Lattice:* Monoclinic, primitive, with $a = 7.51$ Å, $b = 9.70$ Å, $c = 3.53$ Å, $\beta = 93.3°$

Number of formula units per unit cell: 4

Space group: $P2_1/n$ (No. 14)

Atoms all in general positions, with parameters as follows:

	x	y	z
Na	0.285	0.004	0.713
C	0.077	0.238	0.289
O(1)	0.071	0.367	0.260
O(2)	0.205	0.162	0.195
O(3)	0.941	0.170	0.437
H	0.86	0.25	0.53

The structure is shown in Figure 13-14. There are NaO_6 octahedra, considerably distorted, sharing edges to form ribbons parallel to [001], like the AlO_6 octahedra in diaspore (§13.11); unlike diaspore, however, the ribbons do not share corners with each other. Instead, their corners are

---

* The axes of reference used here are those of the original structure determination by W. Zachariasen, followed by Wyckoff in *Crystal Structures*. The unit cell of later workers (Chapter 13, references 19 and 20) is given in terms of Zachariasen's by the matrix 001/010/$\bar{1}0\bar{1}$.

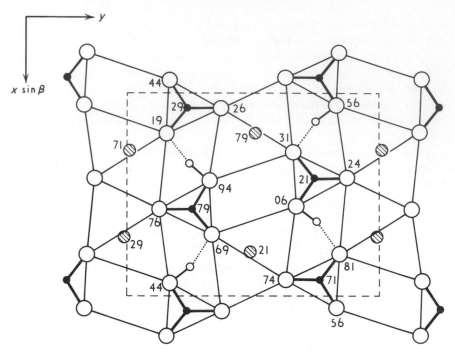

Figure 13-14 NaHCO$_3$, projected down [001]. Large open circles are O; small black circles are C; hatched circles are Na; small open circles are H. Heavy solid lines are C—O bonds; light solid lines are O—O edges of polyhedra, and the O—H links of the hydrogen bonds; dotted lines are the H $\cdots$ O links of the hydrogen bonds; broken lines outline the unit cell. Heights are in units of $c/100$ above (100).

shared with the CO$_3$ groups. The O(1) atom is shared by two octahedra and one triangle, O(2) and O(3) each by one octahedron and one triangle. The CO$_3$ groups are almost exactly planar, and they lie nearly parallel to (101). The bond lengths C—O(1), C—O(2), C—O(3), are 1.25 Å, 1.27 Å, 1.34 Å, respectively. There is only one short O—O distance that is neither an edge of the CO$_3$ triangle nor an edge of the Na octahedron—it is, in fact, shorter than any of the octahedron edges. It is O(2)—O(3), of length 2.61 Å.

Obviously this 2.61 Å length must be the hydrogen bond. Is O(2) or O(3) the donor? Pauling's rule gives no help, because for both O atoms the electrostatic valence sum is $4/3 + 1/6 = 3/2$. However, the difference between C—O(2) and C—O(3) is extremely large, and it clearly points to O(3) as donor. This was confirmed by direct determination of the hydrogen position, giving O(3)—H = 1.07 Å.

From Table 13-1 we notice that, as for H$_2$SO$_4$, the O—O length is greater than would have been predicted from the electrostatic valence; from Figure 13-1, it is also anomalously long relative to O—H. The next shortest distance in the structure is the octahedron edge O(3)—O(1), of length about 3 Å, and it lies at an acute angle to O(3)—O(2). As in H$_2$SO$_4$, it therefore seems that a slight lengthening of the strong hydrogen bond is correlated with the shortening of an adjacent O—O distance.

Instead of considering the structure as built from ribbons of NaO$_6$ octahedra cross-linked by sharing corners with CO$_3$ groups, we might have begun by considering the CO$_3$ groups and the hydrogen bonds only (ignoring the Na—O bonds as weak). They form pleated sheets parallel to (101), the cleavage plane of the crystals. Recognition that the CO$_3$ groups lay nearly in this plane was the clue to the determination of the structure in the first instance. Nevertheless, thinking about it in terms of the NaO$_6$ octahedra probably gives better insight into its general character and relation to other structures.

13.17 POTASSIUM HYDROGEN MALONATE, KHCH$_2$(CO$_2$)$_2$

This is an example of an organic salt whose anion is itself symmetrical, and which incorporates a very short hydrogen bond.

Lattice: Monoclinic, C-face-centred, with $a = 9.473$ Å, $b = 11.56$ Å, $c = 4.73$ Å, $\beta = 91.6°$

Number of formula units per unit cell: 4

Space group: $C2/m$ (No. 12)

Atomic coordinates:

K	in 4(i):	$x, 0, z$, with $x = 0.155$, $z = 0.279$
C (of CH$_2$ group)	in 4(g):	$0, y, 0$, with $y = 0.305$
C (of CO$_2$ group)	in 8(j):	x, y, z, with $x = 0.095$, $y = 0.237$, $z = 0.805$
O(1)	in 8(j):	x, y, z, with $x = 0.177$, $y = 0.305$, $z = 0.666$
O(2)	in 8(j):	x, y, z, with $x = 0.095$, $y = 0.133$, $z = 0.776$
H (of CH$_2$ group)	in 8(j):	x, y, z, with $x = 0.065$, $y = 0.361$, $z = 0.129$
H (bonding)	in 4(e):	$\frac{1}{4}, \frac{1}{4}, 0$

The fact that the general position in this space group is eightfold, while the cell content is 4 formula units, means that K, H, and the C of the CH$_2$ group, must be in special positions. The K atom lies on the mirror plane, and C on the rotation axis. The CO$_2$ groups are in general positions, with one O coordinated to the K atom, the other O linked across the centre of symmetry at 1/4, 1/4, 0 to a similar atom from the next molecule, forming a hydrogen bond. The structure is shown in Figure 13-15.

The hydrogen bond is very short, 2.46 Å. Careful work suggests that it is truly centrosymmetric, with the proton at the centre.

Other compounds with similar short bonds include sodium hydrogen acetate, potassium hydrogen bisphenylacetate, and many other carboxylic acids. In potassium hydrogen bisphenylacetate there is additional evidence that the hydrogen atom is truly central, not only from infrared measurements but also from the fact that the apparent amplitude of thermal vibration, which would be large if there were disorder between two narrowly separated sites, is no greater along the bond than at right angles to it.

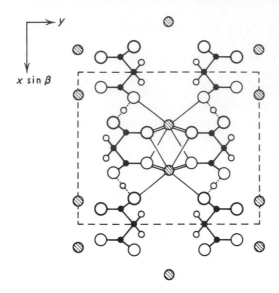

y

$x \sin \beta$

Figure 13-15 KHCH$_2$(CO$_2$)$_2$, projected down [001]. Small black circles are C; small open circles are H; hatched circles are K; large open circles are O (heavy circles indicating a higher level than light ones.) Fine lines show K—O bonds from two of the K atoms; dotted lines are hydrogen bonds. There is a rotation diad at 0, y, 0, a screw diad at 1/4, y, 0, a mirror plane at $y = 0$, and centres of symmetry at 0, 0, 0 and 1/4, 1/4, 0.

13.18 ICE Ih (HEXAGONAL ICE)

We now turn to compounds containing water molecules, and begin with the common form of ice, stable below 0° C and at normal pressures. This is called ice Ih to distinguish it from polymorphs found at different temperatures and pressures. (The numerals II to VII are used to distinguish forms found at increasingly high pressures.)

In spite of the familiarity, importance, and apparent simplicity of ice I, we are not yet sure of all the details of its structure. It is easiest to begin by ignoring the hydrogens and considering only the oxygen atoms. To this approximation it is straightforward, and the formal description is as follows:

> Lattice: Hexagonal, primitive, with $a = 4.523$ Å, $c = 7.367$ Å (at 0° C)
> Number of formula units per unit cell: 4
> Space group: $P6_3/mmc$ (No. 194)
> Atomic coordinates:
> O in 4(f): $\frac{1}{3}$, $\frac{2}{3}$, u, with $u = \frac{1}{16}$ (at $-20°$ C)

The oxygen array is shown in Figure 13-16. It is exactly like that of the Si atoms in high tridymite (§11.13), or like that of all atoms together in wurtzite, the hexagonal form of ZnS(§4.11). Thus, perpendicular to the hexad axis there is a layer of puckered six-rings. Operation of the mirror plane gives a second puckered layer joined to the first by vertical bonds, which help to form other six-rings at right angles to the first set. The vertical six-rings are boat-shaped, and the rings in the layer chair-shaped. Each O atom is surrounded tetrahedrally by four other O atoms. The structure is very open, with a volume of about 32 Å^3 per oxygen. (Its persistence in

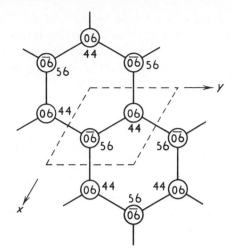

Figure 13-16 Ice I*h* : projection of oxygen array on (0001). The unit cell is shown by broken lines; solid lines are hydrogen bonds. Two layers coincide in projection; heights of atoms in the first layer (in units of *c*/100) are marked inside the open circles representing atoms, and heights of those in the second layer outside the circles.

imperfect form above the melting point explains the contraction of water between 0° C and 4° C; the decrease of volume as the cavities of the framework are gradually filled in outweighs at first the increase due to increasing nearest-neighbour distances.)

The O—O distances in the puckered layer and perpendicular to it, though not required to be equal by symmetry, in fact are not significantly different; both are close to 2.76 Å. Moreover, the two kinds of O—O—O angle, in the puckered layer and in the vertical six-ring, are also equal, within our present accuracy of measurement, to each other and to the ideal tetrahedral angle, 109°28′.

It is obvious that the 2.76 Å distances represent hydrogen bonds. Since there are as many hydrogen atoms as bonds, one might think at first sight that

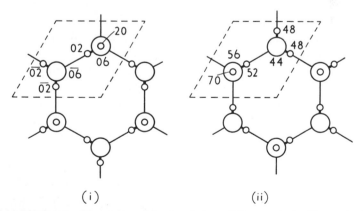

(i)   (ii)

Figure 13-17 Possible scheme of local ordering of H atoms in ice I*h*: projection on (0001). The two puckered layers at different heights are shown separately in (i) and (ii), with heights of atoms marked (in units of *c*/100). Large circles are O, small circles are H. Broken lines outline the unit cell; thick solid lines are O—H bonds; thin solid lines are H · · · O bonds. Axes are as in Figure 13-16.

placing them would be easy. If they could go midway between O atoms, there would be no difficulty. But it was realised from quite an early date that this is impossible; they must belong to one of their two neighbours as donor. Accepting this, if we try to assign two H atoms to every O, and one H to every bond, and keep at least trigonal symmetry in a perfectly ordered structure, it needs a very large number of atoms to make up a repeat unit, and the pattern of the bonds is improbably complicated. We are forced to conclude that disorder of some sort is present.

We may look at the problem in the following way. Each bond contains two possible sites for H, about 1 Å from either end, but only one of them can be occupied. If there is complete disorder, the chance of occupation of either site is the same; hence, statistically, we have half a hydrogen atom in each, and this satisfies the symmetry requirement of the original space group. We can complete our description of the structure as follows:

Half hydrogens in $4(f)$: $\frac{1}{3}, \frac{2}{3}, z_1$, with $z_1 \simeq 0.20$
and in $12(k)$: $x_2, \bar{x}_2, z_2$, with $x_2 \simeq 0.45, z_2 \simeq 0.05$

This model, with completely random distribution of hydrogens among the possible sites, was put forward by Pauling. It can be tested by diffraction methods, which show the average structure over a very large number of unit cells. The results of such work show half-atom peaks near the predicted position.

Accurate location of H atoms by neutron diffraction is always difficult; it is easier to obtain exact measurements with the deuterium analogue, D_2O. We assume that (within present accuracy of measurement) the position parameters of D are the same as those of H; this is reasonable, because the lattice parameters of H_2O and D_2O are almost exactly equal. At $-50°$ C, the O—D distance is 1.01 Å (as compared with O—H $= 0.96$ Å in the H_2O vapour molecule). The D atoms in the puckered layer perpendicular to the hexad axis lie very slightly off their O—O lines—perhaps not significantly so. These D atoms and also those in the vertical bonds have peaks which are somewhat drawn out at right angles to their bonds.

It has been suggested that the H—O—H bond angle of the free molecule, $105°$, is retained unchanged (or little changed) in the solid. If so, the H atoms would lie off the O—O joins, with O—H at an angle of about $2°$ to O—O, but in a disordered way; and this disorder would account for the drawing out of the peaks at right angles to the bonds, which otherwise might be attributed to thermal vibration (cf. §14.3).

Without knowing details, we can predict that if there are changing configurations of disorder, they will contribute to the entropy. This can be calculated for a particular model and compared with experimental results. It was done by Pauling for the model described above, and satisfactory agreement was found.

Pauling's model may be accepted as a good approximation. Nevertheless, it still leaves many questions unanswered. Is the arrangement of H atoms truly random down to the finest scale? Or are there small domains within which the arrangement is ordered? Are there places—perhaps on domain boundaries—where there are three H atoms, or only one, attached to a particular O—or two H atoms, or none, on a particular O—O join? If so, are the positions of such imperfections fixed, or do they move by the jumping of H atoms between neighbouring sites, or by switching of orientation of the H₂O molecules through 110°? Is jumping or switching cooperative, so that movement at one atom facilitates movement at a neighbouring atom? Might such cooperation result in movement of domain boundaries? We may guess at answers to such questions, but it is not easy to get clear experimental evidence.

So far, no model assuming local ordering of H atoms has been found to be an improvement on Pauling's model. One of the simplest hypothetical ordered arrangements is that shown in Figure 13-17, which no longer possesses hexagonal symmetry. Moreover, it is polar: all the O—H bonds perpendicular to the plane of the diagram point upwards. Claims to have observed a polar structure in ice have been made on several occasions, but have generally been refuted by later workers. Disordered orientation of domains could of course explain the absence of a net external dipole moment, as well as the hexagonal symmetry of the average structure; but ordering *within* fairly large domains would still affect the intensity of neutron diffraction in a way which has not been observed experimentally. (For very small domains the effect might be masked; possibilities are still open.)

However, even if disorder is complete in ice I*h*, there remains the possibility of a transition to a lower-symmetry ordered form at some lower temperature, analogous to that in KH_2PO_4. It has been suggested that such a transition may occur at about $-130°$ C, and that the low-temperature form is polar. The evidence for the transition has been questioned, but, if it does exist, the ordered form is different from any of the other polymorphs to be described below, because its array of oxygen atoms and its linkage pattern of hydrogen bonds are the same as those of ice I*h*, whereas all the other polymorphs have different linkage patterns.

13.19 ICE I*c* (CUBIC ICE)

At temperatures below $-80°$ C, a new form of ice can be obtained, which is cubic. It has to be grown directly from the vapour, and cannot be formed by cooling hexagonal ice. Its structure is related to that of hexagonal ice as ideal cristobalite is to ideal tridymite, or as zinc blende is to wurtzite. This means that each puckered layer of oxygen atoms is displaced relative to the one below it in such a way that, while the three lower corners of its rings are vertically above the three upper corners of the previous rings, its three upper corners are vertically above the *centres* of the previous rings. There are still vertical six-rings between layers, but they are now chair-shaped like those in the layers, instead of boat-shaped. The oxygen array is in fact identical with the array of carbons in diamond. As regards the hydrogens, a detailed study by electron diffraction has shown that a model corresponding to Pauling's, with half-hydrogens nearly on the O—O joins, gives the best fit.

The formal description is as follows:

Lattice: Cubic, face-centred, with $a = 6.350$ Å
Number of formula units per unit cell: 8
Space group: $Fd3m$ (No. 227)
Atomic coordinates (with origin chosen so that centre of symmetry is at $\frac{1}{8}, \frac{1}{8}, \frac{1}{8}$):
    O   in $8(a)$:   0, 0, 0
    Half-hydrogens in $32(e)$: x, x, x,   with $x \simeq 0.09$

The O—O distance, 2.75 Å, is almost exactly the same as in hexagonal ice. The O—O—O angles are exactly tetrahedral; but, as in hexagonal ice, there is a spread of H positions normal to the O—O bond which would be compatible with an H—O—H angle of about 105° and orientation disorder of the molecules. The O—H distance is 0.96 ± 0.03 Å.

13.20 ICE II

This is the first of the polymorphs of ice formed with increasing pressure. It is öf interest as the first high-pressure polymorph in which the hydrogen atoms were found to be ordered. A study of their different environments has given useful general information about the bonding requirements of water molecules.

The lattice is rhombohedrally-centred hexagonal, with $a = 13.0$ Å, $c = 6.2$ Å; the space group is $R\bar{3}$ (No. 148), and there are 36 molecules per unit cell, in positions very closely resembling those of the higher space group $R\bar{3}c$ (No. 167). In $R\bar{3}$, the general position is 18-fold; there are thus two independent molecules in the unit cell.

The structure contains puckered six-rings like those of ice I, but instead of being joined into horizontal layers they are cross-connected to others at higher and lower levels, leaving much less empty space between them—as we should expect since it is a high-pressure form. The volume per oxygen atom is about 25 Å³ as compared with 32 Å³ for ice Ih, though the O—O distances in the two structures are 2.81 ± 0.03 Å and 2.75 Å respectively.

If the space group had been $R\bar{3}c$, all rings would have been alike; as it is, they are of two independent shapes, one flattened and one steeper than in ice Ih. This allows much more freedom of adjustment to the O—O—O angles; since there are two independent O atoms in general positions, there are twelve independent angles. The water molecules can thus find ways of orienting themselves which are energetically satisfactory and yet capable of ordered repetition. From inspection of the angles, it proved possible to pick out two angles from the twelve with which to associate the H—O—H angles of the water molecules, and from this to explain the detailed character of the distortion from higher symmetry. The deduction of ordering of hydrogens, made thus on structural grounds, was confirmed by the unusually low value of the entropy.

Detailed examination of the surroundings of the water molecules brought out a point of general importance. The energy of the water molecule depends not only on the configuration of the bonds to which it is donor, but also appreciably on those to which it is acceptor. What is found to be important is the angle made by these acceptor bonds with the symmetry plane of the molecule through the bisector of the H—O—H angle: if the bonds lie in or near this plane the arrangement is energetically favourable, their particular direction in the plane making much less difference. There are plenty of examples illustrating this same effect among the hydrates to be described later in this chapter.

13.21 ICE STRUCTURES IN GENERAL, AND H_2O FRAMEWORKS

There is an important analogy between polymorphs of ice and polymorphs of silica (§11.11). The structural unit in both cases is the corner-sharing tetrahedron, and the variety of linkage patterns which can be built from such a unit is responsible for the variety of polymorphs. The examples of different ice structures we have considered above are only a few of the most interesting or important. Some of the ice polymorphs, like some polymorphs of silica, are high-pressure forms. Some are exactly the same as certain silica structures: we noted the correspondence of ice Ih (hexagonal ice) to tridymite and cubic ice to cristobalite, and another example is the correspondence of ice III (tetragonal) to keatite. Others, such as ice II and ice V, have no silica analogues, nor has an ice analogous to quartz yet been found. Ice VII, with regular 8-coordination of oxygen atoms, and ice VIII, with irregular 9-coordination, have again no silica analogues, though they both retain the tetrahedral arrangement of hydrogen bonds.

In the same way, we should not be surprised to find structures comparable to felspars and zeolites, in which particular framework patterns are stable only when their cavities are filled by suitable spacers—which, for H_2O frameworks, may be neutral molecules. The most striking examples are the gas hydrates, described in the next section.

13.22 GAS HYDRATES

Certain gases with small, relatively inert molecules form hydrates with about six to eight molecules of water per gas molecule. The structure in fact consists of a symmetrical framework of tetrahedral water molecules, with large cavities holding the gas molecules. The framework, taken by itself, is actually isomorphous with the low-density form of silica, melanophlogite (§11.11); the whole arrangement, with its filled cavities, is reminiscent of a zeolite (§12.18).

> Lattice: Primitive, cubic, with $a \simeq 12$ Å
> Contents of unit cell: $6M \cdot 46H_2O$, or $8M \cdot 46H_2O$ (where M stands for the gas molecule)
> Space group: $Pm3n$ (No. 223)
> Atomic coordinates
> $\quad$ O(1) in 6(c): $\frac{1}{4}, 0, \frac{1}{2}$
> $\quad$ O(2) in 16(i): x, x, x, with $x = 0.18$
> $\quad$ O(3) in 24(k): $0, y, z$, with $y = 0.31$, $z = 0.12$
> $\quad$ M(1) in 6(d): $\frac{1}{4}, \frac{1}{2}, 0$
> $\quad$ M(2) in 2(a): $0, 0, 0$

All the O—O bond lengths are about 2.75 Å, and most of the angles nearly tetrahedral, though a few are distorted to about 125°. The volume per oxygen is about 36 Å^3, which is greater than in ice Ih; because of this, and the distorted angles, we should expect the framework to be unstable if all the cavities were left empty. However, provided the 6(d) cavities are filled, it does not matter so much about the 2(a) cavities. In chlorine hydrate, for example, only the 6(d) cavities are filled, and the composition is therefore $(Cl_2)_6 \cdot (H_2O)_{46}$, sometimes written $Cl_2 \cdot 7\frac{2}{3}H_2O$. Bromine hydrate is similar, but in SO_2 hydrate all eight cavities can be filled, giving the composition $(SO_2)_8 \cdot (H_2O)_{46}$ or $SO_2 \cdot 5\frac{3}{4}H_2O$.

13.23 HYDRATES

Water molecules can play different roles in a structure according to whether their negative corners serve as acceptors for hydrogen bonds or are bonded directly to cations. Hitherto, in discussing ice and the gas hydrates, we have considered only the first-mentioned case. A typical example of the second is $MgSO_4 \cdot 4H_2O$, where Mg is octahedrally coordinated by four water molecules and two O atoms from sulphate groups, the O—H bonds pointing outwards away from Mg and linking their water molecules to sulphate O atoms as acceptors. In this sort of configuration, which is typical of many hydrates, the water molecule acts rather like a charge distributor. In other hydrates there may be water molecules playing different roles, some coordinated round cations and some acting as acceptors to hydrogen bonds; an example is $CuSO_4 \cdot 5H_2O$, where four water molecules are of the first kind and one is of the second.

An interesting test has been made (Chapter 13, reference 5) to show whether the hydrogen bonds in such salt hydrates are electrostatic in character. The water molecule was assumed to consist of a charge of -1 at the centre of its oxygen atom and two charges of $\pm 1/2$ at distances of 0.97 Å from the centre in directions making an angle of 109.5° with each other. This molecule was placed with its oxygen atom at the correct position in an accurately determined structure, and allowed to vary its orientation until its electrostatic energy with respect to the rest of the structure was a minimum. The hydrogen positions thus predicted agreed excellently with those observed, confirming the hypothesis that the energy is essentially electrostatic.

When we want to consider the complete environment of the water molecule—the bonds it accepts as well as the bonds it donates—the model just described is not as helpful as the tetrahedral model discussed in §13.3 and illustrated in the structures of ice (though it may be regarded as an approximation to the tetrahedral model). Configurations which are not tetrahedral do occur quite commonly, but they are generally recognisable either as distorted tetrahedra, or (when there are three bonds only) as flattened tetrahedra with one corner missing, or as plane triangles. The three-bond configurations in particular illustrate the effect mentioned in §13.20—that it is the symmetry plane of the molecule bisecting the H—O—H angle which is important, bonds to the negative end of the molecule tending to be in this plane whether or not they make tetrahedral angles with the other bonds. These generalisations apply whether the water molecule is bonded to a cation or acts as acceptor to a hydrogen bond from an acidic or alkaline OH group or another water molecule.

The bonds to which the water molecule is donor are generally medium-long, 2.7 Å to 2.9 Å, but they may be bifurcated (and rather longer in consequence), or there may even be a particular hydrogen atom which does not participate in any bond. The possibility of existence of a particular hydrate depends on whether a packing can be found for the water molecules which satisfies the coordination requirement of the cations and is reasonably satisfactory for their own bond-length and bond-angle requirements. The network of hydrogen bonds thus shows a great variety of patterns in different structures. Though it is generally fairly simple in its principles, it is often complicated to describe in detail.

The same rules seem to apply to water in the cavities of zeolites as to other hydrates. There is evidence that in various structures, including chabazite, gismondite, and brewsterite, the cavity cations—Na, Ca, Sr—tend to be surrounded as far as possible by water molecules, which are linked by hydrogen bonds to other water molecules and to the O atoms of the framework.

Hydrates of strong acids have special features, which will be discussed in §13.27.

13.24 ALUM, MAl(SO$_4$)$_2$·I2H$_2$O

The alums are a group of hydrated double sulphates; M is a fairly large monovalent cation, for example, Na, K, Rb, NH$_4$, or Cs; and Al may be replaced by other trivalent cations, such as Cr, Fe, Ga (though we shall consider only compounds of Al here). They are fairly typical examples of

Table 13-5 Atomic Position Parameters in Alums (In Units of 10^{-2})

	Na ALUM (γ)			Rb ALUM (α)			Cs ALUM (β)		
u_s	27			31			33		
u_1	33			25			26		
	x	y	z	x	y	z	x	y	z
O(2)	30	28	15	32	26	42	28	34	44
O(W1)	08	04	32	04	14	30	−16	05	28
O(W2)	−04	06	14	02	−01	16	00	00	15
H(1)	12	08	32				−18	14	28
H(2)	12	−01	31				−22	01	27
H(3)	00	05	20				06	−01	19
H(4)	−11	09	16				−05	03	20

hydrated salts, and are also of interest because of differences due to the size of the M cation.

Though all alums are cubic and possess the same space group, there are nevertheless three different structures, which have been called α, β, γ; the order of increasing M-cation size* is γ, α, β. Representative examples are sodium aluminium alum, rubidium aluminium alum, and caesium aluminium alum. Their description is as follows:

> Lattice: Cubic, primitive, with $a \simeq 12$ Å
> Number of formula units per unit cell: 4
> Space group: $Pa3$ (No. 205)
> Atomic coordinates:
>   Al   in $4(a)$: 0, 0, 0
> M in $4(b)$: $\frac{1}{2}, \frac{1}{2}, \frac{1}{2}$
>   S   in $8(c)$: u_s, u_s, u_s
>   O(1) in $8(c)$: u_1, u_1, u_1
>   O(2), O(W1), O(W2), and H(1) to (4): all in general positions $24(d)$:
> x, y, z.

Values of the parameters are given in Table 13-5.

In all the alums, Al is surrounded octahedrally by water molecules O(W2). Water molecules O(W1) are grouped round M; in α-alum and γ-alum they form the complete environment of M, a somewhat distorted octahedron, but in β-alum they lie almost in a plane hexagon, above and below which sulphate oxygens make contact with the large cation. On the other hand, α-alum and β-alum differ from γ-alum in the orientation of the SO_4 groups; in the α and β forms, the S—O(1) bonds point away

---

* Although this is true for the examples listed here, it may not hold more generally, e.g., for alums in which Cr replaces Al.

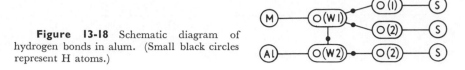

Figure 13-18 Schematic diagram of hydrogen bonds in alum. (Small black circles represent H atoms.)

from the nearest M atom, in the γ form they point towards it. The β-alums and γ-alums are ordered, but all α-alums studied so far are thought to show some disorder. This, if true, is explained by the fact that an octahedral environment is too loosely fitting for large ions such as Rb or K. What has been suggested is that some SO_4 groups are inverted, displacing the nearest M and thereby giving it additional (though slightly more distant) neighbours. The disordered fraction is about 10 percent for Rb alum and 30 percent for K alum.

Another interesting difference is that, though in Na alum the $Al(H_2O)_6$ octahedron is very regular, its orientation is not parallel to that of the unit cell but rotated by about 40° round the cube body diagonal; by contrast, in the α-alums and β-alums, it is oriented nearly parallel to the unit cell, but it is less regular—particularly in α-alums. This sort of difference, where regularity of the octahedron shape is associated with freedom to tilt, and symmetrical orientation with some distortion of shape, is reminiscent of what we have noted in perovskite structures in comparing $NaNbO_3$ and $KNbO_3$ (§12.10).

The hydrogen bonds in the three structures can all be represented by the same general scheme, shown in Figure 13-18. The individual atoms linked by them are, however, differently selected from the equipoint in the different structures, and the *directions* of the bonds and their whole patterns in space are therefore different. In spite of this, the bond *lengths* between corresponding types of atoms are closely similar in all three. They are listed in Table 13-2. The two bonds emanating from $O(W2)$, coordinated round Al, are markedly shorter than those from $O(W1)$, coordinated round M. This is an example of the increasing strength of the hydrogen bond with increasing electrostatic valence of the donor's cation neighbour, which was discussed in §13.2.

The O—D bond lengths as determined by neutron diffraction do not vary significantly with O—O length (Table 13-6); the difference between Na alum and Cs alum is perhaps significant, and is in the direction we should expect from the lower polarising power of Cs as compared with Na. On the other hand, the X-ray values for the distance to the maximum of electron density associated with the proton are shorter than the neutron values for the distance to the proton itself, and most obviously so for $O(W1)$, which has the cation neighbour of lower electrostatic valence (Table 13-1), and is therefore less strongly polarised. These results—and others like them, for many hydrates—suggest that when polarisation by the donor's neighbours is low, the electron cloud of H is more strongly distorted by the donor itself, and displaced away from the proton towards the centre of the donor; higher

Table 13-6 O—H Distances in Alums

DONOR ATOM	CATION NEIGHBOUR	MEAN $O(W)$—O	MEAN $O(W)$—H or $O(W)$—D	
			BOND LENGTHS (Å)	
			Neutron	X-ray
Na alum				
$O(W1)$	Na	2.78	1.03	0.80
$O(W2)$	Al	2.63	1.01	0.97
Cs alum				
$O(W1)$	Cs	2.79	0.96	—
$O(W2)$	Al	2.63	0.98	—

polarisation not only repels the proton away from the donor but also tends to separate the electron clouds of donor and H and bring the electron-density maximum of the latter into closer association with its own nucleus.

The interbond angles in Cs alum are representative of effects in many hydrates. Water molecule (1) has a nearly regular tetrahedral configuration, comprising its bond to Cs, the two hydrogen bonds to which it is donor, and the one it accepts from water molecule (2). Water molecule (2) has a nearly plane triangular environment, its bond to Al lying almost in the same line as the bisector of the angle between the two hydrogen bonds to which it is donor. O(2) has also a plane triangle for the two hydrogen bonds to which it is acceptor and its bond to S, whereas O(1) has a slightly elongated tetrahedron for three hydrogen bonds (related by the triad axis) and its bond to S. Both water molecules have H—O—H angles of about 107°, and the O—H directions do not differ by more than a few degrees from the O—O directions of their bonds.

13.25 TRONA (SODIUM SESQUICARBONATE), $Na_2CO_3 \cdot NaHCO_3 \cdot 2H_2O$

This material (common in domestic use as bath salts) is of interest for illustrating the occurrence of different kinds of hydrogen bonds in the same material, and also in comparison with sodium bicarbonate, $NaHCO_3$.

Lattice: Monoclinic, C-face-centred, with $a = 20.3$ Å, $b = 3.5$ Å, $c = 10.3$ Å, $\beta = 106\frac{1}{2}°$
Number of formula units per unit cell: 4
Space group: $C2/c$ (No. 15)
Atomic coordinates:
  Na(1) in 4(e): 0, y, 1/4, with $y = 0.75$
  H(3) in 4(a): 0, 0, 0

All other atoms in general positions 8(*f*), with parameters as follows:

	x	y	z
Na(2)	0.15	0.16	0.43
C	0.09	0.26	0.10
O(1)	0.15	0.37	0.10
O(2)	0.05	0.14	0.99
O(3)	0.07	0.26	0.21
O(*W*)	0.21	0.67	0.35
H(1)	0.19	(0.57)	0.26
H(2)	0.26	(0.75)	0.36

(The *y* parameters of H(1) and H(2) are estimated by interpolation, not measured).

The atoms O(1), O(2), and O(3) are the three corners of the carbonate group, which is effectively planar. As with $NaHCO_3$ (§13.15), however, we may more usefully begin by considering the coordination of Na. The Na(1) atom, lying on the rotation diad axis, is octahedrally coordinated by O(2) and O(3). These octahedra, shown at *A* in Figure 13-19(b), share edges between corners marked 3 to form the central chain of a ribbon running parallel to [010]. The Na(2) atom, in a general position, has also six neighbours, but arranged in a trigonal prism, as at *B* or *C* in Figure 13-19(b). Prisms share the edges marked 1—*W* with each other to form a chain parallel to [010]; the chains at *B* and *C* are related by the rotation diad axis, and share corners 2 and 3 (but not edges) with the chain at *A* to form the triple ribbon shown. The heights of the atoms are shown in Figure 13-19(c). Both the octahedra and the prisms have unshared edges parallel to [010] of length *b*, which is 3.5 Å.

Through the carbonate groups, which share edges 1—2 with the prisms, the ribbons are connected into sheets parallel to (100), as shown in Figure 13-19(a). Adjacent sheets are related by diad screw axes such as that at $x = 1/4$, $z = 1/4$.

Two of the six neighbours of Na(2) (coinciding in projection) are water molecules (shown as circles). Each water molecule is tetrahedrally surrounded, as shown in Figure 13-19(d), by two Na(2) atoms (hatched circles) and two O(1) atoms (small black circles). The O(*W*)—O(1) distances are 2.74 Å and 2.78 Å, and obviously represent hydrogen bonds. The four bonds round the diad screw axis are not coplanar, but form a spiral. One pair lies within the tightly bonded sheet, the other serves as the only link between sheets (apart from van der Waals forces). The hydrogen positions, found independently by X-ray and neutron diffraction, confirm this deduction. The two independent O—H lengths are nearly equal, with mean values of 1.01 Å from neutron measurements and 0.90 Å from X-ray measurements; this difference is another example of the effect, discussed in §13.9 and §13.24, of the displacement of the electron-density maximum from its proton when the H_2O molecule is not highly polarised by its cation neighbours. The angles,

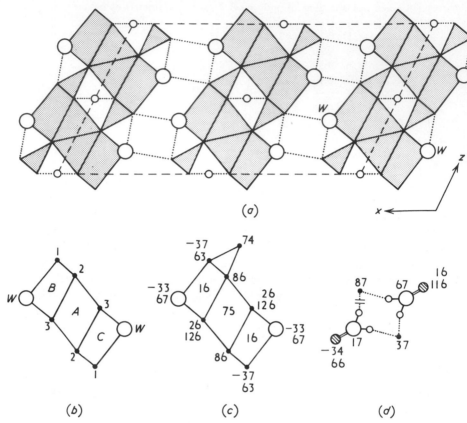

Figure 13-19 (a) Projection of trona on (010). Broken lines outline the unit cell. Shaded areas are polyhedra round Na, and CO_3 groups; large circles are water molecules; small circles are H atoms not belonging to water molecules; dotted lines are hydrogen bonds. (b) Detail of part of (a) near $x = 0$, $z = 1/4$, showing triple ribbon of Na polyhedra. The area marked A is an octahedron round Na (1), those marked B and C trigonal prisms round Na(2). Points marked 1, 2, 3, and W are the atoms O(1), O(2), O(3) of carbonate groups, and a water molecule, respectively. (c) Same part of structure as (b), showing also one CO_3 group, with heights of atoms in units of $b/100$. (Note that some atoms separated by translation-repeat b form part of the same polyhedron.) (d) Detail of part of (a) near $x = 1/4$, $z = 1/4$. Heights of atoms are marked as in (c). Hatched circles are Na(2), (two being superposed in projection); large circles are O(W); small circles are H atoms of water molecules. Bars across a dotted line indicate that the bond ends on the atom in the next unit cell above or Below. Small black circles in (b), (c), and (d) are centres of O atoms; C atoms are not marked.

like the lengths, are in the usual range, with 114° for O(1)—O(W)—O(1) and 109° for H—O—H.

So far, there are no new features in this structure—but we have not yet considered how to place the acidic hydrogen H(3). Since there are only four such H atoms per unit cell, they must be in special positions. The only short O—O contact hitherto unmentioned is that between O(2) atoms through the centres of symmetry, of length 2.50 Å.; obviously it carries the H(3) atoms.

As with KH_2PO_4 (§13.12), the question arises whether they are truly central or only statistically so. Neutron diffraction in this case shows that they are actually off-centre, the O—H distance being 1.12 Å. The donor O(2) has, as its other neighbours, one C, one Na(1), and one Na(2) atom, arranged roughly tetrahedrally. The relations between the electrostatic valence of the C—O bond (4/3), the O—O distance, and the O—H distance, fit smoothly into Table 13-1 and Figure 13-1, without any of the anomalies which appeared for $NaHCO_3$.

13.26   OXALIC ACID DIHYDRATE, (COOH)₂·2H₂O

This structure has for many years been a "crystallographer's guinea-pig"—an important structure which is re-investigated every few years when improving techniques suggest that better refinement can be obtained. More recently its fully deuterated analogue, $(COOD)_2 \cdot 2D_2O$, has also been studied. It occurs in two polymorphous forms, α and β, of which α is isomorphous with the protonated compound, $(COOH)_2 \cdot 2H_2O$. For brevity, we may call the three structures α-POX, α-DOX, and β-DOX. The two α structures resemble one another very closely; and though there are rather greater differences between α-DOX and β-DOX, they are chiefly in the lattice parameters and bond angles, and hardly at all in the interatomic distances. We shall see below how this comes about.

The formal description of all three forms is the same; they differ only in the actual values of the parameters.*

Lattice: Monoclinic, primitive, with parameters as in Table 13-7
Number of formula units per unit cell: 2
Space group: $P2_1/n$ (No. 14)
Atomic positions: All atoms in general positions 4(e), with parameters as in Table 13-7

Since there are only two oxalic acid molecules per cell, and the general position is fourfold, they must lie at the only special positions, centres of symmetry; the molecules are therefore centrosymmetric. They are also very nearly planar. They lie with their C—C bonds approximately in the [10$\bar{1}$] direction. The two molecules are related by the n-glide plane at height 1/4,

---

* Here we keep the conventional (morphological) choice of unit cell and the numbering of atoms used in all the earlier work on $(COOH)_2 \cdot 2H_2O$. For the original work on $(COOD)_2 \cdot 2D_2O$ a different unit cell was used, derived from the conventional one by the matrix 101/010/100, and giving the space-group symbol as $P2_1/a$; also the origin was moved by $\frac{1}{2} c_{new}$, and the numbering of atoms O(1) and O(2) was interchanged.

There is, however, no valid reason for the change. The relationships of the α and β structures are shown up more clearly if the conventional unit cell is retained. There is no special virtue in choosing a unit cell so that the space group has the symbol $P2_1/a$ rather than $P2_1/n$. The rather large β angle of the conventional cell is not a serious disadvantage.

Table 13-7 Comparison of Structures of Oxalic Acid Dihydrate

LATTICE PARAMETERS

	α-POX	α-DOX	β-DOX
a (Å)	6.12	6.15	5.15
b (Å)	3.61	3.61	5.05
c (Å)	12.06	12.10	11.98
β (°)	106.3	106.6	124.4
Long diagonal of unit cell (Å)	14.9	15.0	15.5
Angle between long diagonal and c-axis	22.8°		15.9°

ATOMIC POSITION PARAMETERS

	α-DOX			β-DOX		
	x	y	z	x	y	z
C	$\bar{1}$.955	0.055	0.051	$\bar{1}$.955	0.022	0.050
O(1)	0.085	$\bar{1}$.939	0.148	0.082	$\bar{1}$.848	0.146
O(2)	$\bar{1}$.782	0.230	0.036	$\bar{1}$.785	0.200	0.038
O(W)	$\bar{1}$.550	0.614	0.181	$\bar{1}$.520	0.420	0.172
D(1)	0.026	0.008	0.217	0.028	$\bar{1}$.880	0.215
D(2)*	0.429	0.685	0.115	0.648	0.328	0.151
D(3)*	0.641	0.445	0.152	0.406	0.553	0.106

* In the published work, the numbering of corresponding atoms in the two structures, determined by different authors, has unfortunately been interchanged.

and are thus identical in the (010) projection, shown in Figure 13-20(a). The water molecules are in general positions, and all are symmetry-equivalent. The b axis (as in trona, §13.25) is so short that no atoms overlap in projection.

Consider first α-POX. There are three short distances between the water molecule O(W) and the carboxylic oxygens O(1) and O(2), shown by dotted lines; O(1)—O(W) is 2.51 Å, and the two O(2)—O(W) distances are 2.86 Å and 2.88 Å. We have two reasons for expecting the acidic hydrogen, H(1), to lie on O(1)—O(W): not only is this bond very much shorter than the others, but also C—O(1) is significantly longer than C—O(2), 1.29 Å as against 1.21 Å. (Early workers had considerable difficulty in measuring the C—O distances accurately enough to establish this point. It is much harder to obtain accurate interatomic distances in soft organic materials like this than in the hard oxides of which we had many more examples in this

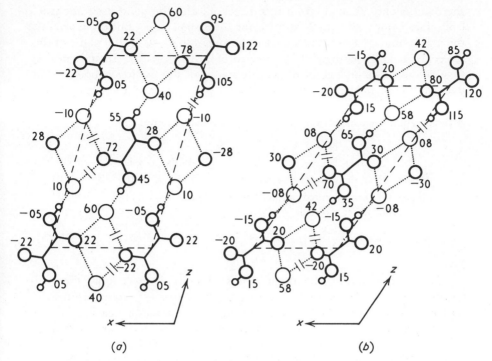

(a)   (b)

Figure 13-20 Oxalic acid dihydrate, projected on (010): (a) α-(COOH)₂·2H₂O, (b) β-(COOD)₂·2D₂O, on the same scale. Broken lines outline the unit cell. Oxalic acid molecules are shown by heavy lines, with medium large circles marking O atoms, and small circles H(1) atoms; large light circles are water molecules.

Heights of atoms are marked in units of $b/100$. Dotted lines are hydrogen bonds; bars crossing them indicate a bond ending on an atom in the unit cell above or below that whose height is marked. The placing of H(1) atoms is only schematic and not to exact scale; H(2) and H(3) are not shown.

book. This is because of the much larger thermal vibrations of the soft materials—see Chapter 14.)

The water molecule provides the hydrogens, H(2) and H(3), for the two longer bonds of 2.86 Å and 2.88 Å, as well as acting as acceptor for the short 2.51 Å bond. The O(1)—H(1) distance, 1.03 Å, is significantly longer than either of the two O(W)—H distances, both equal to 0.95 Å.

This association of high polarising power of the donor's neighbour with long O—H and short O—O distances is again in agreement with the generalisations of Figure 13-1 and Table 13-1(a).

The isomorphous deuterium compound, α-DOX, is very closely similar, with O—D bond lengths of 2.52 Å, 2.88 Å, and 2.91 Å—marginally longer than in α-POX.

Turning to β-DOX, shown in Figure 13-20(b), we find that the oxalic acid molecule is just the same as in α-DOX and α-POX, but its orientation has changed. Its C—C axis lies even closer to [10Ī]—the (010) projection of the bond is only about 1° from this direction, and the bond itself about 9° from it, as compared with 2° and 15° for the α structures. But the tilt about the axis is much greater, the line at right angles to C—C in the plane of the molecule making an angle of 53° instead of 26° with (010). In spite of this very considerable change of tilt, the topology of the hydrogen bonds—the pattern of their linkages—remains the same as before, and their lengths are only slightly different—2.54 Å, 2.85 Å, and 2.82 Å. What has changed most is the angular relationship of the bonds, as shown below:

	α-POX	α-DOX	β-DOX
H—O—H angle of water molecule	106°	106°	110°
Angles between accepted and donated bonds of water molecule	⎰113° ⎱118°	114° 119°	130° 119°
Sum of angles	337°	339°	359°

In the α-structure, the three bonds to O(W) are roughly directed towards three corners of a tetrahedron; in the β-structure, they are nearly coplanar, as shown by the fact that the sum of the angles is nearly 360°.

Why should there be these two different forms of DOX and why does POX never adopt the β form? The answer to the first of these questions is not known. Under apparently identical conditions, sometimes the α form is obtained, sometimes the β. All we can say with certainty is that the energy difference between the two cannot be great, and once crystallisation in a particular form has begun, it continues in that form. To the question why POX never adopts the β form we can give a tentative answer by trying to see why the planar bond configuration should more easily be adopted by D_2O than by H_2O. An O—D · · · O bond is generally slightly weaker than the corresponding O—H · · · O bond, because of its different zero-point energy; this fits in with the observation of longer bonds in α-DOX than in α-POX. Weaker accepted bonds mean less polarisation of the water molecule and less concentration of negative charge at the corners of a tetrahedron, hence less tendency to direct the accepted bond towards one such corner when the other is not involved in a bond. It is an accidental feature of the particular linkage pattern that very nearly the same bond lengths can be achieved by minor adjustments of position of the water molecule and a change of orientation of the oxalic acid molecule, and this allows two different angular environments to the water molecule without destroying the overall pattern.

The changes in lattice parameters between α-DOX and β-DOX, striking though they are, are a simple consequence of the large change of molecular tilt. The changes between α-POX and α-DOX are very much smaller, but even these can be affected by minor changes of tilt, and so do not measure directly the differences of bond lengths.

13.27 HYDRATES OF STRONG ACIDS

In these compounds, the acidic hydrogen is no longer attached to the anion as donor, but adds itself to a water molecule. The oxonium ion, H_3O^+, behaves structually in the same way as H_2O or OH^-, except that, being highly polarised by its anion neighbours, it tends to form shorter hydrogen bonds than does the water molecule. One of the first compounds in which its presence was predicted was perchloric acid hydrate, $HClO_4 \cdot H_2O$. Since the H atoms are all directly attached to one O which does not belong to the ClO_4 group, the formula ought really to be written $(H_3O)^+(ClO_4)^-$, and the compound renamed oxonium perchlorate. Other examples where this sort of renaming is needed will occur later in the discussion. It is analogous, for the acids, to the sort of renaming which has become accepted for alkaline and for amphoteric compounds; for example, we no longer use $Al_2O_3 \cdot H_2O$ or $AlO_3 \cdot 3H_2O$ as a structural formula, but rather $AlOOH$ or $Al(OH)_3$, to match the structural reality.

In oxonium perchlorate, none of the (H_3O)—O distances is particularly short, and it is suggested that the group may be rotating or may have some kind of orientation disorder; alternatively, all the bonds from it might be bifurcated. More commonly, as already noted, the H_3O group participates in very short bonds. In some cases, which we shall consider in detail later, each of two O atoms joined by a very short bond carries two H atoms, which they donate to slightly longer bonds; if the H atom in the very short bond is central, as seems likely, the group constitutes in effect an $(H_5O_2)^+$ molecule-ion. It is possible that molecule-ions more complex than this can be built up in the same way, by the formation of very short bonds between the highly polarised O atoms.

13.28 NITRIC ACID TRIHYDRATE, HNO₃·3H₂O

This structure is the first in which the presence of an oxonium ion, H_3O^+, was directly observed in an electron-density map. The lattice is orthorhombic, primitive, with $a = 9.5$ Å, $b = 14.7$ Å, $c = 3.4$ Å, and there are four formula units per cell. The space group is $P2_12_12_1$ (No. 19). There are no special positions in this space group; the general position is fourfold; hence the formula unit is the asymmetric unit, and we have to locate one N atom, six O atoms, and seven H atoms. Figure 13-21 shows the structure in projection on (001). The c axis is so short that there are no overlapping atoms. Three O atoms (numbered 1, 2, 3) form with N the nitrate group, which is planar, but tilted with respect to the plane of projection. The other three O atoms (4, 5, 6) are connected by a three-dimensional network of hydrogen bonds to each other and to the O atoms of the nitrate group. There are

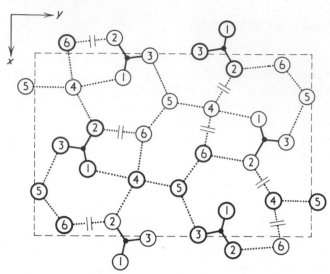

Figure 13-21 Nitric acid trihydrate, projected on (001). The O atoms of NO_3 and H_2O are shown by large open circles, with symmetry-related atoms numbered alike; heavily outlined circles are at heights above $c/2$, and light circles below $c/2$. The N atoms are small black circles, and H atoms are omitted. Dotted lines are hydrogen bonds, interrupted by bars when they end on atoms in the next unit cell above or below.

seven such bonds in the asymmetric unit, four of them ending on nitrate O atoms.

Even though the positions of the hydrogen bonds are easily identifiable from the O—O distances, it would not be easy, in this structure, to predict the positions of the hydrogen atoms on the bonds. Experimentally, it was found that none of them are attached to nitrate O atoms. Both O(4) and O(6) represent water molecules, and O(5) an oxonium ion. The three bonds from O(5), of length 2.49 Å, 2.57 Å, and 2.62 Å, are significantly shorter than those from O(4) and O(6), which range from 2.73 Å to 2.81 Å. The O—H distances—or rather, the distances from O to the electron-density maxima associated with H—have a mean value of 0.9 Å, and though some differences were observed, they were not significant when compared with possible experimental error. The O(4) atom has a tetrahedral environment, with two donated bonds and two accepted bonds; O(6), with two donated bonds and one accepted bond, and O(5), with three donated bonds, may both be classified as "tetrahedral-with-one-corner-missing." The structural formula is $(H_3O)^+(NO_3)^-\cdot2H_2O$.

13.29 HYDRATES OF HYDROCHLORIC ACID, HCl·H₂O, HCl·2H₂O, AND HCl·3H₂O

This series of compounds illustrates some of the effects mentioned in §13.27.

The *monohydrate*, HCl·H₂O, has a rhombohedral lattice, with $a = 4.05$ Å, $\alpha = 73°30'$, and one formula unit per unit cell. The structure is disordered, with space group $R\bar{3}m$ (No. 166) for the average structure; it is easiest to describe it, however, by picking out regions of local order which, if repeated perfectly, would lead to the space group $R3m$ (No. 160). Then the atomic coordinates are as follows: Cl at 0, 0, 0; O at u, u, u, with $u = 0.44$; H at x, x, z; x, z, x; z, x, x; with $x = 0.31$, $z = 0.65$. The three hydrogen atoms thus form the base of a rather flat trigonal pyramid, with O at the apex, having H—O—H angles of 117°, and O—H distances of about 0.96 Å. This group must be interpreted as an H_3O^+ group, and the formula of the compound should therefore be written $(H_3O)^+Cl^-$. In this ideal ordered structure, all H_3O groups are oriented pointing in the same direction, bonded to a layer of Cl atoms on one side only, and nearer to them than to Cl atoms on the other side. In the disordered structure the H_3O groups point in either direction along the triad axis at random, displaced towards the Cl layer to which they form bonds.

In the *dihydrate*, $HCl \cdot 2H_2O$, there are again layers of Cl atoms bound to the oxygens by hydrogen bonds, but the detailed linkage pattern is very different. The structure is monoclinic, with $a = 3.99$ Å, $b = 12.06$ Å, $c = 6.70$ Å, $\beta = 100.5°$, space group $P2_1/c$ (No. 14), and four formula units per unit cell; the formula unit is thus the asymmetric unit. The atomic coordinates are as follows: Cl at 0.02, 0.33, 0.15; O(1) at 0.55, 0.13, 0.02; O(2) at 0.29, 0.06, 0.30. There are thus slightly puckered layers of Cl atoms parallel to (100), at about 4 Å separation, with the oxygen atoms lying between them.

The projection on the plane of the layer is shown in Figure 13-22. There is one short O—O distance, of 2.41 Å, between O(1) and O(2), and four short O—Cl distances, in the range 3.04 Å to 3.10 Å, two from O(1) and two from O(2). These five distances must represent the hydrogen bonds. But if O(1) and O(2) are each donors of H to two O—H $\cdots$ Cl bonds, to which of them will the H on the O(1)—O(2) bond belong? This is a very short bond, and the two O atoms on which it terminates are so much alike in their environment, though not related by symmetry, that it seems the H should be equally associated with both, and hence it should be central (or nearly so) in the bond. These predictions of the hydrogen positions were confirmed by experiment.

Here, therefore, we have found that we must write the group of oxygens with their associated hydrogens as $(H_5O_2)^+$. One oxygen, O(1), has its three bonds at nearly tetrahedral angles—the configuration is "tetrahedral-with-a missing-corner" (cf. §13.3)—while the other set, from O(2), is much more like a plane triangle. The group as a whole lies between two puckered layers of Cl atoms. Three of its four O—Cl bonds attach it to the layer on one side (the layer below it, in the part of the structure shown in Figure 13-22), while the fourth attaches it to the layer on the other side (above it in the figure). The structural formula should be written $(H_5O_2)^+Cl^-$.

The *trihydrate*, $HCl \cdot 3H_2O$, has yet a different pattern of bonds. It is also monoclinic, with $a = 7.58$ Å, $b = 10.15$ Å, $c = 6.71$ Å, $\beta = 123.0°$, and

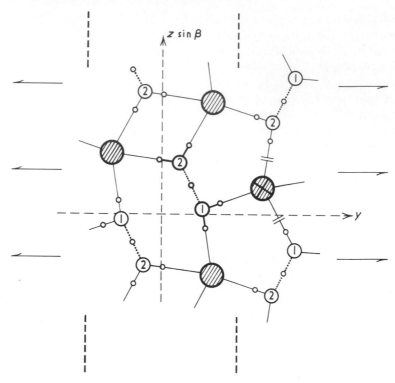

Figure 13-22 Part of the structure of HCl·2H$_2$O, or (H$_5$O$_2$)$^+$Cl$^-$, projected down [100]. Large shaded circles represent Cl atoms near height zero, except the one with a cross added, which is near height 1. Medium and small open circles represent O and H atoms respectively; those with heavy outlines are between heights zero and 1, those with light outlines between zero and −1. Solid lines are O—Cl bonds, dotted lines are O—O bonds; bonds crossed with double bars join O(1) and O(2) between heights zero and −1 to Cl atoms near heights zero and −1 respectively. Thin broken lines show part of the outline of the unit cell with axial directions marked; heavy broken lines and arrows outside the diagram indicate the positions of the symmetry operators.

the formula unit is thus, as before, the asymmetric unit. The atomic co-ordinates are as follows: Cl at 0, 0.16, 0; O(1) at 0.75, 0.42, 0.04; O(2) at 0.63, $\bar{1}$.99, 0.13; O(3) at 0.74, 0.21, 0.30. Atoms Cl, O(2), and O(3) lie close to (10$\bar{1}$) planes passing through the points 0, 0, 0, and 1/2, 0, 0; O(1) atoms lie between these planes. There is a set of three different short O—O distances joining atoms in the sequence O(1)—O(3)—O(2)—O(1)′—O(3)′—O(2)′..., where the first three atoms mentioned are prototype atoms with the coordinates given above, and the second three are related to them by the lattice face-centring translation **a**/2 + **b**/2. In addition, there are four short O—Cl distances, one joining O(2) to a Cl in the same layer, one joining O(3) to a Cl in the next layer, and two joining O(1) to Cl atoms in layers on either side of it. These seven short distances represent the hydrogen bonds; together they form zigzag O—O chains whose overall course is parallel to [100], with side connections linking all the Cl atoms three-dimensionally.

The position of the hydrogen atoms in the bonds can be predicted in the same way as for the dihydrate. For the O—Cl bonds, O must be donor; hence, for the two O—O bonds to O(1), O(1) must be acceptor and O(2) and O(3) donors. This allocation leaves only the hydrogen in the O(2)—O(3) bond to be placed. As in the dihydrate, this is a short bond, 2.43 Å as compared with 2.65 Å for O(1)—O(2) and 2.75 Å for O(1)—O(3), and the atoms at each end have physically similar environments; hence it is likely that the hydrogen will be central in the bond, or nearly so, giving rise to the same kind of group, $(H_5O_2)^+$, as in the dihydrate. The two oxygens have, as before, three hydrogen bonds each; but here the configuration of both sets approximates to tetrahedral-with-a-missing-corner, instead of a plane triangle. For O(1), which is part of an ordinary water molecule H_2O, the four bonds are more or less tetrahedral. The chains of linked O atoms extending through the structure consist of an alternation of water molecules with $(H_5O_2)^+$ groups. The structural formula should therefore be written $(H_5O_2)^+Cl^-\cdot H_2O$.

13.30 SUMMARY AND COMMENTS

We have seen in this chapter the great variety of ways in which the presence of hydrogen can affect structure building. Even so, the list of examples studied is not exhaustive.

We have not paid much attention, for example, to structures where the hydrogen atoms, all covalently bonded to one kind of atom, form a sort of surface coating over a molecule, and the molecules are held together by van der Waals forces only. Typical examples of such materials are the hydrocarbons, but perhaps $Ca(OH)_2$ (§13.9), where the layers coated with H groups on the outer sides of the O atoms are effectively infinite two-dimensional molecules, might be regarded as also belonging to this category.

On the whole, we have focussed attention on structures in which hydrogen bonds are formed. The examples illustrate the almost continuous range of lengths and strengths, from ultra-short bonds where the hydrogens may be central, through normal medium-length bonds where the hydrogen clearly belongs to one donor and is pointed towards one acceptor, to the very weak and bifurcated bonds where the link between hydrogen and acceptor has become very feeble, hardly more than a van der Waals contact. Medium-length bonds are about equal to the sum of the ionic radii of donor and acceptor atoms in the absence of hydrogen; this gives us the picture of the hydrogen atom embedding itself within the electron shell of the donor without greatly affecting its size. The angular arrangement of the complete set of bonds reaching an atom is important, taking into account not only the hydrogen bonds to which it is donor or acceptor, but also the bonds to its other neighbours, whether they are ionic, semipolar, or homopolar. There is a strong tendency for this arrangement to be tetrahedral, but other variants are possible.

Generalising, we may say that the setting up of a three-dimensional network of hydrogen bonds linking together all the other structural units contributes greatly to the stability of a structure; but whether such a network can be built depends on the possibility of adapting atomic positions to satisfy the required geometry of hydrogen bond lengths and angles (which are flexible but not infinitely so) without at the same time destroying arrangements which are satisfactory for the more rigid building elements of the structure.

We have considered at some length the role of the H_2O molecule, distinguishing its occurrence from that of $(OH)^-$ on the one hand and the much rarer $(OH_3)^+$ and $(O_2H_5)^+$ on the other. We have found no examples where the water molecule must be considered as rotating, with its hydrogen atom positions averaged out over a spherical surface. This is partly because we have chosen, as far as possible, to describe well-defined structures with relatively little disorder in other respects; and if rotation of H_2O occurs at all it is likely to be associated with other disorder in structures where its detection is difficult. The general problem of rotation of molecules will be considered in §15.4.

In contrast to the attention paid to H_2O, little has been said here about the other small hydrogen-rich group, the ammonium ion, NH_4^+. This was because we chose to concentrate attention on $O—H \cdots O$ bonds, by comparison with which other hydrogen bonds, for example $N—H \cdots O$ bonds, can easily be understood later. In the NH_4^+ group, the H atoms are bonded tetrahedrally to their donor N. Formerly it was assumed that the ion was likely to rotate when built into crystal structures, and thus in effect become spherically symmetrical. However, later evidence has shown that rotation is the exception rather than the rule, and that formation of a tetrahedrally arranged set of hydrogen bonds is much more usual. One example of this is $NH_4H_2PO_4$, described in §13.14; other illustrations will be given in Chapter 15. Like H_2O, however, NH_4^+ sometimes fits into positions where its ideal tetrahedral array of bonds cannot be formed, and where some may be missing or so much distorted that their identification as bonds is doubtful.

Since structures containing hydrogen bonds, whether $O—H \cdots O$ bonds, $N—H \cdots O$ bonds, or others, often represent a rather delicate compromise between the geometrical requirements for lowest energy of the hydrogen bond system and of the other structure-building elements, it is not surprising that temperature changes may upset the balance and bring about a transition to a different structure. Many of the transitions to be described in Chapter 15 will provide further examples of structures in which hydrogen bonds are involved.

PROBLEMS AND EXERCISES

1. Zinc hydroxide, $Zn(OH)_2$, is orthorhombic, with $a = 5.16$ Å, $b = 5.13$ Å, $c = 4.92$ Å. There are four formula units per unit cell, and the space group is $P2_12_12_1$. Given the atomic positions of Zn and O as set out below, (a) describe the environments of Zn and O, (b) predict possible positions for H, and describe the system of bonds to which they give rise.

   Zn at 0.100, 0.125, 0.175
   O(1) at 0.430, 0.025, 0.085
   O(2) at 0.125, 0.325, 0.370

2. A mineral of composition $Fe_2Te_2O_{10}H_6$ (sonoraite) has Fe^{3+} atoms in 6-coordination and Te^{4+} atoms in 3-coordination. The oxygen atoms, numbered (1) to (10), have cation neighbours as follows: O(1), one Fe; O(2), O(4), two Fe; O(3), O(5), O(6), O(9), O(10), one Te and one Fe; O(8), one Te and two Fe; O(7), no cation neighbour. Using electrostatic valence rules, allocate the H atoms to the O atoms.

3. A mineral with empirical composition $2BaO \cdot Al_2O_3 \cdot 5H_2O$ has all its Al atoms octahedrally coordinated, two octahedra related by a centre of symmetry sharing an edge. The oxygen atoms in the shared edge ("bridging oxygens") have one Ba neighbour, the others have two, and each Ba has nine O neighbours. How should the H atoms be allocated? Rewrite the formula in a way that is related to the structure. Is Pauling's electrostatic valence rule satisfied for all oxygens?

4. A very accurate structure determination of brucite, $Mg(OH)_2$, showed that it is strictly isomorphous with $Ca(OH)_2$, with parameters as follows: $a = 3.142(1)$ Å, $c = 4.766(2)$ Å, $z_O = 0.2216(7)$, $z_H = 0.4303(12)$, the figures in parentheses being possible errors, in units of the last figure quoted. Calculate the distances O—O, O—H, H $\cdots$ O, H—H, and the angles Mg—O—H, O—H $\cdots$ O. Compare them with those calculated for $Ca(OH)_2$, using the parameters in the text. Compare also the H—H distances with what you would expect from the van der Waals radius of 1.2 Å.

   After reading §14.6, correct the O—H distance for thermal motion, on the assumption that H is riding on O. The root-mean-square (r.m.s.) amplitudes of H perpendicular and parallel to z are 0.19(2) Å and 0.11(1) Å respectively, while those of O are about 0.010 Å in both directions.

5. The mineral afwillite, with empirical formula $Ca_3Si_2O_{10}H_6$, has structural features as set out below. Use the information to predict the

arrangement of hydrogen bonds, and show which of the oxygen atoms are O^{2-}, which OH^-, and which H_2O. Make a sketch diagram of the hydrogen bond system.

The asymmetric unit is the formula unit quoted. Eight of the oxygen atoms, numbered (1) to (8), belong to isolated SiO_4 tetrahedra: O(1) to O(4) have each only one Ca neighbour at less than 2.6 Å, O(7) has two, and O(5), O(6), O(8), three each. Of the other two oxygens, O(9) has one Ca neighbour but no Si, and O(10) no cation neighbour at all. Each Ca has either six or seven oxygen neighbours. Bond lengths from Si(1) to O(1), O(4), O(7), O(8), are 1.61 Å, 1.71 Å, 1.61 Å, 1.57 Å respectively; and bond lengths from Si(2) to O(2), O(3), O(5), O(6), are 1.65 Å, 1.61 Å, 1.60 Å, 1.58 Å respectively.

6. Lepidocrocite, γ-FeOOH, is orthorhombic, with $a = 12.4$ Å, $b = 3.87$ Å, $c = 3.08$ Å; the space group is *Bbmm*, and there are four formula units per unit cell. All atoms except H are in special positions 4(c) at u, 1/4, 0, with values of u as follows: Fe, 0.322; O(1), 0.718; O(2), 0.925. Originally H was placed in 4(a) at 0, 1/2, 0; in a later inconclusive study, using neutron diffraction with a powder specimen, it was suggested that the 4(a) site might represent a mean position of two "half hydrogens." Taking into account the symmetry and all the interatomic distances between the other atoms, do you agree with this placing of H, or can you suggest a better alternative? Describe the general linkage pattern of the structure.

7. Ammonium hydrogen sulphate contains two independent SO_4 tetrahedra of which one has bond lengths and angles as follows:

S—O(5)	1.44 Å	O(5)—S—O(6)	113°	O(6)—S—O(7)	108°
S—O(6)	1.44 Å	O(5)—S—O(7)	106°	O(6)—S—O(8)	112°
S—O(7)	1.58 Å	O(5)—S—O(8)	112°	O(7)—S—O(8)	104°
S—O(8)	1.45 Å				

The tetrahedra are oriented so that, in projection on a plane normal to b, two edges are approximately parallel, and the edge most nearly parallel to b is inclined at about 24° to it. The length of b is 4.60 Å. Show that the tetrahedra must be linked by hydrogen bonds, and find the position of the H atoms.

8. Beryllium sulphate tetrahydrate, $BeSO_4 \cdot 4H_2O$, is tetragonal, with $a = 7.99$ Å, $c = 10.69$ Å, space group $I\bar{4}c2$, and four formula units per unit cell. Atomic coordinates are: S at 0, 0, 0; Be at 0, 1/2, 1/4; O(1) at 0.126, 0.084, 0.077; O(2) at 0.138, 0.397, 0.172. Draw a plan of the structure. Show that it is possible, from the length and angular relations of the O—O distances, to decide which of them are hydrogen bonds. Describe the structure, noting the shape, size, and orientation of the

groups round Be and S, as well as the linkage system of the hydrogen bonds. Predict the positions of the hydrogens.

[Note: in this space group, the $\bar{4}$ axis lies in the c-glide plane; the origin is taken at a centre of $\bar{4}$ inversion. A projection on (001) of the part of the structure between $z = -0.40$ and $z = +0.40$ is perhaps the most helpful.]

9. In a compound whose formula is sometimes written $Na_2HAsO_4 \cdot 7H_2O$, and in which this is structurally the asymmetric unit, the linkage scheme is as follows: As is linked to four O atoms, numbered (1) to (4); each Na is linked to six O atoms, of which O(9) is linked only to Na(1), O(10) and O(11) only to Na(2), the others being linked to both Na atoms. (The total of twelve Na—O links is made up by the use of some symmetry-equivalent O atoms.) There are 15 short O—O distances capable of being hydrogen bonds, joining atoms as follows: 1-5, 1-6, 1-8, 1-10, 2-6, 2-7, 2-8, 2-11, 3-5, 3-7, 3-8, 3-11, 4-9, 4-10, 9-11.

Predict the positions of the hydrogen atoms, and rewrite the formula of the compound so that it indicates their structural role.

Do you consider the environments of O(1), O(2), and O(3) a usual type? What features of the compound favour the type?

[Note: In discussing Pauling's electrostatic valence rule it is helpful to use the 5/6, 1/6 distribution suggested in the footnote to §13.4.]

10. "Hydrogarnet," $Ca_3Al_2(OH)_{12}$, closely resembles grossular, $Ca_3Al_2Si_3O_{12}$ (Chapter 11, Question 20); it has lattice parameter $a = 12.57$ Å, and oxygen position parameters 0.03, 0.05, 0.61, but instead of Si in a special position 3/8, 0, 1/4, it has H atoms in general positions 0.15, 0.09, 0.80. Find the O—H distances, and show that the OH groups are "free," i.e., do not take part in hydrogen bonds. Describe qualitatively the orientation of the OH group with respect to (a) the Al octahedron, (b) the tetrahedron of O atoms round 3/8, 0, 1/4.

11. Ice III is tetragonal, with $a = 6.73$ Å, $c = 6.83$ Å, space group $P4_12_12$. There are 12 water molecules per unit cell, the oxygen atoms having coordinates as follows: O(1) in $8(b)$: 0.106, 0.299, 0.286; O(2) in $4(a)$: 0.390, 0.390, 0. Draw a diagram of the structure. From calculations of the O—O distances, locate the hydrogen bonds, and find their average length; find also the shortest distance of approach of nonbonded neighbours. Is it possible to predict a reasonable set of ordered positions for H?

12. The compound of empirical composition $HBr \cdot 3H_2O$ has the space group $Aba2$, with four formula units per unit cell. Show that there are at least two independent water molecules. Assuming that there are no more than two, what are their point symmetries?

The only short O—O and O—Br distances are as follows: O(1)—O(2), 2.68 Å; O(2)—O(2), 2.47 Å; O(1)—Br, 3.39 Å; O(2)—Br, 3.27 Å; O(1) being the atom in the special position. The bond angles are given as follows:

O(2)—O(1)—O(2)	100°	O(1)—O(2)—O(2)	126°
O(2)—O(1)—Br	118°, 98°	O(1)—O(2)—Br	91°
Br—O(1)—Br	123°	O(2)—O(2)—Br	118°

Sketch schematically the environments of O(1) and O(2), and hence find the linkage scheme of the hydrogen bonds. Suggest a possible set of positions for H. Compare the structure with that of HCl·3H$_2$O (said not to be isostructural). Can you suggest any better way of writing the formula?

Chapter 14 THE EFFECTS OF TEMPERATURE

14.1 INTRODUCTION

Hitherto we have not considered explicitly the effects of changes of temperature on structure, though we have noted in passing that some polymorphic forms of crystals are stable only within certain temperature ranges. The present chapter is directly concerned with temperature effects.

We consider first the thermal vibrations which are always present in a crystal, and then go on to look at their direct consequence, thermal expansion; finally, in Chapter 15, we discuss their effect in bringing about thermal transitions.

We begin with the recognition that atoms are not at rest, as we have hitherto pictured them. They are undergoing oscillations, the amplitudes of which increase with increasing temperature. This does not invalidate any of the conclusions of previous chapters, or any of the descriptions of structures contained in them, provided we take the "atomic position" in such descriptions to mean the central position about which the atom oscillates. Diffraction methods of structure analysis give us information about the probability of finding a scattering element at any particular point in the unit cell, the probability being the result of an averaging process both in space—over large numbers of unit cells—and in time—over the whole time of the observation. If the atom were a point scatterer (as its nucleus is for neutrons), it would appear in a diffraction map as a blob whose density at any point represented the probability of finding the atom at that point. The electron cloud surrounding the atom, itself a probability distribution, is spread out by thermal motion in just the same way. When the thermal amplitudes are the same in all directions, the net effect as seen by X-ray diffraction is not very different from that of the atom at rest, except that the spread is greater.

14.2 THERMAL ELLIPSOIDS

Experimentally, it is often found that amplitudes differ not only for different atoms in the same structure but for the same atom in different directions. If we make the assumption that the oscillations are simple harmonic—an approximation which provides a necessary starting point, and is good enough for many purposes, though we shall have to reconsider it later—we can represent the variation of magnitude with direction as follows.

Consider a point atom undergoing three-dimensional simple harmonic vibration about a mean position taken as the origin. The probability that it is displaced to a position x_1, x_2, x_3 is

$$\exp\left[-\left(\frac{x_1^2}{2u_1^2} + \frac{x_2^2}{2u_2^2} + \frac{x_3^2}{2u_3^2}\right)\right] \tag{14.1}$$

where u_1^2, u_2^2, u_3^2 are the mean square displacements along three orthogonal axial directions X_1, X_2, X_3, which have been chosen so that one of them represents the direction of maximum displacement and one that of minimum displacement. The observed electron density at any point in space will be proportional to the probability of the atom's presence at that point. Thus contours of equal electron density are given by the relation

$$\frac{x_1^2}{u_1^2} + \frac{x_1^2}{u_2^2} + \frac{x_3^2}{u_3^2} = \text{constant} \tag{14.2}$$

and this represents an ellipsoid, the lengths of whose principal axes are proportional to the root-mean-square (r.m.s.) values of the displacements along them. The ellipsoid for which the constant is unity is known as the thermal ellipsoid.*

It is important to notice that the principal axes of the thermal ellipsoid need not, in general, coincide with the axes of reference of the crystal—even if these have been chosen along symmetry axes of the structure. The symmetry of the thermal ellipsoid must however satisfy the point symmetry at the atomic site. Only if this has neither rotation nor inversion axis nor mirror plane are the shape and orientation of the ellipsoid unrestricted. If the atom lies on a symmetry axis or a mirror plane, one principal axis of the ellipsoid must coincide with the symmetry axis or plane normal; if the axis is n-fold, with $n > 2$, the thermal ellipsoid must be an ellipsoid of revolution about this axis; and if the site is of cubic point symmetry, the ellipsoid becomes a sphere.

---

* The thermal ellipsoid is still relevant when the atom is not a point atom, since the effective electron density is obtained by convoluting the electron density of the atom at rest with the probability function.

As illustrations, we may consider barium titanate (§12.6) and calcite (§11.7); plenty of others will be found later in the chapter.

In tetragonal $BaTiO_3$ (Figure 12-3) with space group $P4mm$ (No. 99), the atoms Ba, Ti, and O(1) lie on tetrad axes, and have ellipsoids of revolution about these axes; by contrast, O(2) has only *mm* point symmetry, and hence has a triaxial ellipsoid, though with all of its principal axes fixed—one along the diad and hence parallel to the tetrad, and the other two perpendicular to the two mirror planes. It is easy enough, by inspection, to see that vibrations of O(2) along and perpendicular to the Ti—O bonds will be different. Indeed, from this way of looking at it, the differences which might more easily have escaped notice are those between the two kinds of vibration perpendicular to Ti—O, i.e., along and perpendicular to the tetrad axis.

On the other hand, in orthorhombic barium titanate (Figure 12-4) with space group $Bmm2$ (No. 38), all the atoms have triaxial ellipsoids, but that of O(2) is not fixed in orientation. The atom lies on the mirror plane parallel to (010) but not on that parallel to (100); hence one principal axis is parallel to [010], and the other two have an arbitrary orientation in the (010) plane.

In calcite (Figure 11-11), with space group $R\bar{3}c$, atoms Ca and C lie on triad axes, but the oxygen atoms do not; they lie only on diad axes coinciding with the C—O bond direction. Hence, Ca and C possess ellipsoids of revolution, and O a triaxial ellipsoid with one axis lying along the C—O bond.

It is easy, if one is careless, to assume that the thermal ellipsoid satisfies the point symmetry of the *crystal*; the preceding examples are a reminder that in fact it depends on the point symmetry of the atomic site, which is often much lower.

14.3 THERMAL AMPLITUDES AND INTERATOMIC FORCES

Our interest in the shape of the thermal ellipsoid derives from the light it may throw on interatomic forces. For a simple harmonic oscillator of given energy, since the energy is $4\pi^2 m v^2 u^2$ (where m is the mass or effective mass, v the frequency, and u^2 the mean square displacement), the root-mean-square (r.m.s.) displacement is inversely proportional to the frequency. The crystal is an assemblage of oscillators in thermodynamic equilibrium, and the energy distribution follows Boltzmann's law; in consequence, at a given temperature, the r.m.s. amplitudes will be large for low-frequency oscillations, small for high-frequency oscillations. Low frequencies mean low force constants, and hence, in general, weak forces. Thus at a given temperature we have a broad correlation between large amplitudes and weak forces.

Some illustrative examples are now given.

(i) In a series of 10 representative silicates, the Si and Al r.m.s. displacements were found to range from 0.05 Å to 0.07 Å; on the other hand, in

Table 14-I Root Mean Square Displacements (in Å) of the Oxygen Atom in Calcite and Sodium Nitrate

	CaCO$_3$			NaNO$_3$			
	−143° C	−58° C	27° C	25° C	100° C	150° C	200° C
Longest axis of ellipsoid	0.122(4)*	0.137(4)	0.155(4)	0.233(4)	0.274(5)	0.326(9)	0.504(25)
Intermediate axis of ellipsoid	0.051(8)	0.078(7)	0.089(7)	0.160(3)	0.190(4)	0.204(6)	0.222(10)
Shortest axis of ellipsoid	v. small	0.038	0.032	0.135(3)	0.163(4)	0.177(6)	0.177(11)
Angle between longest axis and triad axis	43°(2)	43°(2)	48°(2)	48°(2)	45°(2)	42°(2)	46°(3)

* Figures in parentheses are standard deviations; e.g., 0.122(4) is 0.122 ± 0.004. There is reason to think that the angle of inclination of the longest axis above room temperature in NaNO$_3$ may in addition be subject to systematic error of about −5°.

salicylic acid (typical of many organic compounds) the range of r.m.s. displacements was from 0.17 Å to 0.29 Å, with a mean value of 0.21 Å.

The small amplitudes for the silicates are associated with strong semipolar bonding of the framework, the large amplitudes for salicylic acid with weak van der Waals forces. It may be noted in passing that the large amplitudes characteristic of organic crystals make it hard to carry out structure analyses with the same accuracy as can be achieved quite easily for silicates.

The numerical values quoted here (and elsewhere in this chapter unless the contrary is clearly specified) are r.m.s. displacements in a principal vibration direction. The three-dimensional overall mean square displacement is the sum of the mean square displacements in the three principal directions, or three times any one of them if they are isotropic.

(ii) As a second example, Table 14-1 shows r.m.s. displacements for sodium nitrate and the isomorphous calcite at room temperature. The smaller values in the latter are associated with electrostatic forces of double the strength.

(iii) Our third example is of a large planar molecule, triaminotrinitrobenzene. The r.m.s. displacement of the oxygen in the molecular plane is about 0.15 Å; perpendicular to the plane, where the restoring forces must be much weaker, it is 0.31 Å.

(iv) The fourth example is a comparison of the r.m.s. displacements of the oxygen atom in calcite in directions along and perpendicular to the C—O bond, whose magnitudes are given in Table 14-1. In §14.2, we saw that one principal axis of the ellipsoid must lie along the bond; it turns out (not surprisingly) that this is the shortest axis. In the plane perpendicular to the bond the r.m.s. displacement is more than four times that along the bond. A similar effect is observed in sodium nitrate.

(v) In quartz, our fifth example, a very careful study has shown that the thermal ellipsoid has its shortest principal axis, corresponding to an r.m.s. displacement of 0.09 Å, lying along the Si—O bond, the other principal values being 0.11 Å and 0.14 Å.

All the preceding values are quoted for room temperature. Clearly, amplitudes are expected to increase with temperature. A more detailed discussion will, however, be deferred to §14-14.

A word of warning should be given at this point about the deduction of r.m.s. amplitudes from observational data. As already mentioned, diffraction experiments give us directly the probability of finding the atom in a particular position, by averaging over both time and space; in other words, they determine the average spread of atomic displacements from a mean position. In imperfect crystals where atoms in successive unit cells are not at identical positions with respect to the lattice, there would be a finite spread even if the atoms were at rest. In practice, where this sort of disorder occurs, the r.m.s. displacements due to disorder are generally as large as, or larger than, the r.m.s. thermal displacements we might normally expect. Hence we must not assume, uncritically, that all r.m.s. displacements recorded for particular experimental studies are thermal displacements. When the magnitudes are larger than we should expect for the particular type of material, or when the anisotropies or temperature dependence are anomalous, we must always ask whether disorder could be a contributing factor. It is not easy in all cases to find an unambiguous answer.

One example occurs in a room-temperature study of tridymite—a structure in which disorder can readily occur. The r.m.s. displacements of oxygen are about 0.4 Å, four times their value in quartz; whether they are due to disorder or true thermal motion must await further evidence. Another example occurs in ice (cf. §13.18), where the observed drawing out of the thermal ellipsoids of the "half hydrogens" at right angles to the O—O line can be alternatively explained as a true thermal mode (involving either distortion of the H—O—H bond angle or libration of the H_2O molecule) or as the consequence of disorder due to the difference between the rigid H—O—H angle and the ideal tetrahedral angle. Further examples will be given later in this chapter.

In general, however, when there is no reason to expect disorder and the experimental measurements have been carefully made on good material, it is fair to assume that the observed ellipsoids do represent the thermal displacements, though the experimental errors of the measurements are often considerably underestimated.*

14.4 NORMAL MODES OF MOTION

The diffraction studies which give us our knowledge of thermal ellipsoids can tell us nothing directly of the frequencies. Neither do they tell us whether there are phase relations between the movements of adjacent atoms. A picture in which each atom behaves as an independent oscillator, moving without regard to its neighbours, is of course unrealistic. Where strong interatomic forces exist, the movement of each atom necessarily affects and is affected by its neighbours; they move in a phase-related way, causing wave motions. The analysis of these, and the determination of the frequency spectrum (variation of frequency with wave number) is the province of *lattice dynamics*. A formal introduction to the subject is outside the scope of this book. To apply its methods quantitatively to any but the simplest structures needs a great deal of skill, and is probably impossible for many of the structures described in the preceding chapters. What we are concerned with, however, is to get a qualitative picture of what is going on in the vibrating crystal, and

---

* Measurements of thermal parameters have been called a rubbish dump for all the undetected systematic errors in a structure determination. This does not mean that we should automatically disbelieve all reported values, but rather that we should examine critically the quality of the experimental work in any individual case before making quantitative deductions.

this can often be achieved from our knowledge of the "static" structure and the interatomic forces involved in it.

Each atom in a crystal is capable of vibration in each of three orthogonal directions in space. Thus if there are p atoms per translation-repeat unit and N such units in the crystal, there are $3Np$ *modes of motion*. As we have seen, however, they are generally not independent; but by combining them in appropriate ways a set of $3Np$ *normal modes* can be obtained, which *are* independent of each other (provided the vibrations are *strictly* simple harmonic) and from which all possible motions of the crystal can be built up. They fall into $3p$ subsets or *branches*; if we are content to consider only one representative mode in each branch (generally that for infinite wavelength), and are not inquiring into the dependence of frequency on wavelength, we can limit our discussion to the modes attributable to a single translation-repeat unit.

The simplest kind of mode is one in which (thinking of it as a standing wave) all atoms in the translation-repeat unit move in phase; this is called an *acoustic mode*. In all other modes, called *optic modes*, some atoms in the translation-repeat unit move out of phase with the others.

In either an acoustic mode or an optic mode, the atomic movements may be parallel to the wave vector, when the mode is *longitudinal*, or perpendicular to it, when the mode is *transverse*. For each wave vector there are thus three acoustic modes, one longitudinal and two transverse, and $3(p-1)$ optic modes, also divided into longitudinal and transverse. Some of them may be *degenerate*, for example the two transverse acoustic modes when the wave vector lies along the [001] direction of a cubic or tetragonal crystal, and the two transverse directions are therefore equivalent. Similar degeneracies may occur for the optic modes.

Consider an example where the translation-repeat unit is a linear molecule A—X—A, lying parallel to the x axis. The nine possible modes may be arranged in four groups as follows.

(i) Translation along $\begin{Bmatrix} x \\ y \\ z \end{Bmatrix}$ 3 modes (acoustic)

(ii) Rotation about $\begin{Bmatrix} y \\ z \end{Bmatrix}$ 2 modes

$\left.\begin{array}{l} \\ \\ \\ \\ \end{array}\right\}$ molecule behaves as rigid body

(iii) Bending in plane $\begin{Bmatrix} xy \\ xz \end{Bmatrix}$ 2 modes

(iv) Stretching along x
    (a) homogeneous, keeping AX distances equal
    (b) off-centring of A, keeping X—X constant

    2 modes

$\left.\begin{array}{l} \\ \\ \\ \\ \\ \end{array}\right\}$ molecule is distorted

To classify these modes as longitudinal or transverse we must know the direction of the wave vector. Suppose it is along the x axis. Then translation along x gives the longitudinal acoustic mode, and stretching along x the two longitudinal optic modes; all the others are transverse. The two transverse acoustic modes are degenerate; so are the two rotational modes and the two bending modes; but the two stretching modes differ physically and not merely in their orientation in space. Hence there are six distinguishable branches of the dispersion curve, two acoustic and four optic. For a wave vector parallel to y or z, one rotational mode and one bending mode are longitudinal, and the stretching modes are transverse; the degeneracy disappears.

We shall not, however, have much need to use this classification in what follows.

Not all possible modes are necessarily excited at room temperature. To find out which are fully excited under any given conditions, we note that the quantum condition $kT \gg h\nu$ must be satisfied. At the lowest temperatures, there is only zero-point energy. As the temperature increases, the first mode to begin to be excited is that with the lowest frequency and hence with the lowest force constant. Since intramolecular forces are generally much greater than those between molecules (indeed, this is the criterion used in one definition of a molecule), modes involving distortion of molecules, such as those in groups (iii) and (iv), are likely to have higher frequencies than those involving only translation and rotation (unless the latter are impeded by strong interactions between molecules). Hence at low temperatures we expect to find the molecule vibrating as a rigid body. As the temperature increases the bending modes will probably be the next to be excited, because force constants and frequencies are generally smaller for bending than for stretching (though there may be exceptions). Eventually, at sufficiently high temperatures, all possible modes will be excited—provided the crystal does not decompose or melt first.

For more complicated crystals, and in particular for nonmolecular crystals such as most of those described in Chapters 11 and 12, the complete enumeration of all $3p$ modes is not always so simple. Fortunately, for an understanding of much of the actual behaviour of a crystal, completeness is not necessary. All we have to do is to search for low-frequency modes, because these are the modes with large thermal amplitudes, and therefore the modes with most effect both on the experimentally observable thermal ellipsoids and on the actual character and properties of the structure—in particular, its thermal expansion and its approach to phase transition.

We have to look, on the one hand, for structure-building units—molecules or groups of atoms—which are tightly bonded and might be treated as rigid; and, on the other hand, for "soft" features, where the structure will most easily yield. The latter may be regions of weak forces, like the van der Waals forces in molecular crystals; in framework structures, like some of those in Chapters 11 and 12, the weakness is more likely to be in bond angles.

If, working in this way, we can postulate a low-frequency mode, we can test its correctness by seeing whether it will explain the observed magnitudes and anisotropies of the various thermal ellipsoids. Other tests, involving study of the frequencies by spectroscopic methods, are also of importance, but are outside the scope of this book. Identification of important modes opens up the possibility of explaining the thermal properties of the crystal in terms of the temperature dependence of the energy of the mode and hence of its amplitude.

14.5 EXAMPLES OF CONCERTED MOVEMENTS OF ATOMS

The most obvious application of the ideas outlined in the last paragraph is to molecular crystals. Given a particular set of accurately known thermal ellipsoids, it is possible to show whether they can all be explained in terms of translations and rotations of the molecule as a whole, or whether modes which distort the molecule are also important.

One example, with which we shall be concerned later (§15.17), is thiourea (Figure 14-1). The molecule is planar, except for the two hydrogens H(2) and H(2)', which are slightly out of the plane; both the molecule and the structure have a mirror plane of symmetry bisecting the NCN angle. The molecule moves as a rigid body at 110° K but not at room temperature; however, the assumption of rigid-body motion at room temperature gives a useful approximation. The rotational modes are the most interesting. Their axes correspond closely to the inertia axes of the molecule—the normal to its SCNN' plane, the direction in that plane normal to the SC bond, and the direction of the SC bond. The r.m.s. angular displacements (in degrees) about these axes are 3, $5\frac{1}{2}$, and $4\frac{1}{2}$ respectively at 110° K, 4, 11, and 7 at 293° K.

The smallness of the angular displacement about the normal to the molecular plane at room temperature is explained by the fact that pairs of molecules related by centres of symmetry are held together by intermolecular N—H $\cdots$ S bonds to form a ribbon parallel to the molecular plane and two molecules thick; these bonds clamp the molecules and prevent rotation in their own plane. At 110° K, though the molecules are not strictly parallel, the same kind of linkage occurs. The large angular displacement about the second axis will be considered later in more detail.

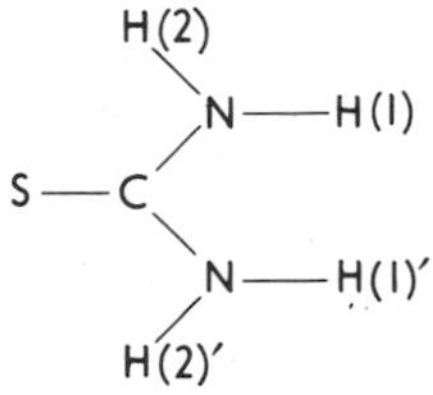

Figure 14-1   Structural formula of thiourea.

Another example of rigid-body motion is given by the SO_4 group in $Li_2SO_4 \cdot H_2O$ at room temperature. Here the SO_4 groups are held to the Li atoms and the water molecule by ionic bonds and hydrogen bonds.

An interesting concerted movement has been found in sodium alum (§13.24). The SO_4 group is placed so that S and one O lie on a triad axis. The other three O atoms have thermal ellipsoids with their major axes at 90° to the S—O bond, but at 65° to the triad axis rather than the 90° one might have expected. This implies that the triangle of oxygen atoms moves as a whole, but its rotation about the triad axis is coupled with translation along the axis. The reason for this screwlike motion is seen if the hydrogen bonds to the three oxygens are considered. Rotation of the triangle without translation would shorten the H . . . O distance considerably, while the coupled movement keeps it nearly constant.

Examples of a different kind come from corner-linked octahedral frameworks of the perovskite family (§12.4 and 12.8). It is plausible to suggest that there will be no great distortion of octahedron shape or size, and that oscillations of the framework will therefore only involve changes of orientation of octahedra. We may test this qualitively by inspection of thermal ellipsoids. In $NaNbO_3$, for example, all the thermal ellipsoids have their two longer axes nearly normal to the Nb—O bond, while the Nb displacements are more nearly isotropic and equal to the displacements of oxygen along the bond (order-of-magnitude values of the small and large r.m.s. displacements of oxygen being 0.05 Å and 0.10 Å respectively). These observations are compatible with the assumption of rigid octahedra, but they are not detailed enough to rule out possible modes in which the octahedra are distorted by change of O—Nb—O bond angle.

The possibility of rigid-body movement of octahedra in this structure calls attention to the important and more general point that parts of a structure which move without internal distortion do not necessarily move independently of one another. In the perovskite framework, alternate octahedra must tilt in opposite directions—just as they do in the "static" structures of the family. More generally, the principle is that the topology of a linkage pattern must be maintained throughout a thermal oscillation.

An example where an important mode of motion can be deduced from the thermal ellipsoids occurs in calcite (cf. §11.7). Rotation of the CO_3 group in its own plane is coupled to translation along the triad axis (like the O_3 group in sodium alum), and the major axis of the oxygen ellipsoid therefore lies at right angles to C—O but at an angle of about 45° to the triad axis. The reason for this is to keep the Ca—O bond lengths unchanged. The shape of the Ca octahedron changes, however, the distortions being an alternate steepening and flattening parallel to the triad axis. The movement of all atoms in the crystal is cooperative, conditioned by the requirements that the CO_3 group shall not change shape, nor C—O nor Ca—O change length; the "soft" feature is the readiness of the O—Ca—O bond angles to depart from their ideal value of 90°.

It is interesting to compare calcite with $LaAlO_3$ (§12.9), which has the same topology and the same symmetry, $R\bar{3}c$. In $LaAlO_3$ (which belongs to the perovskite family) we expect the AlO_6 octahedron to remain undistorted, but the La—O bond to be "soft."

14.6 CORRECTION OF BOND LENGTHS FOR THERMAL MOTION

One important consequence of thermal displacements is their effect on measurements of interatomic distance. If two atoms are moving independently, our best measurement of the interatomic distance is the distance between the two mean positions; but if they are moving so as to keep the true interatomic distance constant at every instant (the phenomenon known as *riding*), this will not be true.

In Figure 14-2, suppose that atom Y moves so that its distance from atom X is constant—and suppose in the first instance that X is at rest. Y then moves on a spherical cap, shown in cross-section by the arc $Y_1Y_0Y_2$. If α is the r.m.s. angular displacement, the r.m.s. displacement in the plane perpendicular to the bond will be AB, and the mean position of the atom will be A. Let the true length of the bond be l_0, the apparent length l', and let $AB = u'$. Then the correction of the apparent length is given by

$$l_0 - l' = \frac{1}{2}\frac{(u')^2}{l_0} \tag{14.3}$$

In evaluating this we must remember that $(u')^2$ is the two-dimensional mean square displacement, i.e., the sum of the mean square displacements in the two principal directions at right angles to XY_0.

If atom X is moving, the movement of Y relative to X is unaffected—Y continues to ride on X. To obtain $(u')^2$, however, we must now take the difference of the (two-dimensional) mean square displacements of Y and X in the plane at right angles to XY_0.

An example is given by calcium hydroxide (§13.9). We assume that the proton moves so as to keep O—H constant. Since the mean positions of both O and H lie on the triad axis, their thermal ellipsoids are ellipsoids of

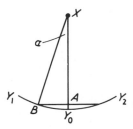

Figure 14-2   Diagram illustrating the shortening of measured bond length.

Table 14-2 Mean Square Amplitudes and Bond-Length Corrections in Calcite and Sodium Nitrate

	CaCO$_3$		NaNO$_3$			
	$-58°$ C	$27°$ C	$25°$ C	$100°$ C	$150°$ C	$200°$ C
Axes of ellipsoid (10^{-2} Å^2)						
Longest, p_1	1.88	2.40(12)	5.4(2)	7.5(3)	10.6(6)	25.4(25)
Intermediate, p_2	0.61	0.79(12)	2.6(1)	3.6(2)	4.2(2)	4.9(5)
Shortest, p_3	0.14	0.10	1.8(1)	2.7(1)	3.2(2)	3.2(4)
Mean square displacements (10^{-2} Å^2)						
$(u')_O^2$ for O $= p_1 + p_2$	2.49	3.19	8.0	11.1	14.8	30.3
u_N^2 for N along triad $= \frac{1}{2}(u')_O^2$	1.25	1.60	4.0	5.6	7.4	15.2
$(u')_N^2$ for N $= u_N^2 + p_3$	1.39	1.70	5.8	8.3	10.6	18.4
Measured bond length (Å)	1.283(2)	1.283(2)	1.245(2)	1.236(3)	1.232(4)	1.215(6)
First model						
$2l_0(l_0 - l') = (u')_O^2 - 2p_3$ (Å^2)	0.022	0.030	0.044	0.057	0.084	0.239
$l_0 - l'$ (Å)	0.009	0.012	0.018	0.023	0.034	0.096
Corrected length (Å)	1.292	1.295	1.263	1.259	1.266	1.311
Second model						
$2l_0(l_0 - l') = (u')_O^2 - (u')_N^2$ (Å^2)	0.011	0.015	0.022	0.028	0.042	0.119
$l_0 - l'$ (Å)	0.004	0.005	0.009	0.011	0.017	0.048
Corrected length (Å)	1.287	1.288	1.254	1.247	1.249	1.263

revolution. Hence, for each, the two principal values in the (0001) plane are equal, and $(u')^2 = 2u^2$, where u^2 is the mean square displacement of O relative to H in any one direction in the (0001) plane. We have information about the mean square displacements from neutron diffraction studies at 20° C and −140° C; the values of $8\pi^2 u^2$ for the proton are 4.23 Å^2 and 3.31 Å^2 at the two temperatures, and for oxygen 0.73 Å^2 and 0.39 Å^2. Hence the bond length corrections are $3.50/(8\pi^2 l_0)$ and $2.92/(8\pi^2 l_0)$, or (taking $l_0 =$ 0.98 Å) 0.045 Å and 0.038 Å. The measured O—H bond lengths are 0.936 Å and 0.944 Å, giving corrected values of 0.981 Å and 0.982 Å. The true bond length is thus significantly longer than the measured bond length, and is independent of the temperature (to the accuracy at present attainable).

A similar correction has been calculated for the N—O bond length in NaNO$_3$. The thermal ellipsoid for oxygen is known, but that for nitrogen has not been studied. Assuming a constant N—O distance, there are two reasonable models for the complete system of motions. In the first model, N moves isotropically, with a mean square displacement equal to that of O along the N—O bond. In the second model, the movement of N in the (0001) plane is the same as in the first, but perpendicular to this its movement is controlled by the requirement that the NO$_3$ group shall remain planar—that N, in fact, rides on its triangle of oxygens. (In the first model, the NO$_3$ group is allowed to become, instantaneously, a trigonal pyramid.) In the second model, the mean square displacements of N and O along the triad axis are equal, and the latter is given by half the sum of the two principal values normal to the bond, since the principal directions concerned are at 45° to the triad axis. Using the data of Table 14-2, the two models then give values for the room-temperature correction as follows:

First model

$$l_0 - l' = \frac{1}{2l_0} [2.6 + 5.4 - 2 \times 1.8] \times 10^{-2} = 0.018 \text{ Å} \qquad \textbf{(14.4a)}$$

Second model

$$l_0 - l' = \frac{1}{2l_0} [2.6 + 5.4 - (1.8 + 4.0)] \times 10^{-2} = 0.009 \text{ Å} \qquad \textbf{(14.4b)}$$

The values of the corrected bond lengths at different temperatures are given for both models, in Table 14-2. Up to 150° C, both models show that there are no significant changes in bond length. The second, more plausible, model indicates that this holds good up to 200° C—even though the largest displacement is then so large that our approximations might well have been too rough. We note, in fact, that at 200° C the r.m.s. angular displacement of NO$_3$ in its own plane is about 20°.

Another bond-length correction for thermal motion, of somewhat similar form but applicable to a different model, will be discussed in §14.10.

These examples illustrate the importance of having some knowledge of thermal motion before attempting accurate bond length calculations. When thermal amplitudes are small, of course, the correction may be negligible. When they are large, however, much depends (as we have seen) on the choice of a correct model to describe the motion. Complete and accurate knowledge of the thermal ellipsoids may allow us to decide between different more or less plausible models; rarely, if ever, will it point uniquely to one particular model without the help of other information or assumptions. Hence, in these matters, we have still to proceed by trial and error.

14.7 SYMMETRY OF LATTICE MODES

When an optic mode is such that all atoms of a continuous framework move together in a phase-related way, the configuration at either end of the oscillation may be regarded as an instantaneous (and perfect) structure. We can then ask, what is the space group of the instantaneous structure? Is it the same as that of the static structure, or different?

Consider, as a hypothetical example, the static structure shown in Figure 14-3(a). Suppose we have a pure stretching mode, in which all the anion neighbours of one cation move towards it. In so doing, each must move away from its other cation neighbour, which therefore ceases to be a translation repeat of the first. The result is as shown in Figure 14-3(b), where alternate octahedra have become large and small (and, of course, each changes from large to small and vice versa during an oscillation). The original space group was $Pm3m$ (No. 221) applying to a unit cell with lattice parameter a; the "instantaneous" space group is $Fm3m$ (No. 225) applying to a unit cell with lattice parameter $2a$. This represents a lowering of symmetry, because there are fewer symmetry elements per unit volume, i.e., per number of atoms, even though there are more symmetry elements per unit cell.

If the mode had been one involving only displacements in the plane of

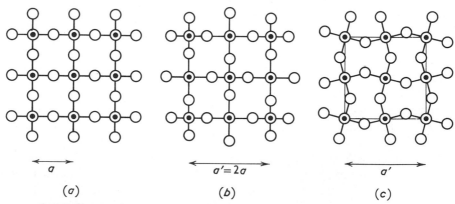

a $a'=2a$ a'

(a)   (b)   (c)

Figure 14-3 Hypothetical structure showing (a) static structure, (b) stretching mode, (c) bending mode.

projection, the instantaneous structure would have been tetragonal, with $a' = 2a$, $c' = c$, space group $C4/mmm$ (No. 123).

If the mode involves bending rather than stretching, similar relations hold. Figure 14-3(c) shows this for rotation about an axis perpendicular to the plane of the diagram. Here again the "instantaneous" cell is larger, with $a' = 2a$, $c' = c$. However, the space group is now $C4/mmb$—No. 127 (the same one as would, with a different choice of axes, be named $P4/mbm$).

Suppose, however, that the static structure had been as shown in Figure 14-3(c). Then the instantaneous structure is derived merely by a change in the tilt parameter, which was already arbitrary. There is no lowering of symmetry. This illustrates a general rule: if, in a lattice mode, atoms move away from special positions on mirror planes, rotation axes, or inversion centres, then the instantaneous structures have a lower symmetry than the static structure; whereas if the only atoms that move are already in general positions, or if they move so that they stay on existing mirror planes or rotation axes, then the instantaneous symmetry is the same as the static symmetry. In neither case does the instantaneous structure gain any symmetry elements not possessed by the static structure.

The examples described above are not completely hypothetical. The structure of Figure 14-3(a) is the ideal perovskite framework, while Figure 14-3(c) represents one layer of the Ti—O framework of low-temperature $SrTiO_3$ (§15.12). In ideal perovskite (if we can ignore the A cations) we therefore expect to find a mode whose instantaneous structures are the same as the static structure of low-temperature $SrTiO_3$. Again, if to the tilts in Figure 14-3(c) were added equal tilts about the two original cube axes in the plane of the diagram, we should have the $LaAlO_3$ structure (§12.9); thus a mode changing only the magnitude of the combined tilt (while keeping the three components equal) will not lower the instantaneous symmetry of $LaAlO_3$, though it would reduce the instantaneous symmetry of ideal perovskite to the actual symmetry of $LaAlO_3$, namely, $R\bar{3}c$. These two structures will be discussed in more detail in §15.12 to 15.13.

The other structure we considered in detail, calcite, starts with the same $R\bar{3}c$ symmetry. Here, however, the movements of oxygen are not along the diad axes of the static structure (on which the C—O bonds lie) but perpendicular to them. Hence the instantaneous structure has lost the diads, and with them the c-glide plane, and is reduced to $R\bar{3}$. The mode responsible for the thermal displacements in calcite is thus different from the important mode in $LaAlO_3$.

14.8 TEMPERATURE VARIATION OF THERMAL AMPLITUDES

In an earlier section, we were working with values of mean square displacements measured at different temperatures; we look next at the actual character of the temperature dependence.

According to classical theory, we should expect the mean energy of a one-dimensional simple harmonic oscillator, $4\pi^2 m\nu^2 u^2$, to be equal to kT, where k is the Boltzmann constant. According to quantum theory, this is true only at temperatures where $kT \gg h\nu$; when $kT \ll h\nu$, the oscillator has only zero-point energy, $\tfrac{1}{2}h\nu$. Hence the mean square amplitude should be proportional to T at high temperatures but level off to a constant value at low temperatures. The levelling off should begin to become noticeable near $T_c = h\nu/k$; for acoustic modes, ν corresponds to ν_{max} of the Debye theory, and T_c to the Debye temperature. The stronger the interatomic forces, the higher the value of ν, and the higher the temperature above which the law $u^2 \propto T$ will apply. If there are optic as well as acoustic modes, they will generally have higher zero-point energies and higher characteristic temperatures; because the effective mass of the moving parts is generally smaller, the mean square displacements tend to be fairly large.

To illustrate some of these points, we may calculate u^2/T for $CaCO_3$ and $NaNO_3$ from the experimental data given in Table 14-2.

For calcite investigated below room temperature we should expect an increase of u^2/T, corresponding to the flattening off of the curve of energy versus T. This is observed for the two larger principal displacements (the third being too small to measure accurately); however, when the standard deviations are considered, the differences are barely significant. On the other hand, for $NaNO_3$ above room temperature, u^2/T for the largest principal displacement *increases* markedly with temperature, and that for the second-largest to a lesser extent. This suggests that here we have some partly-excited optic modes.

A further example is found in potassium dihydrogen phosphate, (§13.12 to 13.13), which has been studied at room temperature, 132° K, and 77° K. For the oxygen atom, the mean square displacements (nearly isotropic) are 0.015 Å^2, 0.007 Å^2, and 0.005 Å^2. We notice that there is a slight increase of u^2/T with decreasing temperature, and this is found, by more detailed work, to be compatible with a Debye curve. The same is true for the phosphorus and potassium atoms. The mean square displacements of hydrogen (again nearly isotropic) are 0.027 Å^2, 0.020 Å^2, and 0.019 Å^2 respectively; relative to the framework (taking the oxygen atom to be representative of this) they are 0.012 Å^2, 0.013 Å^2, and 0.014 Å^2. These values are so nearly constant as to suggest that, even at room temperature, the vibrations of hydrogen relative to the framework possess only zero-point energy.

For organic crystals down to −200° C, the classical approximation generally holds good. Not many structures have been investigated at lower temperatures. One that has been studied at 30° K is ammonium oxalate monohydrate. The average r.m.s. displacement at room temperature is 0.17 Å; at 30° K it is 0.09 Å, instead of the 0.06 Å classical theory would predict. The structure is joined together by hydrogen bonds, and the frequency (and zero-point energy) is therefore likely to be higher than for van der Waals–bonded crystals. A detailed study of DCl (see Chapter 15, reference 97) suggests that, while translational modes are already excited at 4.2° K, one rotational mode does not begin to be excited till about 35° K, and the other still possesses only zero-point energy up to at least 100° K.

14.9 ANHARMONICITY OF VIBRATIONS

So far, we have assumed that all vibrations—whether independent movements of individual atoms or concerted movements in a lattice mode—are simple harmonic. This, though a useful starting point, is not strictly true. Anharmonicity is very important, because strictly harmonic vibrations would give rise to no thermal expansion, and would provide no mechanism for the transference of energy from one mode to another.

Consider a structure whose potential energy V depends only on one structural parameter r (for example, one cation—anion distance), and let us suppose that all conditions except the temperature remain constant. The graph of V versus r has a minimum at r_0, the equilibrium value of r if only potential energy is present (Figure 14-4). But the internal energy of the structure includes also the kinetic energy of the oscillations (with which we may include also, for present purposes, their zero-point energy). At a given temperature the internal energy is constant. Consider what happens at T_1. At either extremity of the oscillation the energy is all potential. Thus the intersections of the horizontal line with the V curve mark the extreme values of the parameter r, and the midpoint of the line marks its mean value.

In general, if for any curve possessing a minimum the origin of coordinates is taken at the minimum, the equation of the curve near the origin can be written in virial form,

$$y = ax^2 + bx^3 + cx^4 + \cdots \tag{14.5}$$

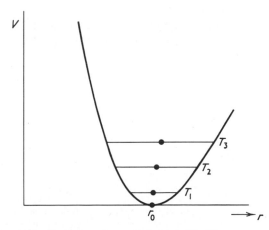

Figure 14-4 Variation of potential energy V with one parameter, r.

where $a > |b| > |c| \ldots$. For small displacements, we can use the approximation,

$$y = ax^2 \tag{14.6}$$

In our case, putting $y = V$, this corresponds to a simple harmonic oscillator with restoring force $2ax$, and centre of oscillation at $r = r_0$. The mean value of r is thus independent of amplitude.

As $|x|$ increases, we can no longer neglect the term bx^3. For any given y, the two roots of the equation of opposite sign are unequal in magnitude, and their mean value—giving the midpoint of the oscillation—is not zero. As T increases, so do V, the two values of $|x|$, and the displacement from $x = 0$ of the midpoint of the line. This can be seen from the figure, which has been drawn giving b a negative sign, so that the static value of the parameter, r_s, increases with temperature.

It must be kept clearly in mind that this diagram does *not* represent a section of crystal space, with an atom moving between certain points in the crystal. The oscillator whose amplitude is being considered is a mode, not an individual atom, and the abscissa represents some important parameter of the crystal, not a lattice spacing. (The amplitude of each individual atom will, of course, be proportional to the amplitude of its mode, but the constant of proportionality may be different for different atoms.) The diagram shows us, schematically, how to relate changes in parameter r_s to changes of temperature; it is a separate step to consider the relation of r_s to lattice spacings. For simple cubic structures, such as rock salt, with no variable parameter except the cation—anion distance, the step is of course obvious, but even in this case to try to read it into Figure 14-4 can be very confusing.

Now consider a structure dependent on two parameters, such as either of those shown in Figure 14-5, in which the distances AA' and MB are geometrically independent (though we must be prepared to find that they may be physically related). Let the parameters chosen to describe the structure be $l = \frac{1}{2}AA'$ and $d = MB$. The potential energy of the structure is a function of both l and d, and to represent it graphically we need three dimensions, replacing the curve of Figure 14-4 by an undulating surface. This is most conveniently represented by a contour map, where l and d are plotted along the x and y axes, and each contour is a curve of constant V. In crystal space, d may lie parallel or perpendicular to l, as in Figure 14-5 (a) or

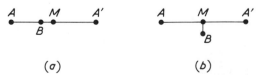

(a) $\qquad\qquad\qquad\qquad$ (b)

Figure 14-5 (a) and (b) Two different structures, each dependent on two geometrical parameters, AA' and MB. Part (b) also illustrates a structure with one geometrical parameter, AA', and a thermal amplitude, MB.

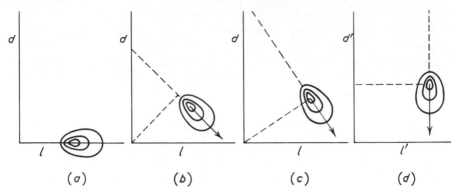

Figure 14-6 Contour maps of potential energy, V, plotted against two parameters d and l, or d' and l'. (See text for details.)

(b) respectively, but in parameter space we have taken them as independent variables and therefore must plot them at right angles.

Suppose in the first instance that the mean value of d is zero. We then have the situation shown in Figure 14-6(a). The area within each contour represents the range of variation of d and l at the temperature concerned. A section of the surface at constant l always gives a symmetrical well, because by symmetry of the structure the potential is the same whether the sign of d is positive or negative. A section at constant d, however, (provided d is small) will have the asymmetrical shape of Figure 14-4. As the temperature increases, successively higher contours represent the internal energy, and the centre of oscillation therefore moves out to higher values of P, along the line $d = 0$.

Now suppose that the mean value of d is not zero. Since the atom is no longer in a special position in crystal space, the change of potential for increase or decrease of d is not necessarily the same, and therefore the well may be asymmetric and contain an anharmonic term in its section at constant l in addition to that in its section at constant d. In general, its shape will be more complicated, but it may have an approximate plane of symmetry in some general direction, as in Figure 14-6(b) or (c). In these examples, the static parameters, given by the coordinates of the centre of the relevant contour, vary together with increasing temperature in the direction shown by the arrow. It is therefore worth considering whether a different choice of parameters would give us a simpler or more informative description. Let us choose d' and l', parallel and perpendicular respectively to the line of the arrow. Then we see that the time-average value of l' is constant, while that of d' varies with temperature. The contours shown in Figure 14-6(b) and (c) have been redrawn with respect to the new axes in Figure 14-6(d)—where, however, the line marked l' has been drawn at an arbitrary negative value of d'. For the conditions represented by Figure 14-6(b), all the contours shown correspond to negative values of d', while for those represented by Figure 14-6(c) the centre of the lowest contour marks the zero value of d'.

To determine the thermal expansion, we now consider the relations between Figure 14-6(b) and (c) and the two models of Figure 14-5. In Figure 14-6(b), the direction of the arrow is equally inclined to the axes of d and l; hence

$$l' = l + d = \text{constant}; \quad -d' = l - d \qquad (14.7)$$

This corresponds to a linear model like Figure 14-5(a), with atom B off-centre between A and A'. Stretching vibrations of the bond $A'B$ are therefore harmonic, and the bond length is independent of temperature, while the AB length and consequently the lattice parameter AA' undergoes thermal expansion. In Figure 14-6(c) we have

$$(l')^2 + (d')^2 = l^2 + d^2 \qquad (14.8)$$

and since the l' axis passes through the position of the minimum, d' is small in the region of interest to us, and is proportional to the change in the angle $\tan^{-1}(d/l)$. This system corresponds to a two-dimensional model like Figure 14-5(b), with parameters l and d at right angles in crystal space. Here AB is given by $l^2 + d^2$, and remains very nearly constant, while the temperature-dependent parameter d' describes changes in the angle MAB and hence in the lattice parameter AA'.

The models we have chosen, and the results we have obtained, are a good approximation to the behaviour of actual structures when l' is the length of a strong bond and d an atomic position parameter controlled by some weaker force or combination of forces. They are still very simple ones; they involve only two initial parameters, and we have not tried to consider examples where the potential well lacks a plane of symmetry in parameter space.

The difficulty of making this kind of approach quantitative lies in the fact that rarely or never do we know the true shape of the potential wells. When there is only one parameter for the structure, a Born-type potential-energy expression might be used; for structures with more than one parameter, this theory can only give a qualitative guide. If there are n independent parameters, the potential energy will be an n-dimensional surface in $(n + 1)$-dimensional space, and our task is to find a convenient and informative section in which only one parameter (or one combination of initially chosen parameters) is changing significantly. This can only be done in one of two ways: either from experimental evidence about parameter changes in the particular structure of interest, or from qualitative prediction based on our knowledge of interatomic forces and thermal changes in similar structures. In fact, we are using empirical knowledge about thermal effects to construct our contour diagrams of potential energy, and not vice versa.

Nevertheless, the approach is of value in helping us to realise what we are doing when we draw illustrative graphs of potential energy versus some one parameter that we intuitively consider to be important—a device which can be very informative if the parameter is wisely chosen.

It is also of use in calling attention to the fact that when a particular position parameter is, by symmetry, zero in the "static" structure, the section involving only that parameter has a symmetrical potential well, and no changes of mean position will occur, however large the amplitude.

There may be potential wells in a structure where the fourth-power term or higher terms in the expression for V as a function of x are not negligible. The extreme case is the *square well*, where the potential energy change is zero for all displacements of less than a certain magnitude, and infinite for any greater displacement. In such wells, of course, the simple harmonic approximation does not hold; the amplitude is constant, independent of the temperature.

14.10 THERMAL EXPANSION: GENERAL IDEAS

In the last section, we have seen how macroscopic thermal expansion is a consequence of anharmonic thermal oscillation of some parameter to whose values the overall lattice parameters are sensitive. On the whole, large anharmonicity is associated with large thermal displacements, and these in turn with weak forces. This general correlation of weak cohesive forces with high thermal expansion can be seen at a glance from inspection of the expansion coefficients listed in Tables 14-3 and 14-4.

It does not necessarily follow that the *direction* of greatest expansion, when this is anisotropic, is the same as that of the greatest thermal displacements of atoms. It is likely to be the same when the displacement is due to a stretching mode, but not when it is due to a bending mode. Figure 14-5(b) illustrates the latter case: changes in the angle MAB cause atomic displacements in the direction MB but lattice parameter changes in the direction AA'. We can, however, make the following generalisations:

- 1. Changes of lattice parameters can be deduced as a geometrical consequence of displacements of atoms in directions in which the vibrations are anharmonic, provided they are all part of the same mode.
2. When a lattice parameter is dependent on the sum of two bond lengths, AX, $A'X$, and the atom X is moving with a large anisotropic amplitude perpendicular to AA', then increase in this amplitude will produce a *contraction* in AA' if the mean square bond lengths are independent of amplitude—and this is true even if the movement of X is strictly harmonic, and the mean positions of AXA' are collinear.
3. When more than one physically independent effect is contributing to the expansion—for example, a bond-length effect and a bond-angle effect—their separate contributions to the expansion coefficient are added to give the macroscopic coefficient.

Table 14-3 Thermal Expansion of Homodesmic Crystals (Ionic or Homopolar) Without Adjustable Angles

MATERIAL	STRUCTURE TYPE	ELECTRO-STATIC VALENCE q	$\dfrac{1}{q^2}$	LINEAR EXPANSION COEFFICIENT NEAR 40° C $(10^{-6}\ deg^{-1})$
(a) Simple cubic crystals				=
Caesium chloride 3 other alkali halides	CsCl CsCl	1/8	64	$\begin{cases}53\\47\ \text{to}\ 62\end{cases}$
Sodium chloride 14 other alkali halides	NaCl NaCl	1/6	36	$\begin{cases}40\\36\ \text{to}\ 50\end{cases}$
Calcium fluoride	CaF_2	2/8	16	19
Magnesium oxide	NaCl	2/6	9	8
Thorium oxide	CaF_2	4/8	4	6
Diamond	Diamond	4/4	1	1
(b) Noncubic crystals and more complex cubic crystals				
Magnesium fluoride, MgF_2 Forsterite, Mg_2SiO_4	Rutile Olivine	2/6	9	$\begin{cases}11\ \text{(mean)}\\11\ \text{(mean)}\end{cases}$
Corundum, Al_2O_3 Spinel, Al_2MgO_4 Gahnite, Al_2ZnO_4 Chrysoberyl, Al_2BeO_4 Hematite, Fe_2O_3	Corundum Normal spinel Normal spinel Olivine Corundum	3/6	4	$\begin{cases}5,\ 6\\6\\6\\5,\ 6,\ 6\\8,\ 8\end{cases}$
Zincite, ZnO	Wurtzite	2/4	4	3, 4
Rutile, TiO_2 Cassiterite, SnO_2	Rutile Rutile	4/6	2	$\begin{cases}7,\ 8\\3,\ 4\end{cases}$

The first generalisation holds because the instantaneous displacement of any atom involved in a particular mode is geometrically related to that of all other atoms in the mode. The third generalization holds in so far as all actual changes of position parameters and lattice parameters are very small. We may demonstrate the second generalisation as follows.

In figure 14-5(b) let B represent the instantaneous position of an atom X, undergoing simple harmonic oscillations about a mean position M; assume A and A' to be at rest. Then, for the squared distances in this position,

$$AM^2 = AB^2 - MB^2 \tag{14.9}$$

The same relation of course holds for their mean values, taken over all positions of B. If l_1 is the r.m.s. length of AB (i.e., the bond length AX) and

Table 14-4 Thermal Expansion of Some Markedly Anisotropic Crystals and Molecular Crystals

MATERIAL	EXPANSION COEFFICIENTS $(10^{-6} \text{ deg}^{-1})$	DIRECTIONS TO WHICH FIGURES QUOTED REFER
Calcite structure		
Calcite	−6, 25	*a, c*
Sodium nitrate	11, 120	*a, c*
Layer hydroxides		
Magnesium hydroxide	11, 45	*a, c*
Calcium hydroxide	10, 33	*a, c*
Aluminium hydroxide	11, 15 38 −6	*b* (in layer); ⊥ to layer −40° to [001] (maximum) +50° to [001] (minimum)
Organic crystals		
Benzene	221, 119, 106	*c, a, b*
Naphthalene	225, 40, 115	*a, b, c*
Anthracene	130, 30, 80	*a, b, c*
Oxalic acid dihydrate	57, −1	Maximum nearly ∥ to C—C bond; minimum nearly ⊥ to C—C bond, in plane (010)
Other molecular crystals		
Chlorine (80° K to 160° K)	66, 36	
Nitrogen (5° K to 20° K)	200	

l'_1 is its contribution to the lattice parameter, and $\overline{u^2}$ is the mean square displacement along MB, then

$$l'^2_1 = l^2_1 - \overline{u^2}$$

or

$$l_1 - l'_1 = \frac{1}{2}\frac{\overline{u^2}}{l_1} \tag{14.10}$$

For the other bond length $A'X$, there is a similar correction, which is to be added to the foregoing, because the instantaneous values of AB and $A'B$ are not independent—whether or not the lengths AM and $A'M$ are equal. Hence, for the total change in AA',

$$\Delta(AA') = -\tfrac{1}{2}\overline{\Delta u^2}\left(\frac{1}{l_1} + \frac{1}{l_2}\right) \tag{14.11}$$

i.e., there is a contraction due to increase in the r.m.s. amplitude of *harmonic* vibration.

Quantitative theories of thermal expansion, so far as they exist, are applicable only to very limited types of structures. We are here concerned with a broader general survey. For this, it is important to distinguish between three sources of macroscopic expansion: bond-length expansion, bond-angle changes, and expansion in the presence of weak residual cohesive forces. These, of course, may all contribute simultaneously to the macroscopic expansion of a particular structure, but it is convenient to consider them separately, with reference to examples where each in turn plays an important part.

14.11 BOND-LENGTH EXPANSION

Experimental evidence about bond-length expansion can most easily be obtained from the macroscopic expansion of cubic structures with only one kind of bond and no arbitrary parameters. Examples are listed in Table 14-3(a).

More complicated structures may still give us reliable information, provided they possess a three-dimensionally linked framework of a kind in which angular distortion is prevented, for example, by the presence of numerous shared edges between polyhedra. Such structures are expected to be nearly isotropic, and the mean linear macroscopic expansion will be that of the cation—anion bond of the framework, irrespective of the nature of the interstitial atoms. Experimental values of expansion coefficients for some such structures are given in Table 14-3(b).

For structures of these kinds, a rough empirical rule may be formulated—that *the mean linear expansion coefficient is inversely proportional to the square of the electrostatic valence.* The extent to which the compounds in the table obey this rule is shown in Figure 14-7. The rule could be justified theoretically for purely ionic crystals, with the modification that the "constant" of proportionality is itself proportional to the interatomic distance and dependent on the Born exponent n. It is, however, more useful to accept it purely empirically as a guide to the normal range of bond-length expansions for different types of atoms, so that any points far off the graph can be recognized as abnormal and given separate attention. Some examples will be considered below. Moreover, for structures which do not qualify for inclusion in Table 14-3 because their macroscopic expansion has obviously other sources than bond-length expansion, the rule allows us to make an order-of-magnitude estimate of the bond-length contribution, and hence, by subtraction, to obtain an estimate of the other contributions.

One compound for which the expansion coefficient is anomalously high is rutile, TiO_2. The effect appears to be characteristic of structures with highly-charged cations, small enough to allow anion—anion contact,

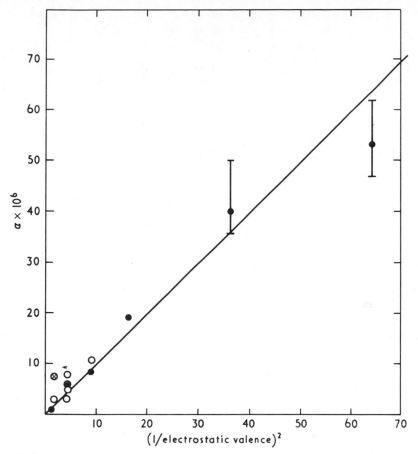

Figure 14-7 Expansion coefficient versus $(1/\text{electrostatic valence})^2$ for structures listed in Table 14.3. Solid circles refer to the simple cubic structures in Table 14.3(a), open circles to the slightly more complex structures in Table 14.3(b). The point marked $\otimes$ is rutile, TiO_2.

centrally placed in their polyhedra. Isomorphous structures with larger cations not allowing close anion—anion contacts obey the empirical rule; this is illustrated by cassiterite, SnO_2, isomorphous with rutile. Other examples with high expansion coefficients due to this cause are the cubic forms of $BaTiO_3$, $KNbO_3$, and $NaNbO_3$ (see §14.14). In materials like these, changes of oxygen—oxygen distances associated with changes of cation—oxygen distances make an important contribution to the energy (§2.8), especially when the cation—oxygen distances are kept equal by symmetry, and it is not surprising that a rule based on consideration of cation—oxygen interactions only should prove inadequate.

In principle it would be desirable to obtain direct information about bond-length expansion in noncubic compounds by measurements of changes in position parameters, but in practice it is very hard to secure the necessary

Table 14-5 Effect of Bond-Angle Changes on Thermal Expansion

STRUCTURES WITH VARIABLE ANGULAR PARAMETERS				STRUCTURES WITH FIXED ANGLES	
Temperature range	Material	Mean linear expansion coefficient (10^{-6} deg^{-1})		Material	Temperature range
0° C to 100° C	Neighbourite NaMgF$_3$	34	11	MgF$_2$	Near 40° C
0° C to 100° C	Perovskite CaTiO$_3$	25	$\{$ 8	Rutile TiO$_2$	Near 40° C
			7	BaTiO$_3$ (tetragonal)	0° C to 100° C
20° C to 200° C	NaNbO$_3$ (Phase P)	11	5	KNbO$_3$ (orthorhombic)	20° C to 200° C
0° C to 100° C	Quartz (low)	12	<1	Beryl Be$_3$Al$_2$Si$_6$O$_{18}$	Near 40° C
Near 40° C	CuSO$_4$·5H$_2$O, MgSO$_4$·7H$_2$O, Na$_2$SO$_4$·10H$_2$O	30 to 50	13	Alum KAl(SO$_4$)$_2$·12H$_2$O	Near 40° C

accuracy. For example, in potassium dihydrogen phosphate (cf. §13.12) the changes in position parameters of the oxygen atom between 293° K and 132° K are not significantly greater than the standard deviation. Taking into account the observed lattice-parameter changes, any expansion of the P—O bond is undetectable, being less than the standard deviation, while that of the K—O bond, though measurable, has estimated error limits of ±50 percent. Subject to this experimental uncertainty, the value of 40×10^{-6} deg^{-1} obtained for the expansion coefficient of K—O agrees satisfactorily with those listed in Table 14-3(a) for electrostatic valence 1/8.

Another example of the limitations of this approach is given by some very careful work on quartz, in which the change in Si—O bond length between 150° K and 300° K was found to be less than 0.001 Å, giving an upper limit of 4×10^{-6} deg^{-1} for the bond-length expansion coefficient.* Similar conditions apply to most studies of organic structures; a statement in such a context that "intramolecular bond lengths remain unaltered" generally means only that any changes which may occur are too small for detection by their effect on measured position parameters.

The special question of the expansion of hydrogen bonds is one to which attention has been directed in the past. For the reasons just mentioned, a direct answer is difficult; the changes such an expansion makes in position parameters are generally barely detectable. For example, in potassium dihydrogen phosphate between 293° K and 132° K there was no measurable change in the O—H distance, and the decrease in the O—O distance was only 0.012 Å—about three times the standard deviation. This does, however, suffice to show that the expansion coefficient for the O—O bond, 30×10^{-6} deg^{-1}, is of the same order of magnitude as for weak ionic bonds. The only structure where the O—O expansion can be deduced directly from the macroscopic expansion is that of ice, which will be considered later (§14.14). In most materials in which hydrogen bonds play a significant part, bond-angle changes are very important and may have much greater macroscopic effects than bond-length changes (cf. oxalic acid dihydrate, described in §13.16 and listed in Table 14-4). Some of these led at one time to the mistaken belief that abnormally large expansion coefficients were always associated with the direction of hydrogen bonds.

14.12 EFFECTS OF CHANGES OF BOND ANGLE

In structures with a three-dimensional network of strong bonds, changes of angle may be brought about either by distortions of the polyhedra, without change of orientation, or by tilts of polyhedra relative to one another without change of shape.

Table 14-5 gives a number of examples of pairs of structures which are

---

* A correction for difference in thermal amplitude, which becomes important between 300° C and 600° C (cf. §15.9), is not likely to be large in this lower temperature range.

much alike except that in the right-hand member of each pair tilts are prevented. Most of the structures have been described in earlier chapters. The first three substances on the left have perovskite structures with corner-linked octahedra; they are compared either with rutile structures or with other perovskite hettotypes in which the symmetry prohibits tilts. Low quartz (§11.14) is compared with beryl, resembling quartz in its corner-linked tetrahedral framework and general openness, but differing in the fact that the orientations of tetrahedra are fixed by the symmetry. The hydrated sulphates, typical of many others, are of fairly low symmetry, with most atoms in general positions; though the bond strengths are qualitatively like those in alum (§13.24), the bond-angle effects predominate because of the difference in symmetry.

An example of the effect of distortion of shape of octahedra is found in calcite. In the mode contributing most to the thermal amplitudes, we saw in §14.5 that the vibration consisted of an alternate steepening and flattening of the octahedron; the effect of anharmonicity is to produce a change in the mean value of the steepness—the angle between the Ca—O bond and the (0001) plane. In consequence, as c increases with temperature, a must decrease. The ratio of their changes can be shown by geometry to be -2.8, assuming no bond-length expansion. The actual ratio of the co-efficients in Table 14-4 can be explained if we add to the bond-angle co-efficient of -15×10^{-6} rad deg^{-1} a uniform bond-length coefficient of $+4 \times 10^{-6}$ deg^{-1}.

Another example of angle effects of a rather different kind is given by oxalic acid dihydrate (see §13.16). Here the maximum expansion coefficient lies in the (010) plane nearly parallel to the C—C bond direction. We saw, by comparison of the hydrogen compound with the β phase of its deuterated analogue, that the tilt of the oxalic acid molecule is controlled by the angles at the water molecules between three hydrogen bonds, and that the weaker bonds in the deuterium compound are associated with a more symmetrical set of bond angles. The marked difference between the two substances indicates a "softness" of the hydrogen bond angles, and therefore of the modes involving tilting of the molecules controlled by them; we may thus expect large thermal amplitudes associated with changes of tilt, and a temperature dependence of the mean value of the tilt angle such that higher temperatures bring about more symmetrical bond angles. The change of tilt to be applied to the α-$(COOD)_2$ molecule to give it the β-$(COOD)_2$ position has two components: a rotation about the C—C line such as to incline the molecular plane more steeply to (010), and a tilt of C—C about a line normal to it in (010), by which the C—C direction becomes more nearly parallel to (010). The latter component produces an expansion in (010) in the C—C direction, the former a contraction per-pendicular to the C—C direction. The observed expansion (Table 14-4) can be explained as the sum of these two tilt effects and a nearly isotropic bond-length expansion.

A further illustration of bond-angle changes occurs in aluminium hydroxide, Al(OH)$_3$. This (see §13.10) consists of layers of edge-sharing octahedra, which we may treat (to a first approximation) as rigid, held together between layers by hydrogen bonds nearly, but not quite, perpendicular to the plane of the layer, (001). Expansion of the layer is (to this approximation) impossible; expansion perpendicular to the layer involves change in hydrogen bond length; *shearing* of one layer over the next only involves changes in the very soft bond-angles of the hydrogen bonds. By symmetry, shearing can only take place in the (010) plane; it is in this plane that we must look for maximum and minimum expansions. They are found to be 38×10^{-6} deg^{-1}, -6×10^{-6} deg^{-1}, inclined at about $-40°$ and $+50°$ respectively to [001]. As in previous examples, the actual expansions are the sum of the shearing effect, the isotropic bond-length expansion, and any effects due to bond-angle changes in the octahedral layer.

14.13   CRYSTALS WITH WEAK COHESIVE FORCES

The crystals considered in the last section were held together by bonds which had a specific directional character. We now consider examples in which there are in effect strongly bonded blocks embedded in a matrix characterised by weak undirected bonding. Typically, the "blocks" may be molecules and the "matrix" consist of a system of van der Waals bonds. Then we can estimate the overall expansion by assigning separate expansion coefficients, α_1, α_2, and separate lengths, l_1, l_2, to the blocks and the matrix, and taking a weighted mean

$$\alpha = \frac{l_1}{l}\alpha_1 + \frac{l_2}{l}\alpha_2 \tag{14.12}$$

where l_1 is the distance between centres of extreme atoms in the molecule, and l_2 the distance between the same extreme atoms in adjacent molecules (Figure 14-8).

In compounds having molecules of different shapes or sizes, but with the same sort of end groups, l_2 remains constant while l_1 increases with molecular size. Now $\alpha_2 \gg \alpha_1$; hence we expect smaller values of α for crystals with large molecules, or, within the same crystal, for a direction in which a long

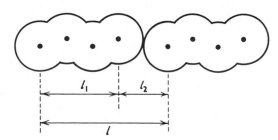

Figure 14-8 Molecules of length l_1, held together by weak cohesive forces across a gap l_2.

axis of the molecule lies. An example is given by the series benzene, naphthalene, anthracene (Table 14-4). The molecules are planar and approximately parallel. The principal expansion coefficient in the direction most nearly perpendicular to the molecular plane, i.e., parallel to the shortest dimension of the molecule (quoted first in the table), is, as expected, the largest. The other coefficients, lying more nearly in the plane of the molecule, decrease as the molecular size increases. (The numerical values quoted are not known accurately enough to allow any more elaborate calculations.)

Another example is calcium hydroxide, $Ca(OH)_2$ (§13.9). Suppose we ignore the hydrogen atoms. Let l_1 be the thickness of the layer in the c direction from oxygen to oxygen, and l_2 that of the interlayer gap; then $l_1 = 2uc$, $l_2 = (l - 2u)c$, where u is the oxygen parameter, of magnitude 0.23. The macroscopic expansion coefficient normal to the layers is given by

$$\alpha_c = 2u\alpha_1 + (1 - 2u)\alpha_2 \qquad (14.13)$$

where α_1 is the expansion coefficient of the layer in the c direction, equal to α_a if the bond angles in the layer do not change, and α_2 is an effective "interlayer coefficient." Substituting values for α_a and α_c from Table 14-5 we have $\alpha_2 = 53 \times 10^{-6}\,deg^{-1}$. This is an order-of-magnitude estimate of the expansion coefficient associated with the weakest cohesive forces in the structure.

In this case, however, direct measurements of parameters have been made at $0°$ and $-140°$ C. They confirm that the $Ca(OH)_2$ layer does not change shape. The Ca—O expansion coefficient is $14(\pm4) \times 10^{-6}\,deg^{-1}$, agreeing within experimental error with α_a and α_1. The O—O (interlayer) expansion coefficient is $32(\pm6) \times 10^{-6}\,deg^{-1}$—less than α_2 because of the inclination of O—O to the c axis.

14.14 TEMPERATURE DEPENDENCE OF THERMAL EXPANSION

In the preceding sections we have quoted thermal expansion coefficients without (in general) specifying the temperature range to which they apply. This is only permissible for very rough working, because expansion coefficients are themselves, in principle, temperature-dependent. A theoretical discussion is outside the scope of this book; but we shall consider a number of examples to illustrate the range of effects.

One practical difficulty concerns the accuracy of the information obtainable. Direct observation gives change of length (or volume) with temperature, and from a smoothed curve through the experimental points the expansion coefficient is deduced. When, however, the coefficient is changing rapidly with temperature, the length of the temperature interval between points and the way the smoothing is done can have quite a big influence on the results.

It is usual for thermal expansion to increase with temperature, but there are exceptions. The rate of increase is not easily predictable, and is not necessarily constant, even when changes of phase at transitions have not to be considered.

Figure 14-9 shows some representative results for strongly bonded cubic materials. Though the rates of increase of the coefficient differ from one material to another—and sometimes for the same material at different temperatures (for example, magnesium oxide)—they are roughly of the same order of magnitude for all, representing an increase in the coefficient of about 1 to 3 $\times$ 10^{-6} deg^{-1} for 100° C increase of temperature.

Materials near their melting points generally undergo much more rapid changes. Figure 14-10(a) gives values of the expansion coefficient of ice—and incidentally illustrates the difficulty of getting accurate and consistent results. The three curves shown are from two careful recent investigations, one of which found significant anisotropy between expansions in the a and c directions, while the other did not. The general characteristics of all sets of results are, however, the same: low values of the coefficient at low temperatures, and a rapid increase to very large values as the melting point is approached.

Remembering that in ice the macroscopic expansion coefficient is a direct measure of the O—O hydrogen bond expansion, we note that at about −100° C, well below the melting point, it is of the same order of magnitude, 30 to 40 $\times$ 10^{-6} deg^{-1}, as the expansion of the hydrogen bond in potassium dihydrogen phosphate in the same temperature range (§14.11). For comparison, the macroscopic expansion of potassium dihydrogen phosphate above room temperature is shown in Figure 14-10(b).

One material of particular interest is quartz, whose expansion is shown in Figure 14-10(c). The coefficients of α_a and α_c both increase with temperature, quite slowly at first and then more and more rapidly up to 575° C, where they suddenly decrease to small negative values. This break actually marks a transition at which there is a change of phase (i.e., a change of structure) from low (α) quartz to high (β) quartz. (cf. §15.9.)

Before going further we may ask: do the lattice parameters increase continuously up to their high-quartz values, or is there a discontinuous jump upward from the highest value reached in low quartz? The curve in Figure 14-10(c) was drawn assuming continuous change, but this is not necessarily true. Figure 14-11 shows the possibilities schematically; ABC represents the variation of spacing with temperature if it is continuous, AEC if it is discontinuous. In practice it is hard to make accurate enough measurements in the region BEC to distinguish between the two. For quartz, there is evidence suggesting that the change is actually discontinuous; i.e., it is represented by AEC.

The rapid increase of spacings and expansion coefficients as the transition point is approached can be understood with the help of the potential energy diagram. Low quartz, as we saw in §11.12, has nearly regular tetrahedra,

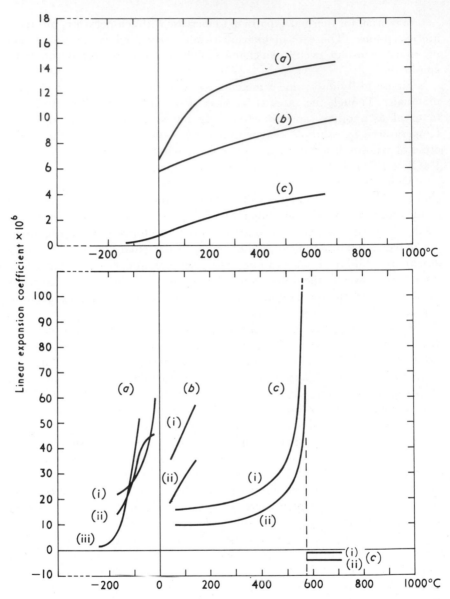

Figure 14-9 Thermal expansion coefficients (in 10^{-6} deg^{-1}) versus temperature for refractory materials: (a) magnesium oxide, MgO: (b) thorium oxide, ThO$_2$; (c) diamond.

Figure 14-10 Thermal expansion coefficients (in 10^{-6} deg^{-1}) versus temperature for materials with rapidly varying coefficients. (a) Ice: (i) α_c (S. La Placa and B. Post, 1960), (ii) α_a (same study, slightly smoothed), (iii) α_a and α_c (R. Brill and A. Tippe, 1967). (b) Potassium dihydrogen phosphate, KH$_2$PO$_4$: (i) α_c, (ii) α_a. (c) Quartz: (i) α_a, (ii) α_c (note the small negative values between 600° and 700° C.)

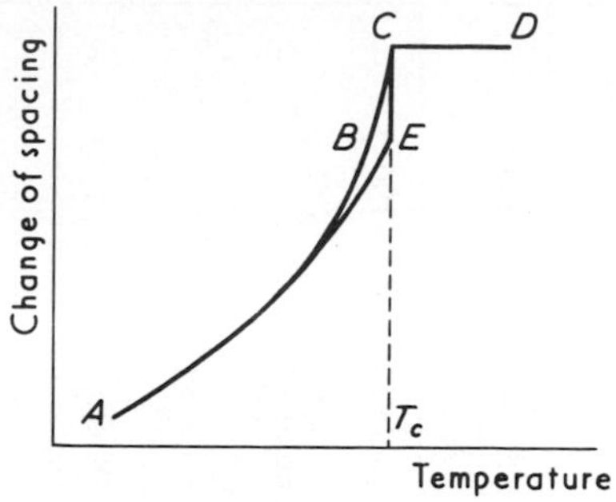

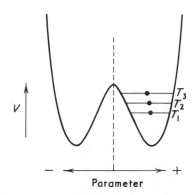

Figure 14-11 Schematic diagram, showing possible variations of spacing with temperature: *ABC*, continuous change below transition at T_c; *AEC*, continuous change to *E*, followed by discontinuity *EC*; *CD*, spacing above transition.

tilted by a small amount from a more symmetrical orientation. We shall show (§15.9) that the tetrahedra are very nearly rigid units, and the tilt angle is the only effectively independent parameter. Whether the tilt is to left or to right of the vertical, the structure is the same—only its orientation depends on the sense of the tilt. Hence the graph of potential energy versus tilt angle (Figure 14-12) must be symmetrical about the line for zero tilt, with a double minimum as shown. Either minimum, though parabolic at the bottom, becomes progressively more unsymmetrical at higher levels: in this temperature region, the anharmonicity of the vibrations increases rapidly, and with it the expansion coefficient. The horizontal line level with the central maximum here represents the transition temperature. (We shall look into this in more detail in §15.3.)

The sudden decrease in expansion coefficient at the transition temperature is easily explained. Changes of tilt and their associated changes in Si—O—Si bond angles contribute to the expansion of low quartz, whereas in high quartz the tilt is zero, and the only possible "static" changes are bond-length expansions and distortion of tetrahedral shape, both likely to be very small. (Contraction of lattice parameters due to increasing thermal amplitudes, as described in §14.10, can be superposed on the other effects in both high and low quartz. This will be considered more fully in §15.9.)

Other silicates undergo changes of expansion coefficients due to changes of tetrahedral tilts, illustrating both the increase before a transition and the

Figure 14-12 Potential energy V versus parameter x for a structure which changes to a higher symmetry when x is zero.

Table 14-6 Expansion Coefficients (in deg^{-1} × 10^{-6}) of Some Structures With Corner-Linked Frameworks

(a) Tetrahedral frameworks

TEMPERATURE RANGE	QUARTZ SiO_2	CRISTOBALITE SiO_2	SODALITE $Na_8(Al_6Si_6O_{24})Cl_2$	NOSEAN $Na_8(Al_6Si_6O_{24})SO_4$	K-LEUCITE $KAlSi_2O_6$	CS-LEUCITE $CsAlSi_2O_6$
0° to 200° C	16, 10		11	11	(mean) 13	0° to 190° C: 16
200° to 500° C	23, 14	220° to 270° C: 9 270° to 330° C: 12 Above 330° C: 1.4	15	19	(mean) 26	3
Above 600° C	−1, −3	1.4	26	4	605° to 820° C: 21 Above 820° C: 3	3
Upper limit of measurement	900° C	1400° C	900° C	900° C	1000° C	800° C

(b) Octahedral frameworks (mean linear coefficients)

LITHIUM NIOBATE $LiNbO_3$	SODIUM NIOBATE $NaNbO_3$	POTASSIUM NIOBATE $KNbO_3$	BARIUM TITANATE $BaTiO_3$
0° to 400° C: 10	0° to 200° C: 10	0° to 200° C: 6	−80° to 0° C: 5 −80° to 20° C: 4 20° to 190° C: 7 20° to 130° C: 4 130° to 200° C: 10
	200° to 360° C: 15	200° to 400° C: 6	
400° to 800° C: 12	360° to 640° C: 21	0° to 420° C: 4	
800° to 1000° C: 15	640° to 800° C: 10	420° to 600° C: 10	

low value after a transition. Values for some zeolite-like compounds are listed, together with those for quartz and cristobalite, in Table 14-6(a). Sodalite and nosean belong to a structure family with one kind of framework, the two leucites to a different one.* All structures in the table are cubic except K-leucite below 605° C, cristobalite below 220° C, and quartz at all temperatures.

An interesting point is that transitions marked by conspicuous changes of expansion coefficient are *not* the transitions at which cubic symmetry is achieved: cristobalite keeps its large (and increasing) expansion coefficient to 330° C, nosean to 525° C, and K-leucite to 820° C. At these temperatures we may safely deduce a change of symmetry by which some possibility of tilting of tetrahedra comes to an end. Cs-leucite, with its larger cation, achieves this at a lower temperature than K-leucite; sodalite, whose "stuffing group" Cl_2 is smaller than the SO_4 group of nosean, has not done so up to the highest temperature studied. It is possible that the cristobalite structure above 330° C may be the idealised high cristobalite described in §11.10.

Structures with corner-linked octahedral frameworks in which tilts are possible may be expected to show the same sort of effect. Sodium niobate (Table 14-6[b]) provides an example. The ideal high-temperature structure with all tilts zero is formed at 640° C. In this, however, the expansion coefficient is still anomalously high. The explanation (as for rutile, §14.11) lies in the small size and central placing of the octahedral cation. Lithium niobate retains its nonzero tilt up to 1200° C.

Other perovskite structures with no octahedral tilts but with off-centre cations provide a contrast to sodium niobate. For the high-temperature cubic phase of barium titanate and potassium niobate (Table 14-6[b]) the expansion coefficient is much the same as for sodium niobate—and has the same explanation. Within each of the lower-temperature phases the expansion coefficient is constant, but close to the transition there is a discontinuous or nearly discontinuous jump. If we take the average change of spacing over a temperature range including such a discontinuity, the corresponding expansion coefficient is even smaller. It is probably these latter values which should be compared with the coefficients for structures with tilts given in the table, because the latter have been calculated as if all spacings changed continuously.

14.15 THERMAL EXPANSION AND TRANSITIONS

The discussion and illustrations of the last section show us how thermal expansion can lead on to transitions. This needs much fuller consideration, and we shall return to further specific examples in Chapter 15, after noting some more general points about transitions.

---

* The names of these minerals, like those of most zeolites, cover a range of compositions differing from the ideal. The samples listed are representative ones fairly close to the ideal.

Chapter 15 PHASE TRANSITIONS

15.1 INTRODUCTION

A *phase*, in the solid state, is characterised by its structure. A solid-state *phase transition* is therefore a transition involving a change of structure, which can be specified in geometrical terms. In this book, we consider only transitions where no change of chemical composition occurs—where a single-phase material changes into another single-phase material of the same composition. The differences between the structures on either side of the transition may be large or small; they may require a partial or complete rearrangement of atoms, or only a slight readjustment of positions. They must, however, be different from thermal expansion, which allows a smooth but not necessarily homogeneous change of all existing variable parameters without creation or destruction of any, i.e., without change of space group.

We shall consider first and foremost the geometrical aspects of transitions. It is not that the geometry is always, in itself, of the greatest practical interest, but that we cannot make convincing studies of the path of a transition unless we know (or can reasonably guess at) the two end states. This is analogous to our need to know the structural details *within* a phase—the atomic positions and variation with temperature and pressure—if we are to understand and predict the physical properties of the phase. Thermodynamic treatments can, of course, be used for a macroscopic approach; but, if they are to be applied to atomic-scale processes, the randomness implicit in them can only be reconciled with the regularities of crystal structure if some physically realistic model can be set up of the ways in which the initial order breaks down and a new order takes shape. We shall refer only briefly to such treatments. Our main concern is with the geometrical relations which must underlie them and allow them to be formulated.

One possible approach to the study of transitions starts from the basic fact that the stable structure under any given conditions is the one with the lowest free energy. This, however, is not very helpful in its application to particular examples, for several reasons. First, calculation of free energies for any but the very simplest structures is still far too difficult: no theoretical treatment using them can be made quantitative in structural terms. Secondly, though measurements of heats of transition give us some useful information, it is hard to get very high accuracy, and at best these measurements give us no details about the separate contributions of different parts of the structure—their evidence is macroscopic, referring to the total effect of the whole unit cell. Thirdly, we know that crystals of the same composition but different structures can coexist for indefinitely long periods at room temperature and pressure (cf. §15.2 for examples). Just because structure A has a higher free energy than structure B does not mean that it will spontaneously undergo a transition to B. There may be need of an initial input of energy to break bonds and give kinetic energy to the atoms, allowing them to find new positions. As they settle into these positions and form new bonds, a greater amount of energy is released. The initial energy needed is called the "energy barrier" or "energy hump," from the analogy of a hill which must be surmounted before a new valley, deeper than the original, can be reached. More insight into the character of transitions can be obtained if we know something of the physical (or structural) origin of such energy humps.

The free energy depends, of course, on other variables besides temperature—for example, on pressure and on electric field. The hill is a multi-dimensional barrier whose height varies with all these thermodynamic parameters. In consequence, transitions may be brought about in different ways. Transitions due to change of pressure are being increasingly studied nowadays, and transitions due to change of electric field are also known. Sometimes a change of one variable, for example, electric field, merely alters the temperature at which a particular transition takes place; sometimes it gives rise to a wholly new phase—just as two passes out of the same alpine valley may lead into one parallel valley or into two different valleys separated from one another by a high mountain range. In studies with large single crystals, or twinned crystals, local pressures may have a considerable effect on the transition temperature; with very small crystals, surface effects (akin to a surface tension) may also impede or encourage a transition. These possibilities have to be kept in mind when detailed studies are being made.

In the next few sections, we shall consider the characteristics of some particular kinds of transitions before attempting any general classification, which follows in §15.8. The remainder of the chapter describes and discusses examples of transitions in particular materials.

15.2 RECONSTRUCTIVE AND DISPLACIVE TRANSITIONS

Most of the older information about solid-state transitions came from mineralogical evidence. The phenomenon of *polymorphism*—the existence of essentially different crystalline species with the same chemical composition—was well known before X-ray analysis was available to determine the nature of the different structures. Examples were the occurrence of SiO_2 as quartz or tridymite or cristobalite, TiO_2 as rutile or anatase or brookite, $CaCO_3$ as calcite or aragonite. Different polymorphs were stable in different temperature ranges—e.g., quartz below 870° C, cristobalite above 1470° C, and tridymite in the intermediate range—but transitions between them were very sluggish, and the high-temperature forms could exist indefinitely at room temperature. On the other hand, pairs of polymorphs had been found which underwent very rapid transitions, and of which the high-temperature form could not be "quenched in." The 575° C high-low transition in quartz was an example. Such transitions could be carried through, reversibly, without destruction of the single crystal. In this they were a contrast to the sluggish transitions, which always involved break-up into small randomly-oriented crystal grains. Again, some low-symmetry minerals, for example perovskite, $CaTiO_3$, gave evidence by their morphology and twinning of having originated in a transition from a single crystal of a higher-symmetry polymorph.

This was the sort of evidence that led to the formulation of a distinction between *reconstructive transitions* and *displacive transitions*. The differences are illustrated in Figure 15-1, a diagram resembling that in the important original paper by M. J. Buerger which contributed much to the subsequent development of the subject. Buerger pointed out that in the sluggish changes the "structures are so different that the only way a transformation can be effected is by distintegrating one structure into small units and constructing a new edifice from the units. Such a transformation is appropriately called a reconstructive transformation." The energy barrier involved in these transitions is high. In the rapid transitions, the displacive transitions, it is low. The distance to be moved by each atom is small, and the position to which it has to move is clearly defined; the change of structure is propagated through the crystal in much the same way as a thermal wave. We shall consider displacive transitions more fully in the next section. Here we need only note their essential features: the atomic displacements are small; no interatomic bonds are broken and no new ones formed; a single crystal of the high-temperature form carefully cooled through the transition and heated again remains a single crystal. In reconstructive transitions, there is extensive breaking of bonds: to achieve suitable rejoining, fragments of structure change orientation in an irregular way, giving crystallites of the new phase with little systematic individual orientation relationship to the old. In displacive transitions the *topology* of the linkage pattern is unchanged; in reconstructive transitions it is drastically altered. This is shown in Figure 15-1. Transitions between the three structures in (a) are displacive, while those between (b) and any in (a) are reconstructive.

A reconstructive transition is thus characterised macroscopically by its sluggishness and by the irregular orientation of crystallites of the new phase, and structurally by large changes in the topology of the linkage. It is generally associated with a large latent heat, and it occurs at a fixed temperature.

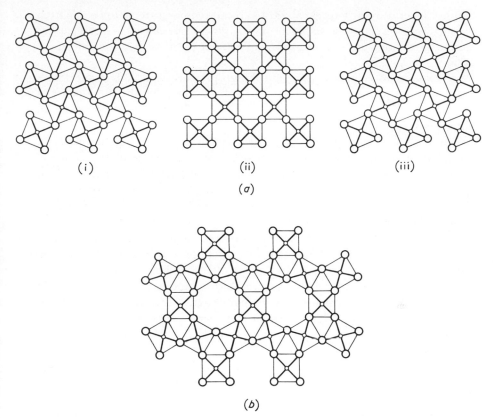

(i)   (ii)   (iii)

(*a*)

(*b*)

Figure 15-1   (a) Diagram of hypothetical structures related by displacive transitions: (i) and (iii) are the low-temperature structure in *obverse* and *reverse* positions; (ii) is the high-temperature structure.

    (b) Diagram of hypothetical structure with different linkage pattern from those in (a); if there were a transition between them, it could only be reconstructive.

Displacive transitions can much more easily escape detection than can reconstructive transitions. Latent heats, if present, are generally small. The transitions are sometimes—not always—spread out over a finite temperature range, and even if they are ideally sharp, they are sensitive to nonuniformities in the conditions, for example to the pressures imposed on them by neighbouring crystals in a large lump of material.

Though reconstructive transitions involve more conspicuous changes than do displacive, it is probably true that they are of less common occurrence, and examples where they have been studied in detail are certainly fewer. Even when the conditions of stability of different polymorphs are known, direct transitions between them are not always clearly established. Conditions which provide the necessary energy for the breakdown of the less stable phase, carrying it over its energy barrier, are not necessarily suitable for the regular crystallisation of the other, and a disordered or glassy phase may result—as in attempts to synthesise quartz by a dry process. Just as high-temperature structures can exist at room temperature, with no apparent tendency to revert to the stable form, so high-pressure structures can exist at atmospheric pressure—for example, coesite, one of the high-pressure forms of silica.

Experiments combining the use of high pressures and high temperatures provide examples of reconstructive transitions. For example, the α-phase of Fe_2SiO_4 or Ni_2SiO_4 (with the olivine structure, §11.9) can be converted in this way to the γ-phase (with the spinel structure, §11.3). On the other hand, α-Mg_2SiO_4 and α-Co_2SiO_4 undergo a transition to a different phase, β, which is based like spinel on a cubic close packing of oxygen atoms, but uses a different selection of octahedral and tetrahedral sites, resulting in Si_2O_7 groups. The olivine → spinel transition (or perhaps the more recently discovered $\alpha \rightarrow \beta$ transition) is believed to occur in deep layers of the earth's crust, where it is geologically important.

A study of the anatase → rutile transition, with a single crystal of anatase as starting material, has shown that the orientations of the new crystallites, though irregularly distributed, are statistically not wholly random, but have preferred directions with respect to the old. Another interesting example occurs in $LiAlSi_2O_6$, which has two reconstructive transitions. The phases are named I, II, and III, in order of increasing temperature, and the two latter have complete Si/Al disorder. Phase I, the mineral spodumene, is monoclinic; Phase II, a "stuffed keatite," is tetragonal; Phase III, a "stuffed quartz," is hexagonal. It has proved possible to study the detailed course of the III → II transition from the high-quartz framework to the keatite framework.

Reconstructive transitions also occur between some of the different forms of ice, Ih, Ic, II, III, V (cf. §13.18 to 13.21). This system provides an illustration of the sort of effect due to energy barriers: ice III prepared at $-40°$ C and 3 kilobars can be quenched to $-196°$ C and atmospheric pressure without formation of ice II, even though this is stable in the intervening range. A similar effect has been observed in the KNO_3 system (§15.22)

A transition between calcite and aragonite is sometimes referred to as a typical example of a reconstructive transition, but experiments at high pressure cast doubt on it. What seems to happen—at least as the first stage—is an orientation-switching transition from calcite to the disordered-calcite structure. The problem is further discussed in connection with the related materials $NaNO_3$ and KNO_3 (§15.22).

A reconstructive transition of technological importance is the $\beta \rightarrow \gamma$ transition in dicalcium silicate, Ca_2SiO_4. The γ phase, isomorphous with olivine (§11.7), has Ca in 6-coordination; the β phase is quite different, with 9-coordinated Ca; the volumes per formula unit are 99 Å^3 and 86 Å^3 respectively. The $\gamma \rightarrow \alpha'$ transition occurs at $800°$ C on heating, missing the β phase altogether, but on cooling the $\alpha' \rightarrow \beta$ transition occurs at $670°$ C, and the $\beta \rightarrow \gamma$ transition not till $520°$ C or below. The volume change and atomic rearrangements are so considerable that even quite small crystals break into fragments. If the crystals form part of a cement clinker, the clinker breaks up; the process is known as "dusting." Since the transition may be delayed by supercooling, it sometimes happens after the specimen has been kept for some time at ordinary temperatures. Fortunately, it is possible to prevent it altogether by the introduction of suitable impurities, which stabilise the β form.

By contrast, the higher transitions in Ca_2SiO_4, $\alpha \rightleftarrows \alpha' \rightleftarrows \beta$, take place with much less drastic rearrangement. They are not reconstructive but displacive: crystals retain their orientation relationship, and the cohesion of the clinker in which they occur is undamaged.

15.3 DISPLACIVE TRANSITIONS

The criterion for a displacive transition is that the new structure should be capable of being derived from the old by a continuous and ordered process of atomic displacement, the largest displacements being so small that no bonds involved in the cohesion of the structure are broken. The path from one structure to the other is reversible for individual atoms. It is not necessary that the actual changes should be continuous; discontinuities are to be expected, but their magnitudes must be small and they must fit into a scheme which is essentially continuous.

The simplest cases are those in which the structure has only one physically-independent variable (apart from bond length, which in the first instance we can treat as constant). Low quartz is an example. As we shall see (§15.9), if we take the SiO_4 tetrahedron as rigid the only physically-independent variable is its tilt about the diad axis. This is small, and in consequence all atoms have only slight displacements from positions of higher point symmetry. Cooperative movement of the atoms towards these special positions, which results from decrease of tilt, does not make any great change in interatomic distances. No bonds are broken; the topology of the structure is unaltered. This statement holds good even if the last stage towards reaching the special positions is a discontinuous jump (a possibility considered in §14.14—cf. Figure 14-11).*

One might almost ask whether this process is a transition at all. The answer lies in the fact that *there is a change of symmetry*, in which the previously variable tilt parameter becomes zero. A transition of this kind is sometimes called *pseudosymmetric* because of the close approximation of the low-symmetry structure to the higher symmetry of the high-temperature form.

It is not necessary to a displacive transition that the changes should concern only one physically-independent variable; we shall have examples

---

* It is important to remember that the large changes in thermal expansion leading up to the transition are not themselves part of the transition; they are merely premonitory symptoms.

to the contrary (§15.11). In practice, however, it often happens that in a given temperature range only one variable is sufficiently temperature-sensitive to need to be considered; in quartz, for example, variations in tetrahedral shape were small enough to be ignored. We shall begin by considering one-variable structures. Here we take up again the line of thought we were following at the end of Chapter 14.

The important variable in any particular case is controlled, below the transition, by forces with a low force-constant. This means that cooperative changes in atomic position depending on this particular variable are easy, and therefore oscillations involving such changes of position are of large amplitude.

What happens to the low-temperature structure when the amplitude, which is increasing with temperature, becomes as large as the mean value of the displacement parameter, which is decreasing? The answer is that, in general, this represents the transition temperature; the discontinuous change of parameter at the transition is equal to the root mean square (r.m.s.) value of thermal amplitude. The evidence in support of this statement is experimental, and will be given in §15.9 and 15.17; meanwhile, its plausibility can be argued as follows.

Consider a two-dimensional structure whose mean atomic positions are as shown in Figure 15-1(a)(i), and suppose that the squares tilt as rigid bodies. All atoms move concertedly, and it is sufficient to consider the movement of one of them, with mean position A, about the centre of the square, O (Figure 15-2). The mean value of the tilt angle is AOQ; the amplitude of oscillation of tilt is represented by AOC. (We do not have to

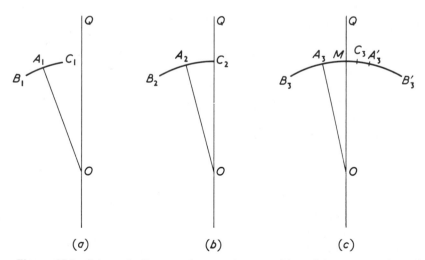

(a)   **(b)**   **(c)**

Figure 15-2 Schematic diagram of approach to transition of the structure shown in Figure 15-1(a) (i). It is assumed that the square moves as a rigid body about its centre O, and hence that every corner atom moves on the arc of a circle about O, as shown here for one of them. Here A is the mean position of the atom, and BC its oscillation range: (a) below the transition; (b) at the transition; (c) above the transition.

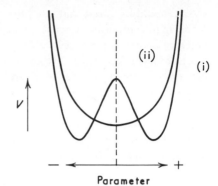

Figure 15-3 Schematic diagram illustrating a change of the potential energy curve at a pure displacive transition: (i) below the transition; (ii) above the transition. (Differences of (i) and (ii) are exaggerated.)

assume that $AOB = AOC$, as would be true for a harmonic oscillation, though it has been drawn in this way.)

Figure 15-2(a) represents the state of affairs below the transition, (b) that at the transition. Suppose now the amplitude still increases, so that, as in (c), $A_3OC_3 > A_3OQ$. The oscillation has obviously ceased to be simple harmonic, because the restoring force at C_3 is not directed backwards to A_3 but onwards to its mirror image A_3', and hence the complete oscillation range becomes B_3OB_3', with mean position M (on OQ) and amplitude MOB_3. The amplitude is thus suddenly nearly doubled. But unless at the same time the frequency changes, this means an abrupt increase of energy, and hence, by the Boltzmann principle, a redistribution of energy out of this particular kind of tilting vibration and into some other vibration which we have hitherto been able to treat as negligible—for example, a stretching vibration of the square. In effect, what happens is that the potential energy curve changes shape, losing its double minimum (Figure 15-3).

This approach suggests that there is an essential discontinuity in a displacive transition. If we had considered only the mean positions of the atoms in the "static" structure, we could have pictured them moving smoothly and continuously to their new positions; allowing for thermal motions, we see that the last step is an abrupt one. Perhaps if we were able to take proper account of the anharmonicity, the break would appear as an anomalous region rather than an actual discontinuous jump. In any case, a discontinuity of the order of magnitude of the thermal amplitude is too small to be easily detected, and may often be ignored without serious harm.

If there is more than one physically-independent parameter which is changing with temperature, then their interaction may be such that at some temperature a new configuration can be found, of lower energy, reached by a discontinuous jump in both parameters. Discontinuities in such cases are likely to be larger than when there is only one parameter. They could conceivably happen without change of symmetry. There might very well be several transitions of this kind occurring in sequence as the highest-symmetry form—the aristotype—is approached. Examples will be found in §15.11.

It should be noted that in a pure displacive transition there is no disorder in the high-temperature form—other than the disorder of thermal motion, which of course is present in all crystals. This is different from the original point of view of Buerger. He supposed that the high symmetry of the high-temperature form such as that shown in Figure 15-1(a)(ii) was only an average effect, made up from the two orientations of the low-temperature form, (i) and (iii). He thought of the crystal above the transition as "a mosaic of small regions (a complexion that is continually changing) of obversed and reversed forms in equal amount." This would correspond to a potential well that had kept its double minimum, but had both sides filled to the same level (Figure 15–4)—representing equal occurrences of the two orientations of the crystal. But, as we have seen, purely displacive transitions are characterised by the conversion of a double minimum to a single minimum at the transition point. The dynamic domain effect envisaged by Buerger in the high-temperature form may occur for related kinds of transitions, such as the orientation-switching transitions and the "hydrogen-hopping" transitions to be dealt with in the next two sections. But the very fact that in them the forms above the transition *necessarily* involve disorder other than ordinary thermal vibration means that they are to be distinguished from the purely displacive transitions discussed in this section.

It is true, of course, that *all* transitions are likely to involve disorder if examined sufficiently closely in a very narrow temperature range near the transition point. This has been observed experimentally for quartz, just below the transition, and (not surprisingly) found to be more conspicuous for less perfect quartz crystals. The concept of a "purely displacive transition" is, in fact, an idealisation which may never be exactly illustrated in practice. But, just as it is useful to ignore thermal motion when classifying crystals as perfect or imperfect, so it is useful to ignore these very short-lived effects— which are of the nature of thermal fluctuations—when classifying transitions as purely displacive or otherwise.*

The atomic movements required for a displacive transition are propagated, as we have seen, by the same sort of mechanism as a thermal wave (though possibly a very anharmonic one), and not by diffusion. This means that the transitions are rapid. It is sometimes supposed, for this reason, that the high-temperature form can never be "quenched in" by rapid cooling through the transition range. Ideally this may be so, but in practice it is sometimes possible because of crystal imperfections, which impede the

---

* When considering possible disorder of atomic positions (as distinct from random occupation of ordered sites by different kinds of atoms), it is wrong to assume disorder, or local order different from the average periodic order, unless there are some facts which cannot equally well be explained by an ordered structure undergoing ordinary thermal vibration—the latter is the simplest assumption, which must be accepted until there is evidence against it. The possibility remains open that transitions at present classed as pure displacive may have to be reclassed as "domain-displacive" if local order different from the average order can be shown to persist well above the transition temperature.

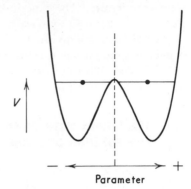

V

Figure 15-4 Oscillation ranges and mean position parameters for the occurrence of two orientations.

$-$ $\longleftarrow$ $\longrightarrow$ $+$

Parameter

propagation of the transition. Nonstoichiometry is particularly effective in this respect.

Although it is easiest to illustrate the idea of displacive transitions by thinking of tilts of octahedra or tetrahedra, as we have been doing, these are not the only possibilities. An independently-variable parameter may, for example, define the shape of a polyhedron (i.e., the degree of its distortion from regularity) rather than its tilt. The elongated octahedron in barium titanate (§12.6) is a case in point; the approach to the transition is marked by the approach of this octahedron to cubic symmetry (§15.10).

We have been discussing these transitions in terms of mean atomic positions and of amplitudes of thermal vibrations, because they are most closely relevant to the geometrical approach used in this book. In addition there are of course changes of *frequency*, which are of great importance, but which lie outside the scope of the book, except for a very brief mention.

Much work, both experimental and theoretical, has been done in this field. The modes most closely associated with displacive transitions have frequencies which are very sensitive to temperature, tending to zero at the transition. The aim is to measure the frequency of these "soft" modes, and, by identifying them with configurations of atoms, to learn something about the forces involved. There are, however, considerable difficulties. At the transition, as we have seen (§14.14), the modes become anharmonic, and this can be formally expressed by treating them as damped oscillations. If the damping is light, they will still give observable resonance effects with incident radiation of the correct frequency; but if it is heavy, the response is smaller and less sharp and harder to recognize with certainty.

The difficulty is not accidental. Physically, it means that if we are so close to the transition that atomic positions are indeterminate and fluctuating, the structure can no longer vibrate as if it were truly periodic. Modes which are compatible with the atomic arrangements common to high-temperature and low-temperature forms, whether these depend on local order or on periodic repeats, may be almost unaffected, but modes depending on oscillation of parameters which become zero at the transition lose their clear-cut identity. The experimental results do not allow of easy generalisation.

In discussing individual transitions, we shall therefore draw on this kind of evidence only when it seems specially relevant to particular points at issue.

We shall illustrate the general ideas of this section by considering, in §15.9 to 15.17, a number of displacive transitions in different materials.

15.4 TRANSITIONS WITH ORIENTATION SWITCHING AND ROTATION OF GROUPS OF ATOMS

In structures containing small molecules or molecule-ions, it may happen that their orientation, either in a plane or in three dimensions, is controlled by rather weak forces. This can have two consequences. Rotational thermal amplitudes (librations) are large, and alternative orientations of not much greater energy are possible (perhaps accompanied by some displacement of neighbouring atoms). With increasing temperature the rotational amplitude increases, and so does the probability of a "wrong" orientation for any molecule. Eventually, if nothing intervenes, we would expect that *either* the limited rotational amplitude passes over into complete rotation (cylindrical or spherical, as the case may be) *or* the increased number of "wrong" molecules produces such extensive distortion of the structure that the distinction between "wrong" and "right" is lost and either orientation is adopted at random. In both cases, the "average molecule" has taken the place of the actual molecule, and is more symmetrical than the latter. The rotating molecule has cylindrical or spherical symmetry, and the two-position average molecule is compounded of the sum of two appropriately-oriented molecules built from half atoms. The corresponding change in the symmetry of the structure is associated with a transition.

In the early days of structure analysis the second possibility was not generally appreciated, and it was assumed that transitions were due to the onset of complete rotation. As the number of refined structure analyses at different temperatures has increased, more and more examples have been found in which the initial transition involves disorder rather than complete rotation. If rotation occurs at all, it does so at much higher temperatures, far above the first transition. It is better to refer to these transitions as orientation-switching transitions rather than rotational transitions.

Looking at it more closely, we find that in the disordered form the choice of molecular orientation is not really random. Each "wrong" molecule displaces its immediate neighbours slightly, and through them influences the orientation of the next molecule. Thus for the crystal as a whole we expect small ordered domains with rapidly shifting boundaries, like those pictured by Buerger for the structure above a displacive transition. Only the average structure is detectable by Bragg diffraction, but studies of diffuse background scattering can sometimes throw light on the character of the dynamic domain texture.

Orientation-switching transitions are fast, not sluggish; but (as for other order-disorder processes) the degree of disorder is a continuous function of temperature, and quite a large temperature interval is needed for the complete change from a state with no wrong orientations to a state with half the orientations wrong. Often the onset of disorder is marked by a sharp change of slope of specific-heat or thermal-expansion curves (cf. Figures 15-16 and 15-17). This, however, is not itself the transition: it is merely the beginning of the premonitory region. The peak at the transition is typically very sharp—nearly as sharp as in a displacive transition. (This need not surprise us, since switching is generally accompanied by some displacement of neighbours.) Compare, for example, the thermal expansion of quartz between 300° C and 600° C, shown in Figure 14-10(c), with that of sodium nitrate between 100° C and 170° C, in Figure 15-16, to be considered in more detail later (§15.20).

15.5 TRANSITIONS INVOLVING HYDROGEN BONDS

Pseudosymmetric transitions very similar to displacive transitions, but with certain distinguishing features, can occur in some compounds containing hydrogen bonds (for example, potassium dihydrogen phosphate, §13.12 and 13.13). In these structures, only small displacements of any atoms—hydrogen atoms or others—are needed to turn the low-symmetry form into a higher-symmetry form, but the higher symmetry would require the hydrogen atoms to lie at the midpoints of hydrogen bonds, on special positions—symmetry centres or rotation diads. Now we know (§13.2) that symmetrical hydrogen bonds are rare, and that in some cases the apparent symmetry is a consequence of disorder, which distributes the hydrogen atoms equally between two possible off-centre sites in the bond. When this happens, the hydrogen atom must somehow be capable of getting from one site to the other with reasonable ease even though its thermal energy is not great enough to surmount the potential energy hill separating the sites. By an extension of the same metaphor we have used in speaking of an energy hill, the process by which the atom gets to the other side of the hill without going over it is called *tunnelling*. It is illustrated in Figure 15-5. By means of it, both wells are filled to the same level, recognisably below the summit of the hill. The thermal amplitude within each well is significantly less than the displacement of the mean position.

The diagram must *not* be taken to imply that the hydrogen atom alone moves in the structure, leaving everything else unchanged. It means rather that with the hydrogen movement the whole piece of structure has switched direction; when the wells are filled to an equal height, there are equal volumes of the crystal in each orientation. The concerted movement of other atoms is brought about by the changed relationship of the hydrogen atom to its anion neighbours in the bond; that which was originally donor becomes

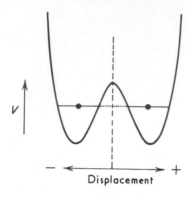

Displacement

Figure 15-5 Schematic diagram showing tunnelling.

acceptor, and vice versa. By the effect of each on its neighbour, the movement is propagated through the crystal as a sort of wave—not an ordinary simple harmonic wave, but one which involves abrupt reversals of direction of small domains of the structure. Waves of this kind can be dealt with mathematically, and are known as *pseudospin waves*.* Within each domain the structure resembles (though it is perhaps not identical with) one of the orientations of the low-symmetry form.

In considering individual examples of pseudosymmetric transitions involving hydrogen bonds, it is thus important to know whether the hydrogen atoms in the high-temperature form are truly central or only statistically so. If the atoms are truly central, the potential well has become single at the transition; and the dynamic displacements of the hydrogen atoms, both above and below the transition, can be described simply as amplitudes of thermal vibration. Transitions of this kind would be classed as pure displacive transitions. On the other hand, statistically-occupied off-centre positions imply the continued existence of a double well above the transition, and a time for the process of switching between orientations which is short compared with the average time spent by a given piece of the structure in the same orientation without switching. It is this latter type which we shall call a "hydrogen-hopping" transition.

There is a considerable resemblance between these "hydrogen-hopping" transitions and the orientation-switching transitions described in the last section. In effect, the hydrogen jump in its bond is equivalent to a reversal of orientation of the O—H group, and, like orientation-switching of other groups, it is accompanied by regular displacements of other atoms. In fact, both types of transition can be thought of as *displacive transitions to which a particular localised geometrical disordering has been added, organised in small domains*. In some cases at least, the "hydrogen-hopping" transitions tend to

---

* This name comes from the fact that the mathematical formulation resembles that of spin waves in magnetic materials. Spin in magnetism is an "either/or" property conveniently pictured as "up" or "down." It is this "up-or-down-ness" which is needed to describe the two orientations of the structure; the word "pseudospin" has nothing to do with anything else that one ordinarily thinks of as "spin."

be sharper than the orientation-switching ones; that is, the temperature range over which the disorder builds up is narrower. Whether the "hydrogen-hopping" is to be thought of as triggering the transition, with other atomic displacements following as a consequence, or whether displacive movements of other atoms provide the conditions which allow "hydrogen-hopping," is a question which would need to be considered individually for any given example, and to which an answer may not always be possible.

Examples of two "hydrogen-hopping" transitions will be described in §15.17 and 15.18. Another possible example occurs in ice III, where the hydrogen atoms appear to be disordered in the bonds at the upper end of its temperature range, but ordered at the lower end. Here, however, further evidence would be needed to distinguish between an orientation-switching and a "hydrogen-hopping" (tunnelling) mechanism.

Not all transitions in structures with hydrogen bonds necessarily involve order-disorder of the hydrogens. If the hydrogen atoms remain attached to the same donor throughout (as in thiourea, §15.17), the transition is a simple displacive one.

15.6 MARTENSITIC TRANSITIONS*

Martensitic transitions were so called because they were first studied in martensite, a compound occurring in steel alloys. They resemble displacive transitions in that the process can be followed as a continuous concerted movement of atoms from the old positions to the new, but differ in the size of the displacements, which are so large that the environments of all (or most) of the atoms are changed in the process. Moreover, they are generally not reversible on the atomic scale—individual atoms do not retrace their paths, and the original single crystal is not re-formed. They are rapid transitions, but extend over a finite temperature range. Generally they involve a considerable shear of the unit cell. They have been given the alternative name of *dilatational changes*.

An example is caesium chloride, whose room-temperature structure (§4.8), changes at 460° C to that of sodium chloride (§4.7). The reverse transformation can be brought about in potassium iodide by the application of pressure. The relationship can best be seen by comparing the primitive unit cell of CsCl—a cube, and therefore a rhombohedron with a 90° interaxial angle—with the primitive unit cell of NaCl, which is a rhombohedron with a 60° interaxial angle. Both have cations at 0, 0, 0, and anions at 1/2, 1/2, 1/2. The structure of NaCl can therefore, ideally, be converted into that of CsCl by a homogeneous compression along the triad axis and extension at right angles to it. The magnitude of this lattice distortion is much greater than

---

* At one time, these transitions were alternatively called *diffusionless transitions*, but this is now too vague a name, because it would also include displacive transitions, "hydrogen-hopping" transitions, and orientation-switching transitions.

that found in displacive transitions. The compression of the octahedron of anions round each cation, and the approach of the two nearest anions on the triad until they are as close to the cation as the other six, thus completing a cube, correspond to a change of coordination number from 6 to 8. This increase of coordination number is here associated with increasing pressure or decreasing temperature.

15.7 SUBSTITUTIONAL ORDER-DISORDER TRANSITIONS

In materials where there are two kinds of atoms sufficiently distinct to occupy separate kinds of sites in the low-temperature structure, but sufficiently alike to cause only a small difference of energy if they go into a "wrong" site, increase of temperature will allow increased disorder, until at high temperatures both kinds of atoms occupy both kinds of site indiscriminately, and blot out any differences between them. The process is cooperative, because an atom in a "wrong" site tends to blur the distinction between "right" and "wrong" in its neighbourhood. To become disordered, atoms must break bonds and move by diffusion to new positions chosen at random from "right" and "wrong" sets. On cooling, the ordered state, though the stable one, will only be re-formed if it is possible to hold the crystal for sufficient time at a temperature high enough to allow diffusion yet low enough to provide an energy incentive for ordering; otherwise the high-temperature form is quenched in. Substitutional order-disorder transitions thus resemble reconstructive transitions in being sluggish and preventable by quenching; they resemble displacive transitions in maintaining complete orientation relationships between old and new structures and involving small changes, if any, in position parameters; they differ from both, and resemble orientation-switching transitions, in having a finite temperature interval from the first onset of disorder to its completion.

An example occurs in copper-gold alloy, Cu_3Au (§4.12). In the high-temperature form, with complete disorder, the average occupant of an atomic site is $Cu_{\frac{3}{4}}Au_{\frac{1}{4}}$; and the four occupied sites, of which one was distinct in the low-temperature form, are now all equivalent. The space group has changed from $Pm3m$ (No. 221) to $Fm3m$ (No. 225); the new structure is that of copper (§4.2).

Many other alloys with order-disorder transitions have equally simple high-temperature forms, but the low-temperature forms are very varied and not always easily predictable.

Order-disorder transitions also occur in ionic and semipolar compounds, the cations exchanging sites. In the felspars, with Si/Al interchanges, the disordering process is extremely sluggish, and the ordering process even more so—indeed, it is doubtful if nearly complete ordering has ever been achieved in the laboratory except when the Si:Al ration is 1:1. The situation is complicated by the fact that there are probably also displacive transitions, and it is by no means certain that the order-disorder changes themselves result in transitions—they may only facilitate or impede the occurrence of the displacive transitions.

A special type of ordering transition (which is better considered in conjunction with displacive transitions) is one in which the electron distribution on the atoms becomes ordered, without any diffusive movement of the cores. This is possible when the atoms concerned have more than one valency state. An example is magnetite, Fe_3O_4, with the inverse spinel structure (§11.3). The high-temperature form has equal numbers of Fe^{3+} and Fe^{2+} distributed at random over sites of the same equipoint; the low-temperature form, by reduction of symmetry, provides separate sites for Fe^{3+} and Fe^{2+}, and the electrons move to satisfy this. The structure will be described in more detail later (§15.25—cf. also §12.13).

15.8 CLASSIFICATION AND SURVEY OF TRANSITIONS

It is perhaps useful to summarise the features of different kinds of transitions in a classification scheme. Though the distinctions drawn are not completely sharp and one may find examples of intermediate character, it can serve as a guide.

(1) No orientation relationships; complete fragmentation of single crystals: *Reconstructive.*

(2) Close orientation relationships; interchange of atoms, randomly, by diffusion: *Substitutional order-disorder.*

(3) Coarse orientation relationships, with large change of shape; no diffusion: *Martensitic.*

(4) Close orientation relationships; changes reversible in single crystal; changes of atomic position rather small, with reversible path: *Reversible.*

   (i) No breaking of bonds; topology completely unchanged: *Pure displacive.*

   (ii) Order-disorder changes of hydrogen atom in a bond; small displacements of other atoms: *Hydrogen-hopping*

   (iii) Order-disorder changes of orientation of small groups of atoms; small displacements of other atoms: *Orientation-switching.*

   (iv) Order-disorder changes of intra-atomic state.

The last-named category, 4(iv), has not been dealt with earlier, and needs a word of explanation. It comprises transitions which are pure displacive as far as the position of atomic centres is concerned, but the displacements are a consequence of ordering of the internal state of the atoms. They include transitions due to ordering of valency electrons, such as that in Fe_3O_4, mentioned at the end of §15.7. They also include magnetic transitions, brought about by ordering of the magnetic spins. The interatomic forces due to the set of ordered magnetic spins are additional to the ordinary structural forces already present, though generally much weaker; if their symmetry is lower, a transition results—characteristically a small homogeneous distortion. For example, NiO is cubic, with the rock-salt structure, above its Néel temperature of 250° C, and rhombohedral when magnetic ordering sets in below this temperature. A transition from cubic to rhombohedral is also reported for Fe_3O_4 (Chapter 15, reference 22) which has the spinel structure. Again, $RbCoF_3$ has the ideal perovskite structure at room temperature, and becomes tetragonal (probably with no change of atomic coordinates) at its Néel temperature of 100° K, its axial ratio decreasing steeply at first, then more gradually, levelling off to a value of $1 - 0.003$ at 4.2° K. There are however examples of crystals, such as $\alpha\text{-}Fe_2O_3$ (hematite—cf. §11.4) at 675° C, $DyAlO_3$ at 3.5° K and perhaps $KCuF_3$ (cf. §12.5) at 243° K, in which magnetic ordering does not lead to a change of structure, though it may give an anomalous thermal expansion.

Transitions involving changes in the character of interatomic bonds are not given a separate category. They include transitions due to onset of the Jahn-Teller effect, where the change in bond system involves a distortion of octahedron shape without off-centring of the cation (as seen for example in $KCuF_3$ at room temperature). In a simple structure, the onset of the effect may result in a displacive transition to a lower-symmetry hettotype—for example, in $DyVO_4$ (zircon family) at 13.8° K.

Two other examples of particular interest may be mentioned, both to be regarded as pure displacive transitions, belonging to category 4(i). One occurs in VO_2 at about 67° C, from a monoclinic hettotype of the rutile family (§12.14) to the aristotype (§11.10). In the room-temperature form the V—O bonds are markedly covalent, forming what are effectively V_2O_4 molecules; in the higher-temperature form there are conducting electrons giving the structure metallic properties. The other example is found in GeTe at about 400° C, though here the nature and extent of bond-character change is still questionable. The high-temperature structure is that of NaCl. At the transition, this becomes extended along one triad axis, and the sheets of unlike atoms perpendicular to the axis move so that the distances between them are no longer all equal, but alternately short and long, in the ratio 0.47:0.53. This gives a polar structure, each close pair of layers constituting a dipole, which can reverse very easily by reversing the sense of displacement of layers from their symmetrical high-temperature arrangement.

We may note that, though the classification scheme does not formally include transitions accompanied by changes in chemical composition, there are obviously close analogies between the structural changes due to minor differences of composition and the structural changes in displacive transitions. One illustration is in the chabazites (§12.20); many others are found among solid solutions of simpler compounds, for example, in the perovskite family.

A question which has aroused some discussion is this: if the coordination number changes at a transition, do we expect it to increase or decrease? Part of the difficulty in making an empirical generalisation arises from the inevitable arbitrariness in deciding on the coordination number when the environment of an atom is irregular (cf. §2.5). Even allowing for this, there is no simple rule. For example, in the reconstructive transition in Ca_2SiO_4 (§15.2) the coordination number of Ca increases from 6 in the low form to 9 in the high form; in displacive transitions in $CaTiO_3$ (§15.14) and $LaAlO_3$ (§15.13) the coordination numbers for Ca and La both increase from 9 in the low form to 12 in the high form; but in the martensitic transition in CsCl (§15.6) the coordination number of Cs decreases from 8 to 6. We note a tendency to make, or keep, the cation environment regular in the high-temperature form—and in displacive transitions this generally implies an increase in coordination number. Beyond this it seems clear that other factors, such as the detailed shape of the polyhedron and its linkage with other polyhedra, are more important in their effect on the energy than the actual coordination number.

For comparison with the classification scheme at the beginning of this section, which is essentially geometrical, we may mention the conventional thermodynamic classification into transitions of first order or of second order—sometimes extended to "transitions of higher order." By definition, a first-order transition is one marked by a discontinuous change of energy—a latent heat—and discontinuities in volume and in lattice parameters; it happens sharply at a particular temperature. A second-order transition is one without a discontinuity in volume, or lattice parameters, and without latent heat, but with peaks in the specific heat and in the thermal expansion extending over a finite temperature range. A third-order transition, by analogy, would be one with a discontinuous rate of change of thermal expansion.

This kind of classification, though much discussed in the literature, is of limited usefulness. Reconstructive transitions are obviously first-order, and substitutional order-disorder transitions are second-order. But what about the smaller transitions in category (4), the reversible transitions? At one time it was thought that they were all second-order, and this is an approximation to the truth in the sense that any discontinuities in energy or volume are small. We know now, however, that discontinuities *do* occur in many of them, including even some which are of the orientation-switching type, and the size of the discontinuity is not obviously correlated with the underlying mechanism of the transition. Discontinuities in all properties are not equally marked; those in energy and specific volume generally go together, but discontinuities in individual lattice parameters—subcell edges and angles— (which are often associated with sharp changes of refractive index) may occur without detectable change of volume. Experimental detection of small discontinuities is difficult, since local nonuniformities and stresses, whether due to defects within a single crystal or to grain boundaries between crystals, may "smear out" a truly discontinuous transition over a finite temperature range.

From the standpoint of theory, when discontinuities are very small it becomes necessary in any case to re-examine the postulates, since variations in thermodynamic coefficients usually treated as constants may now become important—for example, their temperature dependences. The mere statement that a transition is first-order or second-order therefore helps us little. The facts established in attempting to make such a statement are the important thing: the nature of the structural differences on either side of the transition; the temperature variation of the structural parameters and the physical properties; whether a discontinuity is observable in any particular property, and if so how large; whether the structural changes involve any disordering of parts, and if so of which parts.

A word may be said here about "higher-order transitions," which are sometimes said to exist on the evidence of a broad anomaly in the thermal expansion. Now if a transition occurs, there must be a definable difference in the state of the system on either side of it—that is what is meant by a transition. For a structural transition, the difference must lie in the set of parameters needed to describe the structure: this is generally made obvious by a change of symmetry, which necessarily requires the specification of a wholly or partly new set of variables, but discontinuous changes in the numerical values of existing variables, without apparent change of symmetry, would also provide an acceptable criterion. (This agrees with the definition given in §15.1.)

Continuous thermal expansion, without change of symmetry, does not constitute a transition, whether or not the thermal expansion coefficient is changing anomalously. Structurally, therefore, there is no such thing as a "higher-order transition." It is however true that anomalies in thermal expansion may indicate intra-atomic changes, completed over a finite temperature range without bringing about a structural transition. In such cases we need to know the nature of the process causing the anomalies before we can speak of it as a transition. For example, though for α-Fe_2O_3 the anomaly in thermal expansion near 675° C is not accompanied by a structural transition, it is associated with the onset of antiferromagnetic ordering of magnetic spins, and is therefore an antiferromagnetic transition.

In the remaining sections of this chapter, we consider in detail a number of examples of reversible transitions belonging to category (4)—transitions where there is no diffusion or substitutional disorder, but a close geometrical relationship between the initial and final structures, which belong to the same family. When the transition is directly from the aristotype to a hettotype, the change of symmetry always represents a loss of some symmetry elements and no gain of new elements; if the transition is between two hettotypes, there may be gains as well as losses, but never a gain of symmetry elements not possessed by the aristotype. There are of course many other illustrations. Earlier in the book, for example, we mentioned various structures which have polymorphic forms related by displacive transitions— octahedral structures such as cryolite (§12.11) or the 5-4-3 bronzes (§12.21), tetrahedral structures such as cristobalite (§11.12) and anorthite (§12.17). The examples chosen for individual discussion in the following sections are for the most part those on which detailed work has been carried out, allowing us to see not only the geometrical relationships between the end members but something of the mechanism of the transition. They are chosen to illustrate the variety of principles that may apply to these transitions in general.

15.9 THE DISPLACIVE TRANSITION IN QUARTZ

This is one of the few transitions where measurements have been made over a wide temperature range not only of the lattice parameters but also of the atomic position parameters and thermal amplitudes.

It is easiest to see what is happening if we describe the structure in terms of a tetrahedron which is not regular but has orthorhombic symmetry, as it must have in high quartz. In Figure 15-6(a), the symmetry axes of the tetrahedron are the lines through T parallel to EF and FH and perpendicular to the plane of the diagram. Let the distances between midpoints of opposite edges, measured in these three directions, be h_1, h_2, h_3 respectively. We want to find the parameters of high quartz—the lattice parameters a and c, and the position parameter x—in terms of h_1, h_2, h_3. The relations can be found from the diagram:

$$a = h_2 + \sqrt{3}h_1, \qquad c = 3h_3, \qquad x = \frac{1}{3}\left(1 - \frac{h_2}{a}\right) \qquad (15.1)$$

For a regular tetrahedron, with $h_1 = h_2 = h_3$, we would expect an axial ratio $c/a = 1.098$, and a position parameter $x = 0.211$. These are very close to the observed values, $1.093(\pm 3)$ and $0.209(\pm 1)$; the departure from regularity is barely outside the limits of experimental error.

For low quartz, let us assume that the tetrahedron keeps its same shape and size, but is tilted about the x axis by a small angle φ, as in Figure 15-6(b) and (c). Then $PQ = h_1 \cos \varphi$, $MN = h_2$, $RS = h_3 \sin \varphi$. Moreover, a is the sum of the projections on the x axis of HF, FJ, and JK; and the projection of HF is h_2, while the sum of the two others is $2PQ \cos 30°$, which is $\sqrt{3}h_1 \cos \varphi$. Hence

$$a = h_2 + \sqrt{3}h_1 \cos \varphi, \qquad c = 3h_3 \cos \varphi \qquad (15.2)$$

Next we must find the atomic position parameters. Using the notation of §11.14, we take oxygen atom E as the prototype, x, y, z, or (more conveniently) $x_0 - u_0$, $-x_0 - u_0$, $-1/6 + w_0$; a fourth parameter is needed for the silicon atom, which is at $1/2 - u_1, 0, 0$.

The effect of tilt is most easily seen by considering how atom F moves— cf. Figure 15.6(d). Since EF and FJ are perpendicular to diad axes of the structure, and project at 120° to each other, F must move, in projection, in a direction equally inclined to both, i.e., parallel to the y axis. Hence it keeps unchanged its x coordinate, $2x_0$, but its y and z coordinates depend on φ.

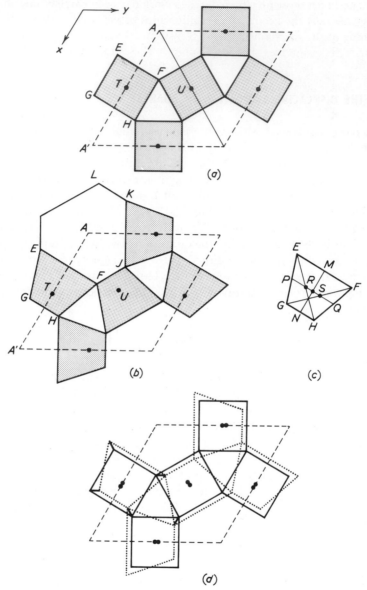

Figure 15-6 Projection (schematic) of quartz on (0001). (a) High quartz. Black dots are Si; squares are projections of tetrahedra, triangles are projections of a spiral arrangement of oxygens. Broken lines outline the unit cell. (b) Low quartz, lettered to show the resemblance to (a). (c) Detail of tetrahedron $EFGH$; MN, RQ, and RS are projections of h_1, h_2, and h_3. (d) Superposition of (a) and (b). Short heavy black lines show the displacements of oxygen atoms during the transition.

Let us write down the coordinates of atoms E, F, G, H, first in terms of x, y, z, and then in terms of x_0, u_0, w_0—see Table 15-1(a). Atoms G and H are

Table 15-1 Geometric Relations in Low Quartz

	(a) Atomic coordinates					
ATOM*	COORDINATES IN TERMS OF x, y, z			COORDINATES IN TERMS OF x_0, u_0, w_0		
E	x	y	z	$x_0 - u_0$	$-x_0 - u_0$	$-\frac{1}{6} + w_0$
F	$x - y$	$-y$	$-z$	$2x_0$	$x_0 + u_0$	$\frac{1}{6} - w_0$
G	$1 + y - x$	$-x$	$\frac{1}{3} + z$	$1 - 2x_0$	$-x_0 + u_0$	$\frac{1}{6} + w_0$
H	$1 + y$	x	$-1 + \frac{2}{3} - z$	$1 - x_0 - u_0$	$x_0 - u_0$	$-\frac{1}{6} - w_0$

	(b) Geometry of the tetrahedron							
	COORDINATES OF MIDPOINTS			COMPONENTS OF h's IN AXIAL DIRECTIONS			LENGTH OF PROJECTION	
*				x	y	z	$\perp$ to c	$\parallel$ to c
M	$\frac{3}{2}x_0 - \frac{1}{2}u_0$	0	0	$1 - 3x_0$	0	0	$(1 - 3x_0)a$	0	h_1
N	$1 - \frac{3}{2}x_0 - \frac{1}{2}u_0$	0	0					
P	$\frac{1}{2} - \frac{1}{2}x_0 - \frac{1}{2}u_0$	$-x_0$	w_0	x_0	$2x_0$	$2w_0$	$\sqrt{3}x_0a$	$2w_0c$	h_2
Q	$\frac{1}{2} + \frac{1}{2}x_0 - \frac{1}{2}u_0$	x_0	$-w_0$					
R	$\frac{1}{2} - u_0$	$-u_0$	$-\frac{1}{6}$	u_0	$2u_0$	$\frac{1}{3}$	$\sqrt{3}u_0a$	$\frac{1}{3}c$	h_3
S	$\frac{1}{2}$	u_0	$\frac{1}{6}$					

* The labelling of points refers to Figure 15-6(c).

translation repeats of L and K; L is derived from E by operation of the triad screw axis 3_2, and F and K are derived from E and L by the diad axis parallel to AA' at height zero.

Next, we find the coordinates of the midpoints of the edges—see first two columns in Table 15-1(b)—and hence, by subtraction, the components along the axial directions of lengths h_1, h_2, h_3, which project as PQ, MN, RS (third column). The projected length in the plane of the diagram and the height difference in the c direction are recorded in the fourth and fifth columns. Hence we obtain three relations involving x_0, u_0, w_0:

$$h_2 = (1 - 3x_0)a, \qquad \tan \varphi = \frac{2w_0c}{\sqrt{3}\,x_0a}, \qquad \tan \varphi = \frac{\sqrt{3}\,u_0a}{c/3} \qquad (15.3)$$

The remaining parameter, u_1, is found from the midpoint of all three lines MN, PQ, RS, which is seen from the table to be $\frac{1}{2} - \frac{1}{2}u_0$, 0, 0; hence $u_1 = -\frac{1}{2}u_0$.

To find changes in parameters, changes in cos φ can to a first approximation be neglected; hence a and c may be treated as constant while changes in u_0 and w_0 are being considered. To this approximation, x_0 is constant (as we saw it should be by considering the movement of atom F), and $u_0/w_0 = 2c^2/(9a^2x_0)$, which is constant, while $u_0/u_1 = -2$. Table 15-2 shows how

Table 15-2 Atomic Position Parameters and Tilt Angles in Quartz

°C	x_0	u_0	w_0	u_1	u_0/w_0	u_0/u_1	tan φ	φ
20	0.207	0.060	0.048	0.030	1.25	2.0	0.28	15.6°
450	0.208	0.043	0.036	0.022	1.19	2.0	0.20	11.3
520	0.209	0.040	0.029	0.018	1.38	2.2	0.19	10.8
545	0.210	0.038	0.028	0.016	1.36	2.4	0.18	10.2
560	0.209	0.035	0.027	0.014	1.30	2.5	0.16_5	9.4
570	0.208	0.032	0.021	0.013	1.52	2.5	0.15	8.5
600	0.207	0	0	0	—	—	0	0
	±0.002	±0.002	±0.002	±0.001	±0.07	±0.1	±0.01	±0.7
Predicted					1.27	2.0		

good the agreement is (though, not surprisingly, it becomes slightly coarse at higher temperatures), and therefore confirms our hypothesis of a rigid tetrahedron with changing tilt.

We may use either u_0 or w_0 to calculate φ, and this is included in the table. Its room-temperature value is about 16°; just below the transition, it has dropped to about half this. Obviously, the last step at the transition is either discontinuous or extremely abrupt compared to what has gone before.

The only structurally important angle that changes with tilt is Si—O—Si, the angle projecting as TFU in Figure 15-6(a). Let this be α, and let us calculate the changes expected if the rigid-tetrahedron model holds. We can show that (measuring changes from the high-quartz end)

$$\frac{\Delta \sin \frac{1}{2}\alpha}{\sin \frac{1}{2}\alpha} = \frac{\Delta(TU)}{TU} = \frac{\Delta a}{a} + \frac{3}{2}\left(\frac{u_1 a}{TF}\right)^2 \qquad (15.4)$$

(where TU and TF now refer to the three-dimensional interatomic distances, not to their projections). Substituting from earlier equations, we get

$$\frac{\Delta \sin \frac{1}{2}\alpha}{\sin \frac{1}{2}\alpha} = \frac{1}{2}\sin^2 \frac{1}{2}\varphi\left[-\frac{4\sqrt{3}}{1+\sqrt{3}} + \frac{1}{1+(3a/2c)^2}\right] = -2.20 \sin^2 \frac{1}{2}\varphi \qquad (15.5)$$

Putting in the room-temperature value of α, 144°, we find

$$\Delta\alpha = -9°$$

whence the predicted high-quartz value of α is 153°. This agrees well with the directly measured value.

What happens meanwhile to the thermal amplitudes? In particular, how does the amplitude of the oxygen atom compare with the displacement of the atom at the midpoint of its oscillation? Displacement is measured from the high-quartz position, and is proportional to the tilt angle φ; it is the amplitude in the direction of the displacement which is relevant for comparison. This is not exactly the direction of the largest amplitude, but it is not far off it. The actual values are given in Table 15-3 together with the displacements, which are equal to $[(u_0a)^2+(w_0c)^2]^{\frac{1}{2}}$. At room temperature the amplitude

Table 15-3 Thermal Amplitudes and Atomic Displacements of the Oxygen Atom in Low Quartz

		20° C	560° C
Thermal amplitudes (in Å)	Greatest	0.133	0.229
	Intermediate	0.099	0.166
	Least	0.052	0.141
	Along line of movement of F	0.110	0.190
Displacement from high-quartz position (in Å)		0.400	0.194

is much less than the displacement, but with increasing temperature the amplitude increases and the displacement decreases, till, just below the transition, they are very nearly equal. This confirms what was suggested in §14.13: the transition "goes" when the r.m.s. amplitude in the direction of the displacement becomes equal to the displacement.

Why should the high-quartz structure not remain stable at lower temperatures? Why is the tilting of the tetrahedra energetically favourable? Its only obvious connection with the potential energy is through the part of it associated with the Si—O—Si angle α. Hence it seems that the potential energy is reduced when α decreases, and that the tilting is a mechanism to bring about the decrease in α. The opening out of α in the change from low quartz to high quartz marks an increase in potential energy though it minimises the free energy.

So far, we have avoided making direct predictions of lattice parameter changes using the rigid-tetrahedron model. If we do so, allowing only φ to change in the right-hand side of equation 15.2, we find very poor agreement with the observed changes. There must be *some* changes in the tetrahedron itself, and the simplest hypothesis is to assume a homogeneous (isotropic) change of size without change of shape. This means that all our previous calculations based on *ratios* of lengths still hold good. To a good approximation we can write

$$a = h(1 + \sqrt{3} \cos \varphi), \qquad c = 3h \cos \varphi \qquad (15.6)$$

and hence

$$\frac{\Delta a}{a} = \frac{\Delta h}{h} + \frac{\sqrt{3}}{1 + \sqrt{3}}\,\Delta\cos\varphi, \qquad \frac{\Delta c}{c} = \frac{\Delta h}{h} + \Delta\cos\varphi \qquad (15.7)$$

(since φ is small and $h_1 \simeq h_2 \simeq h_3$). Taking experimental values of $\Delta a/a$ and $\Delta c/c$ measured from room temperature to $570°$ C, with the corresponding values of φ from Table 15-2, we find that they agree in giving $\Delta h/h$ as about -7×10^3, corresponding to a *negative* expansion coefficient of -12×10^{-6} deg^{-1}. This result, at first sight unexpected, agrees in order of magnitude with the direct measurement of the mean Si—O distance, which drops from 1.608 Å at room temperature to 1.598 Å at $570°$ C.

In arguing thus, however, we have omitted to notice that it is the *root-mean-square* value of the Si—O distance rather than its *mean* value that is physically significant. In Figure 15-7, which represents the situation schematically, the mean square value of AP and AQ is $AM^2 + MP^2$. If p is the r.m.s. value of the thermal amplitude in the plane perpendicular to AB, the correction to be applied to AM, the measured (mean) value of Si—O, is $\Delta l/l = \frac{1}{2}p^2/l^2$, where l is the bond length. Using the measured values of p^2 at room temperature and $560°$ C, we find the correction is $(0.080 - 0.027)/(2 \times 1.6^2) \simeq 10 \times 10^{-3}$, corresponding to an expansion coefficient of about 18×10^{-6} deg^{-1}. Adding it to the coefficient -12×10^{-6} deg^{-1} calculated for AM, we find a net expansion coefficient of the r.m.s. bond length of 6×10^{-6} deg^{-1}, which is of the order of magnitude we should expect. This is a warning against drawing hasty conclusions from measured distances without allowing for the effect of thermal vibrations (cf. §14.10).

There are other points of interest that cannot be discussed here—for example, the fact that near $450°$ C the Si atom begins to move off-centre in its tetrahedron. This would imply a second physically-independent variable, which we have ignored. We are justified in doing so as a first step. What we have done is to pick out one mode of motion, judged from the displacements and the thermal amplitudes to be the most important, and to trace its geometrical consequences. A lattice frequency, of value 207 cm^{-1}, which disappears at the transition, is associated with just this displacement of the oxygen atom and hence with the tetrahedral tilting mode.

One other observation near the transition should be mentioned. Whereas in high quartz there is no disorder—the atoms oscillate about

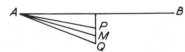

Figure 15-7 Schematic diagram to show the correction of the measured Si—O bond length for thermal oscillation. The line PQ represents the oscillation range of the atom; M is its midpoint, taken as the mean atomic position, and AM is therefore the measured interatomic distance.

positions of high point-symmetry with amplitudes that are not appreciably larger, and may even be smaller, than those in low quartz—there is a region just *below* the transition in which disorder occurs. In this, a single crystal of low quartz breaks up into small domains with the two permissible orientations, corresponding respectively to positive and negative values of u_0; the domains seem to be constantly shifting boundaries, switching from one orientation to the other, with the atoms in each retaining the appropriate low-quartz positions. This is an example of what is often found: a break-up of crystal texture, which *smears out* the sharpness of a transition. The temperature range in which it occurs depends on the perfection or imperfection of the original "single" crystal.

15.10 DISPLACIVE TRANSITIONS IN PEROVSKITES

The perovskite family (§12.4) includes so many hettotypes (variants of lower symmetry, §12.1) that it is not surprising to find many different examples of displacive transitions between them. In particular, we may distinguish cases where the transitions are due to B-cation displacement (cf. §12.6), to tilting of octahedra (§12.8 to 12.10), and to both effects together (§12.10 to 12.11).

15.11 BARIUM TITANATE, BaTiO₃

One of the most extensively studied transitions is that between cubic and tetragonal BaTiO₃ (§12.6). There are several reasons for its importance. (i) Tetragonal BaTiO₃ is ferroelectric (§12.7), and therefore illustrates the interaction of electric field and polarisation. (ii) The structure is one of the simplest showing this kind of effect, and therefore a favourite "guinea pig" for comparison of theory and experiment. (iii) The transition occurs a convenient distance above room temperature, making experimental conditions relatively easy. The fact that the changes are very small makes the relationship between the two forms easy to trace; against this, it means that reliable measurements of the magnitudes of the changes are sometimes difficult to obtain.

Macroscopically, the cubic form is characterised by one parameter, the lattice spacing a. The tetragonal form is characterised by two, which we may take either as the lattice spacings a and c or (more conveniently for our present purposes) as the cube root of the volume, $(a^2c)^{\frac{1}{3}}$, and the difference from unity of the axial ratio, $c/a - 1$. The temperature variation of these is shown in Figure 15-8(a) and (b). We notice the small but distinct discontinuity at the transition at 120° C, and the premonitory region in which changes tending towards the high-temperature form have already begun—a region of different extent in (a) and (b). In (a), if we extrapolate the straight lines above and

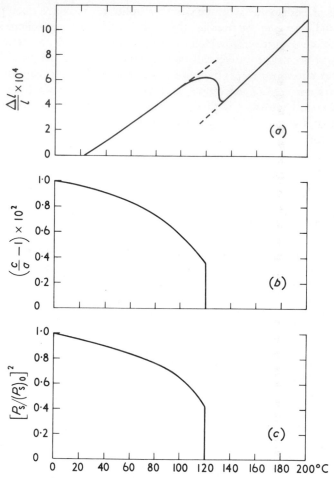

Figure 15-8 Changes in $BaTiO_3$ at and near the 120° C transition. (a) Relative changes in the mean lattice parameter $l = (a^2c)^{\frac{1}{3}}$. (b) Changes in the shape parameter, $c/a - 1$. (c) Relative changes in the square of the spontaneous polarisation.

below the transition (whose slopes give the expansion coefficients recorded in Table 14-6[b]), we see that the premonitory region starts at approximately 110° C and includes a total change of more than 2×10^{-4} in $\Delta l/l$ as compared with less than 2×10^{-4} at the actual transition temperature, 120° C.

Detailed experimental results of this kind are not easily reproducible, because of the difficulty of securing absolutely uniform conditions, not only of temperature but of pressure, to which the transition may be sensitive. Crystal perfection can play a role here; if there are differently oriented domains, the pressure exerted on each by its neighbours may affect the transition temperature significantly. Because of these difficulties, there has been some doubt in the past as to whether the discontinuity was real. It is now generally agreed that a discontinuity does exist if the crystal is unstressed; but it may be weakened or destroyed by stresses applied (whether deliberately or accidentally) to the crystal.

Both the heat of transition and the entropy of transition are very small; they are hard to measure accurately, for the reasons given above. Values obtained in one set of experiments are given in Table 15-4. Comparing the

Table 15-4 Comparison of Effects at the Cubic-Tetragonal Transition in Isomorphous Materials

		BaTiO$_3$	KNbO$_3$	PbTiO$_3$
Thermal effects	Transition temperature	120° C	410° C	490° C
	Energy change (in 10^3 J mol^{-1})	0.21	0.80	4.80
	Entropy change (in J mol^{-1} deg^{-1})	0.5	1.2	6.3
Geometrical effects	Shape factor $(c/a - 1) \times 10^2$			
	Just below transition	0.4	1.2	2
	Well below transition	1.0	1.7	6
	B-cation displacement (in Å)	0.12 (0.125*)	?* (0.171*)	0.30
Electrical effects	Spontaneous polarisation (in C m^{-2})			
	Just below transition	0.18	0.26	Large
	Well below transition	0.24	0.30	

* Since the displacement has not been measured in tetragonal KNbO$_3$, values for orthorhombic BaTiO$_3$ and KNbO$_3$ (in parentheses) are included for comparison.

values for BaTiO$_3$ with those for KNbO$_3$ and PbTiO$_3$, isomorphous structures which depart further from the ideal cubic, we see that the magnitude of the energy and entropy changes is correlated with that of the change in axial ratio.

We now turn to consider the structural details of the transition. There are no variable atomic position parameters in the cubic form. In the tetragonal form, there are three, which may be chosen in various ways, most obviously as the z parameters of any three of the four crystallographically independent atoms Ba, Ti, O(1), or O(2) with respect to the fourth as origin (cf. §12.6). For our purposes, however, it is most informative to take the origin at the geometrical centre of the octahedron, which gives us $2z_{O(2)} = -z_{O(1)}$; our other two parameters are z_{Ti} and $_{Ba}$.

From a physical point of view, we obtain a useful approximate description by considering only z_{Ti} and ignoring the other parameters. The parameters $z_{O(1)}$ and $z_{O(2)}$ are small; if they were zero, it would mean that the octahedron suffered no distortion except a homogeneous change of shape, measured by the change in c/a. The Ba displacement is not only absolutely smaller than the Ti displacement, but relative to the bond lengths it affects it is even smaller. Hence the main features of the transition are the Ti displacement and the homogeneous change of shape.

Unfortunately there are no direct measurements of the Ti displacement as a function of temperature. However, there is good reason to suppose that it is physically closely correlated with the octahedron shape. This would be expected theoretically, from the arguments used in §2.8 to predict the occurrence of off-centring (cf. Figure 2-6); moreover, it is borne out experimentally by comparison of axial ratio and off-centre displacement in isomorphous compounds (Table 15-4) and more generally for other types of off-centring. There is also indirect evidence from the macroscopically-measured spontaneous polarisation P_s, which (from order-of-magnitude arguments) we may take as proportional to the displacement of Ti from the centre of negative charge, (i.e., of the octahedron). Figure 15-8(c) shows the measured variation of P_s^2 with temperature. The close resemblance to the curve of $c/a - 1$ is obvious.

Hence, to a good approximation, the transition in $BaTiO_3$ represents a change in one independent variable—the Ti displacement. It is, however, an effect of bond length rather than bond angle, of stretching rather than of bending (in contrast to the quartz transition).

This discussion has been in terms of the static structure. It would be desirable to consider also the thermal vibrations. Unfortunately it is very hard to measure the thermal amplitudes with sufficient accuracy even at room temperature, and there is not enough information about other temperatures.

Since the difference in the "static" structures consists in a movement of Ti within its octahedron, along its z axis, we should expect, at the approach to the transition from below, an increasing amplitude of oscillation of Ti in this direction, relative to the oxygen framework, which moves as a rigid body. The transition will occur when the thermal amplitude equals the off-centre displacement, which then drops discontinuously to zero. The premonitory region is that in which the thermal amplitude has become anomalously large.

Measurements of thermal amplitudes are not complete enough to provide conclusive evidence about this hypothesis, but they do not contradict it. Though we should expect the thermal amplitude of Ti in the z direction to be large just below the transition, it could well have decreased to a more normal value by room temperature; it is not observed as abnormally large.

The soft mode above the transition will be one whose atomic configuration at an extreme of its oscillation corresponds to the static structure below the transition. Thus it is one in which Ti vibrates relative to a rigid oxygen framework. The slight distortion of the octahedron in the low-temperature form suggests that another mode in which the oxygens vibrate against one another will not be wholly negligible, but it is much smaller than the first-mentioned one. Evidence for the existence of the stretching mode comes from various spectroscopic studies.

Recognising the soft mode associated with a transition does not of course "explain" the transition. If we can explain the low-temperature structure, and can show that there is only one independent parameter sensitive to temperature, then we can predict the soft mode and the high-temperature structure. To predict the soft mode without presupposing the low-temperature structure is no easier than to predict the low-temperature structure.

For KNbO$_3$, with its bigger discontinuities at the transition, it is perhaps less satisfactory to discuss it in terms of a thermal mode, since this assumes that the vibrations can usefully be treated as approximately harmonic. The concept of pseudospin waves can be used instead. Geometrically, the effects are the same, and all the curves for temperature variation of physical properties are very much the same shape for KNbO$_3$ as for BaTiO$_3$.

The other transitions in BaTiO$_3$ and KNbO$_3$ at lower temperatures may be dealt with more briefly—the tetragonal-to-orthorhombic and ortho-rhombic-to-rhombohedral transitions. Geometrically the relationship is more complicated because, although the symmetries of both forms are subgroups of that of the aristotype, neither is a subgroup of the other. In the change from orthorhombic to tetragonal, for example, the diad axis and one mirror plane in which it lies are lost, while a tetrad axis and mirror planes in different directions are gained. The transition is still displacive, but we expect it to be a little harder to achieve—to have a larger energy hump. This is indicated experimentally by a larger thermal hysteresis. For example, in one series of experiments, the tetragonal/orthorhombic transition was observed at 0° C with increasing temperature and at −10° C with decreasing temperature, whereas corresponding differences in the cubic/tetragonal transition were only about one or two degrees C. The bigger energy hump at the transition does not necessarily mean a bigger energy difference between the forms on either side of it, as can be seen from Table 15-5. The latent heat of the orthorhombic/tetragonal transition is actually much less than that of the tetragonal/cubic transition.

Table 15-5 Energy Changes at the Three Transitions (in 10^3 J mol^{-1})

Transition	BaTiO$_3$		KNbO$_3$	
	Temperature	Energy	Temperature	Energy
Cubic/tetragonal	120° C	0.21	410° C	0.80
Tetragonal/orthorhombic	0° C	0.10	210° C	0.36
Orthorhombic/rhombohedral	−90° C	0.05	−40° C	0.13

The single crystal, if taken carefully through such a transition, retains its identity, but it is likely to undergo more complicated twinning, since the choice of orientation for each individual component has to be made not only with decreasing temperature, as at the cubic/tetragonal transition, but also with increasing temperature.

An alternative suggestion about the character of the transitions in $BaTiO_3$ and $KNbO_3$ has been put forward. According to this, the Ti atoms are disordered among different possible off-centre positions displaced towards the eight faces of the octahedron. Only in the lowest form, the rhombohedral form, are they ordered in one such site, the same for every unit cell. In the orthorhombic form they are distributed over two adjacent sites, in the tetragonal form over four, with the same z coordinate, in the cubic form over all eight. The distribution is not wholly random, but is ordered in domains. Reversing the direction of polarisation means switching the choice of occupied site in all appropriately oriented domains; going through a transition with increasing temperature means switching the choice in half the domains, perhaps with the creation of new domains in the process.

Perhaps the two pictures of the transition are not as different as they appear at first sight, because it is hard to distinguish between the effects of shifting domains, as in this picture, and anharmonic lattice modes of large amplitude. The disorder model would, however, predict larger apparent thermal amplitudes for Ti perpendicular to the z axis than along it, for which there is no clear evidence—though the limits of error are too great for certainty.

15.12 STRONTIUM TITANATE, $SrTiO_3$

An interesting example of a pure tilting transition occurs in $SrTiO_3$ at $103°$ K. In the low-temperature form, alternate octahedra are tilted in opposite directions about the tetrad axis. Figure 15-1(a) serves as a diagram of the structure (except that in it the tilt angle is given an arbitrary exaggerated value). Alternate layers of octahedra with the configurations of (i) and (iii) are superposed. Referred to the original cubic axes of reference, the unit cell is (nearly) doubled in all three directions; referred to tetragonal axes rotated by $45°$ from these, the new lattice parameters are $\sqrt{2}a, \sqrt{2}a, 2a$; the cell contains four formula units, and the space group is $I4/mcm$ (No. 140).

With rigid octahedra, and undisplaced B-cations, the only variable is the tilt angle. This has been measured by paramagnetic resonance methods (EPR methods) as a function of temperature, and is shown in Figure 15-9, curve (a). For convenience of comparison, the curve has been normalised so that the values actually plotted are ω/ω_0 and T/T_{tr}, where ω_0 is the tilt angle far below the transition and T_{tr} the transition temperature. Here ω_0 is $1.3°$.

This is an interesting example of a structure determination by resonance methods rather than by the more familiar diffraction methods. It is made

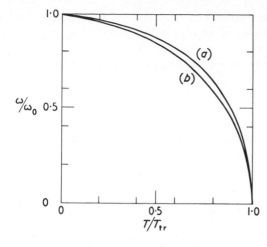

Figure 15-9 Variation of tilt angle with temperature: (a) in SrTiO₃, (b) in LaAlO₃. Both curves are normalised by plotting ω/ω_0 versus T/T_{tr} where ω_0 (the low-temperature tilt angle) and T_{tr} (the transition temperature) are 1.3°, 103° K in (a) and 6.3°, 800° K in (b).

possible by the simplicity of the structure involved, and the fairly extensive knowledge of a wide range of related structures in the perovskite family.

Because there are no cation displacements in the low-temperature form, there are no dipole moments in the octahedra; hence the structure is nonpolar.*

The transition is associated with a *soft mode* of the lattice (a mode whose frequency is strongly temperature-dependent) of such a kind that the instantaneous configuration of atoms at one extreme of the oscillation, above the transition, corresponds to the configuration of their mean positions below the transition. It is thus a mode whose wavelength corresponds to twice the ideal-perovskite lattice vector, and hence, keeping the ideal-perovskite axes of reference, its wave number is half-integral.

It seems likely that SrZrO₃ has a somewhat similar transition near 1170° C. X-ray studies below this temperature show the tetragonal symmetry, and the extra powder lines, which indicate doubling of the lattice parameters. The axial ratio c/a well below the transition is about 1.01; for regular octahedra with a tilt angle ω it is given by $1/\cos\omega$, and on this assumption the tilt angle would therefore be 8°.

It is easy to understand why the tilt should be larger, and should persist to much higher temperatures, in SrZrO₃ than in SrTiO₃. The ZrO₆ octahedron is much larger than the TiO₆ octahedron (cf. Table 2-2, from which we see that Ti—O = 1.96 Å, Zr—O = 2.09 Å) and hence the octahedron has to tilt substantially to make contact with Sr. Higher temperatures are therefore needed to achieve thermal amplitudes large enough to bring about the transition to cubic. A naïve but useful picture is to think of the effective size

---

* In the literature, the transition has sometimes been mistakenly called antiferroelectric, because of the observed doubling of the lattice parameters. What was overlooked by writers doing so was that, while as a general rule an antiferroelectric requires doubling of lattice parameters, doubling of lattice parameters does not require a structure to be antiferroelectric, and this one is not.

of Sr as being increased by thermal vibrations until it is large enough to fill out its cavity and keep the framework from crumpling round it.

At lower temperatures further complexities occur in both these materials. In $SrTiO_3$ it is possible that the changes involve off-centre displacement of Ti. In $SrZrO_3$ the transitions, which occur at 830° C and 700° C, are probably due to changes of tilt; off-centring of Zr is unlikely.

15.13   LANTHANUM ALUMINATE, $LaAlO_3$

Another transition closely resembling that in $SrTiO_3$ at 103° K occurs in $LaAlO_3$ at about 530° C. The structure of $LaAlO_3$ below the transition was described in §12.9 and Figure 12-7(c). In it, as in $SrTiO_3$, the difference from the ideal cubic form involves tilts of adjacent octahedra in opposite directions about one symmetry axis, and a consequent doubling of all lattice parameters if the original axes of reference are retained. With a different choice of axes, a primitive unit cell containing only two formula units can be found. There is no displacement of cations from ideal positions in either material. In $LaAlO_3$, as in $SrTiO_3$, the transition is associated with a single soft mode, whose wave vector, in terms of the ideal-perovskite lattice, is defined by half integers. The one real distinction is that in $LaAlO_3$ the tilts are about a triad rather than a tetrad axis of the octahedron, leading to the space group $R\bar{3}c$ (No. 167).

For this structure as for $SrTiO_3$, the tilt angles have been measured by resonance methods, with the results shown (in normalised form) in Figure 15-9, curve (b). The resemblance to curve (a) for $SrTiO_3$ is very striking. The actual value of the tilt in $LaAlO_3$ at low temperatures is about 6.3°. This agrees well with the angle of about 6° deduced from the measured position parameters of the oxygen atom at room temperature (§12.9).

If, however, assuming that the octahedra remain regular, we try to calculate the tilt angles from the lattice parameters using the geometrical relation $c/a = \sqrt{6}/\cos\omega$, we find ourselves in difficulties. It was noted in §12.9 that the interaxial angle of the primitive rhombohedral cell is *more* than 60°, and of the pseudocubic face-centred rhombohedral cell *more* than 90°. This is true at all temperatures. Geometrically, it implies that c/a is *less* than $\sqrt{6}$, instead of greater, as it would be for a pure tilt (since in that case c would be unchanged while a decreased). Hence there must be an actual distortion of the octahedron—a flattening down the c axis. But this distortion must decrease as the tilt angle decreases, since their net effect, the deviation of the axial ratio from $\sqrt{6}$, decreases smoothly towards the transition. The change of shape is therefore associated with the change of tilt—they are not independent variables.

An explanation of the effect was given in §12.9. With regular octahedra, the three La—O bonds in the (0001) plane are necessarily shorter than the six bonds inclined to it, by an amount which increases with the tilt angle.

Compression along the c axis helps to reduce the inequality, at the expense of a flattening of the octahedron.

We can easily calculate the compression if we know ω, the tilt angle, and ε, the deviation of the interaxial angle from $60°$. (With pseudocubic axes, the corresponding deviation from $90°$ is $\sqrt{3}\,\varepsilon/2$.) It is easily shown geometrically that

$$\frac{2}{3}\sin\frac{\alpha_R}{2} = \left[\left(\frac{c}{a}\right)^2 + 3\right]^{-\frac{1}{2}} \tag{15.8}$$

whence

$$\frac{1}{\sqrt{6}}\left(\frac{c}{a}\right)_{\text{obs}} = 1 - \frac{3}{4}\sqrt{3}\,\varepsilon \tag{15.9}$$

For a tilted structure with regular octahedra,

$$\frac{1}{\sqrt{6}}\left(\frac{c}{a}\right)_{\text{reg}} = \frac{1}{\cos\omega} = 1 + \frac{1}{2}\omega^2 \tag{15.10}$$

The distortion of the octahedron itself is given by the ratio of these two quantities,

$$\left(\frac{c}{a}\right)_{\text{oct}} = 1 - \frac{3\sqrt{3}\,\varepsilon}{4} - \frac{1}{2}\omega^2 \tag{15.11}$$

Substituting the numerical values for ε and ω (in radians), we find for LaAlO₃ that the small terms are -0.002, -0.005, and the compression is therefore 0.007, or 0.7 percent.

Effects in the isomorphous PrAlO₃ confirm the idea that the compression and the tilting increase together. At room temperature the tilt angle, deduced from the measured position parameters of oxygen, is about $8°$, and ε is $0.3°$. In equation 15.11, the terms are therefore -0.006 and -0.008 respectively, and the compression is accordingly 1.4 percent. These are rough calculations, but they show unmistakably that the greater tilt angle in PrAlO₃ is matched by a greater compression, the compression varying roughly as the square of the tilt angle. In PrAlO₃ as in LaAlO₃ the axial ratio decreases towards its ideal cubic value with increasing temperature, and the temperature at which a transition is reported is considerably higher than for LaAlO₃. The same is true for NdAlO₃ with its still greater value of ε.

15.14 PRASEODYMIUM ALUMINATE, PrAlO₃, AND CALCIUM TITANATE, CaTiO₃

We have now considered examples of transitions involving tilting about the tetrad axis and the triad axis; we turn next to one in which tilting occurs about the diad axis.

This is the transition in $PrAlO_3$ to the low-temperature form described in §12.10. The way in which it occurs has been the subject of a detailed experimental study. The high-temperature form is here not cubic but rhombohedral. Comparison is easiest in terms of the large pseudocubic rhombohedral cell, to which the new orthogonal cell is related as follows:

	a_R	b_R	c_R
a_0	$\frac{1}{2}$	0	$-\frac{1}{2}$
b_0	0	1	0
c_0	$\frac{1}{2}$	0	$\frac{1}{2}$

This matrix does more than specify a consistent set of axes with which to display relationships; it actually indicates which directions in the rhombohedron become which in the orthogonal cell. What happens is that (010) planes of the rhombohedron—which remain (010) planes in the new cell—shear along the new [001] direction till [010] becomes normal to the other two axes. Within the plane, a and c repeats change very little, but perpendicular to it, along b, there is considerable compression.

Now consider what happens to the tilts during the transition. The low-temperature form, with its tilt only about a diad axis, is derived from the high-temperature form with tilts about a triad axis. If we think of the triad-axis tilt as made up of equal components about three tetrad axes (as in §12.8), clearly one component must be lost at the transition—the component about [010]. The other two combine to give a tilt about [001], which lies in the plane of [010] and the original triad. If their magnitudes are unchanged, we expect the resultant tilt in the orthogonal form to be $\sqrt{2/3}$ times what it was in the rhombohedral form. The observed tilts are 6°27' and 7°53' respectively, with a ratio of 0.819 as compared with the expected 0.816.

The octahedron changes shape in accordance with the new symmetry. It is now compressed nearly equally in the [010] and [001] directions. Relative to the [100] direction, the compression ratios are $b/(\sqrt{2}a)$ and $(c \cos \omega)/a$, equal to $1-0.012$ and $1-0.010$ respectively.

Besides the changes in framework configuration, there are displacements of Pr atoms from their ideal positions. These are very small in magnitude (§12.10) and their arrangement is a rather low-symmetry one—lower than that of the framework. Nevertheless they probably play a vital and specific role in the transition. It is noteworthy that neither $LaAlO_3$ nor $NdAlO_3$, isomorphous with $PrAlO_3$ in the rhombohedral form, shows any similar transition. The energy associated with the Pr displacements probably sways the balance between the two simplest configurations of the framework, the rhombohedral and the orthogonal. Putting it slightly differently, one might say that the transition only occurs because a change of shape which is energetically satisfactory for the framework can also provide satisfactory (even if low-symmetry) environments for the Pr atoms.

There is a further transition at 151° K, involving even smaller displacive changes and still lower symmetry. Both transitions, like so many others of the kind, are characterised by systematic twinning in small domains.

Other orthorhombic structures we have considered, such as CaTiO$_3$ (§12.10), are based on two independent tilts, ω about the tetrad axis of the octahedron, φ about the diad axis. Do the transitions undergone by such structures involve changes of tilt system?

We cannot answer this unambiguously, because measurements of tilts from position parameters have only been made at room temperature; and the lattice-parameter relations which would hold if the octahedra were regular, cannot be relied on when distortion of the octahedron shape is likely.

$$\frac{c}{a} = \frac{1}{\cos \varphi}, \qquad \frac{b}{\sqrt{2}\, a} = \frac{1}{\cos \omega} \qquad\qquad (15.12)$$

In CaTiO$_3$, for example, the (rather roughly) measured tilt angles at room temperature are $\varphi = 9°$, $\omega = 4°$, as compared with 10° and 0° deduced from the lattice parameters. On the other hand, if we assume that any *changes* of distortion with temperature are smooth and relatively small, so that they do not mask the effects of tilt, we may make useful predictions. Evidence from the rare-earth orthoferrites (§12.10) suggests that this is legitimate.

The observed decrease of c/a for CaTiO$_3$ above room temperature then indicates a decrease in φ, but a complete series of measurements up to the transition at 1200° C is not available. For the isomorphous NaMgF$_3$, both c/a and b/a decrease; b and a become equal at 700° C, suggesting the existence of a tetragonal phase between 700° C and the final transition to cubic at 900° C. Systematic analysis of tilt systems may make it possible to predict such high-temperature structures from their lattice parameters.

15.15 SODIUM NIOBATE, NaNbO$_3$

In the last two sections we have considered separately pure B-off-centring transitions without change of tilt, and pure tilting transitions without B off-centring. As a working rule, we may say that off-centring transitions depend on the effective size of B relative to its octahedron, tilting transitions on the requirements of the A cation.* These are independent factors, and both can operate in the same structure, leading to a complex sequence of transitions. A notable example is NaNbO$_3$.

---

* In structures lacking an A cation, such as WO$_3$ or the trifluorides discussed in §12.11, a limit to tilting must be set by anion—anion contacts between octahedra, supplemented by bond-angle requirements at the anion and electrostatic interactions. These factors could also play a part even when A cations are present. The same generalisations are true for tetrahedral framework structures.

Consider the room-temperature form of NaNbO$_3$ (Phase *P*—see §12.10). There are three independent variables, all likely to change with temperature—the tilt about the tetrad axis, the tilt about the diad axis, and the off-centre displacement of *B*. Qualitatively, the decrease of the tilts is exactly like those we have been considering in §15.14 for pure tilting transitions of the same geometry, for example CaTiO$_3$ and NaMgF$_3$; the decrease of Nb displacement corresponds to that discussed in §15.11 for Ti in BaTiO$_3$, except for its direction—here it is along the diad rather than the tetrad axis.

At room temperature, the tilt angles in NaNbO$_3$ are $\varphi = 8.6°$, $\omega = 8.5°$. The value of φ agrees well with what we should have predicted from the axial ratio c/a assuming undistorted octahedra, but for the other axial ratio $b/(2\sqrt{2}\,a)$ we should have predicted a value greater than 1, whereas it is actually 0.003 *less*. The octahedron must be compressed in the [010] direction by a fractional amount *f*, where $(1-f)\cos \omega = b/(2\sqrt{2}a)$, i.e.,

$$f = \left(\frac{b}{2\sqrt{2}\,a} - 1\right) + \frac{1}{2}\omega^2 = 1.5\,\% \tag{15.13}$$

This is just the compression we find in the [010] direction in orthorhombic KNbO$_3$; in both structures it is associated with the Nb displacement in the (010) plane. Parallel to (010) the octahedron is undistorted at room temperature; we may reasonably assume that it remains so at other temperatures, and hence calculate the tilt angle φ from the axial ratio c/a. Values deduced in this way are plotted in curve (*a*) in Figure 15-10.

The third variable parameter, the *B*-cation displacement, can be measured more easily for this structure than for BaTiO$_3$, because of the nearly antiparallel arrangement of the displacements. (In BaTiO$_3$ they were all in the same direction.) Curve (*b*) in Figure 15-10 shows the experimental values.

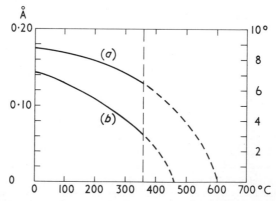

Figure 15-10 Temperature variation of structural parameters in NaNbO$_3$: (*a*) tilt angle about diad axis of octahedron, (*b*) off-centre displacement of Nb. A transition occurs at 360° C.

It is interesting to notice that the extrapolations of the two curves, shown dotted in the figure, cut the temperature axis at different places. There are in fact several different transitions observed in this range, before the final transition to cubic at 640° C.

It is easiest to understand the sequence of transitions in NaNbO₃ if we recognize that changes within the octahedron, involving the type of Nb displacement, occur at roughly the same temperatures as in KNbO₃, but that the detailed effects are influenced by interaction with changes in the tilt system. There is a tendency for the displacement direction to be simultaneously the direction of a tilt axis and a symmetry axis of the structure.

The low-temperature form, Phase N, was described in §12.9. It has a triad-axis tilt and a triad-axis displacement of Nb. We can consider both tilt and displacement as made up of three equal components, the tilt components about the three tetrad axes of the aristotype, and the displacement components along these axes. Near $-100°$ C, one component of displacement becomes zero, and the tilt component about the same axis, [010], is no longer held equal by symmetry to the others; it becomes the independent ω tilt. To describe the transition to Phase P, it is easiest to proceed in two steps, *via* an intermediate Phase Q, though in fact the transition $N \rightleftarrows P$ is direct, and the steps are not separated. In Phase N, alternate layers of octahedra parallel to (010) have ω tilts of opposite senses; in transforming to Phase Q, alternate layers reverse this sense so that all acquire the same sense of ω tilt; but they keep the original relation between their φ tilts, and the Nb displacements do not reverse but remain parallel. There is now a mirror-plane relation between successive layers. In transforming to Phase P, a double layer of octahedra reverses the sense of its ω tilt and also of its Nb displacements.

The transition $P \rightleftarrows Q$ is experimentally known to occur. Formation of Q from P is not brought about by change of temperature but by application of a large electric field—about 90 kV cm⁻¹. The crystal often breaks to pieces in the process.

At about 360° C, another component of the Nb displacement becomes zero. This allows the tilt component about the corresponding axis to become independent: there is no longer a single φ variable. Above the transition, in Phase R, the Nb displacement is along a tetrad axis of the octahedron, and there are three independent tilt components. There is in addition a further complexity: every third layer of octahedra parallel to (010) is different from the intervening pair, and has central Nb atoms.

In Phase R, as in Phase P, the Nb displacement decreases steadily with increasing temperature, becoming zero at about 480° C. Above this, the three independent tilts persist to about 520° C; one becomes zero at this temperature, a second at about 570° C, and the third at 640° C. The highest noncubic phase, T_2, is closely related to, but not identical with, low-temperature SrTiO₃ (§15.12).

In the two upper transitions, a single-phonon mode can be recognized, as in the SrTiO₃ transition. By contrast, the $P \rightleftarrows R$ transition at 360° C, which involves interaction between the Nb displacement and one or both of the independent tilts, is sharply discontinuous, marked by rearrangement into a new geometrical configuration. The same is true for the $N \rightleftarrows P$ transition.

The importance of the interaction between tilt and displacement in determining the temperature sequence of the different symmetries is illustrated by the contrast between $NaNbO_3$, N(rhombohedral) $\rightarrow P$(orthorhombic), and $PrAlO_3$, orthorhombic $\rightarrow$ rhombohedral, both occurring with increasing temperature. Apart from the B-cation displacement, the rhombohedral frameworks are alike, and the orthorhombic frameworks differ only in the absence of an ω tilt in $PrAlO_3$.

The whole $NaNbO_3$ series illustrates the principle that, where there is a sequence of phase transitions, the higher-temperature phase is not necessarily characterised by the higher symmetry. Phase P has lower symmetry than Phase N, Phase R lower than Phase P. Only at the final transition to the aristotype can we be confident in predicting a gain of symmetry.

There is a tremendous variety of small displacive transitions among the perovskites, and they are not all readily predictable. We have looked at some representative examples in considerable detail, to see the different factors which are at work, and to suggest lines along which others can be analysed. Broadly speaking, the factor making most of them possible is the softness of the bond angle at the corner-linking oxygens, which allows for easy tilting. On the other hand, the factors leading to the choice of a particular tilt system are varied; very often the size of the A cation and the kind of bonds it forms—their strength and possible directedness—are the most important, but cation-cation and anion-anion interactions may also be effective. What it amounts to is that the energy differences between hettotypes are generally so small that contributions to the energy which are normally quite small, and which differ in relative magnitude from one material to another, are sufficient to sway the balance and determine which hettotype will be stable.

15.16 MAGNESIUM PYROPHOSPHATE, $Mg_2P_2O_7$

The displacive transitions in perovskites which we considered in the preceding sections were characterised by the fact that the octahedra remain rigid or undergo only slight distortions of shape. An example where this does not hold good occurs in $Mg_2P_2O_7$.

The higher-temperature form is isomorphous with thortveitite, $Sc_2Si_2O_7$. It has the space group $C2/m$ (No. 12) with two formula units per cell; hence one O atom, distinguished as $O(1)$, must occupy a special position at a centre of symmetry. This O atom is the shared corner between two PO_4 tetrahedra forming a discrete P_2O_7 group. All the other O atoms form an array which is very roughly in hexagonal close packing, with Mg atoms at the octahedral interstices. The Mg octahedra together constitute an edge-sharing sheet parallel to (001), the sheets being held together only by bonds through P to O(1). Each O, except O(1), has one P neighbour and two Mg neighbours, and Pauling's rule is thus exactly satisfied.

The one unusual feature of the structure is that, because of the symmetry, the two P—O(1) bonds must be collinear. In the isomorphous Sc$_2$Si$_2$O$_7$ the same requirement applies, and the existence of collinear Si—O(1) bonds was regarded as so unexpected that a special study was made, and the fact was confirmed. In Mg$_2$P$_2$O$_7$ as in Sc$_2$Si$_2$O$_7$, an angle at O(1) of about 130° to 140° would have been more in accordance with expectation.

It is therefore not surprising to find that below about 65° C the symmetry changes to allow a change of shape of the P$_2$O$_7$ group. At the transition, two lattice parameters, a and c, are approximately doubled. (Actually there is a slight increase of volume of about 3 percent per formula unit, due to discontinuous expansion in the a direction.) The new cell contains eight formula units, and the new space group is $B2_1/c$ (No. 14, with an unconventional choice of axes), which still contains only eight general positions in the equipoint; hence O(1) can occupy a general position. The P$_2$O$_7$ group in fact changes shape to give a P—O(1)—P angle of 144°. But this cannot be done, keeping the PO$_4$ groups nearly regular, without disturbing the Mg environment, and the arrangement that results is rather odd and unusual for Mg. The two octahedra per formula unit have now become crystallographically different, and in one of them one of the six O atoms is pushed out to a distance of 3.35 Å, leaving Mg with only five near neighbours.

Clearly the structure of Mg$_2$P$_2$O$_7$ involves a conflict between the preferred environment of Mg, a regular octahedron, and the preferred bent bond in the P$_2$O$_7$ group. It cannot achieve both. The potential-energy contribution is greater for the strong directed P—O bonds, and the potential energy is the determining factor in the low-temperature form. But, as the temperature increases, redistribution of vibrational energy between the different modes increases the entropy contribution to the free energy, and does so in a way which favours the higher-symmetry configuration. It is, however, the contribution from the Mg—O group which tips the balance, and allows a straight P—O—P bond to be formed at a temperature as low as 65° C. An alternative way of putting this is to say that the transition in the direction of increasing temperature is triggered by the formation of the sixth Mg—O bond to the originally 5-coordinated atom.

15.17 THIOUREA, SC(NH$_2$)$_2$

Thiourea provides an example of a displacive transition in a molecular crystal. Strictly speaking, there are believed to be *four* transitions, with three intermediate forms, occurring in quite a narrow temperature range, about

170° to 210° K. We shall however consider only the lowest form, V, and the highest form, I (the room-temperature form), and deal with the transition as if it occurred in a single step from one to the other instead of being broken up into four smaller steps. This does not falsify the overall picture, since the transition as a whole is reversible.

The room-temperature form is orthorhombic with $a = 7.66$ Å, $b = 8.54$ Å, $c = 5.52$ Å; space group *Pnma* (No. 62); $Z = 4$. The molecule (see §14.5 and Figure 14-1) is planar except for two of the hydrogen atoms, and in the structure this plane lies parallel to [010], and therefore projects as a straight line in Figure 15-11(a). The C—S bond of the molecule lies in the (010) mirror plane of the structure. We may think of the S atom as the head of the molecule, and the end carrying the NH_2 groups as the tail. In Figure 15-11(a), which is a schematic diagram of part of the structure, one sees the centrosymmetric head-to-tail pair surrounding the origin. The *n*-glide and *a*-glide planes are at $x = 1/4$ and $z = 1/4$ respectively.

The molecules, as we saw in §14.5, behave nearly like rigid bodies. We have to ask whether any of the relatively weak intermolecular bonds may be hydrogen bonds. They would be N—H $\cdots$ S bonds, a kind we have not hitherto considered, but they are no different in principle from the O—H$\cdots$O bonds dealt with in Chapter 13, though their actual lengths are, of course, different. As with O—H$\cdots$O bonds, it is hard to draw a sharp dividing line between long, weak, hydrogen bonds and non-bonding van der Waals contacts; but since in any case we are not here interested in very weak bonds, we may classify all such as non-bonding contacts.

Examining the structure, we find three pairs of N—S contacts, shown by the dotted and broken lines. Though the hydrogen atoms have been left out of the diagram, for clarity, they lie in fact reasonably near the lines. The first pair are of length 3.394 Å, which is the right length for a bond. They join pairs of molecules which lie, in projection, near the origin into ribbons parallel to [010]. They are the bonds mentioned in §14.5 which clamp the molecules so as to prevent rotation in their own plane. The second pair of contacts are of length 3.696 Å, which is really too long for N—H $\cdots$ S bonds. The third pair are of intermediate length, 3.526 Å. Though still too long to be recognised as bonds, they are nearer the borderline for consideration as such.

Below 169° K, as a result of the transition, the space group becomes $P2_1ma$ (No. 26)—the *n*-glide plane is lost. The resultant structure is shown schematically in Figure 15-11(b). The molecules themselves are unchanged in size and shape—this is what we should expect from their rigid-body behaviour—but they have swung round and changed their orientation. In consequence, the intermolecular contacts have changed their lengths. Of each of the three pairs shown in Figure 15-11(a), one has become longer and one shorter. For the shortest pair this has not altered their structural role; they still continue to tie the molecules together into the [010] ribbons. For the other pairs it makes a big difference: the shorter member of each pair has now become a hydrogen bond. The changes in position of the

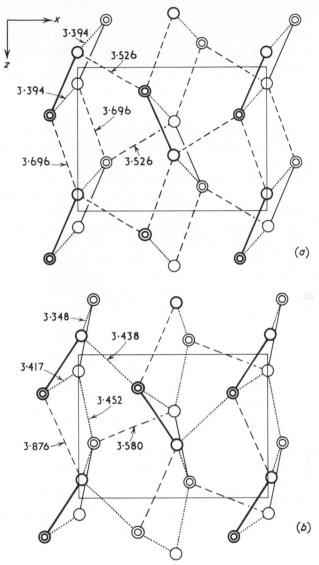

Figure 15-11 Projection of thiourea on (010): (a) room-temperature form; (b) form which occurs below 170° K. The planar SCN$_2$ group projects as a line; S is shown by a single circle, and N by a double circle. The C atom (between S and N) and the H atoms are omitted for clarity. Molecules with S and C at height $y = 1/2$ are shown by thick lines, those at $y = 0$ by thinner lines; the unit cell is outlined. Dotted lines represent N—H · · · S bonds, broken lines other fairly short N—S contacts. Lengths are given in Å.

hydrogen atoms, and the N—H—S bond angles, which we shall not describe in detail, confirm the conclusions drawn from the changes in the N—S lengths. It is important to notice, however, that in none of these bonds does "hydrogen-hopping" take place. The N atom remains the donor throughout, and the S atom the acceptor.

From the diagram, it can be seen that the molecules lying more nearly parallel to the (100) plane are linked into nearly linear chains in the [001] direction by one set of bonds, while those inclined more steeply to (100) are linked by a different set into very zigzag chains with an overall direction along [100]. The only bonds linking one kind of molecule to the other are those originally present in the high-temperature form.

The low-temperature structure, we notice, has polar symmetry, with its polar axis in the x direction (space group $P2_1ma$, point group $2mm$). If there are any dipoles in the structure, their net effect will add up in the x direction. Now the thiourea molecule is a dipole, with a net positive charge on N and a net negative charge on S. Let the magnitude of the dipole be p; its direction, within each molecule, is along the line SC. For a pair of molecules in the [010] ribbon, the dipoles are nearly but not exactly antiparallel; the sums of the x and z components are $p(\cos \alpha_1 - \cos \alpha_2)$, $p(\sin \alpha_1 - \sin \alpha_2)$, respectively (see Figure 15-12). For the next ribbon, at a distance $a/2 + c/2$ from the first, the x components are the same but the z components have changed sign. The net dipole moment per molecule is thus $(p/2)(\cos \alpha_1 - \cos \alpha_2)$, and it lies in the x direction. The value calculated in this way from the structural angles α_1 and α_2 and the dipole moment of thiourea in solution agrees excellently with the measured dipole moment of the crystal.

At the transition, there is no reason why one contact distance of a pair rather than the other should become short, though the choice in one pair determines that of the others. An alternative way of putting it, remembering that the changes of length are consequences of the change of orientation of the molecules, is to say that there is no reason why one particular molecule should turn clockwise rather than anticlockwise, though in the completed transition everything else is determined by this. The two choices are equally possible, and result in the same low-temperature structure in two different

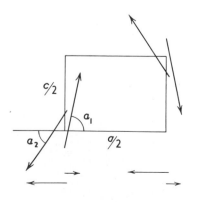

(a)

(b)

Figure 15-12 Thiourea, showing components of dipole moments of individual molecules in the x direction. The molecules are represented by the arrows in (a), a projection corresponding to Figure 15-11(b), and the x-components of their dipole moment are proportional to the x-components of their length, shown by the arrows in (b).

orientations, with oppositely directed x axes. Reversal of direction merely means a slight tilting of the molecules in opposite senses—the same movement that occurs at the transition, but through rather more than twice the angle. Since the reversal of the structure implies the reversal of the net dipole moment, it is not surprising to find that it can be brought about by an applied electric field—in other words, that the low-temperature structure is ferroelectric.

We have described the transition in terms of the "static" structure for atoms at rest at their mean positions, but in fact a good deal is known about the thermal amplitudes. It was noted in §14.5 that the molecule vibrates very nearly as a rigid body even at room temperature, and more exactly so in the transition range; and that the largest amplitude is about the particular inertia axis of the molecule which lies along [010] in the structure. A little below the transition, the r.m.s. amplitude of this motion is 5.5°, while the mean tilt of each molecule from a position parallel to its neighbour is 7°. At room temperature the pairs are of course parallel, and the r.m.s. amplitude is 11°.

This allows us to explain the mechanism of the transition. At low temperatures, potential energy plays the preponderant part in determining the structure, and this is lowest when the N—S contact lengths can adjust themselves to give a simple network of fairly strong bonds and leave other contacts as nonbonding; the bonds hold the molecules in a nonparallel position. The forces opposing tilt about [010] are, however, small, and the mode in which the molecules tilt is therefore a soft one. As thermal energy increases, the amplitude increases and the mean tilt angle decreases till the two become equal with a value a little less than 7°. Thereafter the symmetrical position of the molecules is the mean position of each, though the vibration about it continues soft, with increasingly large amplitude. The transition is exactly analogous to that in quartz (§15.9).

Here we have ignored vibrations of the molecule about its other inertia axes. These, though smaller, are not negligible—4.5° and 3° respectively—and may affect the detailed course of the transition and account for the observed complexities of the intermediate forms. Nevertheless it is useful to see how by considering only the mode with the largest amplitude we can make sense of the transition and predict the overall effect.

The deuterated compound, SC(ND$_2$)$_2$, behaves in an exactly similar way.

15.18 POTASSIUM DIHYDROGEN PHOSPHATE, KH$_2$PO$_4$

The transition here is one involving hydrogen bonds. The forms above and below the transition at 123° K were described in §13.12 and 13.13. It was noted in §13.13 that the symmetry of the high-temperature form requires the hydrogen atom to be central in its bond, but that this may be only statistically satisfied, with the hydrogen actually disordered between two off-centre sites.

The issues raised by such disorder were discussed as a general problem in §15.5. We now consider this particular example in more detail.

Evidence comes from very detailed neutron-diffraction studies made at three temperatures—room temperature, 132° K, and 77° K, the two lower bracketing the transition.

First consider the mean positions of K, P, and O atoms. From room temperature to 132° K there is no change in the size, shape, or orientation of the PO_4 tetrahedron, and only small changes in K—O and O—H · · · O lengths, of the order of magnitude to be expected from thermal expansion. At 77° K, one of the O—O edges perpendicular to [001] has lengthened very slightly, and the other contracted; this change of shape of the tetrahedron is responsible for the shearing of the unit cell in the (001) plane. At the same time, P has moved by 0.08 Å along [001] towards the longer edge, and K by 0.04 Å in the opposite direction. The O atoms of the shorter tetrahedron edge are in fact the donor atoms of the hydrogen bond—the OH groups— and those of the longer edge the acceptor atoms. In the low-temperature form, the displacements of the K, P, and O atoms are all clearly related to the choice of site of the H atom within the bond.

At 77° K the hydrogen peaks are isotropic, and are displaced in an ordered way by 0.20 Å from the centre of the bond. At 132° K and at room temperature they are symmetrically placed in the bond but elongated along its length. Two models account equally well for their shape: a centrally-placed anisotropically-vibrating atom, with mean square amplitude three times as great along the bond as across it; or two half-atoms vibrating isotropically with an r.m.s. amplitude of about 0.17 Å, placed symmetrically at 0.175 Å from the centre. The evidence is thus unfortunately unable to distinguish between the models, but it certainly does not rule out the dis-ordered half-atom model, which is the double-well model of §15.5.

None of the other thermal vibrations are markedly anisotropic or other-wise abnormal. The mean square amplitudes for P, K, and O are approxi-mately equal, and proportional to the absolute temperature. But the mean square amplitude of H (excluding the anomalously large value which might be due to disorder) is much more nearly constant—indeed, if the mean square amplitude of O is subtracted, as representing the effect of the frame-work relative to which H moves, no temperature dependence remains. This (as we saw in §14.8) was interpreted to mean that the hydrogen atom, because of its low mass, has only zero-point energy.

If we assume that the double-well model for the motion of hydrogen atoms is correct, does it follow that the other atoms in the high-temperature form are also statistically distributed between two positions? It is unlikely. For example, the largest displacement at the transition is the displacement of P along [001], but above the transition P has no abnormally large thermal amplitude in that direction—indeed, it is nearly isotropic. For all the other thermal parameters, the r.m.s. amplitudes above the transition are considerably larger than the corresponding displacements between 132° and 77° K, making two-position disorder improbable.

It is of course an important feature of the transition that the low-temperature form has polar symmetry and a net dipole moment. The alternative choice of off-centre site for hydrogen simply reverses the direction of the structure as a whole, and the dipoles with it (cf. Figure 13–12); the low-temperature structure is thus ferroelectric, as we noted in §13.13.

The operation of switching in the ferroelectric form is obviously very similar to the operation of domain reversal in the high-temperature form, assuming the double-well model for the latter. Most physical properties of the material show a continuous relationship across the transition which can be accounted for in this way. For example, it was noted at an early date that the ratio of the spontaneous polarisation to the shape factor $b/a-1$ in the low-temperature form was equal to the ratio of the dielectric constant along c to the geometrically analogous coefficient relating strain to applied field in the high-temperature form. Many detailed studies have been made at the transition, because of the technological importance of the material.

One problem that troubled earlier workers was as follows. The transition, and the ferroelectric reversal process, both appear to be intimately linked with the movement of hydrogen atoms along the O—H $\cdots$ O bond, which is nearly perpendicular to [001]—yet the dipole moment is *along* [001]. Even after it was recognised that the dipoles responsible for the macroscopic effect are in fact the tetrahedra with off-centre P, the strict correlation in the ferroelectric phase between P off-centring and H off-centring was not always realised. We can easily see its nature. Suppose P is originally displaced towards the upper end of its tetrahedron (cf. §13.13 and Figure 13–12). If it moves towards the lower end, towards the OH group, the H atoms are repelled and move across the bond to what were originally acceptor O atoms on other tetrahedra. This occurs cooperatively throughout the structure. Structurally, it is thus obvious that, in ferroelectric reversal, displacement of P implies displacement of H. Similarly, at the thermal transition, we can infer that adoption of a centrosymmetric mean position by P implies adoption of a statistically centrosymmetric position by H, i.e., (for the double-well model) the setting in of "hydrogen-hopping."

What this argument does *not* tell us is which is cause and which consequence. Is it the H atom or the P atom whose structural environment becomes unstable, and thus triggers the transition, forcing a change of environment on the other?

That it is probably the H atom is suggested by observed changes in length in the hydrogen bond. Between 77° and 132° K the O—H $\cdots$ O length decreases by 0.01 Å, the O—H length by -0.02 Å, and the off-centre displacement of H by 0.025 Å. Though too near the limits of error to be conclusive, these changes all point in the same direction, and suggest, for increasing temperature, a weakening (perhaps only a very slight one) of the O—H bond, and a consequent movement of H towards the acceptor O. (A weakening of the O—H bond, it must be remembered, implies a strengthening of the O—H $\cdots$ O bond—cf. §13.2.) Eventually H is near enough to the centre of the bond for tunnelling to set in. This changes the dynamic environment of P, and makes its vibrations take place about the centrosymmetric position as mean position.

Such a picture of the transition process, though accounting plausibly for the diffraction evidence, needs confirmatory evidence from other lines of work before it can be regarded as more than a guess. Clear-cut evidence is not easily obtained, but modern work—whose conclusions are summarised below—seems to support it.

The most successful of the older theories of the transition was that of J. C. Slater. In this it was postulated that, even in the disordered high-temperature form, each tetrahedron had two H atoms attached to it and each bond had one and only one H in it. The possible ways in which a piece of structure, or domain, could be built to satisfy these criteria were enumerated, and for each a count was made of the number of tetrahedra with their dipoles pointing up, down, or sideways. Since the energy of the individual dipole depends on its orientation relative to the field—taken as pointing upwards—the energy of the domain was known, and the number of domains of that particular configuration was therefore given by the Boltzmann distribution. Hence the entropy of the transition could be calculated. Very reasonable agreement was obtained.

Macroscopic changes of properties at the transition on the whole support the idea that it has order-disorder character. They generally show no discontinuity, but are much sharper than is usual for transitions of substitutional disorder. The heat of transition is 0.36×10^3 J mol^{-1}, the entropy 3.1 J mol^{-1} deg^{-1}. The entropy is very much greater than that for $BaTiO_3$ (Table 15–4), though the heats of transition are roughly the same—a result which we should expect for an order-disorder transition.

Attempts to find the frequency and character of a mode associated with the transition have met with considerable difficulties, of the kind mentioned in §15.3. Broadly speaking, there are three possible dynamic models for the structure above the transition. In the first, we have a single-minimum potential well, and an ordinary nearly-harmonic vibration—a phonon mode—of rather large amplitude. In the second, we have a double-minimum potential well, and tunnelling motion of the hydrogens, correlated with each other to give a *pseudospin wave*. In the third, we have again the double-minimum well, but the "hydrogen-hopping" and switching of direction is not periodic—it has a *relaxation time*, like an overdamped mechanical system. Definite evidence for a frequency associated with the transition has been hard to find, but the single-well model is almost certainly ruled out.

It seems likely that what is actually happening can best be described by a mode involving tunnelling, with a characteristic frequency, but heavily damped, so that it is difficult to distinguish from a Debye relaxation. However, the tunnelling mode is coupled to one or more ordinary phonon modes determining the movements of the other atoms in the structure. Direct interaction of H atoms with one another would not produce the transition; it is their indirect linkage through the other atoms that does so. The H atoms make the tunnelling motion, and the other atoms follow them adiabatically.

The corresponding deuterium compound, KD_2PO_4, undergoes a very similar transition, but at a much higher temperature—213° K as compared with 123° K for KH_2PO_4. This great temperature difference has been something of a puzzle, hard to explain by any of the older quantitative theories. It is probably associated with differences in the tunnelling behaviour of H and D atoms. In any case it reinforces the view that the transition is triggered by changes in the hydrogen bond.

15.19 TRIGLYCINE SULPHATE, $(NH_2CH_2COOH)_3 \cdot H_2SO_4$

The transition in this material is very similar in essentials to that in potassium dihydrogen phosphate. It takes place at 49° C, below which the crystal has space group $P2_1$ (No. 4) with three formula units in the unit cell. The three molecular groups are thus crystallographically independent in the low-temperature form. A reliable structure determination is difficult, because the crystal is damaged by the radiation used to investigate it. It is, however, worth describing the results of the first careful low-temperature study and using them to make deductions about the transition. Though some of the details may need reconsideration later, the general principles are likely to remain valid.

The structure determination indicated that the three independent molecules are chemically of two kinds—one glycine molecule or "zwitter ion," $NH_3^+CH_2COO^-$, and two glycinium groups, $NH_3^+CH_2COOH$, analogous to ammonium ions. Together with one SO_4^{2-} ion they constitute the asymmetric unit of structure. In the glycinium group the N, C, and O atoms are coplanar, but in the zwitter ion the N atom sticks out slightly from the plane of the rest. We may call the COO^- or $COOH$ end of the molecule its head, the NH_3^+ group its tail.

The atomic arrangement approximates to one which would satisfy the higher symmetry of the space group $P2_1/m$ (No. 11). It is shown schematically in Figure 15-13(a). Glycinium I lies roughly in the mirror plane—more exactly, at 12° to it—while glycinium III and zwitter ion II face each other, head to head, across the pseudocentre of symmetry, tied together there by a short hydrogen bond of length 2.44 Å (marked 1 in the figure). Other hydrogen bonds join the heads and tails of the three molecular groups to one another and to SO_4 groups to form a continuous network throughout the structure.*

---

* A more recent study (Chapter 15, reference 55) suggests some changes of detail: the NH_3 group of glycine I may be disordered (thus destroying the bonds marked 2 and 3 in the figure), and the hydrogen atom between glycine II and glycine III (marked 1) appears to belong to II rather than III. If this latter finding is substantiated, the original identification of II with the zwitter ion and III with glycinium would need to be altered; however, the general picture of the ferroelectric reversal process given in the text remains unaffected.

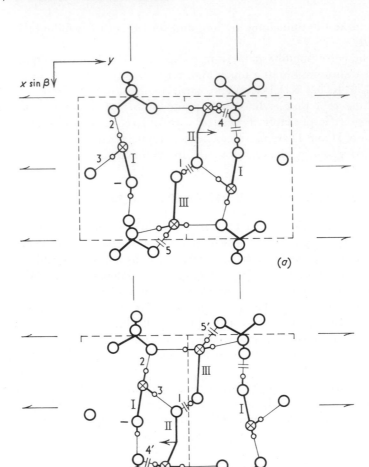

Figure 15-13 Triglycine sulphate: schematic diagram of part of room-temperature structure, projected on (001), illustrating ferroelectric reversal. In (a), the parts of the structure shown lie between $y = 1/4$ and $3/4$; atoms outside this range are included only when they complete the environments of those within it. (Others can be derived by the operation of the diad screw axis shown.) The outline of the unit cell is shown by broken lines. In (b), reversal of the y axis is accompanied by a change of origin, and the same part of the structure now lies between $+1/4$ and $-1/4$.

All atoms and bonds lie within the unit cell, with z coordinates between 0 and 1, except for the OH group marked $-$, which is in the unit cell below. Bonds terminating on atoms in the unit cell above or below are shown crossed by double lines.

Glycinium groups I and III and glycine molecule II (zwitter ion) are shown schematically by thick lines. Only their terminal atoms are marked separately: N, crossed circles; O⁻, or O of OH group, large open circles; H, small circles. The S atoms are omitted, and the S—O bonds are shown by thick lines.

Thin lines are the H $\cdots$ O part of N—H $\cdots$ O or O—H $\cdots$ O bonds. For the numbers attached to some, see text. One H atom attached to the N atom of group I has been omitted since its bond roughly coincides in this projection with the position of group I; it joins N of one group I to OH of the next group I above, thus forming chains of such groups along [001].

If now the hydrogen atom of the COOH group moves across the pseudo-centre of symmetry towards the COO^- group, the group it leaves becomes a zwitter ion and the one it approaches a glycinium group. We can deduce a logical chain of related movements—though, as with potassium dihydrogen phosphate, it remains undetermined whether the movement of hydrogen in bond 1 provides the trigger or is itself triggered by some of the related movements. The shapes of groups II and III are substantially changed at the nitrogen end, and hence also their contacts with SO_4 and group I. But SO_4 and group I lie nearly on mirror planes, and therefore can restore the original set of contacts, in a new orientation, by tilting to a position which is the mirror image of the original one in this plane. The completed operation is shown in Figure 15–13(b).

Geometrically, what has happened is that the whole structure is reflected in the mirror plane at $y = 1/4$ or $y = 3/4$. The direction of the y axis is reversed, and the origin of coordinates is moved, in Figure 15–13(b), to the position that was $(0, 1/2, 0)$ in Figure 15-13(a). Physically speaking, atoms near the pseudo-mirror plane move across it, while atoms not near it do not move bodily to distant positions but make small movements by which they interchange identities with others that were nearly but not quite their mirror images. This interchange of identities can be seen directly for the two sulphate oxygens near $x = 0.1$. To find the same relationship between groups II and III we have to remember that the mirror image of III (which lies outside the region shown in Figure 15-13) nearly coincides with the actual position of a II group related by the 2_1 axis to the II group shown; hence we can combine the operation of the mirror plane and the 2_1 axis, equivalent to a symmetry centre at $(1/2, 1/2, 0)$, to find the new positions shown in Figure 15-13(b).

We may note that, if glycine molecule II has a dipole moment, its y component is reversed with the structure, as shown schematically by the arrows in the figure; the substance is therefore ferroelectric. (A detailed study of the electrical properties would have to consider the contributions of other possible dipoles, but that is outside the scope of this book.)

The space group of the high-temperature form is $P2_1/m$, and the hydrogen atoms near $(1/2, 1/2, 0)$ are statistically at centres of symmetry. It seems that, as in KH_2PO_4, domains of the two low-temperature orientations exist in dynamic equilibrium. Only in the short $O—H \cdots O$ bond(1) can there be tunnelling. In none of the other $N—H \cdots O$ or $O—H \cdots O$ bonds does the hydrogen atom leave its donor during ferroelectric reversal. The bonds numbered 2 to 5 in Figure 15-13 switch direction, to end on a different acceptor (like those in the hydrogen halides, §15.23); the others, not numbered in the figure, undergo displacive changes only (like those in thiourea, §15.17). We may therefore conclude that in the high-temperature form glycinium I and the SO_4 group behave as rigid bodies, whether they are subject to orientation disorder or to large-amplitude libration; glycinium III and zwitter ion II either change shape dynamically as they interchange identities or have both acquired the same shape (as defined by the mean positions of their atoms subject to thermal oscillations only) different from the low-temperature shape of either. If, as assumed for KH_2PO_4, the transition is triggered by tunnelling of hydrogen in the short bond, the sequence of changes set off by it elsewhere in the structure is more complicated than in KH_2PO_4.

There are no observed discontinuities of lattice parameters, heat content, or spontaneous polarisation, at this transition. The specific heat anomaly, studied in a single crystal, is nearly as sharp as that of KH_2PO_4. The heat of transition is 1.4×10^3 J mol^{-1}, and the entropy is 4.5 J mol^{-1} deg^{-1}.

15.20 SODIUM NITRITE, NaNO$_2$

In the preceding section we considered a transition in which, though "hydrogen-hopping" could provide the trigger, switching of molecular orientation was needed to complete the change. In the present section we go on to an example in which orientation-switching is the main feature.

The room-temperature structure of sodium nitrite is as follows:

> Lattice: Orthorhombic, body-centred, with $a = 3.60$ Å, $b = 5.75$ Å, $c = 5.35$ Å.
> Space group: $Im2m$ (No. 44)
> Number of formula units per unit cell: 2
> Atomic positions:
> Na in $2(a)$: $0, \frac{1}{2} + y_1, 0$, with $y_1 = 0.037$
> N in $2(a)$: $0, y_0, 0$, with $y_0 = 0.072$
> O in $4(d)$: $0, y_0, z_0$, with $y_0 = -0.048$, $z_0 = 0.194$

(Since the structure is polar, the choice of origin along the y axis is arbitrary; it has been chosen so as to allow easy comparison with the high-temperature form.)

The structure is shown in Figure 15-14. It is built from alternating Na ions and NO_2 ions. The NO_2 groups are shaped like blunt arrowheads, and all point in the same direction along the b axis, with their planes perpendicular to the a axis. If they were compressed to spheres, the structure would be that of sodium chloride, the axes of reference being taken along one edge and two face diagonals of its cubic unit cell. The fact that the axial ratios b/a and c/a of the actual structure are greater than $\sqrt{2}$ is due to the nonspherical shape of the NO_2 group.

The N—O distance in the NO_2 group is 1.240 Å, and the O—N—O bond angle is 115°. The Na atom has six oxygen neighbours, two from the adjacent NO_2 group in the b direction, at 2.471 Å, and four more, nearly in the [101] and [$\bar{1}$01] directions, at 2.533 Å. The one near-neighbour nitrogen atom is at 2.589 Å.

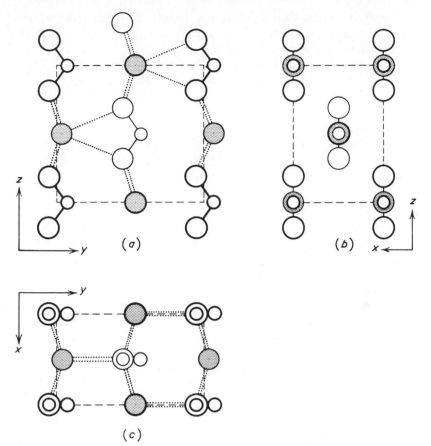

Figure 15-14 Structure of sodium nitrite, NaNO₂, projected (a) on (100); (b) on (010); (c) on (001). Large shaded circles are Na; large open circles are O; small open circles are N; double circles in (c) are two O atoms coinciding in projection. Heavily outlined atoms are at height zero, and lightly outlined atoms at height −1/2. Dotted lines are Na—O bonds, omitted in (b); broken lines outline unit cell.

It is easy to see that this structure is polar, with all dipoles pointing along the *b* axis. Less predictably, it is found experimentally that all dipole directions can be reversed by an electric field—the structure is in fact ferroelectric.

There is a phase transition at about 165° C. In the high-temperature form, the net dipole moment has been lost. The space group is *Immm* (No. 71). This *could* have resulted if the NO₂ groups had begun to rotate freely. However, a detailed study at 185° C has shown that there is not rotation, but disorder: half the NO₂ groups now point in each direction along +*b* and −*b*, and the Na atoms occupy two sites at random to correspond. The structure is, in fact, statistically a superposition of the two orientations of the room-temperature structure, having in common the point chosen above as origin, which now becomes statistically a centre of symmetry.

As usual for order-disorder processes, the setting in of disorder is gradual. At 150° C, still appreciably below the transition, about 10 percent of the NO_2 groups are in the reversed orientation.

We may note here the occurrence of a complexity just above the first transition—a very narrow temperature range, of less than a degree, in which a different phase is found. In it, though equal numbers of NO_2 groups point in either direction, the choice of direction is not random, but represents a modulation in the *a* direction giving a new and longer period. We shall not consider it further, but treat the transition as if it went straight from the low-temperature to the high-temperature form. This (which is what we did in considering thiourea, §15.16) will not falsify the overall picture, though it may obscure some of the details.

An interesting problem posed by the structure is this: just how does the NO_2 group reverse its direction? Does it move in the same way when it reverses spontaneously with increasing temperature as it does when it reverses ferroelectrically under the influence of an applied field?

Three possible methods of reversal can be envisaged: (i) tunnelling by the nitrogen atom through the potential barrier between its two attached oxygens, (ii) rotation about the inertia axis of the group which lies along the *a* axis of the crystal, (iii) rotation about that lying along the *c* axis of the crystal. To get a wider view of the problem, we should list the other modes of motion—rotation about the diad axis of the molecule (the *b* axis of the crystal), and translation in each of the three axial directions—all of which may be active, though they do not directly affect the direction in which the NO_2 group points.

The tunnelling mechanism is considered to be unlikely, because NO_2 is quite a rigid molecule, and the O—N—O angle is not easily distorted. Of the three rotational modes, assuming them independent, quantum mechanics would suggest that the rotation about *a* is most readily excited, because it is associated with the largest moment of inertia, and rotation about *c* the least readily, because its moment of inertia is the smallest. However, this argument is not conclusive, because the rotations are not in fact independent of the translations. This is shown in Figure 15-15. For rotation about the *a* axis, as in Figure 15-15(a), if the Na—O distances are to remain equal, Na must be displaced along *c*; for rotation about the *c* axis, as in Figure 15–15(b), Na must be displaced along *a*. Librations about one axis are therefore coupled to displacements along another axis. Effects like this make predictions much more doubtful. Only for *b*, the axis of diad symmetry, are librations independent of translational movement.

Experimental evidence as to which modes are excited is not wholly conclusive. Interpretation of the thermal parameters measured by diffraction methods is not easy, since librations and linear displacements can contribute to the same observed parameter. Infrared studies suggest that at room temperature libration about *a* has the largest amplitude. Again, in electrical experiments, it is easy to reverse the dipole direction in a

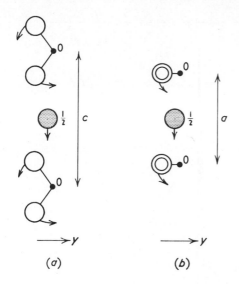

Figure 15-15 Part of NaNO$_2$ structure, showing correlated librational movement of NO$_2$ group and displacement of Na: (a) libration about a—cf. Figure 15-14(a); (b) libration about c—cf. Figure 15-14(c). Large shaded circles are Na; large open circles are O; small black circles are N; heights are marked.

single-crystal plate cut so that the field is applied along c, but not in one with the field along a, and this supports the hypothesis that the NO$_2$ group moves in the bc plane, i.e., about the a axis. Other work, however, suggests that the c-axis libration has an appreciable amplitude.

What seems quite clear, however, is that neither of these librations, nor any of the translational modes, is, separately, the one responsible for the transition. None of the normal modes describing small displacements becomes unstable at the transition. The actual flipping over of the NO$_2$ group involves much bigger movements. Something is known of the times involved for reversal. Well above the transition, where the NO$_2$ groups can be assumed to switch independently, the relaxation time is of the order of 10^{-11} sec, whereas the librational frequency about the a axis is 3×10^{-12} sec; hence the group flips over the barrier only about once in every ten vibrations. Nearer the transition, interaction between NO$_2$ groups makes flipping more difficult and slows down the relaxation time of the reversal process.

The thermal expansion of the crystal has been carefully studied and has interesting features, not yet fully explained. It is shown in Figure 15–16. At room temperature, and up to about 100° C, all three coefficients are of the same order of magnitude, with $\alpha_a > \alpha_c > \alpha_b$, and all increase in the same way. But near 100° C, α_c suddenly starts to decrease, and continues to do so smoothly (becoming zero at about 130° C and increasing in negative magnitude thereafter) right up to the melting point, except for a superposed negative peak at the 163° C transition. Over the same temperature range, α_a and α_b increase in a corresponding way, with large positive peaks at the transition. It is possible that these high, sharp, peaks in all coefficients really represent (at least in part) discontinuous jumps in spacings rather than smooth expansions (cf. §14.14 and Figure 14-11). In any case, they show up very clearly the correlation between all three spacings. If $1/\alpha$ is plotted

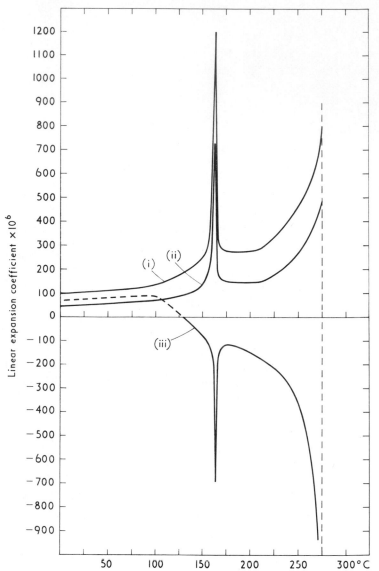

Figure 15-16 Thermal expansion of sodium nitrite: $NaNO_2$; (i) along a axis; (ii) along b axis; (iii) along c axis.

against temperature, it gives straight lines, of the form $1/\alpha = A - BT$, which, if extrapolated beyond the melting point, meet at a point $T = 292°$ C, $1/\alpha \simeq 1 \times 10^3$ deg, $\alpha = 1000 \times 10^{-6}$ deg^{-1}.

The reason for the negative α_c and the positive α_a and α_b can be found in the librations of the NO_2 group. The dimensions of the structure are determined by the projections, along the three axial directions, of the O—O edge of the group and the Na—O bonds. These lengths remain constant except for bond-length expansion (negligible for O—O, fairly small for Na—O), but the O—O edge, whose mean direction is along the c axis, tilts with increasing amplitude on either side of this axis, and hence decreases the r.m.s. value of its component along c and increases it in the other directions. The observed thermal expansion therefore shows that libration about *either a or b*, or more probably both, begins to increase appreciably from about 100° C and continues to do so up to the melting point, irrespective of the 163° C transition. This confirms the conclusion, stated earlier, that the cause of the transition is not, in this case, increasing amplitude of libration, but direct interaction between the dipoles (an Ising model).

If, at high temperatures, librations about all three axes began to have large amplitudes tending to make the NO_2 group spherical, we should expect to find the axial ratios approaching their ideal cubic values, $a:b:c = 1/\sqrt{2}:1:1$. But in fact, though $\sqrt{2}\,a$ and c become more nearly equal with increasing temperature, b is greater than c at room temperature and continues to increase as c decreases; this argues against any approach to free rotation.

To sum up: sodium nitrite is an example of a material in which the transition is due to disordering of dipoles; there is some evidence as to how the NO_2 molecule-ions actually switch orientation, namely by rotation in their own plane; increasing librational amplitudes give a large anisotropic contribution to the thermal expansion, becoming effective above 100° C; but these amplitudes, though they loosen the structure enough to make spontaneous flipping over of NO_2 groups possible at high temperatures, are not themselves the cause of the transition.

15.21   SODIUM NITRATE, NaNO₃

Sodium nitrate is another material with molecule-ions which could conceivably rotate. It has a transition at 275° C which, when first investigated, was thought to be due to the onset of rotation, but which is now known to be another example of orientation-switching.

The room-temperature structure is isomorphous with that of calcite, described in §11.7. The lattice is rhombohedrally-centred hexagonal, with $a = 5.07$ Å, $c = 16.82$ Å; the unit cell contains six formula units. If we prefer to use rhombohedral axes of reference, $a_R = 6.32$ Å, $\alpha = 47°16'$, and the unit cell contains two formula units. The space group is $R\bar{3}c$ (No. 167).

The transition is marked by sharp peaks in the specific heat and mean linear coefficient of thermal expansion, shown in Figure 15-17(a) and (b). The expansion is strongly anisotropic (like that of calcite—cf. §14.12); the separate coefficients along a and c are shown in Figure 15-17(c). It can be seen that the coefficient along a changes very little, but from about 100° C the coefficient along c begins to rise increasingly fast towards the peak at 275° C. The total abnormal increase of volume, above what would have

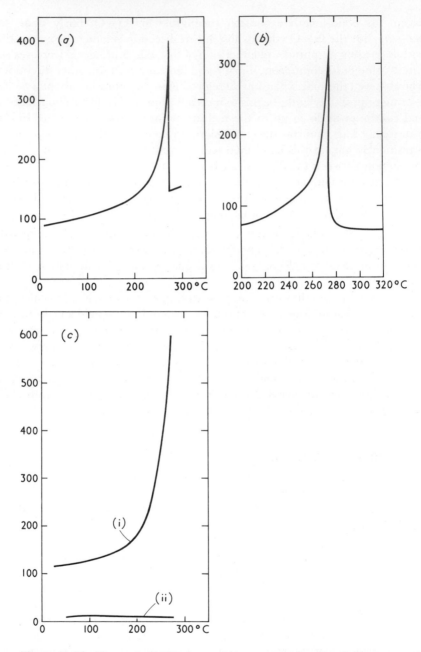

Figure 15-17 Changes in $NaNO_3$ in transition range as a function of temperature: (a) specific heat, in $J\ mol^{-1}$; (b) mean linear expansion coefficient, in $10^{-6}\ deg^{-1}$; (c) linear expansion coefficients along (i) the c direction, (ii) the a direction, in $10^{-6}\ deg^{-1}$.

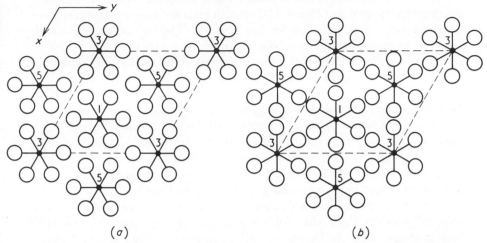

Figure 15-18 Possible disordered structures for high-temperature sodium nitrate. Small black circles represent N, large open circles "half-oxygens," i.e., sites occupied by O with statistical probability 1/2. Heights are marked in units of $c/12$ of the original calcite cell ($R\bar{3}c$) or $c/6$ of the new cell ($R\bar{3}m$). The Na atoms (not shown) lie 3 height units above N atoms; e.g., there is an Na atom at height 4 above the N atom at height 1. The orientation of the primitive rhombohedron of the calcite structure has been retained; this means that the primitive rhombohedron of the disordered structure is shown in *reverse* orientation (cf. §7.6).

been predicted from the steady trend well below the transition, is about 2 percent.

Structurally, the transition is marked by the disappearance of certain diffraction maxima—those that were due to the lack of mirror planes of symmetry through the triad axis. The new space group is thus $R\bar{3}m$, and the identity period along c is halved; hence the hexagonal unit cell contains three formula units, and the rhombohedral unit cell one formula unit. Just above the transition, the lattice parameters are $a = 5.09$ Å, $c = (1/2) \times 17.68$ Å.

The array of Na and N atoms remains unchanged; it is only the O atoms which have altered their positions to achieve the mirror-plane symmetry. Of the possible ways in which they could do so, complete rotation of the NO₃ group as a sphere was ruled out, at an early stage, by diffraction evidence, but rotation in its own plane, about its triad axis, was for some time thought to be the explanation. If we consider disordered orientation of the NO₃ groups between two positions, there are two different models, shown in Figure 15-18(a) and (b). In (a), half the NO₃ groups, at random, have switched through 180° (or 60°) as compared with the original structure, shown in Figure 11-11(a) and (b); there are six "half-oxygens" surrounding each N atom at each level. In Figure 15-18(b), all the NO₃ groups have switched through ±30°, the choice of +30° or −30° being random over the crystal as a whole.

Direct evidence from X-ray investigation of diffuse background diffraction favours the model of Figure 15-18(b). This is the model that seems more likely on geometrical grounds, for two reasons. The first concerns contacts between O atoms of different NO_3 groups. With the model of 15-18(a), or with a rotating NO_3 group, some unit cells would have edges containing straight-line sequences N—O $\cdots$ O—N, of predicted length $1.2 + 2.8 + 1.2 = 5.2$ Å, considerably longer than the observed 5.09 Å; the need to provide for such sequences would suggest a considerable thermal expansion along a, quite contrary to what is observed. With the model of 15-18(b), there are no abnormally short O—O contacts in the (0001) plane, and none in the c-direction either, since the interlayer distance is $c/6 = 2.8$ Å, long enough to allow vertical superposition of O atoms in different layers. The second reason concerns the packing round Na. In Figure 15-18(a), each Na has, in the layers immediately above and below it, 12 equidistant "half-oxygen" neighbours, of which only 6 can actually represent neighbours of any individual Na; in Figure 15-18(b) there are 18 half-atoms, hence 9 true neighbours, and though they are not equidistant they are nearly enough so to suggest that, with minor adjustments, a satisfactory 8- or 9-coordinated environment can be found.

Consideration of thermal amplitudes below the transition supports this. We saw in §14.5 that librations of NO_3 groups in their own plane were geared to translations parallel to the triad axis and correlated with each other so as to maintain the $\bar{3}$ symmetry of the cation, whose octahedron alternately stretched and flattened. At 200° C, the r.m.s. amplitude of libration was 20°; the changes of shape of the octahedron were already very large. Suppose, at 275° C, the r.m.s. amplitude has increased to 30°. The O atoms of the NO_3 groups now reach the position of the possible mirror planes at one extremity of the libration; if they overshoot, they are subject to different restoring forces, and a switch of orientation is likely to occur. But, just as we saw happened for pure displacive transitions (§15.3, Figure 15-2), so in this case the symmetrical position reached at the extremity of vibration (or libration) immediately below the transition is likely to serve as the mean position of vibration (or libration) above the transition. We therefore expect a switch of orientation of $+30°$ or $-30°$ (at random, so far as the present argument goes) from the mean position in the ordered calcite structure.

The high-temperature form is commonly called the *disordered-calcite* structure.

Calcite probably undergoes a similar transition at 975° C, where it is known to have a thermal anomaly.

A recent study has shown that at very high pressures $NaNO_3$ undergoes a different transformation. In the new form, with space group $R3c$, the NO_3 groups tilt about their triad axes in an ordered way, all groups in the same layer having the same tilt, and those in successive layers opposite tilts; there are also small displacements of Na and probably of N perpendicular to the layer. The tilt angle is about 5° when first observed, at 45 kbars, and increases smoothly to 15° at 70 kbars. If it increased to 30°, both layers would be identical; the new structure would be that of KNO$_3$ II (see §15.22). It also is ferroelectric. No change of volume is detected at the transition, which appears to be thermodynamically second order. The relations of this transition to the high-temperature transition in NaNO$_3$, and to the calcite-aragonite transition, need further investigation.

15.22 POTASSIUM NITRATE, KNO$_3$

Potassium nitrate has a more complex set of transitions than sodium nitrate. There are four different forms, whose stability ranges of temperature and pressure are shown in the sketched phase diagram of Figure 15-19. Of these forms, I has a disordered-calcite structure; II, the normal room-temperature form, has the aragonite structure (cf. §11.8); III has a structure to be described later; and though IV has not been analysed in detail, one suggestion is that its K and O atoms together form an approximately cubic-close-packed array, with N occupying triangular interstices in such a way that the NO$_3$ groups are not all parallel to the same plane.

Transitions between the forms are not all of the same character. At atmospheric pressure, the transition II → I occurs at about 130° C, but it is not directly reversible. On cooling, III is formed from I at about 126° C, and only transforms to II at about 115° C. If it is kept dry and in a vacuum, and cooled fairly rapidly, III can be cooled down to room temperature and kept there metastably; it can be reheated to 120° C without reverting to II. On the other hand, high-pressure experiments at room temperature showed direct formation of IV from II, without any trace of III; there was a tendency for IV to remain in existence after reduction of the pressure, after growth in crystallite size at the high pressure, though on standing II was formed again.

All these results illustrate the principle that the formation of a phase at a given temperature and pressure is not determined merely by the value of its free energy, but by the path by which it can be reached from its previous state, and, in particular, whether the energy hump to be surmounted is

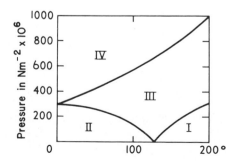

Figure 15-19 Sketched phase diagram of KNO$_3$, showing stability ranges of pressure and temperature for the four phases.

large or small. In KNO_3, it is obvious that the relationship between III and I must be particularly close. We therefore begin by considering the structure of III and the III $\rightleftarrows$ I transition.

Potassium nitrate III has a rhombohedrally-centred hexagonal lattice, with parameters (at 120° C) $a = 5.482$ Å, $c = 9.144$ Å, containing three formula units per unit cell; the rhombohedral axes of reference have $a_R = 4.394$ Å, $\alpha = 77.2°$, and the primitive rhombohedron contains one formula unit. The space group is $R3m$ (No. 160). (At 25° C the parameters are $a = 5.469$ Å, $c = 8.994$ Å, $a_R = 4.354$ Å, $\alpha = 77.8°$.) The atomic positions at 120° C are as follows:

K in $3(a)$: $0, 0, z_K$, with z_K taken as zero to fix the origin
N in $3(a)$: $0, 0, z_N$, with $z_N = 0.56$
O in $9(b)$: $2u, u, z_O$, with $u = 0.11$, $z_O = 0.56$

The structure is shown in Figure 15-20. All NO_3 groups lie in the same plane, between planes of cations; all point in the same direction, with their O atoms lying on mirror planes. With this arrangement, each K has nine oxygen neighbours—three above, lying close to it in projection, and six below, much farther away. To equalise the two sets of distances, the K atom must adjust its height, sinking down relative to the NO_3 groups. Since there is only one K atom in the primitive unit cell, all behave alike. By this movement, the centrosymmetry of the structure is lost, and it becomes polar.

A polar structure is capable of possessing a dipole moment. Can we recognise one here? Because the two K—N distances to the NO_3 groups above and below a given K atom have become unequal, the environments of K and N atoms are one-sided, and this can result in a dipole moment. We

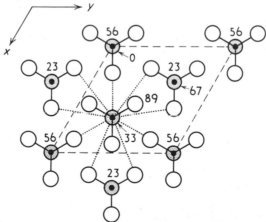

Figure 15-20 Structure of KNO_3, Phase III. Small black circles are N, shaded circles K (coinciding with N in projection), and large open circles O. Heights are indicated in units of $c/100$ for the NO_3 groups and the K atoms. Dotted lines show bonds from the K atom at height 33 to O atoms at heights 23 and 56.

may take the direction of the shorter $N \to K$ distance as indicating the dipole direction.

The next question that suggests itself is: can the dipole direction be reversed? To effect this for the structure as shown in the diagram, all the K atoms must move upward relative to all the N atoms, thus making the short $N \to K$ vectors point downward instead of upward.* But, to allow room for K in its new position, the NO_3 groups must all switch orientation by 60°, cooperatively. Such a movement is not likely to be difficult; we saw in §15.21 that a very similar movement was responsible for the $NaNO_3$ transition. By switching of the NO_3 orientation, and an up-and-down movement of K, the structure as a whole can reverse. It is in fact a ferroelectric, its reversal mechanism consisting of orientation-switching geared to simple atomic displacement—just as the reversal mechanism of potassium dihydrogen phosphate consisted of "hydrogen-hopping" geared to simple atomic displacement.

From the structure of Phase III, it is easy to predict the character of the III ⇌ I transition. It is an order-disorder process of orientation-switching of NO_3 groups in their own plane, with consequent adjustment of K positions. The structure of Phase I is the statistical average of the two orientations of Phase III; it is the same as that of Phase I of $NaNO_3$, shown in Figure †5-18(b).†

Additional information is available about the character of the disorder in Phase I of KNO_3 from its diffuse X-ray diffraction. Immediately above the transition, there are small domains each having the structure of Phase III; in shape they are slabs a very few unit cells thick, their planes lying parallel to the c axis and perpendicular to a, and their polar directions pointing alternately up and down. Their boundaries appear not to be fixed, but to be dynamically moving, like those in potassium dihydrogen phosphate; and, like the latter, they may perhaps be explainable in terms of pseudospin waves. As the temperature rises, their size and shape become more irregular and their identity less clear-cut.

The III ⇌ I transition is completely reversible, not only as regards properties in the region below 126° C, but as regards intensities of diffuse background above it. The premonitory region begins at about 100° C, below which the thermal expansion is not abnormal but resembles that of $NaNO_3$ Phase II (calcite structure); along the triad axis it is 120×10^{-6} deg^{-1} at 25° C and 200×10^{-6} deg^{-1} at 100° C, and in the a direction 36×10^{-6} and 18×10^{-6} deg^{-1} respectively. But thereafter the c-expansion increases rapidly to a value of about 900×10^{-6} deg^{-1} at 126° C, while the a expansion

---

* In reality, of course, both K and NO_3 probably move in opposite directions about their centre of gravity.

† Figure 15-18(b) and Figure 15-20 differ in orientation by 180° in the plane of projection. This is due to the fact that Figure 15-18(b) was derived from the conventional (*obverse*) orientation of a rhombohedron containing two formula units; this is the *reverse* orientation of a rhombohedron of half the height, containing one formula unit.

changes little if at all. At the transition itself there are discontinuities: the
c axis increases from 9.14 Å to 9.64 Å, and the a axis decreases from 5.48 Å to
5.42 Å. Above the transition, the expansion coefficient along c remains
large, 308×10^{-6} deg^{-1}, but constant, while that along a has a small
negative value of about -10×10^{-6} deg^{-1}. Thus the character of the
anisotropy of parameter changes is maintained across the transition. Immediately below, at, and above the transition, changes in c are large and positive,
while changes in a are nearly zero or small and negative.

In contrast to the smooth reversibility of the I $\rightleftarrows$ III transition is the
discontinuity and lack of direct reversibility of the II $\rightarrow$ I transition.
Structurally, the latter involves a complete rearrangement of the cation and
anion positions, to give them new environments. Unlike the I $\rightleftarrows$ III transition, therefore, it cannot be considered as a displacive transition modified by
orientation-switching. On the other hand, it is not a typical reconstructive
transition. Single crystals heated and cooled through the transition in a
carefully controlled cycle keep an orientation related to the original; sometimes the result is a single crystal, slightly less perfect, in the original orientation; more commonly it is a multiple twin, with components slightly
misoriented, but with their c axes nearly as in the original crystal.

We may envisage a possible mechanism for the transition as made up of
the following steps.

(a) In the aragonite structure (Phase II) let the two layers of NO_3 ions
on either side of $z = 0$ merge into a single layer at $z = 0$, and similarly for
the two on either side of $z = 1/2$.

(b) Let the pseudohexagonal lattice parameters become truly hexagonal.

(c) Let the layers of cations and anions parallel to (001) slide across one
another to bring both cations and anions into a sequence of cubic close-packing.

We can study this sliding process in detail as follows. Let Figure
15-21 represent an array of possible atomic sites in projection on

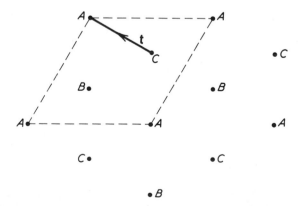

Figure 15-21 Projection on (001) of atomic sites in close-packed array, to illustrate
movements at the II $\rightarrow$ I transition in KNO_3.

(001). In the idealised aragonite structure—after step (b) has been completed—the sequence of occupation in successive equally-spaced layers is $A(C)B(C) \ldots$, where round brackets refer to the N atom of the NO$_3$ group, and the orientation of the group is disregarded. Let **t**, lying in (001), represent the particular C $\rightarrow$ A translation shown in the figure. At the transition, movements of all layers can occur as multiples of **t**, and a regular repetition of six layers is needed:

two cation layers with the intervening anion layer, no change;
next anion layer above, translate by **t**;
next cation layer, and the anion layer above it, translate together by $-$**t**.

The structure formed by regular periodic repetition of these moves has the correct ordered array of K and N—the array found in calcite—but has not the same orientation of its NO$_3$ group. Indeed, if the above procedure for the transition is followed exactly, trigonal symmetry has been lost when the orientations are taken into account. However, our choice of a single vector **t** for all the layer movements was artificial; other vectors at 120° to it would serve equally well for the array of K and N, but would have different consequences for the NO$_3$ orientations, which did not satisfy trigonal symmetry to start with. If the choice of "slip vectors" (**t** translations) is random, the resultant structure is statistically trigonal. It is then in fact the disordered-calcite structure shown in Figure 15-18(b).

There are two points to notice here. First, the orientation of a pair of oppositely directed NO$_3$ groups is *not* that of calcite, shown in 15.18(a), but that of aragonite, Figure 11-14, which differs by 30°. Of course, the groups could undergo a further change of orientation during the transition, but reasons were given in §15.21 why the structure of Phase I is more likely to be that shown in Figure 15-18(b)—and indeed the close relationship to Phase III discussed above confirmed this. Hence, we ought *not* to call the II $\rightarrow$ I transition an "aragonite $\rightarrow$ calcite transition", since the true calcite structure is not involved in it at all.

Secondly, the fact that the path of the transition II $\rightarrow$ I allows choice between numerous combinations of structurally equivalent "slip vectors" means that the reverse transition will not necessarily adopt the same choice—indeed, we have seen that the easiest reverse path is via Phase III. Hence multiple twinning is to be expected. Only if there are good nuclei favouring the growth of one twin component rather than another is anything approximating to a single crystal to be expected after completion of the cycle.

The detailed course of the transition is thus likely to be influenced by the number of growth nuclei present, and this can be altered by external conditions—for example, whether the crystal has been specially dried or is exposed to the atmosphere, or how far above the transition temperature it has been heated. In the II $\rightleftarrows$ I cycle, normally dried crystals of KNO$_3$ kept their orientation better than did those which had undergone special drying.

Effects of nucleation could be seen also in the part of the cycle involving Phase III, which could be supercooled to room temperature in a vacuum, but reverted to Phase II at a much higher temperature when exposed to the atmosphere.

Transitions similar to the II → I transition in KNO_3 are found in the alkaline earth carbonates, in $BaCO_3$ at about 800° C and in $SrCO_3$ at about 912° C. A further transition occurs in $BaCO_3$ at about 970° C, where it becomes cubic, but even here it is unlikely that the CO_3 groups are rotating freely. More probably, they are arranged with their planes perpendicular to the four triad axes, though whether in an ordered or disordered way is unknown.

15.23 THE HYDROGEN HALIDES

Transitions in solid hydrogen halides, discovered from specific-heat studies in the 1920's, were among the first for which rotation of molecules was suggested as an explanation. It has since been conclusively shown that the transitions are due to disordering of orientation rather than to free rotation.

Comparable effects occur in HCl, HBr, and HI, but because HI has been less studied we shall omit it here. The deuterium isotopes, DCl and DBr, behave very similarly, though with slightly different lattice parameters and transition temperatures, shown in Table 15-6. The low-temperature

Table 15-6 Transition Temperatures in the Hydrogen Halides

HCl	III	98° K			I'	120° K	I
DCl	III	105° K					I
HBr	III	90° K	II	114° K	I'(?)	117° K	I
DBr	III	93° K	II	120° K			I

forms (Phase III) are all isomorphous, but as the deuterium compounds have been studied most fully we describe them.

> Lattice: Orthorhombic, *B*-face-centred, with lattice parameters as in Table 15-7
> Space group: $Bb2_1m$ (No. 36)
> Number of formula units per unit cell: 4
> Atomic positions: All atoms in $x, y, 0$, with $x = 0$, $y = 0$ for the halogen, and as in Table 15-7 for D

The origin used for this description is not the conventional one given in the *International Tables for X-ray Crystallography;* instead, it is chosen on a mirror plane, midway between two *b*-glide planes. In the *y* direction, where the choice is arbitrary, it has been taken so as to make the *y* parameter of the halogen atom zero.

Table 15-7 Lattice Parameters (in Å) and Atomic Position Parameters

	a	b	c	x_D	y_D
DCl at 77° K	5.05	5.37	5.82	0.83	0.17
DBr at 84° K	5.44	5.61	6.12	0.82	0.18
	a	b	c		
HCl at 92° K	5.08	5.41	5.83		
DCl at 92° K	5.07	5.40	5.83		
HCl at 118° K	5.48 (assumed cubic)				
DCl at 118° K	5.47 (assumed cubic)				

The structure is as shown in Figure 15-22(a). All atoms lie in special positions on mirror planes. Taken by themselves the halogen atoms form an all-face-centred array, but the D atoms do not conform to this. Each D is attached to one halogen, its donor, and points towards another halogen, the acceptor of its Cl—D · · · Cl or Br—D · · · Br bond. The halogen atoms are thus bonded in zigzag chains, with an angle of about 90° between bonds, the chains extending parallel to the y direction, with the y components of the Cl—D or Br—D vectors all pointing in the same sense. The structure can be seen to be polar. Each molecule is a dipole; since their y components are all parallel, the crystal has a net dipole moment or spontaneous polarisation. The structure as a whole can be reversed by an electric field, and the material is therefore ferroelectric.

Consider the possible ways in which the structure might reverse direction. (i) Each molecule could change direction by 180°; this means that the D atoms move from positions marked by small open circles in Figure 15-22(b) to positions marked by crossed circles. (ii) Each molecule could change direction by 90°, moving its D atoms from the small open circles to the crossed circles of Figure 15-22(c). (iii) Each D atom could jump to a new position in its bond, again moving from the small open circles to the crossed circles of Figure 15-22(c); it is thus indistinguishable from (ii) in its results. In no case do the halogen atoms move, yet in (i) the position of the chain has altered ·by a distance $\mathbf{a}/2 + \mathbf{b}/2$, as shown in Figure 15-22(d), while in (ii) and (iii) it has remained unaltered, as shown in Figure 15-22(e). It is not known for certain which mechanism actually occurs in ferroelectric reversal, but all three possibilities are relevant when we consider what happens at the transition.

The average structure of DBr (Phase II) has been studied at 107° K. It is still orthorhombic, with a unit cell very nearly as in Phase III, containing four formula units, but its space group is now *Bbcm* (No. 64). The structure is that shown in Figure 15-22(b), if we allow small open and crossed circles both to represent statistical half-atoms. The atomic position parameters, and the DBr bond lengths, are almost exactly the same as in the low-temperature form. The thermal amplitudes are considerably greater.

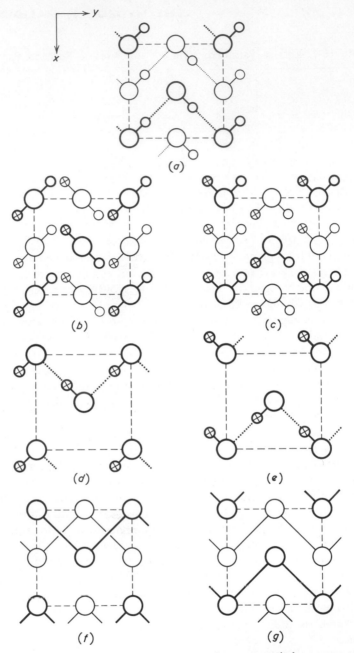

Figure 15-22 Structures of hydrogen halides. Large open circles represent halogen atoms, and small circles H or D, actual or average positions. Atoms shown with heavy line; are at height 0, and those with thin lines at height 1/2. Broken lines outline the unit cells; dotted lines are hydrogen bonds, forming zigzag chains.

(a) Ordered structure of Phase III. (b) and (c) Effect of switching the direction of the molecule through 180° and 90° respectively; the small crossed circles represent the new position of H or D. If the small open and crossed circles are taken as statistical half-atoms, (b) represents the average structure of DBr, Phase II, and (c) that of HCl, Phase I'. (d) and (e) Reversal of direction of the chain at height 0, corresponding to the changes of orientation of the molecule in (b) and (c) respectively. (f) and (g) Relation of the reversed chain at height 0 to the chain in the original direction at height 1/2, corresponding to (b) and (c) respectively.

It seems likely that the zigzag chains, which kept their identity during ferroelectric reversal in Phase III, continue to exist as such above the transition. This means that the possible mechanisms for ferroelectric reversal are also possible mechanisms for thermal disordering. We can then go a little further in analysing the kind of local order that must occur in Phase II. Since its average structure is as shown in Figure 15-22(b), a chain reversed in direction must also be displaced laterally, as in Figure 15-22(d), and hence must be above one in the original direction, as in Figure 15-22(f).

It is assumed that HBr, above its 90° K transition, is isomorphous with DBr above 93° K.

On the other hand, single-crystal studies have shown that HCl, between 98° K and 120° K, is not isomorphous with DBr. It is called Phase I'. Its average structure has the space group *Bbmm* (No. 63) and is represented by Figure 15-22(c). Reversal of chain direction has here not been accompanied by lateral displacement of chains; instead, oppositely directed chains are interleaved in projection, as shown in Figure 15-22(g).

The possibility remains that neither II nor I' is essentially disordered, but that what is observed is an average over domains with a larger true primitive unit cell, built from a periodic sequence of pairs of oppositely directed chains, such as those in Figure 15-22(f) or (g) respectively. Even if there is disorder, the number of chains pointing in each direction must be statistically matched over small areas, and the local order, giving the relationships betwen antiparallel neighbours, must be like that of Figure 15-22(f) for HBr and Figure 15-22(g) for HCl.

The transition at about 114° K in HBr (to which there is nothing corresponding in DBr), may possibly be from II to I'.

The high-temperature form, Phase I, is the same for all the halides. It is cubic, with space group *Fm3m* (No. 207). It seems likely that the zigzag chains continue to exist above the transition, though with more randomness of choice between the different axial directions. A neutron diffraction study of powdered DCl at 111° K showed conclusively that the molecules were not rotating, but that the average structure had its D atoms distributed equally over twelve sites with position parameters very nearly equal to those in Phase III. This, however, may represent an average over three orientations of the orthorhombic Phase I', each twinned in a way which keeps the *b* direction constant but interchanges *a* and *c*—a kind of twinning which has been found in single crystals of HCl between 98° K and 120° K. Nevertheless, the set of orientations possible to twin components, stable in space and time, below a pseudosymmetric transition is likely to be the same as the set possible for local disordered domains with dynamically shifting boundaries above the transition. It is thus probable, though not proved, that even in Phase I there is orientation disorder of molecules, (with some degree of local order) rather than complete rotation.

The differences in lattice parameters, and the thermal expansions, are related to the structures. Differences between HCl and DCl, in Phase III, at the same temperature (Table 15-7) are greatest in the a and b directions, the plane of the zigzag chains. Thermal expansion in DCl, in this phase, is greatest in the same plane, greater along the length of the chain than across it. At the transition, the total change of volume in HCl is 1.9 percent. The transition occurs sharply, over a narrow temperature range of less than one degree; it is thermodynamically first order. In DBr, on the other hand, the transition is gradual, and spread over about 20° C; it is thermodynamically second order. In Phase III, the thermal expansion along a has the greatest anomalous increase, giving a large positive peak at the transition; that along c decreases, giving a fairly large negative peak, while that along b has a small but fairly sharp increase. Above the transition they all have constant values:

$$\alpha_a \simeq \alpha_b \simeq 400 \times 10^{-6} \deg^{-1}, \qquad \alpha_c \simeq -200 \times 10^{-6} \deg^{-1}$$

Like the thermal expansion, the specific heat of DBr shows a characteristic λ-shaped peak at the 90° K transition. Its shape depends on the fraction of reversed chains in thermodynamic equilibrium at any temperature.

The spontaneous polarisation of HBr also falls smoothly to zero, dropping from about 75° K, below which its value is nearly constant at 4×10^{-3} C m^{-2}, along a curve that is roughly parabolic. For HCl, the spontaneous polarisation is greater, 12×10^{-3} C m^{-2}, and it remains constant right up to the transition temperature, where it drops discontinuously to zero.

15.24 AMMONIUM HALIDES

Ammonium, NH$_4^+$, is our only example of a small positively-charged molecule-ion. Many of its compounds are isomorphous with those of potassium, and the interatomic distances early suggested that it behaves as a spherical monovalent cation of radius 1.43 Å.

This, however, is an oversimplification, though a useful starting point. We must consider some representative structures in more detail.

If the ion is *not* rotating, it has tetrahedral point symmetry, and will fit best into an environment where its neighbours are tetrahedrally arranged. This happens in ammonium fluoride, NH$_4$F. If the cation were spherical, we should predict an 8-coordinated CsCl structure from the radius ratio, or, by analogy with RbF, with nearly the same cation radius, the 6-coordinated NaCl structure. Instead, we find the wurtzite structure (§4.11) in which both cations and anions are 4-coordinated. The axial ratio is not perfectly ideal, though close to it—1.614 instead of 1.633—but the atomic position parameter of N is such as to make the four N—F distances almost perfectly equal (of length 2.71 Å), at the cost of a deviation of the F—N—F bond angles of about 0.5° from ideal tetrahedral.

The symmetry does not require the NH_4 groups to be rotating or disordered, and a detailed study in fact shows that they are perfectly ordered. The description of the structure is as follows.

> Lattice: Hexagonal primitive, with $a = 4.439$ Å, $c = 7.165$ Å (at room temperature)
> Number of formula units per unit cell: 2
> Space group: $P6_3mc$ (No. 186)
> Atomic positions:
> F in $2(b)$: $\frac{1}{3}, \frac{2}{3}, z_F$, with $z_F = 0$
> N in $2(b)$: $\frac{1}{3}, \frac{2}{3}, z_N$, with $z_N = 0.3780$
> H(1) in $2(b)$: $\frac{1}{3}, \frac{2}{3}, z_1$, with $z_1 = 0.224$
> H(2) in $6(c)$: $x, \bar{x}, z_2$, with $x = 0.461$, $z_2 = 0.427$

The H—N—H angles remain almost exactly tetrahedral, and the two independent N—H distances are both close to 1.04 Å, not significantly different from one another. The mean square amplitude of vibration of the H atom perpendicular to the bond is nearly twice as great as along it— an effect similar to that in ice (§13.18). If this anisotropy is attributed to a rigid-body libration of the NH_4 group, its angle is only 6°, not suggesting any approach to free rotation.

In fact, it is clear that the structure of NH_4F is determined by the strong N—H $\cdots$ F bonds, rather than by simple packing requirements.

In all other ammonium halides, the hydrogen bonds are weaker than in the fluoride, and the energy advantages of bond formation and of close packing are more evenly balanced. It is therefore not surprising to find a variety of structures and of transitions between them.

At room temperature, and below it, NH_4Cl, NH_4Br, and their deuterium analogues, possess structures closely related to CsCl; at higher temperatures, their average structures are isomorphous with NaCl. The series has been studied in fullest detail for ND_4Br, in which there are three transitions, at about $-104°$ C, $-58°$ C, and 125° C.

Phase IV of ND_4Br, studied at $-195°$ C, is cubic, with $a = 4.00$ Å, two formula units per unit cell, and space group $P\bar{4}3m$ (No. 215). The N atoms are at 0, 0, 0; Br at 1/2, 1/2, 1/2; D at x, x, x, with $x = 0.15$. The ND_4 groups thus lie at positions of tetrahedral point symmetry, and all point in the same direction. Each D atom takes part in one N—D $\cdots$ Br bond, and each Br atom is therefore bonded to four cation neighbours, instead of to eight as in CsBr—the other four neighbours are at the same distance from it, but so oriented that no bonds are formed. The structure is shown in Figure 15–23(a). The angle of libration of the ND_4 group is not more than 5°.

At $-104°$ C there is a transition to Phase III, in which adjacent columns of ND_4 groups extending along [001] reverse their sense, and at the

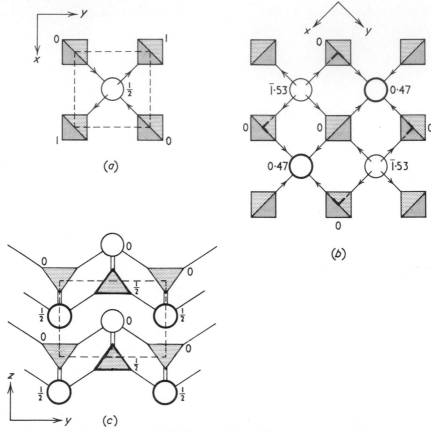

Figure 15-23 Structures of ND_4Br: (a) Phase IV, projected on (001); (b) and (c) Phase III, projected on (001) and (100) respectively. The ND_4 group is represented as a tetrahedron, with positions of D atoms at the corners and N atoms (not shown separately) at the centres; the Br atoms are the large open circles. Heights are marked in (a) and (b) as fractions of c, in (c) as fractions of a. Thin lines drawn from the corners of the tetrahedra represent N—D $\cdots$ Br bonds; in (a) and (b) the arrows point upwards. Bonds to Br atoms not shown in the diagram have been omitted. Unit cells are outlined with broken lines in (a) and (c); in (b) only the corners are marked.

same time the (001) layer of Br atoms becomes puckered. This results in tetragonal symmetry, with a larger primitive unit cell. It is an example of a transition in which increase of temperature *lowers* the symmetry—examples which occur only rarely. At $-78°$ C, Phase III is as follows.

Lattice: Tetragonal, primitive, with $a = 5.83$ Å, $c = 4.14$ Å
Number of formula units per unit cell: 2
Space group: $P\bar{4}/nmm$ (No. 129)
Atomic positions:
  N in $2(a)$: $0, 0, 0$
  Br in $2(c)$: $0, \frac{1}{2}, u$, with $u = 0.47$
  D in $8(i)$: $0, y, z$, with $y = 0.15$, $z = 0.15$

This structure is shown in Figure 15-23(b). The relationship to Phase IV is seen more clearly if we choose new a and b axes at 45° to those for the primitive cell; the new cell has $a = 2 \times 4.12$ Å, $c = 4.14$ Å, and contains four formula units. The system of N—D $\cdots$ Br bonds has changed completely from what it was in IV. There are still four bonds to each Br, but now they are all on one side of it, either above or below—the Br atom has point symmetry 4, not $\bar{4}$. Accordingly, it moves toward the ND_4 layer to which it is bonded, and away from the other. The result is a sandwich with ND_4 groups in the middle and Br atoms alternately on either side; the sandwiches pack together, held by van der Waals forces and residual electrostatic forces, with the upper Br atoms of one lying nearly in the same plane as the lower Br atoms of the next—Figure 15-23(c).

At $-58°$ C there is a further transition to Phase II. This is the temperature at which rotation was formerly supposed to set in. Above it, the average structure has full cubic symmetry and is isomorphous with CsCl. However, it has been conclusively shown that the ND_4 groups are not rotating, but have two-position disorder, which may be thought of as change of orientation by 90° about any of their tetrad axes or as inversion through their centre.

The uppermost transition, at 125° C, differs from all those hitherto considered in requiring a big change in the nitrogen-halogen array—the same that occurs in CsCl at 445° C, when it changes to the NaCl structure. The ND_4 group now has six Br neighbours, not eight, and there is no obvious way of placing it to form four N—D $\cdots$ Br bonds. So we may ask: does it simply lose all its tendencies to bond formation and rotate freely? Or do individual groups set themselves specifically so as to make one or more good N—D $\cdots$ Br contacts, tolerating longer distances for the others? Experiment shows that the latter is the case: there is certainly not free rotation, but disorder between a number of possible orientations. It is hard to distinguish between models in which the preferred settings allow one good contact, two equally good, or three equally good, but infrared evidence suggests the model with one contact. One corner of the tetrahedron points directly at Br, along a $\langle 100 \rangle$ direction—of which there are six choices—and the whole group rotates or is disordered about this line.

In all four phases, the ND_4 group remains tetrahedral and the N—D distance 1.03 Å, within the accuracy of measurement. Libration of the group increases with temperature: at $-195°$ C it is less than 5°, at $-78°$ C about 10°, and at room temperature, in the disordered-CsCl structure, about 12°. But even in the highest form, at about 200° C, rotation is restricted to one axis at most.

It is worth noting that in the two disordered forms, II and I, there seems to be some positional disorder of Br, as we might expect when it finds itself in different local environments, depending on particular choices of orientation of its neighbours. The positional disorder could be (and in the first instance is) interpreted as thermal motion (see §14.3), but a clue is given by the fact that in the two ordered phases the thermal amplitude of Br is much smaller

Table 15-8 "Thermal Amplitudes"* of Atoms in ND_4Br

PHASE	TEMPERATURE	"THERMAL AMPLITUDE" IN Å		
		Br	N	D
IV	−195° C	0.05	0.10	0.12
III	−78° C	0.07	0.12	0.21
II	20° C	0.15	0.15	0.18
I	200° C	0.19	0.19	0.23

* These quantities represent the r.m.s. values of atomic displacements, and therefore include any spread due to disorder as well as true thermal displacement (see text).

than that of N, while in the disordered phases the observed value is equal to that of N. The values at the different temperatures are given in Table 15-8.

The other ammonium halides have structures isomorphous with one or more of those already described, but not the complete sequence. For example, NH_4Br has phases I, II, III, with transitions at 138° C and −38° C (above those for ND_4Br) but Phase IV does not occur down to −196° C. Both NH_4Cl and ND_4Cl have phases I and II, but the transition below this takes them straight into the Phase IV structure—the tetragonal ordered phase does not occur. (There is an interesting analogy here with the hydrogen halides, §15.23.)

It was the transitions in the range below room temperature which attracted most interest in the early days. The specific heats give the typical λ-shaped curves; but it is interesting to notice that if c_v is studied instead of the more usual c_p—that is, if the transition is carried out under external stress—the peak is much less pronounced. The transitions vary considerably in sharpness. The thermal hysteresis breadth—the difference between transition temperatures on heating and cooling—is associated with the volume change rather than with the value of the transition temperature. For example, NH_4Cl, NH_4Br, and ND_4Br, with transition temperatures of −30° C, −39° C, and −58° C respectively, have volume changes of 0.45, 0.15, and 0.20 percent, and thermal hysteresis breadths of 0.35°, 0.03°, and 0.11°.

The variety of behaviour of the ammonium group in ammonium halides is representative of its behaviour in other compounds. There may be cases in which it is rotating freely, but they are very much rarer than was at one time supposed. To a first approximation we may take it as spherical, but when we look more closely into details we must be prepared to find its tetrahedral shape and tetrahedral bonding playing an important role, whether the orientation is ordered in a perfect structure or is disordered between two or more settings. One example of this, discussed in §13.14, is ammonium dihydrogen phosphate, isomorphous with the corresponding potassium and rubidium compounds in its high-temperature form, but differing from them in the low-temperature form both in symmetry and in the orientation of PO_4 tetrahedra, whose arrangement must be influenced by that of the NH_4 tetrahedra.

15.25 MAGNETITE, Fe_3O_4

We conclude with an example of a transition depending on the ordering of valency electrons. Magnetite, as we saw in §12.12, is an inverse spinel, in which the tetrahedral sites (eight per unit cell) are occupied by Fe^{3+} and the octahedral sites (sixteen per unit cell) by equal numbers of Fe^{2+} and Fe^{3+}, distributed at random.

At about 120° K there is a transition marked by a specific-heat anomaly, a decrease in electrical conductivity, and a decrease in ease of magnetisation. The changes in lattice parameters are small, and there has been uncertainty about the symmetry of the low-temperature form. Recent work suggests that it is orthorhombic, with $a = 5.91$ Å, $b = 5.94$ Å, $c = 8.39$ Å, and space group *Imma* (No. 74). The c axis lies along a cube edge of the high-tempera-ture form (space group *Fd3m*, No. 227) and the a and b axes along cube face diagonals. In the absence of a magnetic field, the choice of which cube edge is to be c, and which face diagonal at right angles to it is to be a, is random, and single crystals become, below the transition, polysynthetic twins with components in any of six possible orientations. However, if the crystal is cooled in a magnetic field, the cube edge nearest the field direction becomes the c direction of the orthorhombic phase—this is the direction of easiest magnetisation. By using a suitable inclined field, or by applying lateral pressure in addition to a [100] field, a particular choice of a and b axes can be imposed and an untwinned crystal of the low-temperature form can be obtained. Structure analysis shows that the Fe^{2+} and Fe^{3+} in the octahedral sites order themselves into rows parallel to the b and a axes respectively. In Figure 11-3, for example, all the octahedral cations (Fe instead of Al) would be Fe^{2+} in (a) and (c), Fe^{3+} in (b) and (d). The direction of the Fe^{2+} rows has easier magnetisation than that of the Fe^{3+} rows; it is also the direction of the shorter cell edge, which is slightly surprising because Fe^{2+} is larger than Fe^{3+}.

15.26 SUMMARY AND COMMENTS

(1) The emphasis in this chapter has been on transitions with very small differences between the structures on either side of them. On the whole they are more interesting, and teach us more about the properties of materials, than transitions where there is an extensive breakdown of the structure. They are probably also of much commoner occurrence in nature. On the other hand, the fact that such a high proportion of the examples are of ferro-electric transitions is a reflection of the current interest in ferroelectricity, because of which such transitions have been more extensively studied. They

must, of course, be seen as part of the wider picture; we have to take care neither to describe as peculiar to ferroelectrics effects that in fact are characteristic of all small transitions nor to generalise to the wider category effects that are really due to the unbalanced dipoles of ferroelectrics.

(2) Most small transitions are *pseudosymmetric*, i.e., the low-temperature form has an atomic arrangement which approximates closely to the higher symmetry of the ultimate high-temperature form, the aristotype. This is true whether the higher symmetry is achieved by simple displacements of atoms, without disorder, as in quartz; by orientation-switching, as in sodium nitrite; or by "hydrogen-hopping," as in potassium dihydrogen phosphate. The symmetry of the ultimate high form is therefore predictable, whether its structure is perfect, like high quartz, or disordered, like sodium nitrite and potassium dihydrogen phosphate.

On the other hand, the low-symmetry form cannot be predicted geometrically from the high-symmetry form, even though it satisfies the condition that its symmetry must be a subgroup of that of the high-symmetry form. The symmetry operator lost may be a translation vector rather than an axis or a plane—this is what happens when a lattice parameter is doubled—and for this the number of geometrical possibilities is unlimited. Predictions, if they are possible at all, must be based either on knowledge of the physical character of the structure or empirically by analogy with structures believed to be similar whose transitions are known. Even so, the structures found experimentally are very often quite different from anything one would have thought of beforehand, though they may be understandable when found. Examples occur in sodium niobate, praseodymium aluminate, magnesium pyrophosphate, thiourea, potassium nitrate III—indeed the transitions in barium titanate, now so well known, were completely unexpected and inexplicable when first discovered.

(3) The transition from a given low-symmetry form to the ultimate high-symmetry form does not always occur in a single step. While some materials have only one transition (e.g., quartz), others have several phases intervening between the lowest and the highest (e.g., barium titanate, sodium niobate, thiourea, sodium nitrite, ammonium chloride). In such cases it is not generally true that the higher-temperature form at any intermediate transition gains symmetry elements without losing any already present. Sometimes it even belongs to a lower symmetry class (e.g., Phase III, as compared to Phase IV, of ammonium bromide); sometimes it has a longer repeat period, implying a lower density of symmetry elements (e.g., Phase R in sodium niobate, and the very complex phase just above 163° C in sodium nitrite).

(4) The mechanism of transition varies from one structure to another.

The simplest transitions (e.g., in strontium titanate, or in quartz) are those where the force constants and therefore the frequency of a single lattice mode are temperature-dependent. A transition occurs when the thermal amplitude, in the low-temperature form, of an atom moving according to this mode becomes equal to the displacement of its mean position from the high-symmetry mean position. The mode concerned—the "soft mode"—loses its identity at the transition. (Note that this effect is *not* restricted to ferroelectrics, though chiefly discussed in that context.)

In other structures (e.g., $NaNO_2$), there is direct interaction between dipoles, giving a random distribution between two opposite orientations at high temperatures, an alignment in an ordered structure at low temperatures.

In others again, (e.g., KH_2PO_4), the onset of disorder is described by a pseudospin wave, in which tunnelling of a hydrogen atom in its bond can play a part.

Sometimes two or more physically-independent variables are temperature-sensitive, and their interaction leads to more complex sequences of transitions (e.g., $NaNbO_3$, $CaTiO_3$). Sometimes conditions affecting one particular atom, such as hydrogen in a rather short bond, set off a series of interlocking mechanisms elsewhere in the structure (e.g., KH_2PO_4, triglycine sulphate).

With an ordinary thermal mode, the structure above the transition remains completely ordered; a pseudospin wave allows for dynamic disorder above the transition.

Mechanisms can be studied more easily in ferroelectrics than in other structures, because the mechanism of ferroelectric reversal *below* the transition is valuable in throwing light on the mechanism of structural change *at* the transition.

(5) The structure found at any temperature—even in the immediate neighbourhood of a transition—is not necessarily that of absolute lowest free energy. Energy hills between structures may be too high to allow a passage, even if the valley beyond them is much lower.

This is particularly true for transitions occurring in single crystals. The possibility of keeping orientation relations favours small pseudosymmetric transitions rather than reconstructive transitions. Where there are a number of hettotypes differing very little in energy, the choice of one rather than another may be influenced by clamping pressures, such as those exerted by differently oriented domains within the same composite crystal.

Moreover, the growth of any particular phase at moderate or low temperatures may depend on the occurrence of suitable nuclei, and these may be strongly influenced by very small amounts of impurity. Once started, a crystal of a particular phase can continue to grow, while another of lower free energy may never be able to get a start.

(6) In dealing with these small transitions, conventional thermodynamics cannot take us very far. Its essential assumptions are randomness and statistical homogeneity; against this, in single-crystal transitions, we have to deal primarily with order and regularity. Even when disorder plays a part, local order is often retained—for example, in high-temperature potassium dihydrogen phosphate and potassium nitrate, where domains with

structure very like the low-temperature form in two opposite orientations are recognisable.

An approach which begins by postulating an ordered low-symmetry structure and allows it limited disorder—for example, domain formation of a physically and chemically reasonable kind—is likely to be more fruitful than one which begins with a statistically random structure above the transition and then tries to impose particular correlations on it.

(7) Pseudosymmetric structures are always likely to undergo transition to the high-symmetry form, though it is not necessarily true that the transition is always reached before the melting point. The approach to the transition is marked by an anomalous increase in thermal expansion, and at the transition there is either a sharp peak in the thermal expansion or a discontinuous jump in spacing of a kind suggested by the expansion leading up to it (the two effects not always being easy to distinguish). This statement applies whether the transition is a pure displacive one or involves orientation-switching or "hydrogen-hopping."

(8) The occurrence of pseudosymmetric structures is due to a conflict of energy requirements between different features, notably between good space-filling packing and the specific bond angles of directed bonds. The lower the symmetry per formula unit, the greater the number of independently variable position parameters, by which the environment of each atom can be separately adjusted; hence a low-symmetry structure tends to have low potential energy. If the number of independent lattice parameters and position parameters, taken together, is less than the number of separately specified bond lengths and bond angles to be built into the structure (assuming atoms at rest at the mean positions of their thermal vibrations), some or all of them must be strained; the effective potential energy is increased by this built-in strain energy. On the other hand, if we include the effects of thermal vibrations, a high-symmetry structure tends to have higher entropy (and therefore lower free energy) because of the different possibilities of energy distribution among modes in a perfect structure, or among configurations of local disorder in a disordered structure. When, with increasing temperature, the entropy effect grows large enough to turn the balance, the low-temperature structure either straightens itself out to give the higher symmetry perfectly, or parts of it become disordered to give it statistically. In either case, it is the weakest or "softest" links that suffer change—the N—H $\cdots$ S bonds in thiourea, the oxygen bond angles in quartz, the individual Na—O bonds in sodium nitrite. By observing these changes in geometry we get a clear indication, though a qualitative one, of the relative roles of the different interatomic forces.

This brings us back to the main theme of the book: that geometrical effects in crystals, properly understood, give us the kind of empirical insight into their dynamic behaviour—the character and interaction of their interatomic forces—that is necessary if quantitative theories are to be based on realistic models.

REFERENCES

EXPLANATORY NOTE

This bibliography begins with a short list of books, including some essential reference books. The list has deliberately been restricted to books whose line of thought has something in common with the treatment of the subject in this text.

This is followed by references to original papers, arranged chapter by chapter, in a sequence following that of the text. They are cross referenced among themselves, but not from the text. Each, however, is given either a subject heading or its title—sometimes both. These references are not intended to be completely comprehensive on any topic, or to give an equal coverage to all topics. Each has been chosen for one (or more) of three reasons:

1. It gives an interesting and readable account of a particular study— sometimes by the original discoverer of a new effect, sometimes by a later worker reviewing it.

2. It gives evidence for factual statements made in the text which are not yet familiar from other books, and in particular about structures (mentioned in the text or the exercises) which are either too recent to be included in Wyckoff's *Crystal Structures* or are not easily traced from that work.

3. It describes new results closely relevant to the text but not available in time to be considered in the discussion; such references are marked (n).

No attempt has been made to ensure that the references for any particular structure include either the first or the most precise determination; they do, however, include the sources of the data used in the text and tables, and in the exercises. The references for Chapter 15 are particularly numerous, because this part of the subject is undergoing the most rapid growth. Arbitrariness of selection is, of course, unavoidable, but for a reader interested in following up any particular topic the references quoted should at least serve as a starting point.

BOOKS

General reference

International Tables for X-Ray Crystallography, Vol. I: ed. N. F. M. Henry and K. Lonsdale (1952, Kynoch Press, Birmingham—for the International Union of Crystallography). *Crystal Structures:* R. W. G. Wyckoff, 2nd ed., Vol. 1–4 (1962–1966, Wiley, New York).

History and background

Fifty Years of X-Ray Diffraction: ed. P. P. Ewald (1962, Oosthoek, Utrecht—for the International Union of Crystallography).
Early Papers on Diffraction of X-Rays by Crystals: ed. J. M. Bijvoet, W. G. Burgers, G. Hägg (1969, Oosthoek, Utrecht—for the International Union of Crystallography).
Articles by M. Laue and W. L. Bragg in *International Tables for X-Ray Crystallography*, Vol. I, pp. 1–5, (1952, Kynoch Press, Birmingham—for the International Union of Crystallography).

Reference and further reading for particular chapters

C. H. MacGillavry: *Symmetry Aspects of M. C. Escher's Periodic Drawings* (1965, Oosthoek, Utrecht—for the International Union of Crystallography), for Chapters 3, 7, 8, 9.
F. C. Phillips: *An Introduction to X-Ray Crystallography* (1946, 1963, Longmans, London), for Chapters 5, 6.
M. J. Buerger: *Elementary Crystallography* (1956, Wiley, New York), for Chapter 8.
W. L. Bragg and G. F. Claringbull [and W. H. Taylor]: *Crystal Structures of Minerals* (1965, Bell, London), for Chapters 11, 12, 13.
A. F. Wells: *Structural Inorganic Chemistry*, 3rd ed. (1962, Oxford University Press, London), for Chapters 11, 12, 13.
G. E. Bacon: *Applications of Neutron Diffraction in Chemistry* (1963, Pergamon, Oxford), for Chapters 2, 13.
W. C. Hamilton and J. A. Ibers: *Hydrogen Bonding in Solids* (1968, Benjamin, New York), for Chapter 13.

Additional sources of information

For early structure determinations:
Strukturbericht: Vol. 1, ed. P. P. Ewald and C. Hermann, 1931; Vol. 2, ed. C. Hermann and others, 1937 (Akademische Verlagsgesellschaft, Leipzig; reprinted 1964, Johnson Reprint Corporation, New York).
For more recent structure determinations:
Structure Reports, Vol. 8 and following: ed. A. J. C. Wilson, W. B. Pearson, and others. Vol. 25 (1969) gives a cumulative index (Oosthoek, Utrecht—for the International Union of Crystallography).
For a classified list of crystallographic books to about 1964, with supplement to about 1966:
Crystallographic Book List: ed. H. D. Megaw (1965 and 1967, Oosthoek, Utrecht—for the International Union of Crystallography). Continuations to be published in *Acta Crystallographica*.

CHAPTER 2

Ionic radii

(1) W. L. Bragg (1920) *Phil. Mag.* (6), **40,** 169. The arrangement of atoms in crystals.
(2) W. L. Bragg (1926) *Phil. Mag.* (7), **2,** 258. Interatomic distances in crystals.
(3) V. M. Goldschmidt and coworkers (1923–1938), *Geochemische Verteilungsgesetze der Elemente* (Skrifter utgift av det Norske Videnskaps-Akademie: Jacob Dybwad, Oslo).

(4) L. Pauling (1939) *Nature of the Chemical Bond* (Cornell University Press) Chapter 10.
(5) L. H. Ahrens (1955) *Geochim. Cosmochim. Acta*, **2,** 155. Ionic radii of the elements.
(6) L. Pauling (1957) *Acta Cryst.*, **10,** 685. The use of atomic radii in the discussion of interatomic distances and lattice constants of crystals.
(7) R. D. Shannon and C. T. Prewitt (1969) *Acta Cryst.*, **B25,** 925. Effective ionic radii in oxides and fluorides.
(8) R. D. Shannon and C. T. Prewitt (1970) *Acta Cryst.*, **B26,** 1046. Revised values of effective radii.

Atomic environments in relation to interatomic distances

Off-centring

(9) L. E. Orgel (1958) *Faraday Soc. Discussions*, **26,** 138. Ferroelectricity and the structure of transition-metal oxides.
(10) H. D. Megaw (1968) *Acta Cryst.*, **B24,** 149. A simple theory of the off-centre displacement of cations in octahedral environments.

Directed bonds

(11) D. W. J. Cruickshank (1961) *J. Chem. Soc.*, **1961,** 5486. The role of 3d-orbitals in π-bonds between (a) silicon, phosphorus, sulphur, or chlorine, and (b) oxygen and nitrogen.
(n)(12) W. S. McDonald and D. W. J. Cruickshank (1967) *Acta Cryst.*, **22,** 37. A re-investigation of the structure of sodium metasilicate, Na_2SiO_3.

Systematic variations in radii

(n)(13) W. H. Baur (1970) *Trans. Amer. Cryst. Assoc.*, **6,** 129. Bond length variation and distorted coordination polyhedra in inorganic crystals.
(n)(14) G. E. Brown and G. V. Gibbs (1970) *Amer. Mineral.*, **55,** 1587. Stereochemistry and ordering in the tetrahedral portion of silicates.

Packing and choice of structure

(n)(15) G. O. Brunner (1971) *Acta Cryst.*, **A27,** 388. An unconventional view of the "closest sphere packings."
(n)(16) C. T. Prewitt and R. D. Shannon (1969) *Trans. Amer. Cryst. Assoc.*, **5,** 51. Use of radii as an aid to understanding the crystal chemistry of high-pressure phases.

Exercises and problems

Q.12. Scheelite: R. W. G. Wyckoff, *Crystal Structures*, Vol. 3, p. 19.
Q.13. Thaumasite: R. A. Edge and H. F. W. Taylor (1971) *Acta Cryst.*, **B27,** 584.

CHAPTER 4

(1) S. C. Abrahams and J. L. Bernstein (1969) *Acta Cryst.*, **B25,** 1233. Remeasurement of hexagonal zinc oxide.
(2) T. M. Sabine and S. Hogg (1969) *Acta Cryst.*, **B25,** 2254. The wurtzite z parameter for beryllium oxide and zinc oxide.
(3) C. Frondel and U. Marvin (1967) *Nature*, **214,** 587. Lonsdaleite: a hexagonal polymorph of diamond.
(4) K. Lonsdale (1971) *Amer. Mineral.*, **56,** 333. Formation of lonsdaleite from single-crystal graphite.

CHAPTER 9

(1) E. A. Wood (1964) *Bell System Techn. Journ.*, **43,** 541. The 80 diperiodic groups in three dimensions.

(2) W. Cochran (1952) *Acta Cryst.*, **5**, 620. The symmetry of real periodic two-dimensional functions.

(3) W. Nowacki (1960) *Fortschr. Miner.*, **38**, 96. Überblick über "zweifarbige" Symmetriegruppen.

(4) G. Donnay, L. M. Corliss, J. D. H. Donnay, N. Elliott, and J. M. Hastings (1958) *Phys. Rev.*, **112**, 1917. Symmetry of magnetic structures: magnetic structure of chalcopyrite.

(5) G. S. Pawley (1961) *Acta Cryst.*, **14**, 319. Coloured polyhedra.

CHAPTER II

Spinel, Al₂MgO₄

(1) T. F. W. Barth and E. Posnjak (1931) *Z. Kristallog.*, **82**, 325. Spinel structures with and without variate atomic equipoints.

(2) G. E. Bacon (1952) *Acta Cryst.*, **5**, 684. A neutron diffraction study of magnesium aluminium oxide.

Lithium niobate, LiNbO₃

(3) H. D. Megaw (1968) *Acta Cryst.*, **A24**, 583. A note on the structure of lithium niobate.

Calcite, CaCO₃

(4) H. Chessin, W. C. Hamilton, and B. Post (1965) *Acta Cryst.*, **18**, 689. Position and thermal parameters of O atoms in calcite.

Sodium nitrate, NaNO₃

(5) P. Cherin, W. C. Hamilton, and B. Post (1967) *Acta Cryst.*, **23**, 455. Position and thermal parameters of O atoms in sodium nitrate.

Aragonite, CaCO₃

(6) W. L. Bragg (1924) *Proc. Roy. Soc. A*, **105**, 16. The structure of aragonite.

Olivine, (Mg, Fe)₂SiO₄

(7) J. D. Birle, G. V. Gibbs, P. D. Moore, and J. V. Smith (1968) *Amer. Mineral.*, **53**, 807. Crystal structures of natural olivines.

(n) (7a) W. H. Baur (1972) *Amer. Mineral.*, **57**, 704. Computer simulation of Mg₂SiO₄ polymorphs.

Chrysoberyl, Al₂BeO₄

(8) E. F. Farrell, J. H. Fang, and R. E. Newnham (1963) *Amer. Mineral.*, **48**, 807. Refinement of the chrysoberyl structure.

γ-Dicalcium silicate, γ-Ca₂SiO₄

(9) H. O'Daniel and L. Tscheischwili (1942) *Z. Kristallog.*, **104A**, 124. Zur Struktur von γ-Ca₂SiO₄ und Na₂BeF₄.

(10) D. K. Smith, A. Majumdar, and F. Ordway (1965) *Acta Cryst.*, **18**, 787. The crystal structure of γ-dicalcium silicate.

(11) R. Czaya (1971) *Acta Cryst.*, **B27**, 848. Refinement of the structure of γ-Ca₂SiO₄.

Coesite, SiO₂

(12) T. Araki and T. Zoltai (1969) *Z. Kristallog.*, **129**, 381. Refinement of a coesite structure.

(13) H. D. Megaw (1970) *Acta Cryst.*, **B26**, 261. Structural relations between coesite and felspar.

Keatite, SiO₂

    (14) J. Shropshire, P. P. Keat, and P. A. Vaughan (1959) *Z. Kristallog.*, **112,** 409. The structure of keatite, a new form of SiO_2.

Melanophlogite, SiO₂

    (15) B. J. Skinner and D. E. Appleman (1963) *Amer. Mineral.*, **48,** 854. Melanophlogite, a cubic polymorph of silica.

Quartz, SiO₂

    (16) R. A. Young (1962) Report 2569, Air Force Office Scientific Research, Washington, D.C. Mechanism of the phase transition in quartz.

    (17) A. de Vries (1958) *Nature*, **181,** 1193. Determination of the absolute configuration of α-quartz.

    (18) A. Lang (1965) *Acta Cryst.*, **19,** 290. The orientation of the Miller-Bravais axes of α-quartz.

Sanidine, KAlSi₃O₈

    (19) W. H. Taylor (1933) *Z. Kristallog.*, **85,** 425. The structure of sanidine and other felspars.
    (For other felspar references, see Chapter 12.)

Exercises and problems

    Q.1. KBF_3 series: K. Knox (1961) *Acta Cryst.*, **14,** 583.

    Q.9. δ-Mn_2GeO_4: A. D. Wadsley, A. F. Reid, and A. E. Ringwood (1968) *Acta Cryst.*, **B24,** 740.

    Q.10. MnO_2H: L. S. Dent Glasser and L. Ingram (1968) *Acta Cryst.*, **B24,** 1233.

    Q.11. High-pressure forms of Mn_2GeO_4 and Co_2SiO_4: N. Morimoto, S. Akimoto, K. Koto, and M. Tokonami (1970) *Phys. Earth Planet Interiors*, **3,** 161.

    Q.18. Beryl: W. L. Bragg (1930) *Z. Kristallog.*, **74,** 237.

    Q.20. Garnets
        $Ca_3Al_2Si_3O_{12}$: S. C. Abrahams and S. Geller (1958) *Acta Cryst.*, **11,** 437.
        $Mg_3Al_2Si_3O_{12}$: G. V. Gibbs and J. V. Smith (1965) *Amer. Mineral*, **50,** 2023.
        $Y_3Fe_5O_{12}$: P. Fischer, W. Hälg, E. Stoll, and A. Segmüller (1966) *Acta Cryst.*, **21,** 765.
        $Y_3Al_4GaO_{12}$: M. Marezio, J. P. Remeika, and P. D. Dernier (1968) *Acta Cryst.*, **B24,** 1670.
        $Na_3Al_2Li_3F_{12}$: S. Geller (1971) *Amer. Mineral.*, **56,** 18.

    Q.21. High-pressure $KAlSi_3O_8$: A. E. Ringwood, A. F. Reid, and A. D. Wadsley (1967) *Acta Cryst.*, **23,** 1093.

CHAPTER 12

Perovskites

    [See also Chapter 15, references (16) to (18), (24), (27), (36), and (39) to (46)].

With elongated octahedra

    (1) $KCuF_3$: A. Okazaki and Y. Suemene (1961) *J. Phys. Soc. Japan*, **16,** 176.

With off-centre cations

    (2) $BaTiO_3$, tetragonal: J. Harada, T. Pedersen, and Z. Barnea (1970) *Acta Cryst.*, **A26,** 336.

(3) $BaTiO_3$, orthorhombic: G. Shirane, H. Danner, and R. Pepinsky (1957) *Phys. Rev.*, **105,** 856.
(4) $CsGeCl_3$: A. N. Christansen and S. E. Rasmussen (1965) *Acta Chem. Scand.*, **19,** 421.

With tilted octahedra

(5) $LaAlO_3$: C. de Rango, G. Tsoucaris, and C. Zelwer (1966) *Acta Cryst.*, **20,** 590.
(6) $LiNbO_3$, high-temperature: S. C. Abrahams, H. J. Levinstein, and J. M. Reddy (1966) *J. Phys. Chem. Solids*, **27,** 1019.
(7) $CaTiO_3$: H. F. Kay and P. C. Bailey (1957) *Acta Cryst.*, **10,** 219.
(8) $NaMgF_3$: E. C. T. Chao, H. T. Evans, B. J. Skinner, and C. Milton (1961) *Amer. Mineral.*, **46,** 379.
(9) $GdFeO_3$: S. Geller (1956) *J. Chem. Phys.*, **24,** 1236.
(10) $GdFeO_3$ and $YFeO_3$: P. Coppens and M. Eibschutz (1965) *Acta Cryst.*, **19,** 524.
(11) Rare-earth aluminates: S. Geller and V. B. Bala (1956) *Acta Cryst.*, **9,** 1019.
(12) Ionic radii in rare-earth perovskites: S. Geller (1957) *Acta Cryst.*, **10,** 248. [Cf. Chapter 2, reference (6).]
(13) Rare-earth orthoferrites: M. Marezio, J. P. Remeika, and P. D. Dernier (1970) *Acta Cryst.*, **B26,** 2008.
(14) $PrAlO_3$, low-temperature: R. D. Burbank (1970) *J. Appl. Cryst.*, **3,** 112.

With off-centre cations and tilted octahedra

(15) $NaNbO_3$, Phase *P*: A. C. Sakowski-Cowley, K. Łukascewicz, and H. D. Megaw (1968) *Acta Cryst.*, **B25,** 851.
(16) $NaNbO_3$, Phase *Q*: E. A. Wood, R. E. Miller, and J. P. Remeika (1962) *Acta Cryst.*, **15,** 1273.
(17) $NaNbO_3$, Phase *Q*: M. Wells and H. D. Megaw (1961) *Proc. Phys. Soc. London*, **78,** 1258.
(18) $NaNbO_3$, Phase *R*: A. C. Sakowski-Cowley (1967) Thesis (Cambridge).

(18a) $NaNbo_3$, Phase *N;* C. N. W. Darlington (1971) Thesis (Cambridge).

Cryolite structures

(19) Ba_2CaWO_6: E. G. Steward and H. P. Rooksby (1951) *Acta Cryst.*, **4,** 503.
(20) Cryolite: E. G. Steward and H. P. Rooksby (1953) *Acta Cryst.*, **6,** 49.

Trifluorides

(21) VF_3: K. H. Jack and V. Gutmann (1951) *Acta Cryst.*, **4,** 246.
(22) Series of trifluorides: M. A. Hepworth, K. H. Jack, R. D. Peacock, and G. J. Westland (1957) *Acta Cryst.*, **10,** 63.
(23) CrF_3: K. Knox (1960) *Acta Cryst.*, **13,** 507.

Spinel family

(24) Fe_3O_4, low-temperature: W. C. Hamilton (1958) *Phys. Rev.*, **110,** 1050. [See also Chapter 15, reference (22).]

Rutile family

(25) VO_2, monoclinic: J. M. Longo and P. Kierkegaard (1970) *Acta Chem. Scand.* **24,** 420.

"Stuffed silica" structures

(n)(26) Nepheline: W. A. Dollase (1970) *Z. Kristallog.*, **132,** 27.
(n)(27) $LiAlSi_2O_6$, stuffed high-quartz: C. T. Li (1968) *Z. Kristallog.*, **127,** 327.
(n)(28) $LiAlSi_2O_6$, stuffed keatite: C. T. Li and D. R. Peacor (1968) *Z. Kristallog.*, **126,** 46.
(n)(29) $LiAlSi_2O_6$, $LiGaSi_2O_6$, $LiAlGe_2O_6$, stuffed keatite: T. Hahn and M. Behruzi (1968) *Z. Kristallog.*, **127,** 160. [See also Chapter 15, references (10), (11).]

Felspars

(30) General account: W. H. Taylor (1965) Chapter 14 in *Crystal Structures of minerals,* by W. L. Bragg and G. F. Claringbull (Bell, London).

(31) General account: W. H. Taylor (1965) *Tschermaks mineralog. und petrogr. Mitteilungen,* **10,** 5.

(32) NaAlSi$_3$O$_8$, albite: P. H. Ribbe, H. D. Megaw, and W. H. Taylor (1969) *Acta Cryst.,* **B25,** 1503.

(33) CaAl$_2$Si$_2$O$_8$, anorthite: H. D. Megaw, C. J. E. Kempster, and E. W. Radoslovich (1962) *Acta Cryst.,* **15,** 1017.

(34) NaBSi$_3$O$_8$, reedmergnerite: D. E. Appleman and J. R. Clark (1966) *Amer. Mineral.* **50,** 1827.

(n)(35) NaGaSi$_3$O$_8$: H. Pentinghaus and H. U. Bambauer (1971) *N. Jb. Miner. Mh.* **1971,** 94.

(n)(36) RbAlSi$_3$O$_8$: M. Gasperin (1971) *Acta Cryst.,* **B27,** 854.

(n)(37) SrAlSi$_3$O$_8$: E. Bruno and G. Gazzoni (1971) *Z. Kristallog.,* **132,** 327.

(38) Exchange of large cations: J. Wyart and G. Sabatier (1961) *Cursillos y conferencias,* **8,** 23.

Zeolites

(39) General survey: W. Meier (1968) *Proceedings of conference on molecular sieves* (Soc. Chem. and Industry, London), p. 10.

(40) Natrolite: W. H. Taylor (1930) *Z. Kristallog.,* **74,** 1.

(41) Natrolite: W. Meier (1960) *Z. Kristallog.,* **113,** 430.

(42) Chabazite: J. V. Smith, F. Rinaldi, and L. D. Glasser (1963) *Acta Cryst.,* **16,** 45.

(43) Chabazite: J. V. Smith, C. R. Knowles, and F. Rinaldi (1964) *Acta Cryst.,* **17,** 374.

Bronzes

(44) General survey: A. D. Wadsley (1955) *Rev. Pure and Appl. Chem.,* **5,** 165 (Section 3.2).

(45) General survey: G. Burns (1969) *I.E.E.E. Trans. Electron Dev.,* ED-**16,** 506.

(46) Ba$_3$TiNb$_4$O$_{15}$: N. C. Stephenson (1965) *Acta Cryst.,* **18,** 496.

(47) Ba$_3$TiNb$_4$O$_{15}$: P. B. Jamieson and S. C. Abrahams (1968) *Acta Cryst.,* **B24,** 984.

(48) Ba$_{1.4}$Sr$_{3.6}$Nb$_{10}$O$_{30}$: P. B. Jamieson, S. C. Abrahams, and J. L. Bernstein (1968) *J. Chem. Phys.,* **48,** 5048.

(49) PbNb$_2$O$_6$: M. H. Francombe and B. Lewis (1958) *Acta Cryst.,* **11,** 696.

Other complex oxides and fluorides

(50) PNb$_9$O$_{25}$: R. S. Roth, A. D. Wadsley, and S. Anderson (1965) *Acta Cryst.,* **18,** 643.

(51) BaMnF$_4$: E. T. Keve, S. C. Abrahams, and J. L. Bernstein (1969) *J. Chem. Phys.,* **51,** 4928.

(52) NaNbO$_2$F: S. Anderson and J. Galy (1969) *Acta Cryst.,* **B25,** 847.

(53) Ca$_2$Nb$_2$O$_7$: J. K. Brandon and H. D. Megaw (1970) *Phil. Mag.,* **21,** 189.

(54) K$_2$Zn$_2$V$_{10}$O$_{28}$·16H$_2$O: H. T. Evans (1966) *Inorg. Chem.,* **5,** 967.

Exercises and problems

Q.4. Elpasolite: see reference (20).

Q.6. MnF$_3$: M. A. Hepworth and K. H. Jack (1957) *Acta Cryst.,* **6,** 49.

Q.8. See reference (13).

Q.14. LiAlSiO$_4$ (stuffed quartz structure): H. G. F. Winkler (1948) *Acta Cryst.,* **1,** 27.

Q.15. LiAlSi$_2$O$_6$, LiGaSi$_2$O$_6$, LiAlGe$_2$O$_6$ (stuffed keatite structure): see reference (29).

Q.18. Scapolite: J. J. Papike and T. Zoltai (1965) *Amer. Mineral.,* **50,** 641.

Q.19. 5-4-3 bronzes: see reference (45); also

    (i) T. Fukuda and H. Inuzuka (1969) *Acta Cryst.,* **A25,** Suppl., 49.

    (ii) A. Watanabe, Y. Sato, T. Yano, and I. Kitahiro (1970) *J. Phys. Soc. Japan,* **28,** Suppl. 93.

    (iii) F. W. Ainger, J. A. Beswick, W. P. Bickley, R. Clarke, and G. V. Smith (1971) *Ferroelectrics,* **2,** 183.

CHAPTER 13

General principles

(1) J. D. Bernal and R. H. Fowler (1933) *J. Chem. Phys.*, **1**, 515. A theory of water and ionic solution, with particular reference to hydrogen and hydroxyl ions.

(2) J. D. Bernal and H. D. Megaw (1935) *Proc. Roy. Soc. A.*, **151**, 384. The function of hydrogen in intermolecular forces.

(3) R. Chidambaram (1962) *J. Chem. Phys.*, **36**, 2361. Structure of the hydrogen-bonded water molecule in crystals.

(4) R. Chidambaram, A. Sequeira, and S. K. Sikka (1964) *J. Chem. Phys.*, **41**, 3616. Neutron-diffraction study of the structure of potassium oxalate monohydrate: lone-pair coordination of the hydrogen-bonded water molecule in crystals.

(5) W. H. Baur (1965) *Acta Cryst.*, **19**, 909. On hydrogen bonds in crystalline hydrates.

Interpretation of structures

(6) Hydrogarnets: C. Cohen-Addad, P. Ducros, and E. F. Bertaut (1967) *Acta Cryst.*, **23**, 220.

(7) Afwillite: H. D. Megaw (1952) *Acta Cryst.*, **5**, 477.

(n)(8) Role of electrostatic valence: W. H. Baur (1970) *Trans. Amer. Cryst. Assoc.*, **6**, 129.

KOH

(9) J. A. Ibers, J. Kumamoto, and R. G. Snyder (1960) *J. Chem. Phys.*, **33**, 1164.

Ca(OH)$_2$

(10) W. R. Busing and H. A. Levy (1957) *J. Chem. Phys.*, **26**, 563.

(11) H. E. Petch (1961) *Acta Cryst.*, **14**, 950.

Gibbsite, Al(OH)$_3$

(12) H. D. Megaw (1934) *Z. Kristallog.*, **87**, 185 [with correction in reference (2)].

Diaspore, AlOOH

(13) W. R. Busing and H. A. Levy (1958) *Acta Cryst.*, **11**, 798.

Montroseite, VOOH, and paramontroseite, VO$_2$

(14) H. T. Evans and M. E. Mrose (1955) *Amer. Mineral.*, **40**, 861. (See also Chapter 11, Q.10—MnOOH.)

KH$_2$PO$_4$

(15) G. E. Bacon and R. S. Pease (1953) *Proc. Roy. Soc. A*, **220**, 397.

(16) G. E. Bacon and R. S. Pease (1955) *Proc. Roy. Soc. A*, **230**, 359.

NH$_4$H$_2$PO$_4$

(17) E. A. Wood, W. J. Merz, and B. T. Matthias (1952) *Phys. Rev.*, **87**, 544.

H$_2$SO$_4$

(18) C. Pascard-Billy (1965) *Acta Cryst.*, **18**, 827.

NaHCO$_3$

(19) R. L. Sass and R. F. Scheuerman (1962) *Acta Cryst.*, **15**, 77.

(20) B. D. Sharma (1965) *Acta Cryst.*, **18**, 818.

Potassium hydrogen malonate

(21) J. G. Sime, J. C. Speakman, and R. Parthasaraty (1970) *J. Chem. Soc. London A*, **1970**, 1919.

Ice

(22) Ih: S. W. Peterson and H. A. Levy (1957) *Acta Cryst.*, **10**, 70. A single-crystal neutron-diffraction study of heavy ice.

(23) Ih: R. Chidambaram (1961) *Acta Cryst.*, **14,** 467. A bent hydrogen bond model for the structure of ice I.

(24) Ic: K. Shimaoka (1960) *J. Phys. Soc. Japan*, **15,** 106. Electron diffraction study of ice.

(25) B. Kamb (1964) *Acta Cryst.*, **17,** 1437. Ice II: a proton-ordered form of ice.

(26) B. Kamb and A. Prakash (1968) *Acta Cryst.*, **B24,** 1317. Structure of ice III.

(27) B. Kamb, A. Prakash, and C. Knobler (1967) *Acta Cryst.*, **22,** 706. Structure of ice V.

(28) B. Kamb (1969) *Trans. Amer. Cryst. Assoc.*, **5,** 61. Structural studies of the high-pressure forms of ice.

Gas hydrates

(29) L. Pauling and R. Marsh (1952) *Proc. Nat. Acad. Sci. U.S.A.*, **38,** 112. The structure of chlorine hydrate.

(30) R. K. McMullan and G. A. Jeffrey (1965) *J. Chem. Phys.*, **42,** 2725. Polyhedral clathrate hydrates: structure of ethylene oxide hydrate.

Alums

(31) Cs alum: D. T. Cromer, M. I. Kay, and A. C. Larson (1966) *Acta Cryst.*, **21,** 383.

(32) Na alum: D. T. Cromer, M. I. Kay, and A. C. Larson (1967) *Acta Cryst.*, **22,** 182.

(33) K, Rb, and NH$_4$ alums: A. C. Larson and D. T. Cromer (1967) *Acta Cryst.*, **22,** 793.

Trona, sodium sesquicarbonate hydrate

(34) C. J. Brown, H. S. Peiser, and A. Turner-Jones (1949) *Acta Cryst.*, **2,** 167.

Oxalic acid dihydrate

(35) F. F. Iwasaki and Y. Saito (1967) *Acta Cryst.*, **23,** 56.

(36) F. F. Iwasaki, H. Iwasaki, and Y. Saito (1967) *Acta Cryst.*, **23,** 61.

(37) R. G. Delaplane and J. A. Ibers (1969) *Acta Cryst.*, **B25,** 2423.

(38) T. M. Sabine, G. W. Cox, and B. M. Craven (1969) *Acta Cryst.*, **B25,** 2437.

(39) P. Coppens and T. M. Sabine (1969) *Acta Cryst.*, **B25,** 2442.

(40) P. Coppens, T. M. Sabine, R. G. Delaplane, and J. A. Ibers (1969) *Acta Cryst.*, **B25,** 2451.

HNO$_3$·3H$_2$O

(41) V. Luzzati (1953) *Acta Cryst.*, **6,** 157.

Hydrochloric acid hydrates

(42) Monohydrate: Y. K. Yoon and G. Carpenter (1959) *Acta Cryst.*, **12,** 17.

(43) Dihydrate: J. O. Lundgren and I. Olovsson (1967) *Acta Cryst.*, **23,** 966.

(44) Trihydrate: J. O. Lundgren and I. Olovsson (1967) *Acta Cryst.*, **23,** 971.

Exercises and problems

Q.2. Sonoraite: G. Donnay and R. Allmann (1970) *Amer. Mineral.*, **55,** 1003 (simplified).

Q.3. 2BaO·Al$_2$O$_3$·5H$_2$O: A. H. Moinuddin Ahmed and L. S. Dent Glasser (1970) *Acta Cryst.*, **B26,** 867.

Q.4. Brucite, Mg(OH)$_2$: F. Zigan and R. Rothbauer (1967) *N. Jb. Miner. Mh.*, **1967,** 137.

Q.5. Afwillite: see reference (7).

Q.6. Lepidocrocite, γ-FeOOH: A. Oles, A. Szytuła, and A. Wanic (1970) *Phys. stat. sol.*, **41,** 173.

[see also reference (2).]

Q.7. NH_4HSO_4: R. J. Nelmes (1971) *Acta Cryst.*, **B27,** 272.
Q.8. $BeSO_4 \cdot 4H_2O$: I. G. Dance and H. C. Freeman (1969) *Acta Cryst.*, **B25,** 304.
Q.9. $NaHSO_4 \cdot 7H_2O$: W. H. Baur and A. A. Khan (1970) *Acta Cryst.*, **B26,** 1584.
Q.10. Hydrogarnet: see reference (6).
Q.11. Ice III: see reference (26).
Q.12. $HBr \cdot 3H_2O$: J. O. Lundgren (1970) *Acta Cryst.*, **B26,** 1893.

CHAPTER 14

General survey

(1) K. Lonsdale (1948) *Acta Cryst.*, **1,** 142. Vibration amplitudes of atoms in cubic crystals.
(2) K. Lonsdale (1959) *Z. Kristallog.*, **112,** 188. Experimental studies of atomic vibrations in crystals and of their relationship to thermal expansion.

Examples of thermal amplitudes

(3) C. W. Burnham (1964) *Min. Soc. America*, Summer Meeting, Abstracts.
(4) M. Sunderalingam and L. H. Jensen (1965) *Acta Cryst.*, **18,** 1053. Refinement of the structure of salicylic acid.
(5) H. H. Cady and A. C. Larson (1965) *Acta Cryst.*, **18,** 485. The crystal structure of 1-3-5 triamino-trinitrobenzene.
(6) W. A. Dollase (1967) *Acta Cryst.*, **23,** 617. The crystal structure at 220° C of orthorhombic high tridymite from the Steinbach meteorite.

Rigid-body movements and cooperative vibrations—analyses

(7) D. W. J. Cruickshank (1956) *Acta Cryst.*, **9,** 754. The analysis of the anisotropic thermal motion of molecules in crystals.
(8) V. Schomaker and K. N. Trueblood (1968) *Acta Cryst.*, **B24,** 63. On the rigid-body motion of crystals.
(9) G. S. Pawley (1971) *Acta Cryst.*, **A27,** 80. An extended rigid-body model for molecular crystals.

Examples of rigid-body and cooperative motions

(10) Thiourea: M. M. Elcombe and J. C. Taylor (1968) *Acta Cryst.*, **A24,** 410.
(11) $Li_2SO_4 \cdot H_2O$: A. C. Larson (1965) *Acta Cryst.*, **18,** 717.
(12) Na alum: D. T. Cromer, M. I. Kay, and A. C. Larson (1967) *Acta Cryst.*, **22,** 182.
(13) Calcite: H. D. Megaw (1970) *Acta Cryst.*, **A26,** 235.

Corrections to observed bond lengths

(14) D. W. J. Cruickshank (1956) *Acta Cryst.*, **9,** 757. Errors in bond lengths due to rotational oscillations of molecules.
(15) D. W. J. Cruickshank (1961) *Acta Cryst.*, **14,** 896. Coordinate errors due to rotational oscillations of molecules.
(16) G. M. Brown (1968) *Acta Cryst.*, **B24,** 294. Misuse of the "riding" model in correcting bond lengths for effects of thermal motion.

Temperature variation of thermal amplitudes: examples

$CaCO_3$: see Chapter 11, reference (4).
$NaNO_3$: see Chapter 11, reference (5).
KH_2PO_4: see Chapter 13, reference (16).
(17) Ammonium oxalate monohydrate: J. H. Robertson (1965) *Acta Cryst.*, **18,** 410.
(n)(18) Felspar: S. Quareni and W. H. Taylor (1971) *Acta Cryst.*, **B27,** 281.
DCl: see Chapter 15, reference (97).

Electrostatic valence rule

(19) H. D. Megaw (1938) *Z. Kristallog.*, **100,** 58. The thermal expansion of crystals in relation to their structure.

Effect of tilts

    (20) H. D. Megaw (1968) *Acta Cryst.*, **A24,** 589. The thermal expansion of interatomic bonds, illustrated by experimental evidence from certain niobates. [See also references (39), (40).]

Thermal expansions: numerical values

Table 14-3

    (21) *International Critical Tables* (1928), **3,** 43.

Table 14-4

    Calcite structures: see Chapter 11, references (4) and (5).

    (22) Layer hydroxides: H. D. Megaw (1933) *Proc. Roy. Soc. A*, **142,** 198. [See also reference (19).] Organic crystals: see reference (2).

    (23) Oxalic acid dihydrate: K. Gallagher, A. R. Ubbelohde, and I. Woodward (1955) *Acta Cryst.*, **8,** 561.

 (n)(24) Hexamethylbenzene: I. Woodward (1958) *Acta Cryst.*, **11,** 441. Chlorine and nitrogen: see reference (21).

Table 14-5

    (25) $NaMgF_3$: E. C. T. Chao, H. T. Evans, B. J. Skinner, and C. Milton (1961) *Amer. Mineral.*, **46,** 379.

    (26) $CaTiO_3$: H. F. Kay and P. C. Bailey (1957) *Acta Cryst.*, **10,** 219.

    (27) $BaTiO_3$: G. Shirane and A. Takeda (1952) *J. Phys. Soc. Japan*, **7,** 1.

    (28) $NaNbO_3$: I. Lefkowitz, K. Łukaszewicz, and H. D. Megaw (1966) *Acta Cryst.*, **20,** 670.

    (29) $KNbO_3$: G. Shirane, R. E. Newnham, and R. Pepinsky (1954) *Phys. Rev.*, **96,** 581. Other substances: see reference (21).

Temperature variation of thermal expansion

Figure 14-9

    (30) MgO: B. J. Skinner (1957) *Amer. Mineral.*, **42,** 39.

 (n)(31) MgO: D. K. Smith and H. R. Leider (1968) *J. Appl. Cryst.*, **1,** 246. ThO_2: see reference (30). Diamond: see reference (30).

Figure 14-10

    (32) Ice: S. La Placa and B. Post (1960) *Acta Cryst.*, **13,** 503.

    (33) Ice: R. Brill and A. Tippe (1967) *Acta Cryst.*, **23,** 343.

    (34) KH_2PO_4: D. B. Sirdeshmukh and V. T. Deshpande (1967) *Acta Cryst.*, **22,** 438.

 (n)(35) KH_2PO_4: W. R. Cook (1967) *J. Appl. Phys.*, **38,** 1637.

 (n)(36) KH_2PO_4: J. Kobayashi, Y. Uesu, I. Mizutani, and Y. Enomoto (1970) *Phys. stat. sol. (a)*, **3,** 63.

    (37) Quartz: A. H. Jay (1933) *Proc. Roy. Soc. A*, **142,** 237.

Table 14-6

    Quartz: see reference (37).

    (38) Cristobalite: W. Johnson and K. W. Andrews (1956) *Trans. Brit. Ceramic Soc.*, **55,** 227.

    (39) Sodalite and nosean: D. Taylor (1968) *Miner. Mag.*, **36,** 761.

    (40) Leucites: D. Taylor and C. M. B. Henderson (1968) *Amer. Mineral.*, **53,** 1476.

(41) $LiNbO_3$: S. C. Abrahams, H. J. Levinstein, and J. M. Reddy (1966) *Phys. Chem. Solids*, **27**, 1019 (supplemented by private information).
$NaNbO_3$: see reference (28).
$KNbO_3$: see reference (29).
$BaTiO_3$: see reference (27).
For other examples, see Chapter 15, references (25), (39), (46), (68), (72), (73), (86).

Additional studies of structures at different temperatures

(n)(42) N. Foreman and D. R. Peacor (1970) *Z. Kristallog.*, **132**, 45. Refinement of nepheline at several temperatures.
(n)(43) M. Hospital (1971) *Acta Cryst.*, **B27**, 484. Structure cristalline à -140, 20, 120° C et dilatation thermique de l'azélamide.
(n)(44) E. A. Kellett and B. P. Richards (1971) *J. Appl. Cryst.*, **4**, 1. The c-axis thermal expansion of carbons and graphite. [The thermal expansion depends on stacking faults and imperfections below 0° C but not above.]

CHAPTER 15

General reading on transitions

(1) M. J. Buerger (1951) "Crystallographic aspects of phase transitions": Chapter 6 of *Phase Transformations in Solids*, ed. Smoluchowski, Mayer and Weil (1953, Wiley, New York).
(2) K. Lonsdale (1959) *Z. Kristallog.*, **112**, 188. Experimental studies of atomic vibrations in crystals and of their relationship to thermal expansion.
(3) A. R. Ubbelohde (1956) *Brit. J. Appl. Phys.*, **7**, 313. Crystallography and the phase rule.
(4) M. J. Buerger (1961) *Fortschritte Miner.*, **39**, 9. Polymorphism and phase transformations.
(5) A. R. Ubbelohde (1966) *J. Chim. Phys.*, **62**, 3. Thermodynamic and structural aspects of phase transitions that are wholly or partly continuous.
(6) M. J. Buerger (1971) *Trans. Amer. Cryst. Assoc.*, **7**, 1. Crystal-structure aspects of phase transformations.

Reconstructive transitions

$CaCO_3$: see references (77) to (79).
Graphite and lonsdaleite: see Chapter 4, reference (2).
(7) Mg_2SiO_4, Mg_2GeO_4: F. Dachille and R. Roy (1960) *Amer. J. Sci.*, **258**, 225. High-pressure studies of the system Mg_2GeO_4–Mg_2SiO_4, with special reference to the olivine-spinel transition.
(8) Mn_2GeO_4, Co_2SiO_4: N. Morimoto, S. Akimoto, K. Koto, and M. Tokonami (1970) *Phys. Earth Planet. Interiors.*, **3**, 161. Crystal structures of high-pressure modifications of Mn_2GeO_4 and Co_2SiO_4.
(9) TiO_2: P. Y. Simons and F. Dachille (1970) *Amer. Mineral.*, **55**, 403. Possible topotaxy in the TiO_2 system.
(10) $LiAlSi_2O_6$: C. T. Li (1971) *Acta Cryst.*, **B27**, 1132. Transformation mechanism between high-quartz and keatite phases of $LiAlSi_2O_6$ composition.
(11) $Li_2O \cdot Al_2O_3 \cdot nSiO_2$: C. T. Li (1970) *Z. Kristallog.*, **132**, 118. The crystal structure of $Li_2Al_2Si_3O_{10}$ (high-quartz solid solution).
Ice: see Chapter 13, reference (28)
KNO_3: see reference (88).
(12) Ca_2SiO_4: W. Eysel and T. Hahn (1970) *Z. Kristallog.*, **131**, 322. Polymorphism and solid solution of Ca_2GeO_4 and Ca_2SiO_4.
Ca_2SiO_4: see also Chapter 11, reference (10).

Mechanism of reversible transitions

General

(13) W. Cochran (1969) *Advances in Phys.*, **18,** 157. Dynamical scattering and dielectric properties of ferroelectric crystals.
(14) J. D. Axe (1971) *Trans. Amer. Cryst. Assoc.*, **7,** 89. Neutron studies of displacive structural phase transformations.

Octahedral oxides

(15) W. Cochran and A. Zia (1968) *Phys. stat. sol.*, **25,** 273. Structure and dynamics of perovskite-type crystals.
(16) K. A. Müller, W. Berlinger, and F. Waldner (1968) *Phys. Rev. Letters*, **21,** 814 (C). Characteristic structural phase transitions in perovskite-type compounds.
(17) H. Thomas and K. A. Müller (1968) *Phys. Rev. Letters*, **21,** 1256 (C). Structural phase transitions in perovskite-type crystals.
(18) G. Shirane and Y. Yamada (1969) *Phys. Rev.*, **177,** 858. Lattice-dynamical study of the 110° K phase transition in $SrTiO_3$.

Tunnelling, order-disorder, and pseudospin

(19) G. W. Paul, W. Cochran, W. J. L. Buyers, and R. A. Cowley (1970) *J. Phys. Soc. Japan*, **28,** Suppl., 278. Ferroelectric critical scattering from DKDP.
(20) G. Shirane (1970) *J. Phys. Soc. Japan*, **28,** Suppl., 20. Neutron inelastic scattering study of soft modes.

Magnetic transitions and other small displacive transitions

(21) Fe_3O_4: W. C. Hamilton (1958) *Phys. Rev.*, **110,** 1050. A neutron investigation of the 119° K transition in magnetite.
(22) Fe_3O_4: H. P. Rooksby and B. T. M. Willis (1953) *Acta Cryst.*, **6,** 565. The low-temperature crystal structure of magnetite.
(23) NiO: M. V. Vernon (1970) *Phys. stat. sol.*, **37,** K1. Temperature dependence of rhombohedral distortion in NiO.
(24) $RbCoF_3$: J. Nouet, R. Kleinberger, and R. de Kouchovsky (1969) *C. R. Acad. Sci. Paris*, **269,** 986. Étude radiocristallographique à basse température de la pérovskite fluorée $RbCoF_3$.
(25) Fe_2O_3: B. T. M. Willis and H. P. Rooksby (1952) *Proc. Phys. Soc. London B*, **65,** 950. Crystal structure and antiferromagnetism in haematite.
(26) $DyAlO_3$: R. Bidaux and P. Meriel (1968) *J. de Physique*, **29,** 220. Étude par diffractions de neutrons de la structure nucléaire et magnetique de $DyAlO_3$.
(27) $KCuF_3$: A. Okazaki and Y. Suemene (1961) *J. Phys. Soc. Japan*, **16,** 71. The crystal structures of $KMnF_3$, $KFeF_3$, $KCoF_3$, $KNiF_3$, and $KCuF_3$ above and below their Néel temperatures.
(28) $DyVO_4$: P. J. Becker and J. Langsch (1971) *Phys. stat. sol. (b)*, **44,** K 109. Spectroscopic confirmation of a magnetically controllable Jahn-Teller distortion in $DyVO_4$.
(29) VO_2: Y. Hayashi, J. van Landuyt, and S. Amelinckx (1970) *Phys. stat. sol.*, **39,** 189. Electron microscope study of the substructure of VO_2 due to the monoclinic to tetragonal phase transition.
(30) GeTe: G. S. Pawley, W. Cochran, R. A. Cowley, and G. Dolling (1966) *Phys. Rev. Letters*, **17,** 753. Diatomic ferroelectrics.

Quartz

(31) R. A. Young (1962) Report 2569, Air Force Office Scientific Research, Washington, D.C. Mechanism of the phase transition in quartz.

$BaTiO_3$

Structure

[See Chapter 12, references (2) and (3).]

Lattice parameters and physical properties

(32) G. Shirane and A. Takeda (1952) *J. Phys. Soc. Japan*, **7,** 1. Transition energy and volume change at three transitions in barium titanate.

(33) H. F. Kay and P. Vousden (1949) *Phil. Mag.*, **40,** 1019. Symmetry changes in barium titanate at low temperatures and their relation to its ferroelectric properties.

(34) L. E. Cross (1953) *Phil. Mag.*, **44,** 1161. The dielectric properties of barium titanate single crystals in the region of their upper transition temperature.

Thermal vibrations

(35) J. Harada and G. Honjo (1967) *J. Phys. Soc. Japan*, **22,** 45. X-ray studies of the lattice vibration in tetragonal barium titanate.
[See also references (13), (15), (38).]

$KNbO_3$

(36) L. Katz and H. D. Megaw (1967) *Acta Cryst.*, **22,** 639. The structure of potassium niobate at room temperature.

Thermal vibrations or disorder

(37) R. Comès, M. Lambert, and A. Guinier (1968) *Solid State Comm.*, **6,** 715. The chain structure of $BaTiO_3$ and $KNbO_3$.

(38) R. Comès, M. Lambert, and A. Guinier (1970) *Acta Cryst.*, **A26,** 244. Désordre linéaire dans les cristaux (cas du silicium, du quartz et des pérovskites ferro-électriques).

$SrTiO_3$

[See references (14), (16), (17), (18).]

$SrZrO_3$

(39) L. Carlsson (1967) *Acta Cryst.*, **23,** 901. High-temperature phase transitions in $SrZrO_3$.

$LaAlO_3$

[See references (14), (16), (17).]

(40) J. D. Axe, G. Shirane, and K. A. Müller (1969) *Phys. Rev.*, **183,** 820. Zone-boundary phonon instability in cubic $LaAlO_3$.

$PrAlO_3$

(41) R. D. Burbank (1970) *J. Appl. Cryst.*, **3,** 112. A qualitative X-ray study of the 205° and 151° K phase transitions in praseodymium aluminate.

$NaNbO_3$

(42) *P–Q* transition: A. C. Sakowski-Cowley, K. Łukaszewicz, and H. D. Megaw (1969) *Acta Cryst.*, **B25,** 851. The structure of sodium niobate at room temperature.

(43) *P–R* transition: A. Sakowski-Cowley (1967) Thesis, Cambridge. Structural studies of two antiferroelectric phases of sodium niobate.
P–R transition: see also reference (42).

(44) *P–N* transition: C. N. W. Darlington (1971) Thesis, Cambridge. The low-temperature phase transition in sodium niobate.

(n)(45) T_2-cubic transition: F. Denoyer, R. Comès, and M. Lambert (1971) *Acta Cryst.*, **A27,** 414. X-ray diffuse scattering from $NaNbO_3$ as a function of temperature.

(n)(46) T_2-cubic transition: M. Glazer and H. D. Megaw (1972) *Phil. Mag.* **5,** 1119. The structure of $NaNbO_3$ near 600° C, and its relation to soft-phonon modes at the T_2-cubic transition.

$Mg_2P_2O_7$

    (47) C. Calvo (1967) *Acta Cryst.*, **23,** 289. The crystal structure of α-$Mg_2P_2O_7$.

$Sc_2Si_2O_7$

    (48) D. W. J. Cruickshank, H. Lynton, and G. A. Barclay (1962) *Acta Cryst.*, **15,** 491. A reinvestigation of the crystal structure of thortveitite, $Sc_2Si_2O_7$.

Thiourea

    (49) M. M. Elcombe and J. C. Taylor (1968) *Acta Cryst.*, **A24,** 410. A neutron diffraction determination of the crystal structures of thiourea and deuterated thiourea above and below the ferroelectric transition.

KH_2PO_4

    (50) G. E. Bacon and R. S. Pease (1955) *Proc. Roy. Soc. A,* **230,** 359. A neutron diffraction study of the ferroelectric transition of potassium dihydrogen phosphate.

(n)(51) J. Kobayashi, Y. Uesu, I. Mizutani, and Y. Enomoto (1970) *Phys. stat. sol.,* **(a)3,** 63. X-ray study on thermal expansion of ferroelectric KH_2PO_4.

(n)(52) V. Dvorak (1970) *J. Phys. Soc. Japan,* **28,** Suppl., 252. On the phase transition in KH_2PO_4 and the role of depolarizing energy in it.

(n)(53) H. Grimm, H. Stiller, and T. Plesser (1970) *Phys. stat. sol.,* **42,** 207. Neutron measurements on the hydrogen bond potential in paraelectric KDP. [See also references (13), (19), (20).]

Triglycine sulphate

    (54) S. Hoshino, Y. Okaya, and R. Pepinsky (1959) *Phys. Rev.,* **115,** 323. Crystal structure of the ferroelectric phase of $(glycine)_3 \cdot H_2SO_4$.

(n)(55) K. Itoh and T. Mitsui (1971) *Ferroelectrics,* **2,** 225. Refinement of crystal structure of triglycine sulphate.

    (56) M. J. Tello and J. A. Gonzalo (1970) *J. Phys. Soc. Japan,* **28,** Suppl., 199. Ferroelectric specific heat of triglycine sulphate.

$NaNO_2$

Structure at different temperatures

    (57) M. I. Kay and B. C. Frazer (1961) *Acta Cryst.,* **14,** 56. A neutron diffraction refinement of the low-temperature phase of sodium nitrite.

    (58) M. I. Kay, B. C. Frazer, and R. Ueda (1962) *Acta Cryst.,* **15,** 506. The disordered structure of $NaNO_2$ at 185° C.

    (59) M. I. Kay and J. A. Gonzalo (1970) *J. Phys. Soc. Japan,* **28,** Suppl., 284. Neutron diffraction study of $NaNO_2$ [at 150, 185, and 225° C].

Complexities near transition

    (60) Y. Yamada, I. Shibuya, and S. Hoshino (1963) *J. Phys. Soc. Japan,* **18,** 1594. Phase transition in $NaNO_2$.

Mechanism of transition

    (61) Y. Sato, K. Gesi, and Y. Takagi (1961) *J. Phys. Soc. Japan,* **16,** 2172. Study of the phase transition in $NaNO_2$ by polarised infra-red radiation.

    (62) Y. Ishibashi and Y. Takagi (1970) *J. Phys. Soc. Japan,* **28,** Suppl., 261. Lattice vibration of $NaNO_2$.

    (63) S. Sawada and Y. Tokugawa (1964) *J. Phys. Soc. Japan,* **19,** 2105. A change in the direction of spontaneous polarisation in $NaNO_2$ by a lateral electric field.

    (64) I. Shibuya, Y. Iwata, N. Koyana, S. Fukui, S. Mitani, and M. Tokunaga (1970) *J. Phys. Soc. Japan,* **28,** Suppl., 281. Neutron diffraction studies of ferroelectric polarisation reversals in sodium nitrite.

    (65) G. Dolling, J. Sakurai, and R. A. Cowley (1970) *J. Phys. Soc. Japan,* **28,** Suppl., 258. Crystal dynamics of sodium nitrite.

(66) Y. Yamada and T. Yamada (1966) *J. Phys. Soc. Japan*, **21**, 2167. Interdipolar interaction in NaNO$_2$.
(67) Y. Yamada, Y. Fujii, and I. Hatta (1968) *J. Phys. Soc. Japan*, **24**, 1053. Dielectric relaxation mechanism in NaNO$_2$.

Thermal expansion

(68) N. Maruyama and S. Sawada (1965) *J. Phys. Soc. Japan*, **20**, 811. Thermal expansion in NaNO$_2$ crystals.

NaNO$_3$

Older work

(69) F. C. Kracek (1931) *J. Amer. Chem. Soc.*, **53**, 2609. Gradual transition in sodium nitrate. I: Physico-chemical criteria of the transition.
(70) F. C. Kracek, E. Posnjak, and S. B. Hendricks (1931) *J. Amer. Chem. Soc.*, **53**, 3339. Gradual transition in sodium nitrate. II: Structure.

Specific heat

(71) Quoted from A. R. Ubbelohde, reference (5).

Thermal expansion

(72) J. B. Austen and R. H. H. Pierce (1933) *J. Amer. Chem. Soc.*, **55**, 661. The linear thermal expansion of a single crystal of sodium nitrate.
(73) M. Kantola and E. Vilhonen (1960) *Ann. Acad. Sci. Fenn. A*, VI, No. **54** X-ray measurements of the thermal expansion of sodium nitrate.

Evidence for 30° change of orientation

(74) L. A. Siegel (1949) *J. Chem. Phys.*, **17**, 1146. Molecular rotation in sodium cyanide and sodium nitrate.
(75) K. Kurki-Suonio (1962) *Ann. Acad. Sci. Fenn. A*, VI, No. **94**. The "correct" model for sodium nitrate.

High-pressure transition

(n)(76) J. D. Barnett, J. Pack, and H. T. Hall (1969) *Trans. Amer. Cryst. Assoc.*, **5**, 113. Structure determination of a ferroelectric phase of sodium nitrate above 45 kbar.

Transitions of calcite

(77) J. J. Lander (1949) *J. Chem. Phys.*, **17**, 892. Polymorphism and anion rotational disorder in the alkaline earth carbonates.
(78) G. J. F. MacDonald (1956) *Amer. Mineral.*, **41**, 744. Experimental determination of calcite-aragonite equilibrium relations at elevated temperatures and pressures.
(79) J. C. Jamieson (1957) *J. Geol.*, **65**, 334. Introductory studies of high-pressure polymorphism by X-ray diffraction, with some comments on calcite II.

KNO$_3$

Structures

(80) I: P. E. Tahvonen (1947) *Ann. Acad. Sci. Fenn. A*, I, No. **44**. Röntgenometrische Untersuchung über die Molekulrotation im KNO$_3$-Kristall.
(81) I: Y. Shinnaka (1962) *J. Phys. Soc. Japan*, **17**, 820. X-ray study on the disordered structure above the ferroelectric Curie point in potassium nitrate.
(82) II: D. A. Edwards (1931) *Z. Kristallog.*, **80**, 154. A determination of the complete crystal structure of potassium nitrate.
(83) III: T. F. W. Barth (1939) *Z. Physikal. Chem. B.*, **43**, 448. Die Kristallstruktur der Druckmodifikation des Salpeters.

(84) III: S. Sawada, S. Nomura, and S. Fujii (1958) *J. Phys. Soc. Japan*, **13,** 1549. Ferroelectricity in KNO₃.

(85) IV: J. C. Jamieson (1956) Z. *Kristallog.*, **107,** 65. Some X-ray diffraction data on KNO₃ IV—a high-pressure phase.

Transitions

(86) I–III: M. Kantola and T. Tarna (1970) *Ann. Acad. Sci. Fenn. A, VI*, No. **335.** X-ray measurements of thermal expansion of ferroelectric KNO₃.

(87) I–III, and phase diagram: A. Chen and F. Chernov (1967) *Phys. Rev.*, **154,** 493. Nature of ferroelectricity in KNO₃.

(88) II–I, orientation of crystallites: S. W. Kennedy, A. R. Ubbelohde, and I. Woodward (1953) *Proc. Roy. Soc. A*, **219,** 303. The persistence of crystal axes in a thermal transformation.

Alkaline earth carbonates

BaCO₃, SrCO₃: see reference (77).
CaCO₃ (aragonite) see references (77) to (79).

Hydrogen halides

(89) E. Sandor and R. F. C. Farrow (1967) *Nature* **213,** 171. Crystal structure of solid hydrogen chloride and deuterium chloride.

(90) E. Sandor and R. F. C. Farrow (1967) *Nature*, **215,** 1265. Crystal structure of cubic deuterium chloride.

(91) S. Hoshino, K. Shimaoka, and N. Niimura (1967) *Phys. Rev. Letters*, **19,** 1286. Ferroelectricity in solid hydrogen halides (with additional unpublished information).

(92) E. Sandor and M. W. Johnson (1968) *Nature*, **217,** 541. Crystal structure and the lower phase transition in solid deuterium bromide.

(93) E. Sandor and M. W. Johnson (1969) *Nature*, **223,** 730. Kinetics of the lower phase transition in solid deuterium bromide.

(94) K. Shimaoka, N. Niimura, H. Motegi, and S. Hoshino (1969) *J. Phys. Soc. Japan*, **27,** 1078. A new phase of solid hydrogen chloride.

(95) S. Hoshino, K. Shimaoka, N. Niimura, H. Motegi, and N. Maruyama (1970) *J. Phys. Soc. Japan*, **28,** Suppl., 189. Ferroelectricity and phase transitions in solid hydrogen halides.

(96) E. Hanamura (1970) *J. Phys. Soc. Japan*, **28,** Suppl., 192. On the phase transitions in hydrogen halide crystals.

(97) E. Sandor and R. F. C. Farrow (1969) *Discussions of the Faraday Soc.*, **48,** 78. Neutron diffraction study of molecular motion in solid deuterium chloride.

Ammonium halides

(98) Older work on transitions: L. Pauling (1930) *Phys. Rev.*, **36,** 430. The rotational motion of molecules in crystals.

(99) NH₄F: H. W. Adrian and D. Feil (1969) *Acta Cryst.*, **A25,** 438. The structure of NH₄F as determined by neutron and X-ray diffraction.

(100) ND₄Br: H. A. Levy and S. W. Peterson (1953) *J. Amer. Chem. Soc.*, **75,** 1536. Neutron diffraction determination of the crystal structure of ammonium bromide in four phases.

(n)(101) NH₄Br and NH₄I: R. S. Seymour and A. W. Pryor (1970) *Acta Cryst.*, **B26,** 1487. Neutron diffraction study of NH₄Br and NH₄I.
Volume changes: quoted from reference (5).

Fe₃O₄

[See reference (21).]

AUTHOR INDEX

FORMULA INDEX

SUBJECT INDEX

Page numbers in **bold type** refer to main entries for the topic. Page numbers in round brackets refer to the list of books and published papers dealing with the topic. Authors' names are listed in a separate Author Index. A separate Formula Index is also provided.

Acceptor atom. See Donor or acceptor atom
Accuracy of experimental work, limitations of, 307, 349, 386–387, 403, 428–429, 435, 462
Acids, hydrates of, **385–388, 389–393**
Adjustable parameters, number and structural effect of, 221, 269, 275, 285, 304, 307, 376. See also Physically-independent variables
Afwillite, 200, 345, 348, 395, (517)
Al/Si substitution. See Silicon/aluminium substitution; Ordering of silicon/aluminium
Albite, 318–320, 321, 322, (517)
Alkali halides, 8, 20, 30, 45, 50, 54, 55, **82–85,** 91, 100. See also under separate names
Alkaline earth carbonates, 249, 498, (526). See also Aragonite; Calcite
Alloys, 23, 89–91, 448
Alum, 346, **379–382,** 408, 424, (519, 520)
Aluminium, 75
Aluminium avoidance rule, 314, **317–318,** 323, 324, 328
Aluminium hydroxide, 421, 427. See also Gibbsite
Aluminium-iron alloy. See Iron-aluminium alloy
Aluminosilicate framework. See Framework, aluminosilicate
Ammonium dihydrogen phosphate, 366–369, 394, 506, (518)
Ammonium halides, **502–507,** 508
Ammonium hydrogen sulphate, 396
Ammonium ion, 55, 366, 394, 481, 502, 505

Ammonium ion (*Continued*)
  libration of, 503, 505
  spherical approximation to, 502, 506
Amphibole, 201
Amplitudes, thermal,
  and bond-length corrections, 409–412, 419, 458
  and electron-density cloud, 399–400
  and lattice mode, 416, 420, 439–441, 458, 477
  anisotropy of, 400, 419
  correlation with forces and frequencies, 387, **401–403,** 406, 419, 440
  numerical values of, 395, 401–404, 408, 410, 411, 414, 457, 477, 492, 503
  of libration, 477, 486–487, 489, 492, 503, 505
  relation to "static" displacement, 416, 419, 440, 457, 462, 492, 508–509
  temperature dependence of, 399, **413–415,** 477, (520)
  versus disorder, 362, 371, 403–404, 446, 464, 478, 483, 499, 503, 505–506
Anatase, 257, 277–278, 436, 438
Angles,
  bond. See Bond angles
  interaxial, 60, 66, 319
    of pseudocubic structures, 295, 300
    of rhombohedra, 208, 297, 466
    old mineralogical convention for, 201
  interfacial, 13, 103, 117
    constancy of, 4, 103
  tilt. See Tilt angles
Angular displacements, root-mean-square. See Libration of molecule or molecule-ion
Anharmonicity of oscillations, 415–419